W0038362

Neurogenetic Syndromes

Neurogenetic Syndromes
Behavioral Issues and Their Treatment

edited by

Bruce K. Shapiro, M.D.
The Johns Hopkins University School of Medicine
Kennedy Krieger Institute
Baltimore

and

Pasquale J. Accardo, M.D.
Virginia Commonwealth University
VCU Child Development Clinic
Richmond

·P·A·U·L·H·
BROOKES
PUBLISHING CO. ®

Baltimore • London • Sydney

Paul H. Brookes Publishing Co.
Post Office Box 10624
Baltimore, Maryland 21285-0624
USA

www.brookespublishing.com

Copyright © 2010 by Paul H. Brookes Publishing Co.
All rights reserved.

"Paul H. Brookes Publishing Co." is a registered trademark
of Paul H. Brookes Publishing Co., Inc.

Book design by Mindy Dunn.
Typeset by Broad Books, Baltimore, Maryland.
Manufactured in the United States of America by
Sheridan Books, Inc., Chesea, Michigan.

The individuals described in this book are composites based on the authors' experiences or
real people whose situations are masked and used with permission.

The photos in this book are used by permission of the individuals pictured and/or their
parents/guardians.

The information provided in this book is in no way meant to substitute for a medical or
mental health practitioner's advice or expert opinion. Readers should consult a health or
mental health professional if they are interested in more information. This book is sold
without warranties of any kind, express or implied, and the publisher and authors disclaim
any liability, loss, or damage caused by the contents of this book.

Library of Congress Cataloging-in-Publication Data
Neurogenetic syndromes : behavioral issues and their treatment/edited by
Bruce K. Shapiro and Pasquale J. Accardo.
   p. cm.
  Includes bibliographical references and index.
  ISBN-13: 978-1-59857-017-5 (hardcover)
  ISBN-10: 1-59857-017-X (hardcover)
  1. Nervous system—Diseases—Genetic aspects. 2. Neurogenetics. 3. Behavior genetics.
I. Shapiro, Bruce K. II. Accardo, Pasquale J. III. Title.
  [DNLM: 1. Nervous System Diseases—genetics. 2. Chromosome Disorders—genetics.
  3. Chromosome Disorders—therapy. 4. Genetics, Behavioral. 5. Nervous System
Diseases—therapy. 6. Phenotype. WL 140 N49126 2010]
RC346.4.N47 2010
616.8'0442—dc22   2009031791

British Library Cataloguing in Publication data are available from the British Library.

2013   2012   2011   2010   2009

10   9   8   7   6   5   4   3   2   1

Contents

Editors and Contributors vii
Preface xi
Acknowledgments xv

I Neurogenetic Syndromes

1 Behavioral Phenotypes: Recent Advances
   James C. Harris 3

2 Smith-Magenis Syndrome: A Neurobehavioral Syndrome with
   Genetic Underpinnings
   Andrea L. Gropman and Ann C.M. Smith 15

3 Fragile X: Expansion of a Genetic Disorder
   Walter E. Kaufmann 29

4 Behavioral Phenotypes in Down Syndrome: A Probabilistic Model
   George T. Capone 53

5 Autism and Prader-Willi Syndrome: Searching for Shared Endophenotypes
   Travis Thompson 71

6 Williams Syndrome: Psychological Characteristics
   Carolyn B. Mervis and Angela E. John 81

7 Developmental Influences on Psychological Phenotypes
   Gene S. Fisch 99

8 Is There a Behavioral Phenotype in Children with Fetal Alcohol
   Spectrum Disorders?
   Piyadasa W. Kodituwakku 115

II Treatment of Neurogenetic Syndromes

9 Functional Behavioral Assessment: Its Value in the Treatment of Maladaptive
   Behaviors in Individuals with Neurogenetic Syndromes
   Theodosia R. Paclawskyj 135

10 Psychiatric Diagnosis in Individuals with Neurodevelopmental Disability
    Richard B. Ferrell 153

11 Speech-Language Therapy for Children with Social, Emotional, and
    Behavioral Disorders
    Janet E. Turner and Mary K. Boyle 163

12 A Clinical Approach to the Pharmacological Management of Behavioral
    Disturbance in Intellectual Disability
    Sarah Risen, Pasquale J. Accardo, and Bruce K. Shapiro 185

13 Integrating Behavioral and Pharmacological Interventions for
    Severe Problem Behavior Displayed by Children with Neurogenetic
    and Developmental Disorders
    Louis P. Hagopian and Mary E. Caruso-Anderson 217

III Future Implications

14 New Genetic Techniques: Implications for Neurobehavioral Syndromes
    Lisa T. Emrick 243

15 Genetically Informative Phenotypes: Opportunities for Progress
    and Potential Pitfalls
    Peter Szatmari 263

16 Social Phenotypes in Genetically Based Neurodevelopmental Disorders
    Carl Feinstein and Shivani Verma 277

17 Behavioral Phenotypes: Nature versus Nurture Revisited
    Pasquale J. Accardo and Margie L. Jaworski 293

Index 301

Editors and Contributors

Editors

Bruce K. Shapiro, M.D.
Professor of Pediatrics
The Johns Hopkins University School
    of Medicine
The Arnold J. Capute, M.D., M.P.H. Chair
    in Neurodevelopmental Disabilities
Vice President, Training
Kennedy Krieger Institute
707 North Broadway
Baltimore, MD 21205

Pasquale J. Accardo, M.D.
Professor of Pediatrics
Virginia Commonwealth University
James H. Franklin Professor of
    Developmental Research in Pediatrics
VCU Child Development Clinic
3600 West Broad Street
Richmond, VA 23230

Contributors

Mary K. Boyle, M.A., CCC-SLP
Speech-Language Pathologist
Kennedy Krieger Institute
707 North Broadway
Baltimore, MD 21205

George T. Capone, M.D.
Associate Professor of Pediatrics
The Johns Hopkins University School
    of Medicine
Director
Down Syndrome Program
Research Scientist
Division of Neurology and Developmental
    Medicine
Kennedy Krieger Institute
707 North Broadway
Baltimore, MD 21205

Mary E. Caruso-Anderson, Ph.D., BCBA
Research Associate
The Johns Hopkins University School
    of Medicine
Kennedy Krieger Institute
707 North Broadway
Baltimore, MD 21205

Lisa T. Emrick, M.D.
Neurology and Developmental
    Medicine
Kennedy Krieger Institute
707 North Broadway
Baltimore, MD 21205

Carl Feinstein, M.D.
Professor of Psychiatry and Behavioral
    Sciences
Director, Division of Child and Adolescent
    Psychiatry
Endowed Director of Psychiatry,
    Lucile Packard Children's Hospital
Stanford University School of Medicine
401 Quarry Road
Stanford, CA 94305

Richard B. Ferrell, M.D.
Associate Professor of Psychiatry
Dartmouth Medical School
One Medical Center Drive
Lebanon, NH 03756

Gene S. Fisch, Ph.D.
Senior Research Biostatistician
Research Professor
Department of Epidemiology and Health
    Promotion
Colleges of Dentistry and Nursing
New York University
250 Park Avenue South
New York, NY 10003

Andrea L. Gropman, M.D., FAAP, FACMG
Associate Professor
Pediatrics and Neurology
The George Washington University
    School of Medicine and Health Sciences
Attending Neurologist
Children's National Medical Center
111 Michigan Avenue, NW
Washington, DC 20010

Louis P. Hagopian, Ph.D., BCBA-D
Associate Professor of Psychiatry and
    Behavioral Sciences
The Johns Hopkins University School
    of Medicine
Program Director, Neurobehavioral Unit
Kennedy Krieger Institute
707 North Broadway
Baltimore, MD 21205

James C. Harris, M.D.
Professor of Psychiatry and
    Behavioral Sciences,
    Pediatrics, and Mental Hygiene
The Johns Hopkins University School
    of Medicine
The Johns Hopkins Hospital
600 North Wolfe Street, CMCS 346
Baltimore, MD 21287

Margie L. Jaworski, M.D.
Assistant Professor of Pediatrics
Virginia Commonwealth University
VCU Child Development Clinic
3600 West Broad Street
Richmond, VA 23230

Angela E. John, M.A.
Doctoral Student
Department of Psychological and
    Brain Sciences
University of Louisville
317 Life Sciences Building
 Louisville, KY 40292

Walter E. Kaufmann, M.D., Ph.D.
Professor of Pathology, Neurology,
    Pediatrics, Psychiatry, and Radiology
The Johns Hopkins University School
    of Medicine
Director
Center for Genetic Disorders of Cognition
    and Behavior
Kennedy Krieger Institute
716 North Broadway
Baltimore, MD 21205

Piyadasa W. Kodituwakku, Ph.D.
Associate Professor of Pediatrics and
    Neurosciences
Center for Development and Disability
School of Medicine
The University of New Mexico
2300 Menaul NE
Albuquerque, NM 87107

Carolyn B. Mervis, Ph.D.
Distinguished University Scholar and
    Professor
Department of Psychological and Brain
    Sciences
University of Louisville
317 Life Sciences Building
Louisville, KY 40292

Theodosia R. Paclawskyj, Ph.D., BCBA
Assistant Professor
Department of Psychiatry and Behavioral
    Sciences
The Johns Hopkins University School
    of Medicine
Kennedy Krieger Institute
707 North Broadway
 Baltimore, MD 21205

the syndromic level. New genetic techniques allow us to learn how individual genes affect brain function. We are now able to detect genetic defects that occur in DNA. We are able to selectively add or delete genetic material in animals and observe the behavioral effects (knockout models) Neuroimaging studies increase our understanding of the how physiological and structural factors result in the functional impairments that are clinically observed. Neuroscience techniques enable us to understand how environmental perturbations affect neuronal function, how one neuron communicates with another, and how neurons form networks that produce physiological/cognitive functions.

ABOUT THE BOOK

The objective of *Neurogenetic Syndromes: Behavioral Issues and Their Treatment* is to draw together a body of knowledge that comprises behavioral neurogenetics. It is based on the Spectrum of Developmental Disabilities conference. This conference brings together experts to provide an interdisciplinary focus on neurodevelopmental and related disorders. Presentations address public health aspects, diagnostic issues, neuroscience advances, developmental aspects, and current management strategies. Speakers blend current research with clinical expertise to delineate the boundaries of our knowledge in diagnosis, research, and management.

Neurogenetic Syndromes addresses three areas related to neurobehavioral syndromes, which are discussed across the book's three sections. The anticipated audience for this book is neurodevelopmental pediatricians, pediatric neurologists, child psychiatrists, pediatricians, geneticists, developmental-behavioral pediatricians, psychologists, educators, and those who have interest in intellectual disability.

The first section focuses on established neurobehavioral syndromes and provides current information about them. Harris

defines behavioral phenotypes, addresses the methodological difficulties associated with behavioral phenotyping, and demonstrates the associations between behavior and brain function that result from the study of Lesch-Nyhan syndrome. Gropman and Smith provide a comprehensive review of Smith-Magenis syndrome and note that it is the neurobehavioral and sleep disturbances that usually lead to the diagnosis. Kaufmann reviews fragile X syndrome. He addresses the neurobiology of this disorder, reviews its association with autism spectrum disorders and extends the behavioral spectrum of fragile X syndrome, and speaks to the expanded neural and behavioral pathology of fragile X that extends to carriers and those with premutations. Capone uses Down syndrome to outline the questions that form a genotype-behavioral phenotype research agenda. He draws on his clinical experience and knowledge of neuroscience to conclude that evidence exists supporting several behavioral phenotypes within Down syndrome that appear to reflect underlying differences in brain development and organization. Thompson uses the examples of Prader-Willi syndrome and autism to propose a shared genetic association that is responsible for some of the behaviors noted in each disorder. Mervis and John draw upon their clinical experience with Williams syndrome and present evidence of a neuropsychological profile that supports the clinical finding of stronger language abilities in this syndrome. They further explore the relationship and find that the relative strength in language abilities is not homogeneous and that neuroimaging data support some of the observed patterns of impairment. Fisch uses a cross-sectional study of 108 children with Williams syndrome, Fragile X syndrome, and neurofibromatosis type 1 to draw attention to the influences of development on psychological phenotypes. Kodituwakku seeks to determine whether a behavioral phenotype is present in children with an acquired disorder, fetal alcohol syndrome.

The second section of this book addresses the treatment of behavioral disturbances associated with neurobehavioral syndromes. Paclawskyj uses functional behavioral analysis as a method for addressing maladaptive behavior that is frequently seen in children with neurogenetic disorders. Ferrell reviews the history of psychiatric diagnosis in people with intellectual disability and provides a number of points to consider in the diagnostic process. Turner and Boyle address the issues of limited communication and behavior disturbance in children with social, emotional, and behavior disorders. They show that providing a child with a method of communication may substantially decrease maladaptive behavior. Risen, Accardo, and Shapiro provide a clinical overview of the use of psychopharmacologic agents to treat maladaptive behavior in people with intellectual disability. Hagopian and Caruso-Anderson provide a comprehensive overview of an integrated approach to the treatment of behavior disturbance in children with neurogenetic and developmental disorders.

The final section contains reviews of new techniques and forecasts what the study of neurobehavioral disorders may entail in the future. Emrick reviews some of the new genetic techniques that are being employed in the study of intellectual disability. She notes that we have substantially increased our ability to detect genetic abnormality but also cautions that the new techniques still do not explain the complete clinical picture—either in scope or severity. Szatmari makes note of the lack of success in identifying genetic variants that account for neurobehavioral disorders when compared with the progress made in identifying genetic mutations that account for specific syndromes. He distinguishes disorders from phenotypes and notes that disorders are a proxy for phenotype. He proposes a model of genetically informative phenotypes that may serve as a guide for future studies of the genetics of complex disorders. Feinstein and Verma propose a model of social phenotypes that are distinct from behaviors and mood. Concluding the book, Accardo and Jaworski consider the misapplication of neurobehavioral genetics and suggest that the enthusiasm engendered by the pursuit of gene–behavior interactions should be tempered by the effects of treatment and the environment (including epigenetic phenomena).

Neurogenetic Syndromes

Behavioral Phenotypes

Recent Advances

James C. Harris

A neurodevelopmental disorder may be associated with characteristic patterns of behavior that are specific to that disorder. This was first recognized by Langdon Down (1887/1990), who, in his original description of the syndrome that now bears his name, wrote:

> They have considerable powers of imitation, even bordering on being mimics. . . . Several patients who have been under my care have been wont to convert their pillow cases into surplices (vestments) and to imitate, in tone and gesture, the clergymen or chaplain which they have recently heard.

He commented too on personality traits: "Another feature is their great obstinacy— they can only be guided by consummate tact." Although these stereotypes were not confirmed in subsequent studies (Gath & Gumley, 1986; Gunn, Berry, & Andrews, 1981), Down seems to be the first to propose that there may be behaviors that are characteristic for a particular neurodevelopmental disorder. Later Critchley and Earl (1932) identified several behavior problems specific to children with tuberous sclerosis complex. Despite this early interest in syndrome-specific behaviors research, it was not until William Nyhan (1972) proposed that self-injurious behavior was a behavioral phenotype of Lesch-Nyhan syndrome that a concerted focus on genes and behavior was initiated in neurogenetic syndromes.

Two issues contributed to the lack of interest in behavioral phenotypes following the early proposals by Down and others. First was the generally negative reaction against the eugenics movement's early, and false, claims that people with intellectual disability were inherently antisocial. Claims for the genetic bases of personality in that era were suspect, creating a climate in which academic investigations linking genes and behavior were discouraged. Learning theory models that were applied in the education of people with intellectual disabilities emphasized normalization and tended to deemphasize behavior (Harris, 1998a). Little consideration was given to "unlearned behaviors" associated with behavioral phenotypes. Eventually, however, it became apparent that reports from families and clinical observations by physicians of characteristic patterns of behavior in certain syndromes could not be ignored. The interest of parent groups in improving the lives of their children played a major role in encouraging research into behavioral phenotypes.

Advances in genetics are providing new insights into the extent that genes influence behavior. Moreover, neurophysiology and neuroanatomy provide means of studying brain mechanisms, and neuroimaging studies are contributing to our understanding of the brain regions that may be involved (Harris, 2001). Coupled with better evaluation methodology to measure behavioral phenotypes, there is renewed interest in and focus on behavioral phenotypes in developmental neuropsychiatry. Comprehensive study of children with different developmental disabilities is increasing our appreciation for the relative contribution of genetic variables in the pathogenesis of behavioral disorders.

BEHAVIORAL PHENOTYPE

William Nyhan (1972) introduced the term *behavioral phenotype* in his presidential address to the Society for Research in Child Development in 1971 when describing self-mutilating behavior in the syndrome that came to be known as Lesch-Nyhan disease. For him *behavioral phenotype* referred to outwardly observable behavior that is so characteristic of children with a particular genetic disorder that its presence suggests the underlying genetic condition. He wrote:

> We feel that these children have a pattern of unusual behavior that is unique to them. Stereotypical patterns of behavior occurring in syndromic fashion in sizable numbers of individuals provide the possibility that there is a concrete explanation that is discoverable. In these children, there are so many anatomical abnormalities, from changes in hair and bones to dermatoglyphics, that it is a reasonable hypothesis that their behaviors are determined by an abnormal neuroanatomy that would be discoverable, possibly neurophysiologically, ultimately anatomically . . . these children all seem self-programmed. These stereotypical patterns of unusual behavior could reflect the presence of structural deficits in the central nervous system. (p. 235)

Nyhan highlighted three elements in his description: 1) unique stereotypical patterns of unusual behavior, 2) behavior that seems self-programmed, and 3) possible links to structural deficits in the central nervous system.

Such observations as these have resulted in closer scrutiny being paid to behavior in other neurodevelopmental disorders. Distinct behavior patterns have been reported in a number of syndromes arising from genetic or chromosomal abnormalities. Initially the research goal was descriptive, that is, to document well-defined patterns of behavior specific to a syndrome. Subsequently the focus has shifted to understanding the neurobiological mechanisms underlying characteristic behavioral patterns, cognitive processes, and social interactions. Flint (1996, 1998) cites evidence from transgenic animal models that some features of the pathways from genotype to phenotype may be studied in animal models by careful delineation of individual aspects of the behavioral phenotype. Thus although an animal model may not be a model for the full syndrome, it may be for aspects of it.

Current Definition and Characterization

The study of behavioral phenotypes emphasizes the discovery, among individuals with known chromosomal, genetic, or neurodevelopmental disorders, of behavioral features that may be causally related to the underlying condition. Still, although behavior can sometimes be linked with a syndrome, it is possible that not all individuals with a disorder will exhibit classic behavioral features. However, the probability is greater that they will. The essential issue is that the behavior suggests the genetic diagnosis. Examples are characteristic self-mutilation of fingers and lips in Lesch-Nyhan disease; hyperphagia and compulsive behaviors in Prader-Willi syndrome; gaze aversion in fragile X syndrome; and superficial sociability, hyperlalia, and language disorder in Williams syndrome.

Efforts to define what is meant by a behavioral phenotype are continuing. Over 35 years after Nyhan proposed the term *behavioral phenotype,* the definition has been expanded and refined. Currently the definition suggested by Flint and Yule (1994) is commonly used: "The behavioral phenotype is a characteristic pattern of motor, cognitive, linguistic, and social abnormalities that is consistently associated with a biological

Table 1.1.   Methodological considerations in behavioral phenotype research

Choice of control groups
  Between-syndrome comparisons
  Within-syndrome comparisons (e.g., uniparental dysomy versus deletion in
    Prader-Willi syndrome)
Methodological issues in assessment
  Impact of extent of the intellectual disability on symptom expression
  Gender issues
  Impact of environmental experiences
Developmental change across the life span
  Relationship of psychopathology to specific syndromes
  Positive attributes and strengths in assessment
  Other subject/environmental factors

Source: Dykens and Hodapp (2007).

disorder"(p. 667). The definition has three elements: 1) a characteristic pattern of motor, cognitive, linguistic, and social abnormalities; 2) the probability that these behaviors are increased, but that they may not be fully expressed; and 3) that brain mechanisms may be studied in those with the same syndrome with and without the full phenotype (Jinnah & Friedmann, 2001).

When Nyhan first described the behavioral phenotype of Lesch-Nyhan syndrome he focused on one behavior, compulsive self-mutilation. However, Flint and Yule allow examination of an extended behavioral phenotype that potentially includes motor, cognitive, linguistic, and social abnormalities. A broader and extended behavioral phenotype raises questions about the extent to which it is possible to trace pathways from gene to cognition and complex behaviors. Since Nyhan's original description, each of the elements Flint and Yule proposed for Lesch-Nyhan syndrome and its variants has been investigated (Harris, Wong, Jinnah, Schretlen, & Barker, 2002).

METHODOLOGICAL ISSUES IN THE ASSESSMENT OF INDIVIDUAL DIFFERENCES WITHIN AND BETWEEN SYNDROMES

An individual with a syndrome may not exhibit all of the behaviors associated with that syndrome. A probabilistic definition of behavioral phenotype raises questions regarding both within-syndrome and between-syndrome comparisons in determining the behavioral phenotype. Based on a thorough review of published studies, Dykens and Hodapp (2007) propose several critical methodological issues that must be considered when behavioral phenotypes of a particular syndrome are studied (Table 1.1). Issues raised for consideration are the choice of appropriate control groups, methods of between-syndrome and within-syndrome comparisons (e.g., uniparental dysomy versus deletion etiologies in Prader-Willi syndrome), the impact of the severity of intellectual disability on symptom expression, gender issues, the impact of environmental experiences on phenotypic expression, developmental change (trajectory) across the life span, and the relationship of psychopathology to specific syndromes. Moreover, they point out that positive attributes and strengths in affected individuals must be considered along with other subject/environmental factors.

THE EXPRESSION OF BEHAVIORAL PHENOTYPES

Because the genetic background of an individual may also affect phenotypic expression, it is essential to complete a thorough family history. As noted above, differences in the developmental trajectory that emerge as the child enters adolescence and adult life must be taken into account, along with the possibility that environmental factors may modify behavioral presentation over time. Finally, when animal models are used, it should be remembered that mutant mouse models generally do not replicate the full behavioral phenotype found in humans (Skuse, 2000).

Steinhausen et al. (2002) compared four intellectual disability syndromes on subscales and for individual items on the Developmental Behavior Checklist (DBC). Individuals with Prader-Willi syndrome, fragile X syndrome, tuberous sclerosis complex, and fetal alcohol syndrome were evaluated. There were clear differentiations in the behavioral rating across all four syndromes for individual items and at the subscale level. Fragile X syndrome and fetal alcohol syndrome were most clearly differentiated. Prader-Willi syndrome and tuberous sclerosis complex showed overall lower scores and less atypical behavior profiles. Intelligence level, gender, and age did not contribute to variation in the number of behavior abnormalities identified. The authors stated that the DBC can be used to quantitatively demonstrate significant differences and to differentiate behavioral phenotypes among intellectual disability syndromes.

Developmental Trajectories Differ by Syndrome

A lifetime developmental approach is critical in the evaluation of behavioral phenotypes. Its focus is on establishing the natural history or developmental trajectory for behaviors in each syndrome. When a developmental approach is used, changes may be identified in symptom progression and in the behavioral profile as the person matures. This approach is increasingly used to understand pathways from genes to cognition and complex behavior within a syndrome.

Certain brain systems, perhaps those that have most recently evolved, are most vulnerable to neurobiological perturbation. For example, impairments in attention, executive functioning, intentional planning, and impulse control occur across neurogenetic disorders, suggesting that these systems are particularly vulnerable to developmental insult. Moreover, their natural history varies according to the syndrome. Yet despite the differences found between syndromes, the *Diagnostic Manual-Intellectual Disability* (*DM-ID*; Fletcher, Loschen, Stavrakaki, & First, 2007) does not recommend adaptations in criteria for attention-deficit/hyperactivity disorder (ADHD) when it occurs in a neurogenetic syndrome.

Developmental trajectories in brain development have been undertaken in nonsyndromic attention deficit disorder. One magnetic resonance imaging study (Castellanos et al., 2002) recruited 152 boys and girls (89 male and 63 female patients, ages 5–18) with a diagnosis of ADHD. Their brain scans were compared with those of 139 age- and gender-matched controls, children and adolescents without ADHD. The brains of children and adolescents with ADHD were found to be 3%–4% smaller than those of children without the disorder. Smaller brain volumes, in all regions, were linked to the severity of a patient's ADHD symptoms. Never-medicated patients were found to have smaller white matter volumes. Medication treatment was not the cause of the brain changes. Several brain regions showed changes. The frontal lobes were smaller, and reductions were noted in temporal gray matter, the caudate nucleus, and the cerebellum. In addition, developmental trajectories for all brain structures, other than the caudate nucleus, paralleled one another for ADHD patients and controls during childhood and adolescence. Such parallel trajectories suggest that genetic and/or early environmental experiences influence the development of the brain in those with ADHD.

These changes are fixed and do not progress as the child grows older. The pattern of cortical thinning in ADHD occurs mainly in prefrontal brain regions. These regions are part of the anterior attentional network, which includes the medial prefrontal and cingulate regions. Normalization of the right parietal cortex thickness was found in some cases with better outcome and might represent compensatory change in the posterior attentional network when the anterior network is dysfunctional (Shaw et al., 2006). These findings in children with nonsyndrome attention deficit disorder compared with age-matched controls might be extended to brain findings in specific neurogenetic syndromes in the future to further delineate the relationship between brain function and attention.

Impairments in social understanding and communication may occur in many neurogenetic syndromes. In some syndromes such as tuberous sclerosis complex, criteria may be met for Autistic Disorder. The frequency with which social impairments occur in multiple syndromes suggests that brain systems involved in social communication are particularly vulnerable. When individuals with a neurodevelopmental disorder meet diagnostic criteria for a *Diagnostic and Statistical Manual of Mental Disorders, Fourth Edition, Text Revision* (*DSM-IV-TR;* American Psychiatric Association, 2000) diagnostic category, it is important to remember that *DSM-IV-TR* diagnoses are not behavioral phenotypes. That certain syndromes meet broad diagnostic criteria for *DSM-IV-TR* diagnoses suggests a lack of specificity in *DSM-IV-TR* diagnostic categories and more than one pathway to such diagnoses. Moreover, when social impairment or Autistic Disorder is diagnosed in conjunction with a neurodevelopmental disorder, the developmental trajectory will vary with the syndrome. Thus some children with fragile X syndrome may meet diagnostic criteria for an Autistic Disorder in the preschool years, but by adolescence a social anxiety disorder may be the more accurate diagnosis. Thus it is best to focus on the developmental trajectory of symptoms in a particular syndrome and not generalize from that syndrome more broadly. Understood in this way, behavioral phenotypes in specific neurodevelopmental disorders may have greater specificity in their natural history than a *DSM-IV-TR* diagnosis. Finally, understanding the developmental trajectory of a disorder allows the treating physician to provide prognostic advice to families.

BEHAVIORAL PHENOTYPES: BRAIN AND BEHAVIOR

The study of behavioral phenotypes may contribute to an understanding of developmental psychopathology. As noted, studies of behavior over time conducted within syndromes are important to the establishment of developmental trajectory. Studies between syndromes are important when behavioral profiles are compared. Both approaches may contribute to our understanding of mechanisms underlying behavioral phenotypes. Attention deficits, executive dysfunction, and impairments in social understanding and communication occur in various syndromes. Categorical *DSM-IV-TR* diagnoses of ADHD frequently are made in children and adolescents with neurogenetic syndromes based on *DSM-IV-TR* criteria, as shown in Table 1.2. The type of attention profile in a particular disorder may allow elements of the attention network to be studied among syndromes.

Table 1.2. Attention-deficit/hyperactivity disorder diagnosis in neurogenetic syndromes

Phenylketonuria
Fragile X syndrome
Tuberous sclerosis complex
Marfan syndrome
Tourette syndrome
Velocardiofacial syndrome

PHENYLKETONURIA AND ATTENTION DEFICITS

Studies of neurogenetic syndromes focus not only on brain anatomy but also on neurochemistry. Phenylketonuria (PKU) is a genetic disorder whose neurochemistry may provide clues to the mechanism of attention deficits. Asbjorn Folling originally described PKU in 1934. Folling studied a young mother and her two sons; the sons were typical at birth but later developd intellectual disabilities as they grew older. Folling found elevations in phenylketones in the boys' urine and determined that excessive phenylalanine was the cause for the intellectual disability. Subsequently the metabolic pathway was established, and a low-phenylalanine diet was introduced. It was shown that when the diet was followed from the first months of life, severe intellectual disability could be prevented. Newborn screening is now routinely carried out to identify PKU throughout the United States. However, it was later discovered that although the original low-phenylalanine diet prevented severe intellectual disability, it did not necessarily prevent attention deficits.

Adele Diamond (Diamond, 2007; Diamond, Prevor, Callender, & Druin, 1997) and others documented attention deficits and impairment with prefrontal cortical tests, using a modification of the Stroop test (a test of divided attention) in preschool and school-age children treated with the original low-phenylalanine diet. These authors found that standard dietary treatment was insufficient. One possible mechanism for the attention deficits is the role PKU plays in the functioning of the dopamine system. PKU results in a reduction in blood tyrosine, a dopamine precursor. The lack of tyrosine leads to reduced dopamine and impaired functioning of the prefrontal cortex and attention deficits (Pennington, VanDoornick, McCabe, & McCabe, 1985). The dopamine neurons that project to the prefrontal cortex

have higher rates of firing and dopamine turnover (Tam, Elsworth, Bradberry, & Roth, 1990) than dopamine neurons in the corpus striatum. Diamond and colleagues demonstrated this mechanism both in an animal model of PKU and in a longitudinal study of affected children (Diamond, 2001; Diamond, Ciaramitaro, Donner, Djali, & Robinson, 1994). Moreover, they found that affected individuals showed impairment on visual charts used to document visual contrast sensitivity in low light (Diamond and Herzberg, 1996).

That research, presenting a mechanistic explanation, provided the evidence needed to support a change in the medical guidelines for the treatment of PKU (blood phenylalanine levels should be kept between 120 and 360 micromoles per liter [μmol/L]). Further support for the tyrosine hypothesis comes from studies showing that affected children improve when given tyrosine supplementation (Kalkanoğlu et al., 2005).

Findings of dopamine dysfunction in PKU are consistent with dopaminergic models of inattention and working memory in nonsyndromal ADHD. Such studies in young children provide insight into the development of cognitive control, executive functioning, and attention.

LESCH-NYHAN SYNDROME AND ITS VARIANTS: THE EXTENDED BEHAVIORAL PHENOTYPE

Self-injurious behavior in Lesch-Nyhan syndrome (hypoxanthine-guanine phosphoribosyltransferase [HPRT] deficiency) was the first behavioral phenotype described. However, it is not the only distinguishing feature of this neurogenetic disorder. Affected children have a severe dystonic movement disorder, difficulty with expressive language, aggressive behaviors, and cognitive impairments. To understand the range of neurobiological dysfunction, we have focused our research on understanding pathways from

Table 1.3. Spectrum of hypoxanthine-guanine phosphoribosyltransferase (HPRT) deficiency: From genes to cognition and complex behavior

Gene sequencing
HPRT levels
Cerebrospinal fluid (CSF) measures
Studies of brain structure (magnetic resonance imaging [MRI])
Studies of brain chemistry using spectroscopy (magnetic resonance spectroscopy [MRS])
Combining positron emission tomography (PET) and MRS
Cognitive studies
PET and dystonia/self-injurious behavior
Correlations of MRI studies and cognition
Investigations of personality using the NEO Personality Inventory–Revised (Costa & McCrae, 1995)

genes to cognition and complex behavior in this disorder.[1] Table 1.3 shows the steps we have taken in seeking to understand its neurobiology. Self-injurious behavior is regularly found in classic Lesch-Nyhan syndrome with a HPRT level less than 1%. Lesch-Nyhan variants with 2% enzyme do not self-injure. However, they do have a similar dystonic movement disorder and have cognitive impairments. The cognitive impairments are not as severe as that found in classic Lesch-Nyhan syndrome.

To investigate the effects of HPRT deficiency on brain and behavior, we chose subjects with classic Lesch-Nyhan syndrome and others with up to 8% HPRT deficiency. An understanding of the neurobiological basis might contribute to a better understanding of brain mechanisms involved in self-injurious behavior, compulsive aggression, attentional deficits, and intellectual disability (Schretlen et al., 2005). In our studies we first sequenced the gene and measured the enzyme. Concurrently we carried out studies of brain structure using neuroimaging techniques, conducted studies of cognitive functioning, completed behavioral rating scales, and carried out personality inventories.

Lesch-Nyhan syndrome is a rare (1:380,000; Crawhall, Henderson, & Kelley, 1972), sex-linked recessive disease caused by an inborn error of purine nucleotide metabolism. It is caused by an almost complete deficiency of the enzyme HPRT, which is involved in the purine salvage (purine base recycling) pathway (Harris, 1998b). The clinical features are hyperuricemia, intellectual disability, dystonic movement disorder with early hypotonia (Jinnah et al., 2006), dysarthric speech, aggression, and compulsive self-injury. The HPRT-encoding gene is located on the X chromosome in the q26-q27; over 200 different mutations throughout the coding regions have been recognized. This enzyme is normally present in each cell in the body and is highest in the basal ganglia. Its absence results in excessive uric acid production and gout without specific drug treatment (i.e., allopurinol). The full disease requires the virtual absence of the enzyme. Page and Nyhan (1989) have shown that HPRT levels are linked to the extent of motor impairment, presence or absence of self-injury, and cognitive function. Uric acid is not produced in the brain and does not cross the blood-brain barrier.

---

[1]Collaborators on the Lesch Nyhan Research Project: Hyder Jinnah, M.D., Ph.D. (neurology); David Schretlen, Ph.D. (neuropsychology); Peter Barker, Ph.D. (magnetic resonance spectroscopy); Dean Wong, M.D., Ph.D. (positron emission tomography imaging); George Thomas, Ph.D. (genetics); William Nyhan, M.D., Ph.D. (clinical genetics); and Hugo Moser, M.D. (neurology)

Behavioral Phenotype in Lesch-Nyhan Syndrome

Although self-injurious behavior in Lesch-Nyhan syndrome usually is expressed as self-biting, other patterns of self-injurious behavior may emerge over time. The biting pattern usually involves the fingers, mouth, and buccal mucosa and often is asymmetrical. Other associated maladaptive behaviors include head or limb banging, eye poking, pulling fingernails, and psychogenic vomiting. It is a compulsive behavior that the child tries to control but generally is unable to resist. As he grows older he may be better at finding means of self-control and self-restraint. He may ask the help of others to protect him against these impulses. Vulgar speech is characteristic, as is compulsive aggression. The patient may injure through pinching or grabbing others. He often will apologize for this behavior immediately afterward, saying that the behavior was not under his control.

Etiology of Lesch-Nyhan Syndrome

The cause for the neurological and behavioral symptoms in HPRT deficiency is not fully understood. The behavior is not caused by hyperuricemia or excess hypoxanthine, inasmuch as partial variants with hyperuricemia (HPRT levels > 2.0%) do not self-injure. Infants with the classic presentation who are treated for hyperuricemia from birth do develop self-injury. The dystonic movement disorder suggests basal ganglia involvement. Three autopsied cases in the 1980s (Lloyd et al., 1981) demonstrated dopamine deficiency. When it became possible to image ligands binding to the dopamine system by positron emission tomography (PET), we conducted in vivo studies in adult patients (over 18 years of age) with HPRT deficiency, with a ligand that binds to the presynaptic dopamine transporter (Wong et al., 1996).

The first person studied had an IQ in the normal range and gave his own consent. Subsequently we documented reductions in dopamine transporter density of 68% in the putamen and 42% in the caudate in six people with classic Lesch-Nyhan syndrome (< 1% HPRT) and self-injurious behavior.

To further clarify whether the PET findings were linked to the movement disorder or to both the movement disorder and the self-injurious behavior, we extended our PET studies to patients with HPRT levels ranging from 2% to 8% enzyme. These studies sought to identify neurobiological correlates linked to the movement disorder across the spectrum of HPRT deficiency. We studied Lesch-Nyhan variant patients with HPRT levels of 1.8%–8% and two patients with HPRT levels less than 1.5%, using neuroimaging techniques. None of the variants (age range: 12–37 years) had self-injurious behavior. The extent of their motor involvement was documented by a quantitated neurological examination. The two patients with HPRT levels less than 1.5% and two other patients with HPRT levels of 1.8% and 2.5% had severe movement disorder essentially identical to that of the classic cases. Their dopamine transporter binding studied by PET imaging did not differ from that of classic Lesch-Nyhan patients. We concluded from the study of variant cases with motor symptoms but without self-injurious behavior that reductions in dopamine receptor density are not sufficient to explain the self-injury. Still, the extent of motor deficit was correlated with dopamine transporter binding in the putamen in all cases. When the motor disorder was rated on a dystonia rating scale, the putamen dopamine transporter density was significantly correlated with neurological severity. Thus dopamine reduction is linked to the extent of the movement disorder but may not be a sufficient explanation for self-injurious behavior; other neurotransmitters need to be examined in regard to self-injury. Why partial HPRT deficiency does not lead to self-injury remains unclear; perhaps trophic factors are active during brain development and require only minute amounts of the enzyme to affect brain development.

Magnetic Resonance Spectroscopy Studies

PET imaging of the dopamine system with a ligand that binds to the dopamine transporter is sufficient to study dopamine in the striatum but not to evaluate extrastriate brain systems. Thus we turned to magnetic resonance spectroscopy and studied both classic and Lesch-Nyhan variant cases for this purpose. Using N-acetyl-aspartate (NAA) as the neuronal marker, we compared the classic cases and variant cases of Lesch-Nyhan syndrome with normal control subjects. Similar to our PET findings for ligands, NAA reductions were documented in both the classic cases and the variants of Lesch-Nyhan syndrome when compared with controls. Thus a different neuroimaging approach, the magnetic resonance spectroscopy studies, provided additional documentation for the caudate/putaman dysfunction. Moreover, we found a statistically significant reduction in the orbitomedial cortex between classic and variant cases of Lesch-Nyhan syndrome and control subjects. The findings in the orbitomedial cortex, a brain region linked to emotional control, are consistent with affect dysregulation found in our patients.

Neuropsychological Test Performance

Individuals with classic Lesch-Nyhan syndrome (Lesch and Nyhan, 1964) are reported to have intellectual disabilities. We sought to determine the severity of intellectual disability and to establish their neuropsychological profile. We compared 15 patients with classic Lesch-Nyhan syndrome with 9 Lesch-Nyhan syndrome variants and 13 typical adolescents and adults (Schretlen, Harris, Park, Jinnah, & del Pozo, 2001). The mean IQ was 108 in our control sample, 72 for the Lesch-Nyhan variants, and 59 for our classic Lesch-Nyhan syndrome patients. Although variant subjects had higher average IQ scores, on neuropsychological testing they showed cognitive impairment profiles similar to those found in classic Lesch-Nyhan syndrome. Thus neuropsychological testing revealed qualitatively similar cognitive impairments in the two patient groups. Moreover, the variants produced scores that were intermediate between those of patients with Lesch-Nyhan syndrome and typical participants on nearly every cognitive measure. Family members completed the NEO Personality Inventory–Revised (Costa & McCrae, 1995) for both classic and variant patients. The classic cases and variants again differed from the control subjects, showing a profile that was high on neuroticism and extroversion and low on conscientiousness.

These studies document that there is an extended phenotype for HPRT deficiency. Moreover, there is continuity in cognitive findings throughout HPRT levels that range from less than 1% to 8% enzyme.

Treatment

The treatment of Lesch-Nyhan syndrome is multidisciplinary. It begins with medical management of hyperuricemia with allopurinol, treatment of associated illnesses, and management of self-mutilation associated with self-injurious behavior. Occupational and speech therapies are needed for assistance with activities of daily living and communication. Both supportive psychotherapy and behavior interventions are needed for self-injury and aggression. Pharmacotherapy is used as an adjunctive treatment for mood modulation, aggression, and self-injury. Finally, deep brain stimulation is being evaluated as a means to examine the electrophysiological characteristics of limbic and motor globus pallidus internus in classic cases and as a potential treatment for dystonia and severe self-injury (Pralong et al., 2005).

Summary

Our studies of Lesch-Nyhan syndrome and its variants document links to dysfunction in prefrontal striatal circuits (Visser, Bär, & Jinnah, 2000). These patients have severe motor disability, with extrapyramidal findings characteristic of dysfunction of the motor circuits of the basal ganglia. They demonstrate neuropsychological impairments, affect dysregulation, and difficulties in behavioral control that suggest disruption of other circuits of the basal ganglia. Consistent with these findings is a striking reduction (60%–90%) in the dopamine content of the basal ganglia in classic cases and lesser reductions in Lesch-Nyhan syndrome variants. These findings link the neurobehavioral features of Lesch-Nyhan syndrome to dysfunction of the basal ganglia.

CONCLUSIONS

With advances in assessment methodology and better understanding of behavior and brain functioning, it is increasingly possible to study pathways from genes to cognition and complex behaviors in neurogenetic syndromes. Behavioral phenotypes are increasingly being linked to brain dysfunction in neurogenetic disorders. Moreover, links between behavioral phenotypes and psychopathology are an ongoing area of research. Knowledge of behavioral phenotypes is critical in treatment and essential for anticipatory parental counseling for children with neurodevelopmental disorders.

REFERENCES

American Psychiatric Association. (2000). *Diagnostic and statistical manual of mental disorders* (4th ed., text rev.). Washington, DC: Author.

Castellanos, F.X., Lee, P.P., Sharp, W., Jeffries, N.O., Greenstein, D.K., Clasen, L.S., et al. (2002). Developmental trajectories of brain volume abnormalities in children and adolescents with attention-deficit/hyperactivity disorder. *JAMA, 288*(14), 1740–1748.

Costa, P.T., & McCrae, R.R. (1995). *NEO Personality Inventory–Revised*. Port Huron, MI: SIGMA Assessment Systems.

Crawhall, J.C., Henderson, J.F., & Kelley, W.N. (1972). Diagnosis and treatment of the Lesch-Nyhan syndrome. *Pediatric Research, 6,* 504–513.

Critchley, M., & Earl, C.J.C. (1932). Tuberous sclerosis and allied conditions. *Brain, 55,* 311–346.

Diamond, A. (2001). A model system for studying the role of dopamine in prefrontal cortex during early development in humans. In C. Nelson & M. Luciana (Eds.), *Handbook of developmental cognitive neuroscience* (pp. 433–472). Cambridge, MA: MIT Press.

Diamond, A. (2007). Consequences of variations in genes that affect dopamine in prefrontal cortex. *Cerebral Cortex, 17*(Suppl. 1), i161–i170.

Diamond, A., Ciaramitaro, V., Donner, E., Djali, S., & Robinson, M.B. (1994). An animal model of early-treated PKU. *Journal of Neuroscience, 14*(5, Pt. 2), 3072–3082.

Diamond, A., & Herzberg, C. (1996). Impaired sensitivity to visual contrast in children treated early and continuously for phenylketonuria. *Brain, 119*(Pt. 2), 523–538.

Diamond, A., Prevor, M.B., Callender, G., & Druin, D.P. (1997). Prefrontal cortex cognitive deficits in children treated early and continuously for PKU. *Monographs of the Society for Research in Child Development, 62*(4), i–v, 1–208.

Down, J.L. (1990). *Mental affectations of childhood and youth*. London: Mac Keith Press. (Originally published in 1887)

Dykens, E.M., & Hodapp, R.M. (2007). Three steps toward improving the measurement of behavior in behavioral phenotype research. *Child and Adolescent Psychiatric Clinics of North America, 16*(3), 617–630.

Fletcher, R., Loschen, E., Stavrakaki, C., & First, M. (Eds.). (2007). *Diagnostic manual-intellectual disability (DM-ID): A clinical guide for diagnosis of mental disorders in persons with intellectual disability*. Kingston, NY: National Association of the Dually Diagnosed.

Flint, J. (1996). Annotation: Behavioural phenotypes: A window onto the biology of behaviour. *Journal of Child Psychology and Psychiatry, 37,* 355–367.

Flint, J. (1998). Behavioral phenotypes: Conceptual and methodological issues. *American Journal of Medical Genetics, 81,* 235–240.

Flint, J., & Yule, W. (1994). Behavioural phenotypes. In M. Rutter, E. Taylor, & L. Hersov (Eds.), *Child and adolescent psychiatry* (3rd ed., pp. 666–687). Oxford, England: Blackwell Scientific.

Gath, A., & Gumley, D. (1986). Behavior problems in retarded children with special reference to Down's syndrome. *British Journal of Psychiatry, 149,* 156–161.

Gunn, P., Berry, P., & Andrews, R.J. (1981). The temperament of Down's syndrome in infants: A research note. *Journal of Child Psychology and Psychiatry, 22,* 189–194.

Harris, J.C. (1998a). Behavioral phenotypes. In J.C. Harris, *Assessment, diagnosis and treatment of the developmental disorders* (pp. 251–374). New York: Oxford University Press.

Harris, J.C. (1998b). Lesch-Nyhan disease. In J.C. Harris, *Assessment, diagnosis and treatment of the developmental disorders* (pp. 306–319). New York: Oxford University Press.

Harris, J.C. (2001). Behavioral phenotypes: Portals into the developing brain. In K. Davis, D. Charney, J. T. Coyle, & C. Nemeroff (Eds.), *Neuropsychopharmacology: The 5th generation of progress.* Baltimore: Lippincott Williams & Wilkins.

Harris, J., Wong, D.F., Jinnah, H.A., Schretlen, D., & Barker, P. (2002). Neuroimaging studies in Lesch-Nyhan syndrome and Lesch-Nyhan variants. In S. Schroeder, M. Oster-Granite, and T. Thompson (Eds.), *Self-injurious behavior: Gene-brain-behavior relationships* (pp. 269–278). Washington, DC: American Psychological Association.

Jinnah, H.A., & Friedmann, T. (2001). Hypoxanthine-guanine phosphoribosyltransferase deficiency: Lesch-Nyhan syndrome and gout. In C.R. Scriver, A.L. Beaudet, W.S. Sly, & D. Valle (Eds.), *The metabolic and molecular basis of inherited disease* (8th ed., pp. 2537–2570). New York: McGraw-Hill.

Jinnah, H.A., Visser, J.E., Harris, J.C., Verdu, A., Larovere, L., Ceballos-Picot, I., et al. (2006). Delineation of the motor disorder of Lesch-Nyhan disease. *Brain, 129*(Pt. 5), 1201–1217.

Kalkanoğlu, H.S., Ahring, K.K., Sertkaya, D., Møller, L.B., Romstad, A., Mikkelsen, I., et al. (2005). Behavioural effects of phenylalanine-free amino acid tablet supplementation in intellectually disabled adults with untreated phenylketonuria. *Acta Paediatrica, 94*(9), 1218–1222.

Lesch, M., & Nyhan, W.L. (1964). A familial disorder of uric and acid metabolism and central nervous system function. *American Journal of Medicine, 36,* 561–570.

Lloyd, K.G., Hornykiewicz, O., Davidson, L., Shannak, K., Farley, I., Goldstein, M., et al. (1981). Biochemical evidence of dysfunction of brain neurotransmitters in the Lesch-Nyhan syndrome. *New England Journal of Medicine, 305,* 1106–1111.

Nyhan, W. (1972). Behavioral phenotypes in organic genetic disease. Presidential address to the Society for Pediatric Research, May 1, 1971. *Pediatric Research, 6,* 1–9.

Page, T., & Nyhan, W.L. (1989). The spectrum of HPRT deficiency: An update. *Advances in Experimental Medicine and Biology, 253A,* 129–132.

Pennington, B.F., VanDoornick, W.J., McCabe, L.L., & McCabe, E.R.B. (1985). Neuropsychological deficits in early treated phenylketonuric children. *American Journal of Mental Deficiency, 89,* 467–474.

Pralong, E., Pollo, C., Coubes, P., Bloch, J., Roulet, E., Tétreault, M.H., et al. (2005). Electrophysiological characteristics of limbic and motor globus pallidus internus (GPI) neurons in two cases of Lesch-Nyhan syndrome. *Neurophysiologie Clinique, 35*(5–6), 168–173.

Schretlen, D.J., Harris, J.C., Park, K.S., Jinnah, H.A., & del Pozo, N.O. (2001). Neurocognitive functioning in Lesch-Nyhan disease and partial hypoxanthine-guanine phosphoribosyltransferase deficiency. *Journal of the International Neuropsychological Society, 7*(7), 805–812.

Schretlen, D.J., Ward, J., Meyer, S.M., Yun, J., Puig, J.G., Nyhan, W.L., et al. (2005). Behavioral aspects of Lesch-Nyhan disease and its variants. *Developmental Medicine and Child Neurology, 47,* 673–667.

Shaw, P., Lerch, J., Greenstein, D., Sharp, W., Clasen, L., Evans, A., et al. (2006). Longitudinal mapping of cortical thickness and clinical outcome in children and adolescents with attention-deficit/hyperactivity disorder. *Archives of Genera: Psychiatry, 63*(5), 540–549.

Skuse, D.H. (2000). Behavioural phenotypes: What do they teach us? *Archives of Disease in Childhood, 82,* 222–225.

Steinhausen, H.C., Von Gontard, A., Spohr, H.L., Hauffa, B.P., Eiholzer, U., Backes, M., et al (2002). Behavioral phenotypes in four mental retardation syndromes: Fetal alcohol syndrome, Prader-Willi syndrome, fragile X syndrome, and tuberous sclerosis. *American Journal of Medical Genetics, 111*(4), 381–387.

Tam, S.Y., Elsworth, J.D., Bradberry, C.W., & Roth, R.H. (1990). Mesocortical dopamine neurons: High basal firing frequency predicts tyrosine dependence of dopamine synthesis. *Journal of Neural Transmission, 81*, 97–110.

Visser, J.E., Bär, P.R., & Jinnah, H.A. (2000). Lesch-Nyhan disease and the basal ganglia. *Brain Research Brain Research Reviews, 32*(2–3), 449–475.

Wong, D.F., Harris, J.C., Naidu, S., Yokoi, F., Marenco, S., Dannals, R.F., et al. (1996). Dopamine transporters are markedly reduced in Lesch-Nyhan disease *in vivo. Proceedings of the National Academy of Sciences USA, 93*, 5539–5543.

2

Smith-Magenis Syndrome

A Neurobehavioral
Syndrome with Genetic Underpinnings

Andrea L. Gropman and Ann C.M. Smith

Smith-Magenis syndrome (SMS) is a multiple congenital anomaly and mental retardation syndrome. It is caused by a de novo interstitial deletion of chromosome 17p11.2 (including the *RAI1* gene) or a mutation in the *RAI1* gene (Seranski et al., 2001; Slager, Newton, Vlangos, Finucane, & Elsea, 2003; Smith et al., 1986). This deletion results in haploinsufficiency for the retinoic acid induced 1 (*RAI1*) gene. Heterozygous mutations of *RAI1* also cause the SMS phenotype in a smaller subset of patients without deletions. *RAI1* encodes for a novel gene whose role remains unclear. Studies in mouse models suggest that *RAI1* may act as a transcriptional regulator that plays a role in embryonic and postnatal development.

The diagnosis of SMS is based on the recognition of a unique and complex pattern of clinical findings consisting of physical, developmental, and behavioral features that evolve with age. There is a range of cognitive abilities from impaired to typical IQ; however, neurocognitive testing of people with SMS is challenging because of extreme maladaptive behaviors present in these individuals. The characteristic behavioral phenotype seen in SMS is characterized by maladaptive, self-injurious, and aggressive behaviors; stereotypies; and sensory integration disorders. SMS also features a chronic sleep disturbance associated with an inverted circadian rhythm of melatonin secretion. Sleep deprivation may also modify the neurobehavioral phenotype. The treatment of problem behaviors remains an area of active research.

The estimated prevalence of SMS deletion cases was reported to be 1 in 25,000 (Greenberg et al., 1991). However, new cases identified in the last decade as a result of improved molecular cytogenetic techniques, including microarray technology, now suggest the incidence to be closer to 1 in 15,000 births (Elsea & Girirajan, 2008). Despite this increase in the number of cases in recent years, clinical diagnosis based on phenotypic recognition is often delayed.

DIAGNOSIS

The SMS phenotype encompasses age-specific findings, including distinctive craniofacial and skeletal features, history of infantile hypotonia, significant expressive language delay, cognitive and intellectual disability, stereotypies, neurobehavioral problems, and a sleep disorder due to an inverted pattern of melatonin secretion (De Leersnyder et al., 1999, 2001a; Potocki et al., 2000). The characteristic features of SMS are shown in

---

This chapter is dedicated to the families and individuals with Smith-Magenis syndrome and written in special memory of Frank Greenberg, M.D., a dear friend and genetics colleague.

Table 2.1. Smith-Magenis syndrome (SMS) across the life span: Characteristic neurodevelopmental and behavioral features in infants and toddlers

Infant with SMS	Toddler with SMS

Dysmorphisms—diagnostic pearls	**Dysmorphisms—diagnostic pearls**
Down-turned mouth	Down-turned mouth
Brachycephaly	Brachycephaly
Down-slanting palpebral fissures	Down-slanting palpebral fissures
Flat mid-face	Flat mid-face
Failure to thrive	Pes planus or pes cavus
Lethargic	Engaging personality
Parent perception as "good sleepers"	Frequent night awakenings and daytime naps
"Quiet good babies"	Sleep disturbance: short sleep cycle; early risers
Complacent	(5:30 A.M.–6:30 A.M.)
Diminished vocalizations and crying	Decreased pain sensitivity
Neurodevelopmental examination features/ objective findings	**Neurodevelopmental examination features/ objective findings**
Generalized hypotonia	Developmental delays
Delayed gross/fine motor skills	Gross/fine motor delays
Age-appropriate social skills	Marked speech delay (expressive and receptive)
Decreased total sleep for age	Delayed toilet training
	Sensory integration issues
	Stereotypic behaviors: self-hugging, lick and flip behaviors
	Self-abusive behaviors: head banging, hitting, skin picking

Tables 2.1 and 2.2. There are several clinical features that are common among deletion cases but occur significantly less often in *RAI1* mutation cases, including cardiovascular and genitourinary tract abnormalities, hearing loss, hypotonia, and short stature (Edelman et al., 2007; Girirajan et al., 2006).

DIFFERENTIAL DIAGNOSIS

Delayed diagnosis in SMS syndrome is very common. Infants come to clinical attention because of hypotonia and several facial stigmata resembling Down syndrome, includ-ing flat mid-face and down-slanting palpebral fissures (Allanson, Greenberg, & Smith, 1999; Gropman, Duncan, & Smith, 2006), and this may prompt cytogenetic analysis. Other initial diagnoses considered in infants and children with SMS include Prader-Willi syndrome, because of infantile hypotonia, lethargy, and feeding and sleep disorders; DiGeorge or velocardiofacial syndrome (del 22q11.2), because of marked speech delay and cardiac anomalies; and fragile X syndrome, because of autism-like features and behaviors (Behjati, Mullarkey, Bergbaum, Berry, & Dochery, 1997). In addition, many of the children have been diagnosed with autism/pervasive

Table 2.2. Smith-Magenis syndrome (SMS) across the life span: Characteristic neurodevelopmental and behavioral features in school-age children and adolescents/adults

School-age child with SMS	Adolescent/Adult with SMS

Dysmorphisms—diagnostic pearls	**Dysmorphisms—diagnostic pearls**
Coarsening of facial features	Coarsening of facial features
Relative prognathism	Relative prognathism
Chronic sleep disturbance	Chronic sleep disturbance
	Reports of exercise intolerance
Behavior	**Behavior**
Attention-seeking behaviors	Major behavioral outbursts or rage behaviors, property destruction
Frequent outbursts and tantrums	
Sudden mood shifts	Attention-seeking behavior
Impulsivity/aggression	Aggressive/explosive outbursts
Stereotypic behaviors	Impulsive, disobedient
Self-injurious behaviors: hitting self, nail biting or pulling, object insertion into body orifices	Self-injurious behaviors (hitting self/nail yanking, object insertion)
Pes planus or pes cavus	Mood shifts (rapid) without major provocation; attention deficits; argumentative
Bed-wetting	Self-hug, upper-body spasmodic squeeze
	Mouthing of objects, bruxism, licking and flipping of pages in a book
Neurodevelopmental findings	Body rocking, spinning and twirling of objects
Very communicative	**Neurodevelopmental findings**
Cognitive delays	Very communicative
Excellent long-term memory	Cognitive delays
Sensory integration issues	Poor adaptive function
	Excellent long-term memory

developmental disorder because of significant early speech and language delay and the presence of sensory integration disorders and stereotypic and maladaptive behaviors (Gropman et al., 2006).

GENETICS OF SMITH-MAGENIS SYNDROME

As alluded to, there are two genetic mechanisms leading to SMS syndrome: 1) interstitial deletion of chromosome 17p11.2 or 2) a heterozygous *RAI1* mutation. The deletion associated with SMS was first reported in 1982 in two severe patients (Smith, McGavran, & Waldstein, 1982), and the clinical spectrum was further delineated in 1986 (Smith et al., 1986; Stratton et al., 1986). In the majority of cases the deletion is confirmed by a karyotype of at least 550-band resolution accompanied by fluorescent in situ hybridization (FISH) using an *RAI1*-specific probe. Approximately 95% of deletions are detected with this method (Gropman, Elsea, Duncan, & Smith, 2007;

Vlangos, Wilson, Blancato, Smith, & Elsea, 2005), of which 70% are due to a common deletion spanning 3.5 Mb. The remaining 30% are either smaller or larger deletions (Elsea & Girirajan, 2008). Repeat karyotype/FISH is recommended in individuals with a prior "normal" study that was performed more than 5 years ago or with lower-level chromosomal banding. This is especially relevant when SMS was not in the differential diagnosis or when an early suspected diagnosis of trisomy 21 was not confirmed.

In 2003, Slager et al. identified heterozygous mutations in the *RAI1* gene in three individuals with classic features of SMS who did not have detectable deletions. He showed that the majority (~70%) of features associated with SMS result from a loss of *RAI1* function (Girirajan et al., 2006). Therefore, patients with the SMS phenotype who do not have a chromosome 17p11.2 deletion should next be tested for mutations in *RAI1*.

Newer DNA-based molecular techniques, such as quantitative real-time polymerase chain reaction (qPCR) and multiplex ligation-dependent probe amplification (MLPA) to determine gene copy number (e.g., *RAI1*) and targeted chromosome microarrays for comparative genomic hybridization (CGH), play a role in the diagnosis of clinically suspected cases (Elsea & Girirajan, 2008; Truong et al., 2008). To date, all identified *RAI1* mutations occur within exon 3, including nonsense mutations, single-to-multiple nucleotide deletions, and missense mutations (Elsea & Girirajan, 2008).

SMS is a contiguous gene syndrome (Greenberg et al., 1991). Haploinsufficiency of physically linked but functionally unrelated genes is responsible for phenotypic variability (Figure 2.1). Clinical variability exists among individuals even with the same deletion size, thus suggesting that other gene(s) in the deletion interval account for the phenotypic variability observed (Girirajan et al., 2006; Potocki, Shaw, Stankiewicz, & Lupski, 2003).

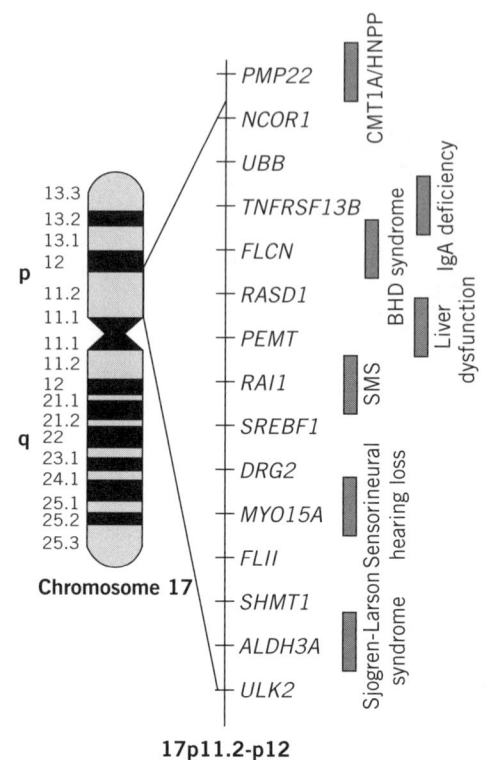

17p11.2-p12

Figure 2.1. Chromosomal region deleted in Smith-Magenis syndrome (SMS), showing the genes in the region and flanking the SMS critical region. Disorders localized to 17p11.2–p12 are indicated. The gene associated with each disorder is indicated to the left. Data on individual genes associated with a particular disorder/syndrome were obtained from published sources and core Smith-Magenis syndrome (SMS) features (*RAI1*). (*Key:* BHD syndrome, Birt-Hogg-Dube syndrome [a rare complex genetic skin disorder—genodermatosis—characterized by the development of benign skin tumors—hamartomas—affecting the head, face, and upper torso.]; CMT1A/HNPP, Charcot-Marie-Tooth disease type 1/hereditary neuropathy with pressure palsies; IgA, immunoglobulin A.) (From Gropman, A.L., Elsea, S., Duncan, W.C., Jr., & Smith, A.C. [2007]. New developments in Smith-Magenis syndrome [del 17p11.2]. *Current Opinion in Neurology 20*[2], 130; adapted by permission.)

Most cases of SMS occur de novo, and hence with low recurrence risk (Howard-Peebles, Friedman, Harrod, Brookshire, & Lockwood, 1985; Zori et al., 1993). At least two cases of monozygotic affected twins (Hicks, Ferguson, Bernier, & Lemay, 2008; Kosaki, Okuyama, Tanaka, Migita, & Kosaki, 2007) and one family with two affected siblings due to maternal mosaicism for the 17p11.2 deletion (Juyal et al., 1995; Smith, Magenis, & Elsea,

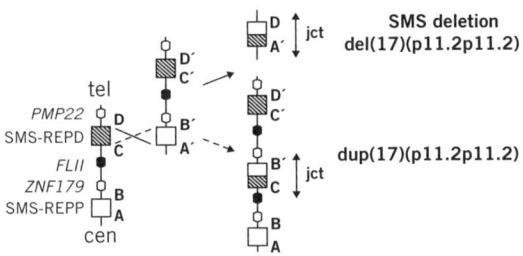

Figure 2.2. The mechanism of the deletion in Smith-Magenis syndrome (SMS) is believed to be due to nonhomologous recombination involving low-copy repeat clusters that flank the SMS interval and is a common mechanism observed in other microdeletion syndromes (Prader-Willi/Angelman syndrome, velocardiofacial syndrome, Williams syndrome).

2005) have been reported. Therefore, parental cytogenetic analyses are highly recommended for all newly diagnosed cases. Random parental origin of the 17p deletion has been documented, suggesting that imprinting does not play a role (Greenberg et al., 1991).

The mechanism leading to SMS involves nonallelic homologous recombination of flanking low-copy repeat gene clusters referred to as SMS-REPs (Chen et al., 1997), which flank genomic regions prone to deletion, duplication, and inversion and act as substrates for inter- and intrachromosomal recombination (Figure 2.2). This mechanism has also been documented in other contiguous gene syndromes such as Williams syndrome, Prader-Willi/Angelman syndrome, and DiGeorge/velocardiofacial syndrome (Lupski, 1998).

CARDINAL FEATURES OF SMITH-MAGENIS SYNDROME

Neurobehavioral Phenotype of Smith-Magenis Syndrome

A striking neurobehavioral pattern consisting of stereotypies, hyperactivity, polyembolokoilamania (insertion of objects into body orifices), onychotillomania (nail yanking), and maladaptive, self-injurious, and aggressive behavior has been recognized in SMS and becomes more apparent with increasing age

(Gropman et al., 2006). These behaviors affect the ability to obtain accurate assessment of the cognitive and developmental status of individuals with SMS. In addition, this is compounded by the presence of marked speech delay, which makes traditional cognitive test batteries inappropriate and difficult to interpret and validate.

Infant Phenotype

Infants with SMS possess a characteristic cherubic facial appearance (100%) and hypotonia (100%). Other common features include delayed gross motor skills, hyporeflexia (84%), generalized lethargy/sleepiness (100%), complacency (100%) and oromotor dysfunction (100%) contributing to feeding difficulties with failure to thrive, and delayed expressive language skills (100%). Social skills appear to be age appropriate (80%) (Gropman et al., 2006). Parents report infrequent crying (95%), and babbling and vocalizations are decreased despite normal hearing tests. Infants with SMS are described by their parents as "good babies."

Sleep disturbances start to appear in infancy, manifested as hypersomnolence and lethargy. Data derived from actigraphy, a noninvasive means of measuring rest/activity, document the emergence of the sleep dysfunction characterized by fragmented sleep with reduced 24-hour sleep time beginning at 6–9 months of age (Duncan, Gropman, Morse, Krasnewich, & Smith, 2003; Gropman et al., 2006) (Figure 2.3).

Feeding difficulties are seen with some frequency in SMS due to hypotonia, lethargy, oral motor dysfunction, gastroesophageal reflux, and/or poor sucking and swallowing (Gropman et al., 2006; Solomon, McCullagh, Krasnewich, & Smith, 2002; Sonies et al., 1997). Nasogastric or gastrostomy tube feedings may be required. Additional findings that can affect early feeding include weak bilabial seal (64%), palatal anomalies including velopharyngeal insufficiency (75%),

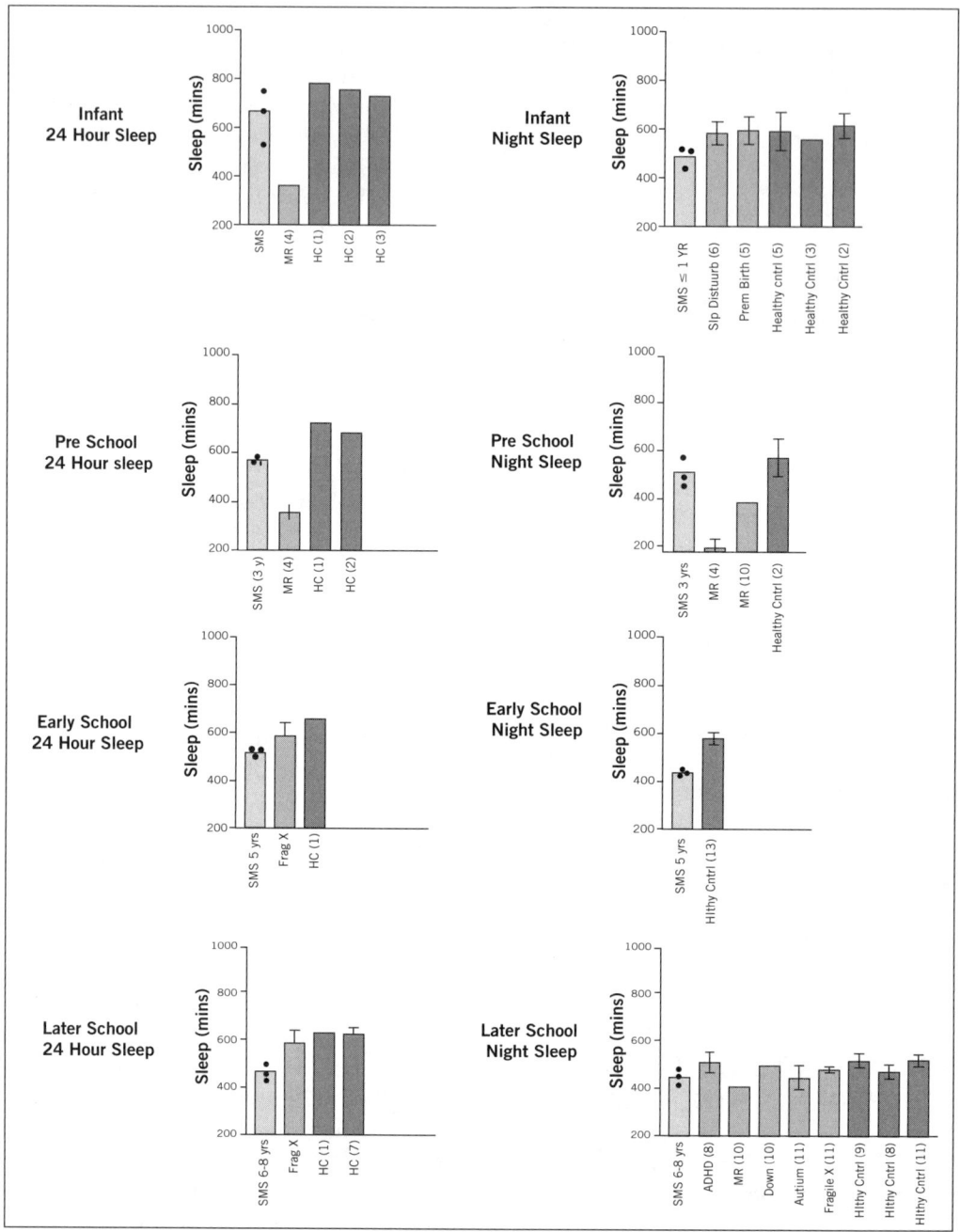

Figure 2.3. Twenty-four-hour sleep actigraphy recording in children with Smith-Magenis syndrome (SMS) (*light gray*) as compared with other patient groups showing estimated sleep (*left*) and nighttime sleep (*right*). SMS children are compared against a group with intellectual disability (*gray*) and typically developing age-matched controls (*dark gray*). Four age groups are arranged from top to bottom: infants (1 year, row one), preschool (3 years, row two), early school (5 years, row three), and later school (6–8 years, row four). Note the lower estimated 24-hour and night sleep in children with SMS, especially compared with healthy control subjects of the same age. Children with SMS often get less sleep than other children with developmental disabilities, although some children with more severe intellectual disability appear to have more severe sleep problems. The estimated sleep from the comparison child groups was derived from actigraphy, video, sleep logs, or electroencephalography. (Reprinted from *Pediatric Neurology 34*[5], Gropman, A.L., Duncan, W.C., & Smith, A.C., Neurologic and developmental features of the Smith-Magenis syndrome [del 17p11.2], 346, Copyright 2006, with permission from Elsevier.)

and open mouth posture with tongue protrusion (less than 30%) (Solomon et al., 2002). Oral-motor dysfunction may result in refusal of highly textured foods and contribute to failure to thrive.

Childhood Phenotype

The physical features and behavioral phenotype of SMS become more recognizable and pronounced with advancing age. Clinicians are most knowledgeable about this stage of SMS, and a diagnosis is more likely to be made once the typical behaviors of SMS emerge. Unusual maladaptive, self-injurious, and stereotypic behaviors occur in 40%–100% of both children and adults with SMS. The prevalence of self-injurious behaviors is nearly universal (96%) and is directly correlated with age and level of intellectual functioning (Finucane, Dirrigl, & Simon, 2001).

Onychotillomania and polyembolokoilamania (Finucane et al., 2001; Greenberg et al., 1991) appear to be unique to SMS, as are two stereotypic behaviors: the spasmodic upper-body squeeze or "self-hug" (Finucane, Konar, Givler, Kurtz, & Scott, 1994) and a hand-licking and page flipping or "lick and flip" behavior (Dykens, Finucane, & Gayley, 1997; Dykens & Smith, 1998). In older children, developmental delay, particularly of expressive language, as well as emerging sleep and neurobehavioral difficulties, dominates the picture (Gropman et al., 2006; Madduri et al., 2006; Martin, Wolters, & Smith, 2006) and may bring patients to clinical attention.

Although sleep disturbance is present at 1 year, the stereotypies and self-injurious behaviors generally do not begin to emerge until after the first 18 months of life. The sleep disturbance and self-abusive behaviors also appear to escalate with age; they are specifically recognizable and reportable at 18–24 months, at school age, and again with the onset of puberty. Reduced 24-hour and total night sleep for age leads to a chronic daytime sleep debt that is compensated for by increased daytime somnolence (napping);

settling difficulties appear to be more prevalent after the age of 10 years (Gropman et al., 2007).

DEVELOPMENT AND BEHAVIOR

Developmental delay or intellectual disability is found in all affected individuals, ranging from profound to borderline functioning. Cross-study comparisons are difficult to directly compare, because of the differences in ages and instruments used. Greenberg et al. (1996) found IQs ranging from 20 to 78 in a series of 27 people, with the majority in the moderate mental retardation range (at 40–54), using the Stanford-Binet (Thorndike, Hagen, & Sattler, 1986). However, 6 of 25 evaluated had IQs in the mild and one in the borderline range. Udwin, Webber, and Horn (2001) examined a larger British cohort of 29 children (younger than 16 years of age) and 21 adults (over 16 years) and reported IQs of 50 in 75% of the children on the Wechsler Intelligence Scale for Children, Third Edition (WISC–III; Wechsler, 1991). Overall, the adult group had higher IQs, with 16 scoring in the range of 50–69 and only 5 scoring below 50 on the Wechsler Adult Intelligence Scale–Revised (WAIS–R; Wechsler, 1981), but showed poor adaptive skills with increased reliance on caregivers, especially in daily living skills. No decline in cognitive function was found among adults up to age 50 years.

Children with SMS uniformly demonstrate significant delays in adaptive behavior, including communication and daily living skills and socialization skills. Age was found to be inversely related to Daily Living Skills on the Vineland Adaptive Behavior Scales (Sparrow, Balla, & Cicchetti, 1984) ($r = -0.68$, $p < .001$) in at least one study, suggesting that proficiency in activities of daily living may plateau in early adolescence, leading to increased dependency on others for support (Martin et al., 2006).

Martin et al. (2006) studied 19 children between the ages of 2 and 12 years, using either the Bayley Scales or the Stanford-Binet Intelligence Scales, Fourth Edition (SBIS–IV)

(Thorndike et al., 1986), and found the majority function in the mild (67%) to moderate mental retardation (MR)/intellectual disability range and 28% in the borderline range of intellectual functioning; only one scored in the low average range of cognitive ability.

Madduri studied 58 subjects (age 1.5 years to 29 years) and reported that cognitive and adaptive abilities were inversely related to deletion size (Madduri et al., 2006). She used a combination of neuropsychological batteries, including the Clinical Linguistic and Auditory Milestone Scale Developmental Quotient (DQ) (Capute et al., 1986) in individuals whose developmental age was less than 3 years, the SBIS-IV, the WISC-III, or the WAIS–R. Intelligence testing was determined based on developmental age. Individuals with larger deletions were found to have significantly lower IQs/DQs (severe to profound MR range) and lower adaptive behavior composite scores compared with individuals with smaller or common deletions. There is no apparent correlation of patients with *RAI1* mutations and cognitive performance (Girirajan et al., 2006).

The neurocognitive profile for SMS is beginning to be recognized. Dykens et al. (1997) described specific cognitive profiles with relative strengths in long-term memory and perceptual closure and relative weaknesses observed in sequential processing and short-term memory. These early findings were confirmed in a larger English cohort of 40 individuals (Udwin et al., 2001) who demonstrated strengths in long-term memory, computer skills, and perceptual skills, in contrast to areas of weakness in visual motor coordination, sequencing, and response speed. Deficits in sensory processing and modulation appear to be prevalent in SMS (personal experience; Hildenbrand & Smith, 2008). Many children exhibit tactile and auditory defensiveness and appear to have problems with depth perception and gravitational insecurity.

Significant speech/language delay, with or without associated hearing loss, occurs in over 90% of individuals with SMS. To the clinician, the absence of age-appropriate babbling or vocalizations should be a major clue leading to further examination of the child for other minor features consistent with SMS. Expressive language skills in early childhood consist primarily of the use of gestures and signs in children younger than 4 years, with some verbal language emerging about 4–5 years of age (Smith et al., 2005; Solomon et al., 2002). With aggressive speech/language therapy, including sign language and a total communication program, fairly understandable expressive language is usually present by school age. Speech parameters include hypernasality with a harsh, hoarse vocal quality. Speech intensity may be mildly elevated, with a rapid rate and moderate explosiveness (Solomon et al., 2002; Sonies et al., 1997).

Dykens and Smith (1998) examined the distinctiveness and correlates of maladaptive behavior as well as the prevalence of self-injurious and stereotypical behaviors in this population. They evaluated 35 children with SMS, using the Childhood Behavior Checklist (Achenbach, 1991) score, and compared these children with age- and gender-matched subjects with Prader-Willi syndrome or mixed mental retardation. All but four subjects with SMS (89%) demonstrated significantly elevated maladaptive behavior scores compared with their counterparts; 12 behaviors differentiated the groups with 100% accuracy. The most frequent stereotypies seen in the SMS group involved the mouth in some way.

Additional diagnosis of co-morbid psychopathologies/conditions, including obsessive-compulsive disorder (OCD), attention deficit and/or hyperactivity disorder (ADD/ADHD), autism/pervasive developmental disorder (PDD), is common in SMS (Finucane et al., 2001; Smith et al., 2005).

Because of significant early speech/language delays, several young children have been diagnosed with an autism spectrum disorder (ASD) prior to confirmation of SMS. Prospective assessment with the Childhood Autism Rating Scale (CARS) (Schopler,

Reichler, & Renner, 1986) among 19 children with SMS (Martin et al., 2006) revealed scores ranging from 24 to 37 (mean total score 31.43, *SD* 4.3), which fall within the mild to moderate range of autism (total scores 30–36.5). Hicks et al. (2008) recently reported monozygous twins with SMS with scores consistent with ASDs.

TREATMENT OF BEHAVIOR IN SMITH-MAGENIS SYNDROME

A behavioral treatment plan must be developed as soon as problem behaviors arise. This is not straightforward, however, as there is generally not one medication that is effective in the majority of children with SMS, or across the developmental stages. A search for organic causes of adverse behavior should be investigated, especially in the child with SMS who is nonverbal or has moderate impairments, as gastroesophageal reflux disease, constipation, and otitis media are present with some frequency in this syndrome. The use of medications to control behaviors has had mixed results in this population (Smith, Dykens, & Greenberg, 1998a), and this is an active area of interest. Adverse reactions to some medications have also been reported (A.C.M. Smith & A. Gropman, unpublished observations).

Unpublished medication history data on 12 children with SMS ages 3–16 years yield a median number of five medication trials; only two children were not on medication therapy, and one of these was enrolled in a strict behavior-modification program (W. Allen & A.C.M. Smith, unpublished observations, 1997). Older stimulant drugs, in our experience, are not particularly useful in controlling behavior or increasing attention span in patients with SMS (W. Allen & A.C.M. Smith, unpublished observations, 1997).

Greenberg et al. (1996) reported some or only transient behavior improvement for several individuals with SMS, both with and without seizures, with carbamazepine. Similar unpublished results (personal experience) were found for eight individuals with SMS. With anticonvulsant treatment, improvements were seen in three individuals on carbamazepine and valproic acid. Among the 12 unpublished SMS medication histories, of 3 children tried on benzodiazepines, clonazepam showed improvement (1) or no change (1), and lorazepam showed improvement (1). Use of tricyclic antidepressants (clomipramine or imipramine) was associated with worsening of behavior. In at least two recent cases, older teens treated with atomoxetine exhibited adverse effects that included a significant decline in sleep time coupled with major escalation of behaviors, especially agitation, and self-injurious and aggressive outbursts leading to psychiatric admission in one case (authors' experience). Recent use of specific serotonin reuptake inhibitors (specifically sertraline and fluoxetine) has shown considerable improvement with respect to behavioral outbursts and sleep for at least three individuals with SMS (Smith et al., 1998a).

SLEEP DISORDER IN SMITH-MAGENIS SYNDROME

An inverted melatonin secretion pattern in which daytime levels are high and nighttime levels are low (i.e., opposite the normal pattern) is pathognomonic of SMS (Figure 2.4). Past studies have consistently documented fragmented and shortened sleep cycles characterized by frequent and prolonged nocturnal awakenings, early sleep offset (mean 5:30 A.M.), and excessive daytime sleepiness (De Leersnyder et al., 2001a; Greenberg et al., 1996; Potocki et al., 2000; Smith et al., 1998b). Data derived from objective sleep measures (i.e., 24-hour polysomnography, multiple sleep latency test, wrist actigraphy, and/or sleep log diaries) are consistent with an advanced sleep phase that is a recognized circadian sleep disorder (De Leersnyder et al., 2001b; Potocki et al., 2000; Smith & Duncan, 2005).

The presence of elevated daytime melatonin levels with low nighttime levels suggests

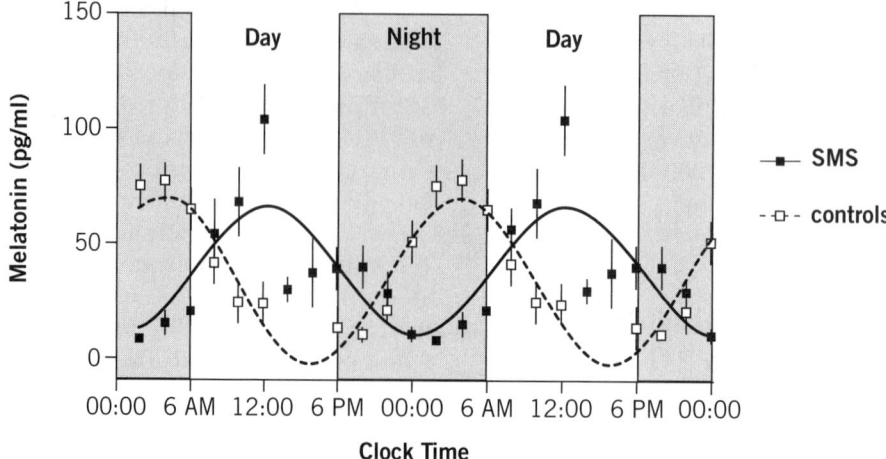

Figure 2.4. The inverted pattern of plasma melatonin is shown in eight children with Smith-Magenis syndrome (SMS) (*solid line, filled squares*) as compared with 15 control subjects without disabilities (*dotted line, open squares*). The lines represent the best-fit sine curves to each data set based on minimal least-squares criteria. The peak of the 24-hour curve is at night (~3 A.M.–4 A.M.) in the control subjects; the peak occurs during the day (noon) in patients with SMS. (*Key:* pg/ml, picograms per milliliter.) (Reprinted from *Pediatric Neurology 34*[5], Gropman, A.L., Duncan, W.C., & Smith, A.C., Neurologic and developmental features of the Smith-Magenis syndrome [del 17p11.2], 347, Copyright 2006, with permission from Elsevier.)

possible therapeutic approaches. De Leersnyder et al. (2001b, 2003) used the daytime β_1-adrenergic antagonist acebutolol (10 milligrams [mg]/kilograms [kg] at 8:00 A.M.) to reduce daytime melatonin, coupled with an evening oral dose of controlled-release melatonin (6 mg at 8 P.M.), to restore nocturnal plasma melatonin levels. Although this uncontrolled trial demonstrated a more normal circadian rhythm of melatonin as well as improved behavior in nine children with SMS, the results may have been biased because of parent expectations. The time of administration is important, because melatonin can have phase-shifting properties when taken at different times. In addition, low therapeutic doses of 0.5–2.5 mg should be used, because higher dosages (5–10 mg) can result in increased daytime levels of melatonin.

NEUROLOGICAL ASPECTS OF SMITH-MAGENIS SYNDROME

Individuals with SMS have demonstrated both central and peripheral nervous system symptoms. Seizures occur in only 11%–30% of individuals with SMS (Goldman et al., 2006; Greenberg et al., 1991, 1996; Gropman et al., 2006; Potocki, Lynch, Glaze, & Walz, 2002). Electroencephalogram abnormalities have been reported in approximately 25% of affected individuals in the absence of a clinical history of seizures (Goldman et al., 2006; Greenberg et al., 1996; Gropman et al., 2006). There is no single seizure type or electroencephalogram finding that is characteristic of SMS. An isolated case of infantile spasms was reported in a 9-month-old girl with SMS (Roccella & Parisi, 1999). Recognition and treatment of seizures is important and may improve attention, behavior, sleep, and overall cognitive functioning. However, careful selection of anticonvulsant is necessary, as adverse side effects of medications such as excessive lethargy, hyperactivity, and irritability have been reported in children with SMS.

Nonspecific central nervous system structural abnormalities have been documented by neuroimaging in over half of affected individuals (Masuno, Asano, Arai, Kuwahara, & Orii, 1992). Computed tomography (CT) scans performed on 25 individuals demonstrated ventriculomegaly in nine, enlargement of cisterna magna in two, and partial absence of the cerebellar vermis in one (Greenberg et al., 1996). Similar findings were seen among a group of 10 children who had undergone previous magnetic resonance imaging: five had ventriculomegaly, two had enlarged posterior fossa, and three had normal scans (Gropman et al., 2006). Despite the clinical finding of oromotor dysfunction, to date no structural abnormalities of the opercular cortex, which subserves these functions, have been reported. Studies employing magnetic resonance imaging volume-based morphometrics and positron emission tomography have reported significant bilateral decrease of gray matter volumes in the insular and lenticular nuclei in SMS children (Boddaert et al., 2004), with hypoperfusion observed in the same regions, based on a small series of patients studied.

Neuropathological study of the initial patient reported by Smith et al. (1982, 1986), who was deleted for the entire 17p11.2 band, showed microcephaly and foreshortened frontal lobes with depletion of neurons frontally. A small choroid plexus hemangioma was also noted in the lateral ventricle (Smith et al., 1986). Moya Moya disease has been reported in one individual (Girirajan et al., 2007). Moya Moya disease is a rare disorder characterized by progressive intracranial vascular stenoses of the circle of Willis, resulting in successive ischemic events.

Stroke has been reported in at least three individuals with SMS; thus, its true prevalence remains undetermined at present (Chaudhry, Schwartz, & Singh, 2007). In this isolated patient, the stroke was believed to be due to accelerated hypercholesterolemia. Cholesterol abnormalities have been reported previously in SMS. Determination of the risk factors for and true incidence of stroke in SMS requires additional study.

Clinical signs of peripheral neuropathy are reported in approximately 75% of individuals with SMS (Greenberg et al., 1996; Gropman et al., 2006). People with SMS have a characteristic appearance of the leg muscles similar to that observed in peripheral nerve syndromes or neuropathies, and either pes cavus or pes planus deformity. Hammer toes are also frequently seen in individuals with SMS, and decreased sensitivity to pain is suspected based on clinical behavior and appearance of increased pain tolerance. Markedly flat or highly arched feet (pes planus or cavus) and unusual gait (foot flap) are generally appreciated in childhood. During early infancy and childhood, signs of peripheral nervous system involvement include hyporeflexia (84%) and decreased sensitivity to pain (Gropman et al., 2006), and hypotonia (100%) is common. Despite suggestion of involvement of the peripheral nervous system, the hypotonia observed in SMS is likely due to a central abnormality. Affected individuals with SMS tend to toe walk despite the absence of tightened heel cords. In one series, distal muscle weakness was present in over half of the individuals examined, and a previously undescribed peripheral neuropathy tremor in the upper extremity (6–8 hertz) was evident in 21% (Gropman et al., 2006).

Peroneal motor nerve conduction velocities are generally normal in childhood. Delayed motor nerve conduction velocities due to biopsy-confirmed segmental demyelination and remyelination similar to that seen in hereditary neuropathy with liability to pressure palsy occur rarely (Greenberg et al., 1996; Smith et al., 1986; Zori et al., 1993). PMP22 located at 17p12 (distal to the SMS critical region) is usually not deleted in SMS (Chevillard et al., 1993; Greenberg et al., 1991; Moncla et al., 1993). It is possible that other genes in the critical region may play a role. It is not believed that the neuropathy observed in SMS is progressive.

Because of their relative insensitivity to pain, individuals with SMS may cause injury

to themselves by object insertion or persistent picking, self-biting, nail yanking, or self-
hitting during uncontrolled rages (Smith
et al., 1998a).

CONCLUSIONS

The incidence of SMS is reported to be 1 in
25,000, but with increasing awareness and
proper diagnosis, improved cytogenetics, and
new molecular techniques, it is expected to
be found to be higher. Most patients are
initially evaluated for other conditions; SMS
is considered later only as the neurobehavioral and sleep patterns become more
recognizable. The features change with age,
and the clinician needs to be aware of this
when evaluating the mildly dysmorphic, hy-
potonic infant who may not yet show the full
spectrum of behaviors. Although haploinsufficiency of *RAI1* is responsible for most of the
SMS features, the involvement of other genes
cannot be ruled out, as severity of the phenotype increases with increased deletion size.
It is possible that other genes or genetic background plays a role in altering the functional
availability of *RAI1* for downstream effects.
Further investigation into additional genes in
the 17p11.2 region is needed to determine
any role they may play in modification of the
features and/or severity of the SMS phenotype in deletion cases. Further studies
evaluating the effects of various classes of
medication on behavior need to be explored.

RESOURCES

Further information can be obtained from
Parents and Researchers Interested in Smith-
Magenis Syndrome (PRISMS) and United
PRISMS (www.prisms.org).

REFERENCES

Achenbach, T.M. (1991). *Childhood Behavior Checklist.* Burlington, VT: Department of Psychiatry,
University of Vermont.
Allanson, J.E., Greenberg, F., & Smith, A.C.M.
(1999). The face of Smith-Magenis syndrome:
A subjective and objective study. *Journal of Medical Genetics, 36,* 394–397.
Behjati, F., Mullarkey, M., Bergbaum, A., Berry,
A.C., & Dochery, Z. (1997). Chromosome deletion 17p11.2 (Smith-Magenis syndrome) in
seven new patients, four of whom had been
referred for fragile-X investigation. *Clinical Genetics, 51,* 71–74.
Boddaert, N., De Leersnyder, H., Bourgeois, M.,
Munnich, A., Brunelle, F., & Zilbovicius, M.
(2004). Anatomical and functional brain imaging evidence of lenticulo-insular anomalies
in Smith-Magenis syndrome. *Neuroimage, 21,*
1021–1025.
Caputo, A.J., Palmer, F.B., Shapiro, B.K., Wachtel,
R.C., Schmidt, S., & Ross, A. (1986). Clinical
Linguistic and Auditory Milestone Scale:
Prediction of cognition in infancy. *Developmental Medicine and Child Neurology, 28,* 762–771.
Chaudhry, A.P., Schwartz, C., & Singh, A.K.
(2007). Stroke after cardiac surgery in a
patient with Smith-Magenis syndrome. *Texas
Heart Institute Journal, 34(2),* 247–249.
Chen, K.S., Manian, P., Koeuth, T., Potocki, L.,
Zhao, Q., Chinault, C.A., et al. (1997). Ho-
mologous recombination of a flanking repeat
gene cluster is a mechanism for a common
contiguous gene deletion syndrome. *Nature Ge-
netics, 17,* 154–163.
Chevillard, C., Le Paslier, D., Passage, E., Ougen,
P., Billault, A., Boyer, S., et al. (1993). Relation-
ship between Charcot-Marie-Tooth 1A and
Smith-Magenis regions. snU3 may be a candi-
date gene for the Smith-Magenis syndrome.
Human Molecular Genetics, 2(8), 1235–1243.
De Leersnyder, H., Bresson, J.L., deBlois, M.C.,
Souberbielle, J.C., Mogenet, A., Delhotal-
Landes, B., et al. (2003). ß1-adrenergic antag-
onists and melatonin reset the clock and
restore sleep in a circadian disorder, Smith-
Magenis syndrome. *Journal of Medical Genetics
40,* 74–78.
De Leersnyder, H., deBlois, M.C., Claustrat, B.,
Roman, S., Albrecth, U., VonKleist-Retzow,
J.C., et al. (2001a). Inversion of the circadian
rhythm of melatonin in Smith-Magenis
syndrome. *Journal of Pediatrics, 139,* 111–116.
De Leersnyder, H., deBlois, M.C., Vekemans, M.,
Sidi, D., Villain, E., Kindermans, C., et al.
(2001b). ß-Adrenergic antagonists improve
sleep and behavioral disturbances in a
circadian disorder, Smith-Magenis syndrome.
Journal of Medical Genetics, 38, 586–590.
De Leersnyder, H., Von Kleist-Retzow, J. C.,
Munnich, A., Claustrat, B., Lyonnet, S.,
Vekemans, M., et al. (1999). Inversion of the
circadian rhythm of melatonin in Smith-
Magenis syndrome. *American Journal of Human
Genetics, 65*(Suppl.), A2.

Duncan, W.C., Gropman, A., Morse, R., Krasnewich, D., & Smith, A.C.M. (2003). Good babies sleeping poorly: Insufficient sleep in infants with Smith-Magenis syndrome (SMS). *American Journal of Human Genetics 73*(5, Suppl.), A896.

Dykens, E., Finucane, B., & Gayley, C. (1997). Cognitive and behavioral profiles in persons with Smith-Magenis syndrome. *Journal of Autism and Developmental Disorders, 27,* 203–211.

Dykens, E., & Smith, A.C.M. (1998). Distinctiveness and correlates of maladaptive behavior in children and adolescents with Smith-Magenis syndrome. *Journal of Intellectual Disability Research, 42,* 481–489.

Edelman, E.A., Girirajan, S., Finucane, B., Patel, P.I., Lupski, J.R., Smith, A.C., et al. (2007). Gender, genotype, and phenotype differences in Smith-Magenis syndrome: a meta-analysis of 105 cases. *Clinical Genetics, 71*(6), 540–550.

Elsea, S.H., & Girirajan, S. (2008). Smith-Magenis syndrome. *European Journal of Human Genetics, 4,* 412–421.

Finucane, B., Dirrigl, K.H., & Simon, E.W. (2001). Characterization of self-injurious behaviors in children and adults with Smith-Magenis syndrome. *American Journal of Mental Retardation, 106*(1), 52–58.

Finucane, B.M., Konar, D., Givler, B.H., Kurtz, M.B., & Scott, L.I. (1994). The spasmodic upper-body squeeze: A characteristic behavior in Smith-Magenis syndrome. *Developmental Medicine and Child Neurology, 36,* 70–83.

Girirajan, S., Mendoza-Londono, R., Vlangos, C.N., Dupuis, L., Nowak, N.J., Bunyan, D.J., et al. (2007). Smith-Magenis syndrome and Moyamoya disease in a patient with del(17)(p11.2p13.1). *American Journal of Medical Genetics A, 143A*(9), 999–1008.

Girirajan, S., Vlangos, C.N., Szomju, B.B., Edelman, E., Trevors, C.D., Dupuis, L., et al. (2006). Genotype-phenotype correlation in Smith-Magenis syndrome: Evidence that multiple genes in 17p11.2 contribute to the clinical spectrum. *Genetics in Medicine, 8*(7), 417–427.

Goldman, A.M., Potocki, L., Walz, K., Lynch, J.K., Glaze, D.G., Lupski, J.R., et al. (2006). Epilepsy and chromosomal rearrangements in Smith-Magenis syndrome [del(17)(p11.2p11.2)]. *Journal of Child Neurology, 21*(2), 93–98

Greenberg, F., Guzzetta, V., De Oca-Luna, R.M., Magenis, R.E., Smith, A.C.M., Richter, S.F., et al. (1991). Molecular analysis of the Smith-Magenis syndrome: A possible contiguous-gene syndrome associated with del(17)(p11.2). *American Journal of Human Genetics, 49,* 1207–1218.

Greenberg, F., Lewis, R.A., Potocki, L., Glaze, D., Parke, J., Killian, J., et al. (1996). Multidisciplinary clinical study of Smith-Magenis syndrome (deletion 17p11.2). *American Journal of Medical Genetics, 62,* 247–254.

Gropman, A.L., Duncan, W.C., & Smith, A.C. (2006). Neurologic and developmental features of the Smith-Magenis syndrome (del 17p11.2). *Pediatric Neurology, 34*(5), 337–350.

Gropman, A.L., Elsea, S., Duncan, W.C. Jr., & Smith, A.C. (2007). New developments in Smith-Magenis syndrome (Del 17p11.2). *Current Opinion in Neurology, 20*(2), 125–134.

Hicks, M., Ferguson, S., Bernier, F., & Lemay, J.F. (2008). A case report of monozygotic twins with Smith-Magenis syndrome. *Journal of Developmental and Behavioral Pediatrics* (1), 42–46.

Hildenbrand, H., & Smith, A.C.M. (2008). *Analysis of the Sensory Profile in children and adolescents with Smith-Magenis syndrome (SMS).* Manuscript in preparation.

Howard-Peebles, P.H., Friedman, J.M., Harrod, M.J., Brookshire, G.C., & Lockwood, J.E. (1985). A stable supernumerary chromosome derived from a deleted segment of 17p. *American Journal of Human Genetics, 37*(Suppl.), A97.

Juyal, R.C., Finucane, B., Shaffer, L.G., Lupski, J.R., Greenberg, F., Scott, C.I., et al. (1995). Letter to the Editor: Apparent mosaicism for del(17)(p11.2) ruled out by fluorescence in situ hybridization in a Smith-Magenis syndrome patient. *American Journal of Medical Genetics, 59,* 406–407.

Kosaki, R., Okuyama, T., Tanaka, T., Migita, O., & Kosaki, K. (2007). Monozygotic twins of Smith-Magenis syndrome. *American Journal of Medical Genetics A, 143*(7), 768–769.

Lupski, J.R. (1998). Genomic disorders: Structural features of the genome can lead to DNA rearrangements and human disease traits. *Trends in Genetics, 14,* 417–422.

Madduri, N., Peters, S.U., Voigt, R.G., Llorente, A.M., Lupski, J.R., & Potocki, L. (2006). Cognitive and adaptive behavior profiles in Smith-Magenis syndrome. *Journal of Developmental and Behavioral Pediatrics, 27*(3), 188–192.

Martin, S.C., Wolters, P.L., & Smith, A.C.M. (2006). Behavior in Smith-Magenis syndrome. *Journal of Autism and Developmental Disorders, 36*(4), 541–552.

Masuno, M., Asano, J., Arai, M., Kuwahara, T., & Orii, T. (1992). Interstitial deletion of 17p11.2 with brain abnormalities. *Clinical Genetics, 41,* 278–280.

Moncla, A., Prias, L., Arbex, O.F., Muscatelli, F., Mattei, M.-G., Mattei, J.-F., et al. (1993). Physical mapping of microdeletions of the chromosome 17 short arm associated with Smith-Magenis syndrome. *Human Genetics, 90,* 657–660.

Potocki, L., Glaze, D., Tan, D.X., Park, S.S., Kas-hork, C.D., Shaffer, L.G., et al. (2000).

Circadian rhythm abnormalities of melatonin in Smith-Magenis syndrome. *Journal of Medical Genetics, 37,* 428–433.

Potocki, L., Lynch, J.K., Glaze, D.G., Walz, K., Noebels, J.L., & Lupski, J.R. (2002). EEG abnormalities and epilepsy in Smith Magenis syndrome. *American Journal of Human Genetics, 71*(4, Suppl.), 260A.

Potocki, L., Shaw, C.J., Stankiewicz, P., & Lupski, J.R. (2003). Variability in clinical phenotype despite common chromosomal deletion in Smith-Magenis syndrome [del(17)(p11.2 p11.2)]. *Genetics in Medicine, 5*(6), 430–434.

Roccella, M., & Parisi, L. (1999). The Smith-Magenis syndrome: A new case with infantile spasms. *Minerva Pediatrica, 51,* 65–71.

Schopler, E., Reichler, R.J., & Renner, B.R. (1986). *The Childhood Autism Rating Scale (CARS) for diagnostic screening and classification of autism.* New York: Irvington.

Seranski, P., Hoff, C., Radelof, U., Hennig, S., Reinhardt, R., Schwartz, C.E.H., et al. (2001). *RAI1* is a novel polyglutamine encoding gene that is deleted in Smith-Magenis syndrome patients. *Gene, 270,* 69–76.

Slager, R.E., Newton, T.L., Vlangos, C.N., Finucane, B., & Elsea, S.H. (2003). Mutations in *RAI1* associated with Smith-Magenis syndrome. *Nature Genetics, 33,* 466–468.

Smith, A.C.M., & Duncan, W.C. (2005). Smith-Magenis syndrome: A developmental disorder with circadian dysfunction. In M.G. Butler & F.J. Meaney (Eds.), *Genetics of developmental disabilities.* Boca Raton, FL: Taylor and Francis Group.

Smith, A.C.M., Dykens, E., & Greenberg, F. (1998a). Behavioral phenotype of Smith-Magenis syndrome (del 17p11.2). *American Journal of Medical Genetics, 81,* 179–185.

Smith, A.C.M., Dykens, E., & Greenberg, F. (1998b). Sleep disturbance in Smith-Magenis syndrome (del 17p11.2). *American Journal of Medical Genetics, 81,* 186–191.

Smith, A.C., Magenis, R.E., & Elsea, S.H. (2005). Overview of Smith-Magenis syndrome. *Journal of the Association of Genetic Technologists, 31*(4), 163–167.

Smith, A.C.M., McGavran, L., Robinson, J., Waldstein, G., Macfarlane, J., Zonona, J., et al. (1986). Interstitial deletion of (17) (p11.2 p11.2) in nine patients. *American Journal of Medical Genetics, 24,* 393–414.

Smith, A.C.M., McGavran, L., & Waldstein, G. (1982). Deletion of the 17 short arm in two patients with facial clefts. *American Journal of Human Genetics, 34*(Suppl.), A410.

Solomon, B., McCullagh, L., Krasnewich, D., & Smith, A.C.M. (2002). Oral motor, speech and voice functions in Smith-Magenis syndrome children: A research update. *American Journal of Human Genetics, 71,* 271

Sonies, B.C., Solomon, B., Ondrey, F., McCullah, L., Greenberg, F., & Smith, A.C.M. (1997). Oral-motor and otolaryngologic findings in 14 patients with Smith-Magenis syndrome (17p11.2): Results of an interdisciplinary study. *American Journal of Human Genetics, 61*(Suppl.), A5.

Sparrow, S., Balla, D., & Cicchetti, D. (1984). *Vineland Adaptive Behavior Scales (VABS).* Circle Pines, MN: American Guidance Service.

Stratton, R.F., Dobyns, W.B., Greenberg, F., De Sana, J.B., Moore, C., Fidone, G., et al. (1986). Interstitial deletion of (17) (p11.2p11.2): Report of six additional patients with a new chromosome deletion syndrome. *American Journal of Medical Genetics, 24,* 421–432.

Thorndike, R.L., Hagen, E.P., & Sattler, J.M. (1986). *Stanford-Binet Intelligence Scale* (4th ed.). Itasca, IL: Riverside.

Truong, H.T., Solaymani-Kohal, S., Baker, K.R., Girirajan, S., Williams, S.R., & Vlangos, C.N., (2008). Diagnosing Smith-Magenis syndrome and duplication 17p11.2 syndrome by RAI1 gene copy number variation using quantitative real-time PCR. *Genetic Testing, 12*(1), 67–73.

Udwin, O., Webber, C., & Horn, I. (2001). Abilities and attainment in Smith-Magenis syndrome. *Developmental Medicine and Child Neurology, 43*(12), 823–828.

Wechsler, D. (1981). *Wechsler Adult Intelligence Scale–Revised.* San Antonio, TX: Harcourt Assessment.

Wechsler, D. (1991). *Wechsler Intelligence Scale for Children* (3rd ed.). San Antonio, TX: Harcourt Assessment.

Vlangos, C.N., Wilson, M., Blancato, J., Smith, A.C., & Elsea, S.H. (2005). Diagnostic FISH probes for del(17)(p11.2p11.2) associated with Smith-Magenis syndrome should contain the RAI1 gene. *American Journal of Medical Genetics A, 132A*(3), 278–282.

Zori, R.T., Lupski, J.R., Heju, Z., Greenberg, F., Killian, J.M., Gray, B.A., et al. (1993). Clinical, cytogenetic, and molecular evidence for an infant with Smith-Magenis syndrome born from a mother having a mosaic 17p11.2p12 deletion. *American Journal of Medical Genetics, 47,* 504–511.

3

Fragile X

Expansion of a Genetic Disorder

Walter E. Kaufmann

During the last decade, our knowledge of neurobehavioral abnormalities associated with mutations affecting the fragile X (*FMR1*) gene has expanded considerably. On one hand, there has been a refinement of the behavioral syndromes, particularly autism spectrum disorders, observed in individuals with *FMR1* full mutation. On the other, it has been well established that the intermediate *FMR1* premutation is associated with a range of neurologic and behavioral problems that vary according to age and gender. This chapter focuses on these multiple neurobehavioral abnormalities, following introductory sections on fragile X-associated disorders (FXDs), and concludes with a discussion on the unresolved issues in the field.

FMR1 MUTATIONS

The *FMR1* gene (fragile X mental retardation gene 1) was identified in a search for the etiology of fragile X syndrome (FXS) (Verkerk et al., 1991). However, it is currently accepted that *FMR1* mutations are associated with at least two additional disorders, termed here in conjunction with FXS as *FMR1*- or FXS-related disorders. The vast majority of mutations affecting *FMR1* involve a polymorphic CGG repeat located in the 5' untranslated region of the gene, which extends into exon 1. Over generations, this CGG repeat is expanded from the typical 4–45 repeats range to 55–200 repeats, termed premutation or carrier status, and eventually

to >200 repeats, termed full mutation or FXS (Kaufmann & Reiss, 1999). CGG repeats in the 46–54 range are referred to as a gray zone or intermediate allele (Dombrowski et al., 2002) (Figure 3.1). The precise boundaries between these CGG expansion categories vary according to the reference population; nonetheless, there is agreement that premutation and full-mutation alleles are linked to phenotypical manifestations (Loesch, Huggins, & Hagerman, 2004; Sherman, 2000). Gray zone expansion mutations are not clearly associated with a clinical profile, nor are they linked with increased instability of the CGG repeat, contrasting with premutation-level repeats that frequently expand to full mutation. Although premutation expansions result in mild, if any, reductions in the levels of *FMR1*'s product (fragile X mental retardation protein or FMRP), they are associated with increases in *FMR1* mRNA (2–8 times normal levels) (Kenneson, Zhang, Hagedorn, & Warren,

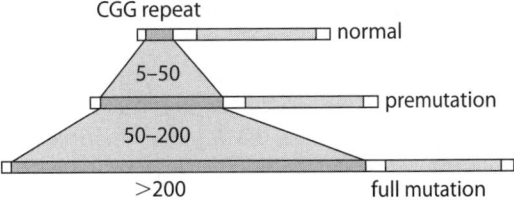

Figure 3.1. Expansion of CGG from 4–45 repeats range to 55–200 repeats (premutation or carrier status) and eventually to >200 repeats (full mutation or fragile X syndrome). (From Korf, R.B. [2000]. *Human genetics: A problem-based approach* [2nd ed., p. 221]. Malden, MA: Wiley-Blackwell; reprinted by permission.)

2001; Tassone et al., 2000). Full-mutation expansions result in marked decreases in FMRP because when the expansion reaches the 200 CGG repeats threshold, an atypical methylation of the CGG region and the upstream (promoter) CpG island occurs (Kaufmann, Abrams, Chen, & Reiss, 1999; Kaufmann & Reiss, 1999; Loesch et al., 2004; Pieretti et al., 1991; Tassone et al., 1999). As commonly seen in promoter CpG island methylation of genes, this phenomenon leads to transcriptional silencing (no mRNA) and therefore no protein (FMRP) synthesis (Sulewska et al., 2007). Thus, phenotypes observed in individuals with premutation are postulated to be the consequence of mRNA accumulation and associated toxicity and, to a lesser extent, mild FMRP deficits, whereas full mutation of FXS status would be the result of a severe FMRP deficit (Hagerman et al., 2009). The prevalence of FXS is approximately 1:4,000 in males and 1:6,000 in females (Hagerman, 2008; Sherman, 2002), making FXS the second most common genetic etiology of intellectual disability and the most frequent inherited severe neurodevelopmental disorder (Kaufmann & Moser, 2000). A small percentage of individuals with FXS present with a combination of premutation and full-mutation alleles, a pattern termed allele size mosaicism. An even smaller FXS subgroup shows full-mutation alleles that are incompletely methylated, a pattern termed methylation mosaicism. Because both forms of mosaicism lead to FMRP production, the phenotype of these individuals is a milder form of FXS. Premutation alleles are quite frequent, with conservative estimates of 1:800 in males and 1:250 in females (Dombrowski et al., 2002; Sherman, 2002).

Transmission of *FMR1* expansions follows the rules of X-linked inheritance, in that half of the offspring of the carrier mother and all of the females, but not the males, of the carrier father will inherit the mutation (Hagerman et al., 2009). Risk of expansion across generations depends on the gender of the carrier and the size of the CGG repeats, with the most frequent clinically relevant situation being a mother with premutation

and her son with full mutation (Hagerman et al., 2009). The two diagnostic assays employed to estimate the numbers of CGG repeats and the methylation status of *FMR1* are polymerase chain reaction and Southern Blot analysis (Kaufmann & Reiss, 1999; Stoyanova & Oostra, 2004). In a few cases, deletions or point mutations, rather than CGG repeat expansions, have been reported to be the cause of FXS, although in these situations the clinical presentation is slightly different from that of typical FXS (Coffee et al., 2008; De Boulle et al., 1993). Following classical X-linked inheritance, FXDs tend to be more severe in males.

FRAGILE X SYNDROME

The prototypical FXD, FXS, is characterized by a severe deficit (in many instances, virtual absence) of FMRP, a ubiquitous RNA-binding protein that predominantly inhibits protein synthesis. Therefore, it is not surprising that the FXS phenotype includes a wide range of physical and neurobehavioral abnormalities. Among the former are dysmorphic features, such as a long and narrow face, large ears, a prominent jaw; connective tissue abnormalities, prominent joint laxity, and other non–central nervous system anomalies like testicular enlargement (Table 3.1) (Oostra & Halley, 1995). Dysmorphic features are more prominent in postpubertal males, whereas young boys may only have large heads or no dysmorphia at all (Lachiewicz, Dawson, & Spiridigliozzi, 2000). For this reason, these physical characteristics should not constitute a diagnostic requirement. In terms of central nervous system involvement, most males and ~25% of females present with developmental delay or intellectual disability. Whereas females with FXS are typically in the borderline to mild intellectual disability range, most boys are more affected in the mild to moderate intellectual disability range (Hagerman et al., 1992). In addition to cognitive impairment, a large proportion of individuals with FXS display behavioral abnormalities. Other neurologic

Table 3.1. Physical and neurobehavioral features of fragile X syndrome phenotype

Physical features	Neurobehavioral features
Large ears	Variable intellectual disability
Thick nasal bridge	Visuospatial impairment
Prominent jaw	Decline in IQ with age
High-arched/narrow palate	Attentional-organizational dysfunction
Pale blue irises	Hyperactivity
Strabismus	Anxiety
Pectus excavatum	Autism-like features
Kyphoscoliosis	Mood instability with aggression and depression (adolescence)
Lax joints	Repetitive/stereotypic movements
Single palmar crease	Rapid/burst-like speech
Flat feet	Hypotonia
Cutis laxa	Nystagmus
Mitral valve prolapse	Seizures
Macroorchidism	

From Kaufmann, W.E., & Reiss, A.L. (1999). Molecular and cellular genetics of fragile X syndrome. *American Journal of Medical Genetics, 88,* 12; adapted by permission.

features include hypotonia, nystagmus, and seizures (Table 3.1). Nevertheless, the spectrum of cognitive and behavioral impairment in FXS is quite wide. Therefore, the diagnosis of the disorder is based on the presence of *FMR1* full mutation (including any mosaicism) and not on clinical features. The following sections discuss in detail the neurobehavioral features of FXS.

FRAGILE X–ASSOCIATED TREMOR/ATAXIA SYNDROME

Fragile X–associated tremor/ataxia syndrome (FXTAS) was recently described as a progressive degenerative disorder characterized by intention tremor, parkinsonism, and generalized brain atrophy in males with *FMR1* premutation (Hagerman et al., 2001). FXTAS is a relatively newly discovered disorder, and its clinical features have been refined over the last few years. Additional manifestations include gait difficulties (ataxia), resting tremor, neuropathic pain, peripheral neuropathy, autonomic dysfunction, dementia, and psychiatric disorders (mainly anxiety and mood disorders) (Hagerman & Hagerman, 2004). Neuroimaging is particularly informative about FXTAS; in addition to brain atrophy,

affecting posterior cerebral regions and the cerebellum, there may be periventricular white matter changes, and the characteristic T2-weighted or fluid-attenuated inversion recovery hyperintensities affecting the middle cerebellar peduncles (MCP sign) (Brunberg et al., 2002). Distinctive neuropathologic features of FXTAS include, in addition to atrophy (neuronal loss) and variable spongy white matter changes, eosinophilic, ubiquitin-positive intranuclear inclusions in neurons and astrocytes, but not oligodendrocytes, throughout the cortex and brainstem (Greco et al., 2002).

Current diagnostic criteria for FXTAS (Table 3.2) follow the format of those for other neurogenerative disorders: 1) possible FXTAS: *FMR1* premutation, intention tremor *or* gait ataxia *and* white matter lesions in the cerebrum *or* moderate generalized brain atrophy; 2) probable FXTAS: *FMR1* premutation, intention tremor *and* gait ataxia *or* MCP sign *and* a minor clinical feature: parkinsonism, executive function deficits, or moderate short-term memory deficiency; and 3) definite FXTAS: intention tremor *or* gait ataxia *and* MCP sign *or* intranuclear inclusions on postmortem examination (Jacquemont et al., 2003).

Unquestionably the most severe *FMR1*-related disorder, FXTAS is predominantly

Table 3.2. Diagnostic criteria of fragile X–associated tremor/ataxia syndrome (FXTAS)[a]

Definite FXTAS	Probable FXTAS	Possible FXTAS
Intention tremor *or* gait ataxia *and* Middle cerebellar peduncle (MCP) sign[b] *or* intranuclear inclusions on postmortem examination	Intention tremor *and* gait ataxia *or* MCP sign and a minor clinical feature: parkinsonism, executive function deficits, moderate short-term memory deficiency	Intention tremor *or* gait ataxia *and* White matter lesions in the cerebrum *or* moderate generalized brain atrophy

From Amiri, K., Hagerman, R.J., & Hagerman, P.J. (2008). Fragile X–associated tremor/ataxia syndrome: An aging face of the fragile X gene. *Archives of Neurology, 65*(1), 21; reprinted by permission. Copyright © 2008 American Medical Association. All rights reserved.

[a]Must be premutation carrier (55–200 CGG repeats).

[b]MCP: symmetric hyperintensities of the MCPs on T2-weighted or fluid-attenuated inversion recovery magnetic resonance imaging.

seen in males who are premutation carriers and older than 50 years (~40%). The median age of major motor manifestations in FXTAS is approximately 60 years. The penetrance of FXTAS movement disorder increases with age, with >50% males with premutation older than 70 displaying FXTAS features. Approximately 4–8% of females with premutation develop FXTAS, though the clinical severity tends to be milder. As a disorder associated with *FMR1* premutation, FXTAS pathogenesis is linked to increased *FMR1* mRNA. The toxicity of the latter would be reflected in the aforementioned inclusions, whose quantity is highly correlated with the number of CGG repeats, which represent protein sequestration leading to cellular dysfunction (Hagerman & Hagerman, 2004).

PRIMARY OVARIAN INSUFFICIENCY

Primary ovarian insufficiency (POI), previously termed premature ovarian failure, is historically the first disorder linked to *FMR1* premutation (Cronister et al., 1991; Table 3.3). POI affects approximately 20% of females with premutation and is defined as cessation of menses before age 40 years (Schwartz et al., 1994; Sherman, 2000; Vianna-Morgante & Costa, 2000). The prevalence of POI is nonlinearly correlated with the number of repeats within the premutation range (Sullivan et al., 2005). POI is characterized by elevated serum follicle-stimulating hormone

levels throughout the menstrual cycle and impaired ovarian function, which has led to the hypothesis that POI is the consequence of abnormal follicular function. Like FXTAS and perhaps other abnormalities associated with *FMR1* premutation, the pathogenesis of POI is linked to increased *FMR1* mRNA and its potential toxic effect. The fact that 0.8–7.5% of females with sporadic POI have *FMR1* premutation, and that this proportion increases to 13% in familial POI, underscores the conclusion that *FMR1* premutation should be excluded in women with POI (Conway et al., 1998; Mallolas et al., 2001; Marozzi et al., 2000; Murray, Webb, Grimley, Conway, & Jacobs, 1998; Sherman, Pletcher, & Driscoll, 2005).

NON-FXTAS FEATURES IN *FMR1* PREMUTATION

One of the most controversial issues in *FMR1*-related disorders is whether individuals with premutation could have distinctive disorders other than FXTAS and POI. Initial reports focused on females with premutation, ascertained because they were mothers of boys with full mutation (FXS), and mainly described emotional-behavioral problems in the anxiety and depression spectrum. Two main groups continue to raise the possibility of a non-FXTAS neurobehavioral profile in *FMR1* premutation: young males with intellectual disability, attention-deficit/hyperactivity disorder (ADHD), and/or autism spectrum

Table 3.3. Features of primary ovarian insufficiency (POI)

Prevalence	20% females with *FMR1* premutation
Impact of *FMR1* premutation	0.8%–7.5% sporadic POI
	13% familial POI
Clinical features	Menopause before 40 years
	Prevalence correlated with number of CGG repeats
	Elevated follicle-stimulating hormone (FSH) levels throughout menstrual cycle
	Other indices of impaired ovarian function

disorders (ASDs); and adults without FXTAS, but with a variety of neurologic and behavioral problems. Because the main focus of this chapter is the neurobehavioral problems associated with FXS and related disorders, these clinical presentations will be discussed in more detail in the following sections.

GENERAL NEUROBEHAVIORAL FEATURES IN FRAGILE X SYNDROME

As mentioned in the introductory section on FXS, the disorder is associated with a wide range of cognitive and behavioral manifestations in correspondence with the critical role of FMRP in synaptic development and function (Bassell & Warren, 2008). Global developmental delay, which evolves into intellectual disability, is the hallmark of FXS in males, though a small proportion of individuals score in the borderline range of intelligence (Hagerman et al., 2009). In our research studies, most boys with FXS without ASDs are in the 55–70 IQ range, whereas those with ASD comorbidity are in the 40–54 IQ range (Budimirovic et al., 2006; Kaufmann, Cortell, & Kau, 2004). This contrasts with females with FXS, a group in which only ~25% have an IQ below 70, although the majority exhibit learning problems (de Vries et al., 1996; Hagerman et al., 2009). Most boys and a small fraction of girls with FXS have a history of developmental delay, usually in multiple domains, and abnormal muscle tone (Kau, Meyer, & Kaufmann, 2002). Delay in gross motor skills is common, as is delay in language development (Kau et al., 2002).

Expressive language is typically more impaired than receptive language (Budimirovic et al., 2006); both are frequently associated with speech disturbances (Hagerman, 2002).

The behavioral phenotype of FXS reaches its maximum expression in boys, who display poor eye contact, excessive shyness, anxious behavior, hand flapping, hand biting, tactile defensiveness, tendency to aggressive behavior, hyperarousal in response to sensory stimuli, attention deficits, hyperactivity, impulsivity, and behavior characteristic of autism (Hagerman et al., 2009). In females with FXS, these features are milder and better described as shyness, social anxiety, features of ADHD combined with language and other learning deficits (e.g., math disability), and mood lability (Hagerman et al., 2009). In many individuals, these behavioral traits reach the diagnostic threshold of one or more categories listed by the *Diagnostic and Statistical Manual of Mental Disorders, Fourth Edition* (*DSM-IV;* American Psychiatric Association [APA], 1994) and its text revision (*DSM-IV-TR;* APA, 2000). The most common *DSM-IV* diagnoses in FXS are ADHD and its associated disruptive behavior disorder; a variety of anxiety disorders, most prominently Generalized Anxiety Disorder and social anxiety; autism spectrum disorders (Autistic Disorder and Pervasive Developmental Disorder-Not otherwise Specified [PDD-NOS]); and mood disorders (Bailey, Raspa, Olmsted, & Holiday, 2008; Hagerman, 2002; Hagerman et al., 2009).

Despite its reported high prevalence (84% attention problems and 66% hyperactivity in males, 67% attention problems and 30% hyperactivity in females, according to the National Parent Survey; Backes et al., 2000; Bailey et al., 2008), which is higher than that in the general population (Sullivan et al., 2006) and in other individuals with intellectual disability (Munir, Cornish, & Wilding, 2000), relatively limited research has been conducted on ADHD in FXS. All three subtypes of ADHD, inattentive, hyperactive, and combined, are recognized in boys with FXS. Boys receiving the diagnosis tend to have younger mental ages and lower FMRP levels (Sullivan et al., 2006). Moreover, boys with FXS with hyperactive subtype of ADHD have greater impairment in response inhibition, which appears to be a relative weakness of boys with FXS (Sullivan, Hatton, et al., 2007). Disturbance in the early development of executive functions (Cornish et al., 2004) and attentional selection (Scerif, Cornish, Wilding, Driver, & Karmiloff-Smith, 2007) seems to be contributory to the high frequency and severity of ADHD in boys with FXS.

Similarly, anxiety disorders are reported as very common conditions in FXS (70% in males, 56% in females, according to the National Parent Survey; Bailey et al., 2008; Hagerman et al., 2009). Nonetheless, because of the difficulties in evaluating anxiety symptoms in young individuals who have cognitive impairments, diagnosis and characterization of anxiety are limited (Sullivan, Hooper, & Hatton, 2007). The following section reviews in more detail social anxiety in the context of social interaction disorders. Mood disorders, particularly depression, are also relatively prevalent in individuals with FXS (12% in males, 22% in females, according to the National Parent Survey; Bailey et al., 2008; Hagerman, 2002). Similar to complex manifestations of anxiety (e.g., symptoms of general anxiety disorder), depression is more prominent in older and higher functioning individuals with FXS (e.g., females). Studies delineating behavioral profiles and factors influencing depressive symptoms emphasize their close relationship with anxiety and social withdrawal (Hessl et al., 2001; Sullivan, Hooper, & Hatton, 2007). As discussed

in the next section, further delineation of behavioral and neurobiological correlates of anxiety and social withdrawal is critical for more accurate diagnoses of anxiety and mood disorders in FXS. The emphasis of this chapter on ASDs and social anxiety reflects the importance of these behavioral syndromes in the management of FXS and FXDs, the increasing body of research in this area, and our contribution to FXS research.

SOCIAL INTERACTION DISORDERS IN FRAGILE X SYNDROME

Disorders affecting social interaction are among the most prevalent and severe in FXS. Social anxiety and ASDs are also among the most challenging FXS-associated disorders from the therapeutic viewpoint (Hagerman et al., 2009). Although most individuals with FXS show shyness or anxious features, and a high proportion of males display features characteristic of autism, only a fraction meets DSM-IV criteria for social anxiety and/or ASDs. During the last decade, ASDs in FXS have received considerable attention. Recent data confirm the high prevalence of Autistic Disorder and PDD-NOS in FXS (here termed ASDs), particularly among young males (46% in males, 16% in females, according to the National Parent Survey; Bailey et al., 2008; Hall, Lightbody, & Reiss, 2008; Harris et al., 2008). Studies have reported a wide range of ASD prevalence rates, from 15% to 60% (Bailey, Hatton, Skinner, & Mesibov, 2001; Bailey et al., 1998; Borghgraef, Fryns, Dielkens, Pyck, & Van den Berghe, 1987; Clifford et al., 2007; Dykens & Volkmar, 1997; Hagerman, Jackson, Levitas, Rimland, & Braden, 1986; Hall et al., 2008; Harris et al., 2008; Hatton et al., 2006; Kaufmann et al., 2004; Philofsky Hepburn, Hayes, Hagerman, & Rogers, 2004; Rogers, Wehner, & Hagerman, 2001), with approximately half of the cases meeting DSM-IV criteria for autism and the other half the DSM-IV criteria for PDD-NOS. Our work has shown

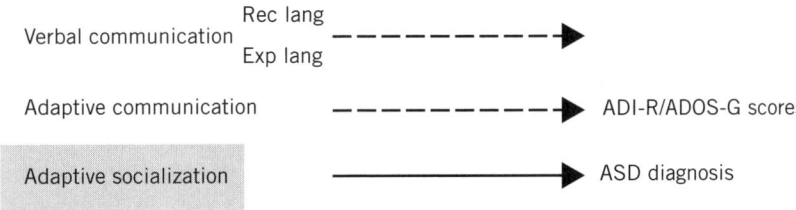

Figure 3.2. Diagram of the relationship between skills and autism spectrum disorder (ASD) in fragile X syndrome (FXS). Note that delay in socialization skills is a selective contributor to the diagnosis and severity (measure as ADI-R/ADOS-G scores) of ASD in FXS. (From Kaufmann, W.E., Capone, G.T., Clarke, M., & Budimirovic, D.B. [2008]. Autism in genetic intellectual disability: Insights into idiopathic autism. In A.W. Zimmerman [Ed], *Autism: Current theories and evidence* [p. 87]. Totowa, NJ: The Humana Press Inc.; reprinted by permission. Copyright © 2008 Humana Press. With kind permission of Springer Science+Business Media.) (*Key:* Rec, receptive; Exp, expressive; lang, language skills; ADI-R, Autism Diagnostic Interview–Revised [Lord, Rutter, & Le Couteur, 1994]; ADOS-G, Observation Schedule–Generic [Lord et al, 2000].)

figures of 25% and 20%, respectively (Kau et al., 2004; Kaufmann et al., 2004). These discrepancies reflect differences on ascertainment strategies and supportive diagnostic methods (Autism Diagnostic Interview–Revised [ADI–R], Lord, Rutter, & Le Couteur, 1994; Autism Diagnostic Observation Schedule [ADOS], Lord et al., 2000) employed in the study. We and others have delineated the behavioral profile associated with the diagnosis of ASDs in boys with FXS, because the significance of a *DSM-IV* diagnosis of ASDs in FXS and other genetic disorders is still unclear, considering that those *DSM-IV* criteria were developed for individuals without known genetic conditions. We have shown that, like atypical social behaviors (e.g., avoidant eye contact, social withdrawal, social anxiety) and repetitive and stereotyped behaviors (e.g., perseveration, hand flapping, self-injury) long recognized in FXS (Bailey et al., 1998; Budimirovic et al., 2006; Hagerman, 2002; Hall et al., 2008; Hatton, Bailey, Hargett-Beck, Skinner, & Clark, 1999; Kau et al., 2004; Merenstein et al., 1996), other more specific behaviors characteristic of autism (i.e., impairment in play) are also present in most young males with FXS. However, there is a clear differentiation between boys with FXS with and without ASDs in terms of selective deficits in peer interactions and socioemotional reciprocity (Kaufmann et al., 2004).

Interestingly, impairment in nonverbal basic social behaviors (e.g., impaired eye contact or social smile) and stereotypies (e.g., hand flapping) are not contributors to the *DSM-IV* diagnosis of ASDs in FXS, emphasizing its core deficit in complex social interactions. Overall, ASD in FXS shows striking similarities with that in the general population, particularly with groups that have language delays (Kau et al., 2004; Rogers et al., 2001).

Longitudinal data for ASDs are scarce but valuable, in that they support the distinctiveness and uniqueness of the behavioral syndrome. This is also the case for FXS; initial studies demonstrated that behavior characteristic of autism is relatively stable over time (Hatton et al., 2006; Sabaratnam, Murthy, Wijeratne, Buckingham, & Payne, 2003). In our recent 3-year longitudinal evaluation of boys with FXS, we demonstrated the stability of the *DSM-IV* diagnosis of autism, but not of PDD-NOS, and corroborated the aforementioned selective deficit in peer relationships (Hernandez et al., 2009). Furthermore, we confirmed our previous finding of specific deficit in adaptive socialization skills, and not in overall cognitive or language skills, in boys with FXS and ASDs (Hernandez et al., 2009; Kaufmann et al., 2004; Figure 3.2), which further supports ASDs as a distinctive social behavior subphenotype of FXS. We also reproduced our finding of severe social withdrawal in most boys with FXS and ASDs, particularly severe social avoidant behavior in late childhood (Budimirovic et al., 2006; Hernandez et al., 2009). Finally, the previously reported decline in the rate of acquisition of cognitive skills throughout childhood in FXS (Fisch et al., 1999; Fisch, Simensen, & Schroer, 2002) was observed only in boys without ASDs. Though lower, overall cognitive function remained relatively stable in boys with FXS and ASDs (Hernandez et al., 2009).

The progressive contribution of social avoidance to ASDs in FXS over time highlights the complex relationship between the shyness trait and anxious features and the diagnosis of ASDs in FXS. Although excessive shyness, anxious behavior, tactile defensiveness, and hyperarousal in response to sensory stimuli are well-accepted key features of the FXS neurobehavioral phenotype (Hagerman, 2002; Hagerman et al., 2009), their distinction as traits from clinically relevant problems is unclear. As exemplified by the results of the National Parent Survey, anxiety is the second most common behavioral abnormality (after ADHD) in individuals older than 6 years with FXS (Bailey et al., 2008). A few studies have surveyed the clinical significance of social avoidance or withdrawal. Merenstein and colleagues (1996) reported that 75% of young males with FXS display excessive shyness and anxiety and 50% report panic attacks, whereas Freund, Reiss, and Abrams (1993) reported that females with FXS also present with social anxiety, shyness, and avoidant personality. Although these publications describe the general features of individuals with FXS who meet *DSM-IV* criteria for different anxiety disorders, they are not specific regarding social anxiety and the delineation between the latter diagnosis and ASD. Studies by Cohen and colleagues (1988) and Bailey et al. (1998) emphasize that individuals with FXS are interested in social interaction but often display social anxiety-like and social withdrawal-like behavior in response to unfamiliar people and novel situations. Considering that the familiarity of people and places plays a critical role in the response of an

individual with FXS, current diagnostic criteria including rating scales appear to be inadequate, inasmuch as these approaches emphasize behavioral styles and not the dynamics of social interaction. Moreover, contrary to the availability of complementary instruments for ASD diagnosis (e.g., ADI–R, ADOS) applicable to FXS and similar disorders (Harris et al., 2008; Hepburn et al., 2008; Kaufmann et al., 2004; Molloy et al., 2009), there are no adequate instruments for assessing anxiety in individuals with low mental age (Sullivan, Hooper, & Hatton, 2007). Recent data on ASDs in the general population, demonstrating that social anxiety is the most common comorbidity (Simonoff et al., 2008), also raise the possibility that avoidance and not indifference (closely linked to core ASD features; Budimirovic et al., 2006) contributes to the diagnosis of both ASD and social anxiety.

Obviously, this is a complicated issue that may not be resolved only by studies of FXS. However, we have begun to approach the relationship between ASDs and social anxiety by in-depth analyses of standardized measures of social withdrawal and novel behavioral laboratory paradigms. We have classified items of the social withdrawal scales of the Aberrant Behavior Checklist (Aman, Singh, Stewart, & Field, 1985) and the Child Behavior Checklist (Achenbach & Rescorla, 2000) as either avoidance or indifference and demonstrated that the severity of both types of behaviors influences the diagnosis and severity of Autistic Disorder in boys with FXS in an age-dependent fashion, with avoidant behaviors being a strong correlate of ASDs in FXS after age 5 (Budimirovic et al., 2006; Hernandez et al., 2009). Furthermore, by applying clinically relevant cutoffs of the social withdrawal scales and confirmatory factor analyses, we have identified two groups of boys with FXS and severe social withdrawal (Kaufmann, Capone, Clarke, & Budimirovic, 2008). Boys with intermediate severe social withdrawal (SSW-I) display high scores on a wide range of behaviors with a predominance of avoidance items. On the other hand, boys in the high severe

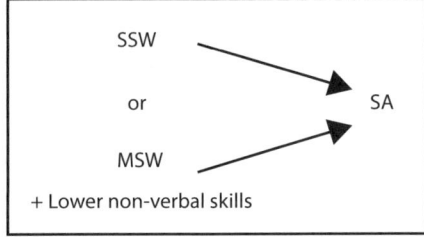 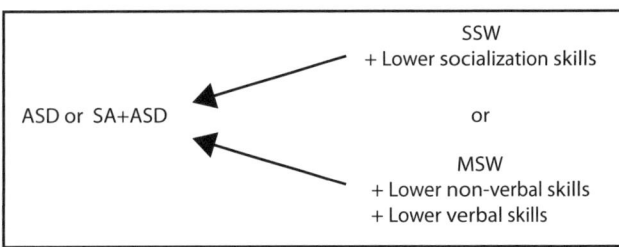

Figure 3.3. Model of the relationships among social withdrawal, cognitive impairment, social anxiety, and autism spectrum disorder (ASD) in fragile X syndrome (FXS). Left panel: Note that either severe social withdrawal (SSW) *per se* or mild social withdrawal (MSW) in conjunction with lower nonverbal skills would lead to social anxiety (SA). Right panel: A more complex combination of impairments, specifically the addition of lower socialization or verbal skills, is required for ASD alone or comorbid with SA. (From Kaufmann, W.E., Capone, G.T., Clarke, M., & Budimirovic, D.B. [2008]. Autism in genetic intellectual disability: Insights into idiopathic autism. In A.W. Zimmerman [Ed], *Autism: Current theories and evidence* [p. 88]. Totowa, NJ: The Humana Press Inc.; reprinted by permission. Copyright © 2008 Humana Press. With kind permission of Science+Business Media.)

social withdrawal (SSW-H) group show high scores on both avoidance and indifference items, and they are distinguished by their diverse and severe indifferent behavior profile. The SSW-H and SSW-I groups are selectively associated with ASDs (SSW-H group) or social anxiety (SSW-I group) (Kaufmann et al., 2008). Thus, these social withdrawal classifications provide a unified view of the relationship between ASDs and social anxiety in FXS and their underlying behavioral styles (Figure 3.3). Complementing these social withdrawal rating scale analyses, we began to apply the Social Approach Scale, a measure of multiple forms of social approach behavior in FXS (Roberts, Weisenfeld, Hatton, Heath, & Kaufmann, 2007). We found that the increase in stranger-approaching behaviors over time, which is observed in typically developing boys, is diminished and restricted mainly to physical movement in boys with FXS. Furthermore, boys with FXS and severe autism-like behavior showed persistent gaze avoidance, whereas those with high scores on shyness and social avoidance displayed particularly fearful facial expression during the initial social encounter. Related to these dynamic profiles, several lines of research connect the hypothalamic-pituitary-adrenocortical (HPA) axis with social withdrawal behaviors in the general population (Mathew & Ho, 2006) and in people with FXS

(Hessl et al., 2002, Hessl, Glaser, Dyer-Friedman, Reiss, 2006). Early work by Hessl and colleagues (2002) showed that individuals with FXS, particularly boys, have higher salivary cortisol levels at baseline and after social and cognitive challenges (i.e., delay in return to baseline) that correlate with more severe problem behavior, including higher scores on the social withdrawal. More recently, the same group found that cortisol reactivity in response to a social challenge at home was inversely correlated with gaze aversion (Hessl et al., 2006), an index of Autistic Behavior in our Social Approach Scale data (Roberts et al., 2007). These data suggest that in ASDs in FXS there is a profound disturbance in cortisol physiology, an apparent "up-regulation" with decreased reactivity that parallels the diminished behavioral response to social interaction, and that cortisol or other HPA axis measures may be valuable in delineating the relationship between ASDs and social anxiety in FXS.

NEUROBEHAVIORAL ABNORMALITIES IN *FMR1* PREMUTATION

The relationship between *FMR1* premutation and clinical manifestations, particularly neurologic and psychiatric features, has been one of the most controversial ones in the FXS literature. At present, it is well accepted that two disorders reviewed here, FXTAS and POI, are linked to *FMR1* premutation. However, there are multiple publications, from small series of cases to more structured evaluations of cohorts, reporting a wide variety of neurobehavioral manifestations. The recent National Parent Survey confirmed the presence of neurobehavioral abnormalities in a significant proportion of individuals with premutation older than 6 years (Bailey et al., 2008). Nonetheless, for most diagnoses, the proportion of affected subjects is lower than that reported for individuals with FXS (full mutation) (Bailey et al., 2008). This literature, divided into pediatric and adult cases, is briefly reviewed next.

Most publications dealing with children with premutation report on small groups of patients who are compared with siblings or other reference groups. The most frequent diagnoses are intellectual disability, ADHD, and ASDs (Aziz et al., 2003; Clifford et al., 2007; Farzin et al., 2006; Goodlin-Jones, Tassone, Gane, & Hagerman, 2004). Although most of these publications report on boys with premutation, girls with premutation can also present with cognitive impairment or an ASD. The National Parent Survey corroborated these publications: 32% of males and 6% of females had developmental delay; 45% of males and 14% of females, attention problems; 30% of males and 3% of females, hyperactivity; and 19% of males and 1% of females, autism. Other typically pediatric problems included aggressiveness (recorded in 19% of males and 8% of females) and self-injurious behavior (recorded in 4% of males and 3% of females) (Bailey et al., 2008). An interesting subtype of boys with premutation is the one that presents with physical and/or behavioral features of FXS, including large ears, hand flapping, poor eye contact, and an ASD (Aziz et al., 2003; Farzin et al., 2006; Goodlin-Jones et al., 2004). It is assumed that this resemblance to

FXS is due to mild deficits in FMRP synthesis (Hagerman et al., 2009).

Despite the pediatric data on developmental delay, the conclusion is that most individuals with *FMR1* premutation have a normal IQ. However, mild and selective cognitive impairments and behavioral problems, better characterized in adults, are likely. Salient among cognitive problems is executive dysfunction, with selective deficit in executive working memory (Cornish et al., 2009; Kogan, Turk, Hagerman, & Cornish, 2008; Loesch et al., 2003; Moore et al., 2004). Arithmetic difficulties have also been reported (Lachiewicz, Dawson, Spiridigliozzi, & McConkie-Rosell, 2006); nonetheless, this association remains debatable (Fisch, 2006). Less controversial is the higher frequency of mild emotional and behavioral problems in adults with premutation; these include social deficits, anxiety, and obsessive-compulsive behavior, which tend to be more prevalent in males (Cornish et al., 2005; Hessl et al., 2005; Tassone et al., 2000). The National Parent Survey reported a relatively high prevalence of anxiety, which, in contrast to ADHD and ASD manifestations, was present in approximately equal proportions in males and females with premutation: 36% of males and 31% of females. The phenotype of females with *FMR1* premutation has recently been expanded to include increased frequency of autoimmune disorders, including hypothyroidism and fibromyalgia, and other, less specific symptoms, such as chronic muscle pain, paresthesias, and history of tremor (Coffey et al., 2008). In spite of this literature, the extent of the neurobehavioral involvement of adults with premutation without FXTAS is still questionable, partly because of the methodology employed in recent studies that rely on adults ascertained for investigations of FXTAS (i.e., younger adults with concerns about FXTAS, older adults without FXTAS). This is best exemplified by the study by Cornish, Turk, and Hagerman (2008) of cognitive trajectories in males with premutation that demonstrated progressive difficulty with inhibitory control over time, which was more prominent in males presenting with FXTAS. This finding, which suggests that deficit in inhibitory control might be a precursor of dementia in FXTAS, could have been interpreted differently (e.g., as an impairment in males with premutation without FXTAS) if data on evolution into FXTAS had not been available. A recent critical review of the literature on adult men and women with premutation who do not meet diagnostic criteria for FXTAS concluded that there is a need for further research using standardized protocols and larger sample sizes (Hunter, Abramowitz, Rusin, & Sherman, 2009). Among the criticisms raised in the review are the focus on female carriers, who are more frequent in the population but less likely to be affected because of X-linked inheritance patterns; older individuals, which makes it difficult to distinguish early FXTAS manifestations from non-FXTAS features; and mothers of children with FXS, whose psychological problems could be linked to the burden of raising an affected child (a major issue in early studies of *FMR1* premutation) (Hunter et al., 2009).

NEUROBIOLOGY OF FRAGILE X-ASSOCIATED DISORDERS

Research on the neurobiological bases of the neurobehavioral phenotype of FXDs has been limited. Three main sources of information have been used: molecular and biochemical analyses of biological samples from affected subjects, neuroimaging studies of affected subjects, and investigations of animal models of *FMR1* premutation and FXS.

The first level of analysis has included measurements of *FMR1* mRNA and protein levels in peripheral leukocyte samples, as well as the number of CGG repeats. In FXS, the size of the CGG repeat expansion is irrelevant because *FMR1* silencing is independent

of expansion when the 200 CGG repeat threshold is reached and the gene is hypermethylated. On the other hand, the severity of the FXS physical phenotype and intellectual impairment is correlated with the magnitude of the FMRP deficit (Bailey, Hatton, Skinner, & Mesibov, 2001; Kaufmann & Reiss, 1999; Loesch et al., 2004; Tassone et al., 1999). However, multiple studies have failed to correlate FMRP levels with behavioral abnormalities, particularly ASDs (Bailey, Roberts, Missett, & Hatton, 2001; Harris et al., 2008). With regard to *FMR1* premutation, the literature has been consistent in demonstrating a relationship between increased *FMR1* mRNA levels and a variety of neurologic and behavioral manifestations (Allen, He, Yadav-Shah, & Sherman, 2004; Hessl et al., 2005; Tassone, Hagerman, Taylor, Gane, et al., 2000). Repeat size has a role in certain aspects of FXTAS (e.g., age of onset, number of intranuclear inclusions; Amiri, Hagerman, & Hagerman, 2008) and POI (age of onset, ovarian dysfunction; Sullivan et al., 2005; Wittenberger et al., 2007).

Sources for neurobiologically relevant data for *FMR1* premutation include neuroimaging studies of subjects with and without FXTAS and an increasing body of literature on a mouse model of FXTAS. The latter shows cognitive decline and increasing behavioral (equivalent to anxiety) and motor difficulties over time (Van Dam et al., 2005). Although these findings have validated the mouse model for FXTAS (e.g., intranuclear inclusions), they have only begun to advance our understanding of FXTAS and neurobehavioral impairment of non-FXTAS *FMR1* premutation, partly because large repeat expansions are also observed (Brouwer et al., 2007; Entezam et al., 2007). In conjunction with this, these studies suggest that transcription of the CGG repeat expansion leads to neurodegeneration (*Drosophila* model; Jin et al., 2003; Jin et al., 2007); that mRNA accumulation may not be the only physiological mechanism (Brouwer, Huizer, Severijnen,

Hukema, et al., 2008); that there may be HPA axis involvement, including elevated serum cortisone levels and ubiquitin-positive inclusions in the pituitary, adrenal gland, and amygdala, particularly in aged mice (Brouwer, Severijnen, de Jong, et al., 2008); and that selective up-regulation of GABA-ergic components in the cerebellum may be a contributing factor (D'Hulst et al., 2009).

As for the animal model data, most neuroimaging studies in *FMR1* premutation focus on individuals with FXTAS. In addition to data dealing with diagnostic features in FXTAS (reviewed by Amiri et al., 2008), morphometric magnetic resonance imaging (MRI) studies have demonstrated reductions in cerebellar and cerebral volumes, where the latter are related to the number of CGG repeats (Loesch et al., 2005). Interestingly, these authors found increased hippocampal volumes that suggest this could be a neurodevelopmental feature of *FMR1* premutation. Another study demonstrated brainstem volumetric reductions in both FXTAS-affected and unaffected premutation subjects and that CGG repeat size correlates with overall atrophy, cerebellar volume, and white matter hyperintensities (a nonspecific feature of FXTAS; see respective section in this chapter) (Cohen et al., 2006). Finally, Hessl and colleagues (2007) showed that males with premutation, without FXTAS, have reduced brain activation in the amygdala and social cognition–related brain regions on functional MRI when they view fearful faces, in correspondence with their psychological symptoms and with a lack of startle potentiation in the same situation, as well as decreased skin conductance when they greet an unfamiliar experimenter. Altogether, these data suggest that *FMR1* premutation may be associated with selective developmental changes in limbic structures, which may explain the high prevalence of emotional behavioral problems in this type of mutation, which may become more pronounced over time with aging. Unquestionably, more research is needed to delineate

the neurobiology of the neurodevelopmental conditions associated with *FMR1* premutation.

The body of literature on the neurobiology of FXS is significantly larger than that on *FMR1* premutation. However, most of the animal model and neuroimaging studies have aimed at characterizing general features of the cognitive and behavioral phenotype of FXS. For this reason, in this section, we focus on research of relevance to social interaction disorders in FXS, with an emphasis on ASDs.

The mouse model of FXS (*FMR1* knock-out mouse) has reproduced the most distinctive anatomical abnormality in the disorder, an aberrant configuration of dendritic spines (i.e., long, tortuous, immature appearance; Kaufmann & Moser, 2000). Nonetheless, the most significant finding, which has opened the possibility of specific treatments, is the demonstration of enhanced activity of class I metabotropic glutamate receptors leading to increased hippocampal long-term depression (Bear, 2005). Long-term depression enhancement has been linked to the postulated negative regulatory role of FMRP in protein synthesis, triggered by metabotropic glutamate receptor activation (Bear, 2005; Hagerman et al., 2009). A general disturbance in events downstream from metabotropic glutamate receptor activation could explain a wide variety of cognitive and behavioral problems in FXS (Bear, 2005), including anxiety and ASDs that might be related to disturbances in metabotropic glutamate receptor-dependent long-term potentiation in the lateral amygdala (Mathew & Ho, 2006; Pelphrey, Adolphs, & Morris, 2004), reported to be reduced in the FXS mouse model (Zhao et al., 2005). Although most behavioral analyses of the FXS mouse model demonstrate that it reproduces some features of FXS (Chen & Toth, 2001; Frankland et al., 2004; Moon et al., 2006; Moon, Ota, Driscoll, Levitsky, & Strupp, 2008; Paradee et al., 1999), at a relatively mild level of severity, recent data suggest that features compatible with social anxiety (i.e., high levels of grooming and shorter duration of nose contact with an unfamiliar mouse) and ASDs (i.e., blunted negative reaction to a more aggressive "non-preferred" unfamiliar mouse) are exhibited by these animals (McNaughton et al., 2008; Mineur, Huynh, & Crusio, 2006). Furthermore, comparisons between different mouse strains carrying the FXS defect suggest that genetic background is critical for the association of the FMRP deficit with behavior characteristic of autism. FXS mice on an FVB/129 background failed to demonstrate significant preference for spending time in the social-partner side in a sociability task, which could not be attributed to low exploration, low activity, or higher levels of anxiety-like behavior, but not FXS mice on a C57BL/6J background. The FVB/129 FXS mouse behavior was similar to that of mice with disruption in the serotonin transporter gene (Moy et al., 2009). In support of a secondary gene modulation of the neurobehavioral phenotype of FXS, including ASDs in FXS, Hessl and colleagues (2008) reported that polymorphisms of the serotonin transporter gene, but not of the monoamine oxidase A gene, influenced aberrant behavior in males with FXS. Individuals who were homozygous for the high-transcribing long genotype exhibited the most aggressive and destructive behavior and the highest levels of stereotypic behavior (Hessl et al., 2008).

In line with the data reviewed for the FXS mouse model, neuroimaging and neuroendocrine studies suggest a preferential involvement of the limbic system and the HPA axis in subjects with FXS. A pioneer functional MRI study of females with FXS demonstrated abnormal activation of key regions involved in social cognition, namely the fusiform gyrus, superior temporal sulcus region, and amygdala, which correlated or suggested abnormal gaze aversion and aberrant processing of gaze features (Garrett, Menon, MacKenzie, & Reiss, 2004). A follow-up study in boys with FXS corroborated abnormal brain activation during gaze processing, specifically lower prefrontal activation, and elevated left insula activation in

response to direct eye gaze stimuli (Watson, Hoeft, Garrett, Hall, & Reiss, 2008). Another functional MRI investigation of individuals with FXS (both males and females) expanded on frontal abnormalities. Holsen et al. (2008) found that subjects with FXS have decreased activation of the medial and superior frontal cortical subregions of the prefrontal cortex (both implicated in social cognition) during successful face encoding, as well as inverse correlations between measures of social anxiety and activation of frontal regions and the hippocampus, which contrasted with mainly direct correlations in controls. A fourth functional MRI study expanded on the abnormalities in brain circuitry involved in emotional regulation. In high-functioning females with FXS, despite preserved emotion recognition, there is reduced activation to facial stimuli in the anterior cingulate cortex (neutral faces) and the caudate (sad faces) (Hagan, Hoeft, Mackey, Mobbs, & Reiss, 2008). In conjunction, these functional neuroimaging studies indicate that in FXS there are widespread abnormalities in the activity of brain regions involved in processing and response to emotional and social stimuli, which could underlie the shyness trait and high prevalence of social anxiety in this disorder.

Structural MRI studies have demonstrated, in connection with FXS, volumetric abnormalities in the social cognition–implicated regions discussed in the previous paragraph. However, limited data are available on the contribution of these anomalies to social anxiety and ASDs in FXS. In addition to overall larger cerebral volumes, several studies have demonstrated increased caudate volume (Gothelf et al., 2008; Hoeft et al., 2008; Lee et al., 2007; Reiss, Abrams, Greenlaw, Freund, & Denckla, 1995), which correlates with FMRP levels, and selective reductions of the superior temporal gyrus and amygdala (Gothelf et al., 2008; Hoeft et al., 2008; Kates et al., 2002). White matter volumetric abnormalities reported in boys and girls with FXS, mainly mild increases in temporal and parietal cortices (Kates et al., 2002; Lee et al., 2007), and reductions in medial prefrontal cortex (Hoeft et al., 2008), the latter in line with diffusion tensor imaging abnormalities in females with FXS (Barnea-Goraly et al., 2003), could be of relevance to white matter changes reported in idiopathic ASDs (Minshew & Williams, 2007). Interestingly, most of these abnormalities have also been described in children with FXS younger than 3 years (Hoeft et al., 2008) and are specific to the disorder, inasmuch as they are not observed in children without FXS who have developmental delays (Hoeft et al., 2008; Kates et al., 2002). Although no formal comparisons between individuals with FXS with and without diagnosis of social interaction disorder have been made, Gothelf and colleagues (2008) reported correlations between increased caudate nucleus volume and scores on several subscales of the Autism Behavior Checklist and the Stereotypy subscale of the Aberrant Behavior Checklist and between reduced posterior vermis volume and the Body Object Use subscale of the Autism Behavior Checklist (Krug, Arick, & Almond, 1980). These correlations suggest that abnormal size of these brain regions may constitute a substrate of ASDs and related behaviors in ASDs. A neuroimaging finding of more direct relevance to ASDs in FXS was reported by us; we found that boys with FXS who met *DSM-IV* criteria for Autistic Disorder have relatively larger posterior–superior cerebellar vermii than their counterparts without autism, although they are smaller than those of typically developing controls (Kaufmann et al., 2003). This result is intriguing because the abnormal region (i.e., vermian lobules VI–VII) is the same one that has consistently been shown to be relatively smaller in individuals with ASDs of unknown cause (Stanfield et al., 2008), a finding confirmed in our own study (Kaufmann et al., 2003).

As mentioned in the section on social interaction disorders in FXS, there are several lines of evidence that anxious and autism-like traits in the disorder are associated with HPA axis dysfunction. Cortisol reactivity (i.e., variability in cortisol levels) to a social challenge is decreased in children with FXS and prominent autism features (i.e., persistent eye gaze avoidance; Hessl et al., 2006). The opposite seems to be true for boys with FXS with prominent social avoidance, who demonstrate a markedly slow return to baseline cortisol levels after a cognitive/social challenge (Hessl et al., 2002). This pattern suggesting abnormal cortisol regulation is also observed in the FXS mouse in response to stress (Markham, Beckel-Mitchener, Estrada, & Greenough, 2006). Although intuitively, decreased hormonal and behavioral responses to social stimuli are compatible with ASDs, and the opposite pattern is consistent with anxiety, the precise mechanism of these cortisol changes in FXS is unknown. We have reported a higher frequency of acetylation of the glucocorticoid-negative regulator annexin-1 in males with FXS (Sun, Cohen, & Kaufmann, 2001), particularly in those with severe social withdrawal (Kaufmann et al., 2003). However, annexin-1 is involved in the acute phase of cortisol modulation (Buckingham et al., 2003), not the process of regulation presumably affected in FXS. Another candidate for abnormal cortisol regulation in FXS is the glucocorticoid receptor alpha, whose synthesis is directly regulated by FMRP (Brown et al., 2001) and whose levels are decreased in dendrites of hippocampal neurons in the FXS mouse (Miyashiro et al., 2003). Autonomic abnormalities that may be linked to the HPA axis and other limbic components have also been reported in FXS. They include increased sympathetic and decreased parasympathetic activities, which appear to be more severe in children with severe autism-like behavior (Roberts, Mirrett, & Burchinal, 2001). These autonomic abnormalities seem to be persistent features correlated with temperament traits (i.e., less persistent; Roberts, Boccia, Hatton,

Skinner, & Sideris, 2006) and present both at baseline and during social interaction (Hall, Lightbody, Huffman, Lazzeroni, & Reiss, 2009). Therefore, their differential contributions to social anxiety and ASDs are unknown.

Although it is clear that FMRP is a key regulator of molecular events secondary to synaptic activity, particularly activation of class I metabotropic glutamate receptors, the role of FMRP-dependent cell signaling in specific features of the FXS phenotype is unknown. A recent publication by Nishimura and colleagues (2007) sheds some light on the bases of ASDs in FXS. By examining gene expression profiles in lymphoblasts from boys with FXS and ASDs, which were compared with those of typically developing boys in a control group and of boys with autism and duplication of chromosome 15 (dup15q), another major cause of ASDs (Cohen et al., 2005), they found 68 genes to be abnormally regulated in both groups of boys with autism. Among them was G protein-coupled receptor 155 (GPR155), a gene regulated by the cytoplasmic FMR1-interacting protein 1 (CYFIP1), an antagonist and binding partner of FMRP that is a member of the RacGTPase system involved in neurite development (Schenck et al., 2003). Because CYFIP1 and another one of its targets, the Janus kinase and microtubule-interacting protein 1 (JAKMIP1 or MARLIN-1), were also dysregulated in patients with dup15q, the authors concluded that the CYFIP1 signaling pathway is implicated in these two genetic forms of ASDs. These data are the first to demonstrate common molecular events in different etiologies of ASDs, further emphasizing the importance of the study of ASDs in FXS for the field of social interaction disorders.

We can conclude that our knowledge of the neurobiology of social anxiety and ASDs and other neurobehavioral features of *FMR1*-related disorders is limited. However, some themes are emerging: 1) The limbic system is particularly susceptible to FMRP deficit and *FMR1* accumulation, and its complex and variable involvement along other cerebral circuits implicated in emotional regulation may lead to either social anxiety or ASDs, or both, in FXS. 2) Secondary genetic events (e.g., modifier gene polymorphisms) may be responsible for the considerable phenotypical variability in this monogenic disorder. 3) FXS shares molecular and neural circuitry features with other genetic etiologies of ASDs and, perhaps, social anxiety. 4) Knowledge of the neurobiology of FXS will lead to more specific "targeted" treatments (e.g., class I metabotropic glutamate receptor antagonists, Berry-Kravis et al., 2009; CRH 1 receptor antagonists, Ising & Holsber, 2007). For an overview on current predominantly symptomatic treatments and future targeted drug therapies, the reader is directed to our recent review on the subject (Hagerman et al., 2009).

CONCLUDING REMARKS

During the almost 18 years since the discovery of *FMR1* as the gene responsible for FXS, remarkable progress has been made in delineating the neurobehavioral features associated with FXS. Clinical "wisdom" has been replaced by hypothesis-driven, research cohort–based studies, leading to probably one of the best-characterized forms of ASDs. Still, many issues remain unresolved in FXS; two of the main ones are the clinical delineation of ADHD and anxiety, including social anxiety. One of the big surprises in the field of developmental disabilities is the discovery of a novel neurodegenerative disorder, FXTAS, in grandparents and great uncles of children with FXS. How was this prevalent and severe disorder hidden for so many years? The review by Amiri and colleagues (2008) discusses some of the possible causes. The recognition of FXTAS has also led to a revision of the clinical significance of *FMR1* premutation. Most data suggest that it is a risk factor, per se insufficient as etiology, for some common developmental disorders (e.g., ADHD,

ASDs) and behavioral problems in adulthood. Nonetheless, data on adults with premutation should be taken with caution, because neurologic and psychiatric findings may represent early stages of FXTAS and not developmental-origin abnormalities. The explosion of knowledge of the neurobiology of FXS has led to probably the first defect-targeted treatments for neurodevelopmental disorders. Similar progress on FXTAS, in conjunction with other advances in neurodegenerative disorders, may soon bring promising therapies for this devastating disorder.

Although the success story of FXS may stimulate more research and the consideration of this disorder as a model for intellectual disability, pediatric anxiety, and ASDs, we should not lose sight of the fact that FXS is a major cause of these behavioral syndromes that frequently goes undiagnosed (Schaefer, Mendelsohn, & Professional Practice and Guidelines Committee, 2008; Shevell et al., 2003). Late diagnosis of FXS is linked to the misconception that physical features (i.e., dysmorphia) and severe neurologic involvement are diagnostic requirements. This and the fact that *FMR1* premutation is a possible cause of intellectual disability and ASDs lead to the strong recommendation of testing for *FMR1* mutations in these disorders, particularly in boys. Finally, the recognition of FXTAS and POI has led to a different view of FXS as a disorder affecting families, not individuals, and corresponding diagnostic and counseling guidelines have been developed (McConkie-Rosell et al., 2007).

REFERENCES

Achenbach, T.M., & Rescorla, L.A. (2000). *Child Behavior Checklist.* Burlington, VT: ASEBA/Research Center for Children, Youth and Families.

Allen, E.G., He, W., Yadav-Shah, M., & Sherman S.L. (2004). A study of the distributional characteristics of FMR1 transcript levels in 238 individuals. *Human Genetics, 114*(5), 439–447.

Aman, M., Singh, N., Stewart, A., & Field, C. (1985). The Aberrant Behavior Checklist: A behavior rating scale for the assessment of treatment effects. *American Journal of Mental Deficiency, 89*(5), 485–491.

American Psychiatric Association. (1994). *Diagnostic and statistical manual of mental disorders* (4th ed.). Washington, DC: Author.

American Psychiatric Association. (2000). *Diagnostic and statistical manual of mental disorders* (4th ed., text rev.). Washington, DC: Author.

Amiri, K., Hagerman, R.J., & Hagerman, P.J. (2008). Fragile X–associated tremor/ataxia syndrome: An aging face of the fragile X gene. *Archives of Neurology, 65*(1), 19–25.

Aziz, M., Stathopulu, E., Callias, M., Taylor, C., Turk, J., Oostra, B., et al. (2003). Clinical features of boys with fragile X premutation and intermediate alleles. *American Journal of Medical Genetics B Neuropsychiatric Genetics, 121B*(1), 119–127.

Backes, M., Genç, B., Schreck, J., Doerfler, W., Lekhmukhl, G., & von Gontard, A. (2000). Cognitive and behavior profiles of fragile X boys: Correlations to molecular data. *American Journal of Medical Genetics, 95*(2), 150–156.

Bailey, D.B., Jr., Hatton, D.D., Skinner, M., & Mesibov, G. (2001). Autistic behavior, FMR1 protein and developmental trajectories in young males with fragile X syndrome. *Journal of Autism and Developmental Disorders, 31*(2) 165–174.

Bailey, D.B., Jr., Mesibov, G.B., Hatton, D.D., Clark, R.D., Roberts, J.E., & Mayhew, L. (1998). Autistic behavior in young boys with fragile X syndrome. *Journal of Autism and Developmental Disorders, 28*(6), 499–508.

Bailey, D.B., Jr., Raspa, M., Olmsted, M., & Holiday, D.B. (2008). Co-occurring conditions associated with FMR1 gene variations: Findings from a national parent survey. *American Journal of Medical Genetics A, 146A*(16), 2060–2069.

Barnea-Goraly, N., Eliez, S., Hedeus, M., Menon, V., White, C.D., Mosley, M., et al. (2003). White matter tract alterations in fragile X syndrome: Preliminary evidence from diffusion tensor imaging. *American Journal of Medical Genetics B. Neuropsychiatric Genetics, 118B*(1), 81–88.

Bassell, G.J., & Warren, S.T. (2008). Fragile X syndrome: Loss of local mRNA regulation alters synaptic development and function. *Neuron, 60*(2), 201–214.

Bear, M.F. (2005). Therapeutic implications of the mGluR theory of fragile X mental retardation. *Genes Brain Behavior, 4*(6), 393–398.

Berry-Kravis, E., Hessl, D., Coffey, S., Hervey, C., Schneider, A., Yuhas, J., et al. (2009) A pilot open label, single dose trial of fenobam in adults with fragile X syndrome. *Journal of Medical Genetics, 46*(4), 266–271.

Borghgraef, M., Fryns, J.P., Dielkens, A., Pyck, K., & Van den Berghe, H. (1987). Fragile X syndrome: A study of the psychological profile in 23 prepubertal patients. *Clinical Genetics, 32*(3), 179–186.

Brown, V., Jin, P., Ceman, S., Darnell, J.C., O'Donnell, W.T., Tenenbaum, S.A., et al. (2001). Microarray identification of FMRP-associated brain mRNAs and altered mRNA translational profiles in fragile X syndrome. *Cell, 107*(4) 477–487.

Brouwer, J.R., Huizer, K., Severijnen, L.A., Hukema, R.K., Berman, R.F., Oostra, B.A., et al. (2008). CGG-repeat length and neuropathological and molecular correlates in a mouse model for fragile X–associated tremor/ataxia syndrome. *Journal of Neurochemistry, 107*(6), 1671–1682. [Epub 2008 Nov 10.]

Brouwer, J.R., Mientjes, E.J., Bakker, C.E., Nieuwenhuizen, I.M., Severijnen, L.A., Van der Linde, H.C., et al. (2007). Elevated Fmr1 mRNA levels and reduced protein expression in a mouse model with unmethylated fragile X full mutation. *Experimental Cell Research, 313*(2), 244–253. [Epub 2006 Oct 13.]

Brouwer, J.R., Severijnen, E., de Jong, F.H., Hessl, D., Hagerman, R.J., Oostra, B.A., et al. (2008). Altered hypothalamus-pitutary-adrenal gland axis regulation in the expanded CGG-repeat mouse model for fragile X-associated tremor/ataxia syndrome. *Psychoneuroendocrinology, 33*(6), 863–873.

Brunberg, J.A., Jacquemont, S., Hagerman, R.J. Berry-Kravis, E.M., Grigsby, J., Leehey, M.A., et al. (2002). Fragile X premutation carriers: Characteristic MR imaging findings of adult male patients with progressive cerebellar and cognitive dysfunction. *American Journal of Neuroradiology, 23*(10), 1757–1766.

Buckingham, J.C., Solito, E., John, C., Tierney, T., Taylor, A., Flower, R., et al. (2003). Annexin 1: A paracrine/juxtacrine mediator of glucocoid action in the neuroendocrine system. *Cell Biochemistry and Function, 21*(3), 217–221.

Budimirovic, D., Bukelis, I., Cox, C., Gray, R.M., Tierney, E., & Kaufmann, W.E. (2006). Autism spectrum disorder in fragile X syndrome: Differential contribution of adaptive socialization and social withdrawal. *American Journal of Medical Genetics, 140A*(17),1814–1826.

Chen, L., & Toth, M. (2001). Fragile X mice develop sensory hyperactivity to auditory stimuli. *Journal of Neuroscience, 103*(4), 1043–1050.

Clifford, S., Dissanayake, C., Bui, Q.M., Huggins, R., Taylor, A.K., & Loesch, D.Z. (2007). Autism spectrum phenotype in males and females with fragile X full mutation and premutation. *Journal of Autism and Developmental Disorders, 37*(4), 738–747.

Coffee, B., Ikeda, M., Budimirovic, D.J., Hjelm, L.N., Kaufmann, W.E., & Warren, S.T. (2008). Mosaic FMR1 deletion causes Fragile X syndrome and can lead to molecular misdiagnosis: A case report and review of the literature. *American Journal of Medical Genetics, 146A*(10), 1358–1367.

Coffey, S.M., Cook, K., Tartaglia, N., Tassone, F., Nguyen, D.V., Pan, R., et al. (2008). Expanded clinical phenotype of women with FMR1 premutation. *American Journal of Medical Genetics, 146A*(8), 1009–1016.

Cohen, D., Pichard, N., Tordjman, S., Baumann, C., Burglen, L., Excoffier, E., et al. (2005). Specific genetic disorders and autism: Clinical contribution towards their identification. *Journal of Autism and Related Disorders, 35*(1), 103–116.

Cohen, I.L., Fisch, G.S., Sudhalter, V., Wolf-Schein, E.G., Hanson, D., Hagerman, R. et al. (1988). Social gaze, social avoidance, and repetitive behavior in fragile X males: A controlled study. *American Journal of Mental Retardation, 92*(5), 436–446.

Cohen, S., Masyn, K., Adams, J., Hessl, D., Rivera, S., Tassone, F., et al. (2006). Molecular and imaging correlates of the fragile X-associated tremor/ataxia syndrome. *Neurology, 67*(8), 1426–1431.

Conway, G.S., Payne, N.N., Webb, J., Murray, A., & Jacobs, P.A. (1998). Fragile X premutation screening in women with premature ovarian failure. *Human Reproduction, 13*(5), 1184–1187.

Cornish, K.M., Kogan, C.S., Li, L., Turk, J., Jacquemont, S., & Hagerman, R.J. (2009). *Lifespan changes in working memory in fragile X premutation males, 69*(3), 551–558. [Epub 2008 Dec 27.]

Cornish, K., Kogan, C., Turk, J., Manly, T., James, N., Mills, A., et al. (2005). The emerging fragile X premutation phenotype: Evidence from the domain of social cognition. *Brain and Cognition, 57*(1), 53–60.

Cornish, K., Turk, J., & Hagerman, R. (2008). The fragile X continuum: New advances and perspectives. *Journal of Intellectual Disability Research, 52*(6), 469–482.

Cornish, K.M., Turk, J., Wilding, J., Sudhalter, V., Munir, F., Kooy, F., et al. (2004). Annotation: Deconstructing the attention deficit in fragile X syndrome: A developmental neuropsychological approach. *Journal of Child Psychology & Psychiatry, 45*(6), 1042–1053.

Cronister, A., Schreiner, R., Wittenberger, M., Amiri, K., Harris, K., & Hagerman, R.J. (1991).

Heterozygous fragile X female: Historical, physical, cognitive and cytogenetic features. *American Journal of Medical Genetics, 38*(2–3), 269–274.

De Boulle, K., Verkerk, A.J., Reyneirs, E., Vits, L., Hendrickx, J., Van Roy, B., et al. (1993). A point mutation in the FMR-1 gene associated with fragile X mental retardation. *Nature Genetics, 3*(1), 31–35.

de Vries, B.B., Wiegers, A.M., Smits, A.P., Mohkamsing, S., Duivenvoorden, H.J., Fryns, J.P., et al. (1996). Mental status of females with an FMR1 gene full mutation. *American Journal of Human Genetics, 58*(5), 1025–1032.

D'Hulst, C., Heulens, I., Brouwer, J.R., Willemsen, R., De Geest, N., Reeve, S.P., et al. (2009). Expression of the GABAergic system in animal models for fragile X syndrome and fragile X associated tremor/ataxia syndrome (FXTAS). *Brain Research, 1253*, 176–183.

Dombrowski, C., Levesque, M.L., Morel, M.L., Rouillard, P., Morgan, K., & Rousseau, F. (2002). Premutation and intermediate-size FMR1 alleles in 10 572 males from the general population: Loss of an AGG interruption is a late event in the generation of fragile X syndrome alleles. *Human Molecular Genetics, 11*(4), 371–378.

Dykens, E.M. & Volkmar, F.R. (1997). Medical conditions associated with autism. In D.J. Cohen & F.R. Volkmar (Eds.), *Handbook of autism and pervasive developmental disorders* (pp. 388–407). New York: John Wiley and Sons.

Entezam, A., Biacsi, R., Orrison, B., Saha, T., Hoffman, G.E., Grabczyk, E., et al. (2007). Regional FMRP deficits and large repeat expansions into the full mutation range in a new Fragile X premutation mouse model. *Gene, 395*(1–2), 125–134.

Farzin, F., Perry, H., Hessl, D., Loesch, D., Cohen, J., Bacalman, S., et al. (2006). Autism spectrum disorders and attention deficit/hyperactivity disorder in boys with the fragile X premutation. *Journal of Developmental and Behavioral Pediatrics, 27*(2, Suppl.), S137–S144.

Fisch, G.S. (2006). Cognitive-behavioral profiles of females with the fragile X mutation. *American Journal of Medical Genetics A, 140*(7), 665–672.

Fisch, G.S., Carpenter, N.J., Simensen, R., Smits, A.P., van Roosmalen, T., & Hamel, B.C. (1999). Longitudinal changes in cognitive-behavioral levels in three children with FRAXE. *American Journal of Medical Genetics, 84*(3), 291–292.

Fisch, G.S., Simensen, R.J., & Schroer, R.J. (2002). Longitudinal changes in cognitive and adaptive behavior scores in children and adolescents with the fragile X mutation or autism. *Journal of Autism and Developmental Disorders, 32*(2), 107–114.

Frankland, P.W., Wang, Y., Rosner, B., Shimizu, T., Balleine, B.W., Dykens, E.M., et al. (2004). Sensorimotor gating abnormalities in young males with fragile X syndrome and Fmr1-knockout mice. *Molecular Psychiatry, 9*(4), 417–425.

Freund, L.S., Reiss, A.L., & Abrams, M.T. (1993). Psychiatric disorders associated with fragile X in the young female. *Pediatrics, 91*(2), 321–329.

Garrett, A.S., Menon, V., MacKenzie, K., & Reiss, A.L. (2004). Here's looking at you kid: Neural systems underlying face and gaze processing in fragile X syndrome. *Archives of General Psychiatry, 61*(3), 281–288.

Goodlin-Jones, B.L., Tassone, F., Gane, L.W., & Hagerman, R.J. (2004). Autism spectrum disorder and the fragile X premutation. *Journal of Developmental Behavior Pediatrics, 25*(6), 392–398.

Gothelf, D., Furaro, J.A., Hoeft, F., Eckert, M.A., Hall, S.S., O'Hara, R., et al. (2008). Neuroanatomy of fragile X syndrome is associated with aberrant behavior and the fragile X mental retardation protein (FMRP). *Annals of Neurology, 63*(1), 40–51.

Greco, C.M., Hagerman, R.J., Tassone, F., Chudley, A.E., Del Bigio, M.R., Jacquemont, S., et al. (2002). Neuronal intranuclear inclusions in a new cerebellar tremor/ataxia syndrome among fragile X carriers. *Brain, 125*, 1760–1771.

Hagan, C.C., Hoeft, F., Mackey, A., Mobbs, D., & Reiss, A.L. (2008). Aberrant neural function during emotional attribution in female subjects with fragile X syndrome. *Journal of American Academy of Child and Adolescent Psychiatry, 47*(12), 1443–1454.

Hagerman, P.J. (2008). The fragile X prevalence paradox. *Journal of Medical Genetics, 45*(8), 498–499.

Hagerman, P.J., & Hagerman, R.J. (2004). The fragile X premutation: A maturing perspective. *American Journal of Human Genetics, 74*(5), 805–816.

Hagerman, R.J. (2002). The physical and behavioral phenotype. In R.J. Hagerman & P.J. Hagerman (Eds.), *Fragile X syndrome: Diagnosis, treatment, and research* (pp. 3–109). Baltimore: The Johns Hopkins University Press.

Hagerman, R.J., Berry-Kravis, E., Kaufmann, W.E., Ono, M.Y., Tartaglia, N., Lachiewicz, A., et al. (2009). Advances in the treatment of fragile X syndrome. *Pediatrics, 123*(1), 378–390.

Hagerman, R.J., Jackson, A.W., III, Levitas, A., Rimland, B., & Braden, M. (1986). An analysis of autism in fifty males with fragile X syndrome. *American Journal of Medical Genetics, 23*(1–2), 359–374.

Hagerman, R.J., Jackson, C., Amiri, K., Silverman, A.C., O'Connor, R., & Sobesky, W. (1992). Girls with fragile X syndrome: Physical and neurocognitive status and outcome. *Pediatrics, 89*(3), 395–400.

Hagerman, R.J., Leehey, M., Heinrichs, W., Tassone, F., Wilson, R., Hills, J., et al. (2001). Intention tremor, parkinsonism, and generalized brain atrophy in male carriers of fragile X. *Neurology, 57*(1), 127–130.

Hall, S.S., Lightbody, A.A., Huffman, L.C., Lazzeroni, L.C., & Reiss. A.L. (2009). Physiological correlates of social avoidance behavior in children and adolescents with fragile X syndrome. *Journal of the American Academy of Child and Adolescent Psychiatry, 48*(3), 320–329.

Hall, S.S., Lightbody, A.A., & Reiss, A.L. (2008). Compulsive, self injurious and autistic behavior in children and adolescents with fragile X syndrome. *American Journal of Mental Retardation, 113*(1), 44–53.

Harris, S.W., Hessl, D., Goodlin-Jones, B., Ferranti, J., Bacalman, S., Barbato, I., et al. (2008). Autism profiles of males with fragile X syndrome. *American Journal of Mental Retardation, 113*(6), 427–438.

Hatton, D.D., Bailey, D.B., Jr., Hargett-Beck, M.Q., Skinner, M., & Clark, R.D. (1999). Behavioral style of young boys with fragile X syndrome. *Developmental Medicine and Child Neurology, 41*(9), 625–632.

Hatton, D.D., Sideris, J., Skinner, M., Mankowski, J., Bailey, D.B., Jr., Roberts, J., et al. (2006). Autistic behavior in children with fragile X syndrome: Prevalence, stability and the impact of FMRP. *American Journal of Medical Genetics A, 140A*(17), 1804–1813.

Hepburn, S.L., DiGuiseppi, C., Rosenberg, S., Kaparich, K., Robinson, C., & Miller, L. (2008). Use of a teacher nomination strategy to screen for autism spectrum disorders in general education classrooms: A pilot study. *Journal of Autism and Developmental Disorders, 38*(2), 373–382.

Hernandez, R.N., Feinberg, R.L., Vaurio, R., Passanante, N.M., Thompson, R.E., & Kaufmann, W.E. (2009). Autism spectrum disorder in fragile X syndrome: A longitudinal evaluation. *American Journal of Medical Genetics, 149A*(6), 1125–1137.

Hessl, D., Dyer-Friedman, J., Glaser, B., Wisbeck, J., Barajas, R.G., Taylor, A., et al. (2001). The influence of environmental and genetic factors on behavior problems and autistic symptoms in boys and girls with fragile X syndrome. *Pediatrics, 108*(5), E88.

Hessl, D., Glaser, B., Dyer-Friedman, J., Blasey, C., Hastie, T., Gunnar, M., et al. (2002). Cortisol and behavior in fragile X syndrome. *Psychoneuroendocrinology, 27*(7), 855–872.

Hessl, D., Glaser, B., Dyer-Friedman, J., & Reiss, A.L. (2006). Social behavior and cortisol reactivity in children with fragile X syndrome. *Journal of Child Psychology and Psychiatry, 47*(6), 602–610.

Hessl, D., Rivera, S., Koldewyn, K., Cordeiro, L., Adams, J., Tassone, F., et al. (2007). Amygdala dysfunction in men with the fragile X premutation. *Brain, 130*(2), 404–416.

Hessl, D., Tassone, F., Cordeiro, L., Koldewyn, K., McCormick, C., Green, C., et al. (2008). Brief report: Aggression and stereotypic behaviors in males with fragile X syndrome-moderating secondary genes in a "single gene" disorder. *Journal of Autism and Developmental Disorders, 38*(1), 184–189.

Hessl, D., Tassone, F., Loesch, D.Z., Berry-Kravis, E., Leehey, M.A., Gane, L.W., et al. (2005). Abnormal elevation of FMR1 mRNA is associated with psychological symptoms in individuals with the fragile X premutation. *American Journal of Medical Genetics B, 139B*(1), 115–121.

Hoeft, F., Lightbody, A.A., Hazlett, H.C., Patnaik, S., Piven, J., & Reiss, A.L. (2008). Morphometric spatial patterns differentiating boys with fragile X syndrome, typically developing boys and developmentally delayed boys aged 1 to 3 years. *Archives of General Psychiatry, 65*(9), 1087–1097.

Holsen, L.M., Dalton, K.M., Johnstone, T., & Davidson, R.J. (2008). Prefrontal social cognition network dysfunction underlying face encoding and social anxiety in fragile X syndrome. *Neuroimage 43*(3), 592–604.

Hunter, J.E., Abramowitz, A., Rusin, M., & Sherman, S.L. (2009). Is there evidence for neuropsychological and neurobehavioral phenotypes among adults with FXTAS who carry the FMR1 premutation? A review of current literature. *Genetics in Medicine, 11*(2), 79–89.

Ising, M., & Holsber, F. (2007). CRH-sub-1 receptor antagonists for the treatment of depression and anxiety. *Experimental & Clinical Psychopharmacology, 15*(6), 519–528.

Jacquemont, S., Hagerman, R.J., Leehey, M., Grigsby, J., Zhang, L., Brunberg, J.A., et al.

(2003). Fragile X premutation tremor/ataxia syndrome: Molecular, clinical and neuroimaging correlates. *American Journal of Human Genetics, 72*(4), 869–878.

Jin, P., Duan, R., Qurashi, A., Qin, Y., Tian, D., Rosser, T.C., Liu, H., et al. (2007). Pur alpha binds to rCGG repeats and modulates repeat—mediated neurodegeneration in Drosophila model of fragile X tremor/ataxia syndrome. *Neuron, 55*(4), 555–564.

Jin, P., Zarnescu, D.C., Zhang, F., Pearson, C.E., Lucchesi, J.C., Moses, K., et al. (2003). RNA-mediated neurodegeneration caused by the fragile X premutation rCGG repeats in Drosophila. *Neuron, 39*(5), 739–747.

Kates, W.R., Fredrikse, M., Mostofsky, S.H., Folley, B.S., Cooper, K., Mazur-Hopkins, P., et al. (2002). MRI parcellation of the frontal lobe in boys with attention deficit hyperactivity disorder or Tourette syndrome. *Psychiatry Research, 116*(1-2), 63-81.

Kau, A.S., Meyer, W.A., & Kaufmann, W.E. (2002). Early development in males with fragile X syndrome: A review of the literature. *Microscopic Research Techniques, 57*(3), 174–178

Kau, A.S., Tierney, E., Bukelis, I., Stump, M.H., Kates, W.R., Trescher, W.H., et al. (2004). Social behavior profile in young males with fragile X syndrome: Characteristics and specificity. *American Journal of Medical Genetics, 126A*(1), 9–17.

Kaufmann, W.E., Abrams, M.T., Chen, W., & Reiss, A.L. (1999). Genotype, molecular phenotype and cognitive phenotype. Correlations in fragile X syndrome. *American Journal of Medical Genetics, 83*(4), 286–295.

Kaufmann, W.E., Capone, G.T., Clarke, M., & Budimirovic, D.B. (2008). Autism in genetic intellectual disability: Insights into idiopathic autism. In A.W. Zimmerman (Ed.), *Autism: Current theories and evidence* (pp. 81–108). Totowa, NJ: Humana Press.

Kaufmann, W.E., Cooper, K.L., Mostofsky, S.H., Capone, G.T., Kates, W.R., Newschaffer, C.J., et al. (2003). Specificity of cerebellar vermian abnormalities in autism: A quantitative magnetic resonance imaging study. *Journal of Child Neurology, 18*(7), 463–470.

Kaufmann, W.E., Cortell, R., Kau, A.S., Bukelis, I., Tierney, E., Gray, R.M., et al. (2004). Autism spectrum disorder in fragile X syndrome: Communication, social interaction, and specific behaviors. *American Journal of Medical Genetics, 129A*(3), 225–234.

Kaufmann, W.E., & Moser, H.W. (2000). Dendritic anomalies in disorders associated with mental retardation. *Cerebral Cortex, 10*(10), 981–991.

Kaufmann, W.E., & Reiss, A.L. (1999). Molecular and cellular genetics of fragile X syndrome. *American Journal of Medical Genetics, 88*(1), 11–24.

Kennson, A., Zhang, F., Hagedorn, C.H., & Warren, S.T. (2001). Reduced FMRP and increased FMR1 transcription is proportionally associated with CGG repeat number in intermediate-length and premutation carriers. *Human Molecular Genetics, 10*(14), 1449–1454.

Kogan, C.S., Turk, J., Hagerman, R.J., & Cornish, K.M. (2008). Impact of the fragile X mental retardation 1(FMR1) gene premutation on neuropsychiatric functioning in adult males without fragile X–associated Tremor/Ataxia syndrome: A controlled study. *American Journal of Medical Genetics, 147B*(6), 859–872.

Korf, R.B. (2000). *Human genetics: A problem-based approach* (2nd ed.). Malden, MA: Blackwell Science.

Krug, D.A., Arick, J., & Almond, P. (1980). Behavior checklist for identifying severely handicapped individuals with high levels of autistic behavior. *Journal of Child Psychology and Psychiatry, 21*(3), 221–229.

Lachiewicz, A.M., Dawson, D.V., & Spiridigliozzi, G.A. (2000). Physical characteristics of young boys with Fragile X syndrome: Reasons for difficulties in making a diagnosis in young males. *American Journal of Medical Genetics, 92*(4), 229–236.

Lachiewicz, A.M., Dawson, D.V., Spiridigliozzi, G.A., & McConkie-Rossel, A. (2006). Arithmetic difficulties in females with the fragile X premutation. *American Journal of Medical Genetics, 140*(7), 665–672.

Lee, A.D., Leow, A.D., Lu, A., Reiss, A.L., Hall, S., Chiang, M.C., et al. (2007). 3D pattern of brain abnormalities in fragile X syndrome visualized using tensor-based morphometry. *NeuroImage, 34*(3), 924–938.

Loesch, D.Z., Bui, Q.M., Grigsby, J., Butler, E., Epstein, J., Huggins, R.M., et al. (2003). Effect of fragile X status categories and fragile X mental retardation protein levels on executive functioning in males and females with fragile X. *Neuropsychology, 17*(4), 646–657.

Loesch, D.Z., Huggins, R.M., & Hagerman, R.J. (2004). Phenotypic variation and FMRP levels in fragile X. *Mental Retardation and Developmental Disabilities Research Reviews, 10*(1) 31–41.

Loesch, D.Z., Litewka, L., Brotchie, P., Huggins, R.M., Tassone, F., & Cook, M. (2005). Magnetic resonance imaging study in older fragile X premutation male carriers. *Annals of Neurology, 58*(2), 226–230.

Lord, C., Risi, S., Lambrecht, L., Cook, E., Jr., Leventhal, B., DiLavore, P., et al. (2000). The

Autism Diagnostic Observation Schedule–Generic: A standard measure of social and communication deficits associated with the spectrum of autism. *Journal of Autism and Developmental Disorders, 30,* 205–223.

Lord, C., Rutter, M., & Le Couteur, A. (1994). Autism Diagnostic Interview–Revised: A revised version of a diagnostic interview for caregivers of individuals with possible pervasive developmental disorders. *Journal of Autism and Related Disorders, 24*(5), 659–685.

Mallolas, J., Duran, M., Sanchez, A., Jiménez, D., Castellvi-Bel, S., Rife, M., et al. (2001). Implications of the FMR1 gene in menopause: Study of 147 Spanish women. *Menopause, 8*(2), 106–110.

Markham, J.A., Beckel-Mitchener, A.C., Estrada, C.M., & Greenough, W.T. (2006). Corticosterone response to acute stress in a mouse model of fragile X syndrome. *Psychoneuroendocrinology, 31*(6), 781–785.

Marozzi, A., Vegetti, W., Manfredini, E., Tibiletti, M.G., Testa, G., Crosignani, P.G., et al. (2000). Association between idiopathic premature ovarian failure and fragile X premutation. *Human Reproduction, 15*(1), 197–202.

Mathew, S.J., & Ho, S. (2006). Etiology and neurobiology of social anxiety disorder. *Journal of Clinical Psychiatry, 67*(Suppl. 12), 9–13.

McConkie-Rosell, A., Abrams, L., Finucane, B., Cronister, A., Gane, L.W., Coffey, S.M., et al. (2007). Recommendations from multidisciplinary focus groups on cascade testing and genetic counseling for fragile X–associated disorders. *Journal of Genetic Counseling, 16*(5), 593–606.

McNaughton, C.H., Moon, J., Strawderman, M.S., Maclean, K.N., Evans, J., & Strupp, B.J. (2008). Evidence of social anxiety and impaired cognition in a mouse model of fragile X syndrome. *Behavioral Neuroscience, 122*(2), 293–300.

Merenstein, S.A., Sobesky, W.E., Taylor, A.K., Riddle, J.E., Tran, H.X., & Hagerman, R.J. (1996). Molecular-clinical correlation in males with the expanded FMR1 mutation. *American Journal of Medical Genetics, 64*(2), 388–394.

Mineur, Y.S., Huynh, L.X., & Crusio, W.E. (2006). Social behavior deficits in the Fmr1 mutant mouse. *Behavioral Brain Research, 168*(1), 172–175.

Minshew, N.J., & Williams, D.L. (2007) The new neurobiology of autism: Cortex, connectivity and neural organization. *Archives of Neurology, 64*(7), 945–950.

Miyashiro, K.Y., Beckel-Mitchener, A., Purk, T.P., Becker, K.G., Barrett, T., Liu, L., et al. (2003). RNA cargoes associating with FMRP reveal deficits in cellular functioning in Fmr1 null mice. *Neuron, 37*(3), 417–431.

Molloy, C.A., Murray, D.S., Kinsman, A., Castillo, H., Mitchell, T., Hickey, F.J., et al. (2009). Differences in the clinical presentation of trisomy 21 with and without autism. *Journal of Intellectual Disabilities Research, 53*(2), 143–151.

Moon, J., Beaudin, A.E., Verosky, S., Driscoll, L.L., Weiskopf, M., Levitsky, D.A., et al. (2006). Attentional dysfunction, impulsivity and resistance to change in a mouse model of fragile X syndrome. *Behavioral Neuroscience, 120*(6), 1367–1379.

Moon, J., Ota, K.T., Driscoll, L.L., Levitsky, D.A., & Strupp, B.J. (2008). A mouse model of fragile X syndrome exhibits heightened arousal and/or emotion following errors or reversal of contingencies. *Developmental Psychobiology, 50*(5), 473–485.

Moore, C.J., Daly, E.M., Schmitz, N., Tassone, F., Tysoe, C., Hagerman, R.J., et al (2004). A neuropsychological investigation of male premutation carriers of fragile X syndrome. *Neuropsychologia, 42*(14), 1934–1947.

Moy, S.S., Nadler, J.J., Young, N.B., Nonneman, R.J., Grossman, A.W., Murphy, D.L., et al. (2009). Social approach in genetically-engineered mouse lines relevant to autism. *Genes, Brain and Behavior, 8*(2), 129–142.

Munir, F., Cornish, K.M., & Wilding, J. (2000). A neuropsychological profile of attention deficits in young males with fragile X syndrome. *Neuropsychologia, 38*(9), 1261–1270.

Murray, A., Webb, J., Grimley, S., Conway, G., & Jacobs, P. (1998). Studies of FRAXA and FRAXE in women with premature ovarian failure. *Journal of Medical Genetics, 35*(8), 637–640.

Nishimura, Y., Martin, C.L., Vazquez-Lopez, A., Spence, S.J., Alvarez-Retuerto, A.I., Sigman, M., et al. (2007). Genome-wide expression profiling of lymphoblastoid cell lines distinguishes different forms of autism and reveals shared pathways. *Human Molecular Genetics, 16*(14), 168–198.

Oostra, B.A., & Halley, D.J.J. (1995). Complex behavior of simple repeats: The fragile X syndrome. *Pediatric Research, 38*(5), 629–637.

Paradee, W., Melikian, H.E., Rasmussen, D.L., Kenneson, A., Conn, P.J., & Warren, S.T. (1999). Fragile X mouse: Strain effects of knockout phenotype and evidence suggesting deficient amygdala function. *Neuroscience, 94*(1), 185–192.

Pelphrey, K., Adolphs, R., & Morris, J.P. (2004). Neuroanatomical substrates of social cognition dysfunction in autism. *Mental Retardation and*

Developmental Disabilities Research Reviews, 10(4), 259–271.

Philofsky, A., Hepburn, S.L., Hayes, A., Hagerman, R., & Rogers, S.J. (2004). Linguistic and cognitive functioning and austim symptoms in young children with fragile X syndrome. *American Journal of Mental Retardation, 109*(3), 208–218.

Pieretti, M., Zhang, F.P., Fu, Y.H., Warren, S.T., Oostra, B.A., Caskey T.C., et al. (1991). Absence of expression of the FMR-1 gene in fragile X syndrome. *Cell, 66*(4), 817–822.

Reiss, A.L., Abrams, M.T., Greenlaw, R., Freund, L., & Denckla, M.B., (1995). Neurodevelopmental effects of the FMR-1 full mutation in humans. *Nature Medicine, 1*(2), 159–167.

Roberts, J.E., Boccia, M.L., Hatton, D.D., Skinner, M.L., & Sideris, J. (2006). Temperament and vagal tone in boys with fragile X syndrome. *Journal of Developmental and Behavioral Pediatrics, 27*(3), 193–201.

Roberts, J.E., Mirrett, P., & Burchinal, M. (2001). Receptive and expressive communication development of young males with fragile X syndrome. *American Journal of Mental Retardation, 106*(3), 216–230.

Roberts, J.E., Weisenfeld, A.A., Hatton, D.D., Heath, M., & Kaufmann, W.E. (2007). Social approach and autistic behavior in children with fragile X syndrome. *Journal of Autism and Developmental Disorders, 37*(9), 1748–1760.

Rogers, S.J., Wehner, E.A., & Hagerman, R.J. (2001). The behavioral phenotype in fragile X: Symptoms of autism in very young children with fragile X syndrome, idiopathic autism and other developmental disorders. *Journal of Developmental and Behavioral Pediatrics, 22*, 409–417.

Sabaratnam, M., Murthy, N.V., Wijeratne, A., Buckingham, A., & Payne, S. (2003). Autistic-like behaviour profile and psychiatric morbidity in fragile X syndrome: A prospective ten-year follow-up study. *European Child and Adolescent Psychiatry, 12*(4), 172–177.

Scerif, G., Cornish, K., Wilding, J., Driver, J., & Karmiloff-Smith, A. (2007). Delineation of early attentional control difficulties in fragile X syndrome: Focus on neurocomputational changes. *Neuropsychologia, 45*(8), 1889–1898.

Schaefer, G.B., Mendelsohn, N.J., & Professional Practice and Guidelines Committee. (2008). Clinical genetics evaluation in identifying the etiology of autism spectrum disorders. *Genetics in Medicine, 10*(4), 301–305.

Schenck, A., Bardoni, B., Langmann, C., Harden, N., Mandel, J.L., & Giangrande, A. (2003). CYFIP/Sra-1 controls neuronal connectivity in Drosophila and links the Rac1 GTPase pathway to the fragile X protein. *Neuron, 38*(6) 887–898.

Schwartz, C.E., Dean, J., Howard-Peebles, P.N., Bugge, M., Mikkelsen, M., Tommerup, N., et al. (1994). Obstetrical and gynecological complications in fragile X carriers: A multicenter study. *American Journal of Medical Genetics, 51*(4), 400–402.

Sherman, S.L. (2000). Premature ovarian failure in the fragile X premutation syndrome. *American Journal of Medical Genetics, 97*(3), 189–194.

Sherman, S. (2002). Epidemiology. In R.J. Hagerman & P.J. Hagerman (Eds.), *Fragile X syndrome: Diagnosis, treatment, and research* (pp. 136–168). Baltimore: The Johns Hopkins University Press.

Sherman, S., Pletcher, B.A., & Driscoll, D.A. (2005). Fragile X syndrome: Diagnostic and carrier testing. *Genetics in Medicine, 7*(8) 584–587.

Shevell, M., Ashwal, S., Donley, D., Flint, J., Gingold, M., Hirtz, D., et al. (2003). Practice parameter: Evaluation of the child with global developmental delay: Report of the Quality Standards Subcommittee of the American Academy of Neurology and the Practice Committee of the Child Neurology Society. *Neurology, 60*(3), 367–80.

Simonoff, E., Pickles, A., Charman, T., Chandler, S., Loucas, T., & Baird, G. (2008). Psychiatric disorders in children with autism spectrum disorders: Prevalence, comorbidity and associated factors in a population-derived sample. *Journal of the American Academy of Child and Adolescent Psychiatry, 47*(8), 921–929.

Stanfield, A.C., McIntosh, A.M., Spencer, M.D., Phillip, R., Gaur, S., & Lawrie, S.M. (2008). Towards a neuroanatomy of autism: A systemic review and meta-analysis of structural magnetic resonance imaging studies. *European Psychiatry, 23*(4), 289–299.

Stoyanova, V., & Oostra, B.A. (2004). The CGG repeat and the *FMR1* gene. *Methods in Molecular Biology, 277*, 173–184.

Sulewska, A., Niklinska, W., Kozlowski, M., Minarowski, L., Naumnik, W., Niklinski, J., et al. (2007). DNA methylation in states of cell physiology and pathology. *Folia Histochemica et Cytobiologica, 45*(3), 149–158.

Sullivan, K., Hatton, D., Hammer, J., Sideris, J., Hooper, S., Ornstein, P., et al. (2006). ADHD symptoms in children with FXS. *American Journal of Medical Genetics, 140A*(21), 2275–2288.

Sullivan, K., Hatton, D., Hammer, J., Sideris, J., Hooper, S., Ornstein, P., et al. (2007). Sustained attention and response inhibition in boys with fragile X syndrome: Measures of continuous performance. *American Journal of Medical Genetics, 144B*(4), 517–532.

Sullivan, K., Hooper, S., & Hatton, D. (2007). Behavioural equivalents of anxiety in children with fragile X syndrome: Parent and teacher

report. *Journal of Intellectual Disabilities Research,*
51(Pt. 1), 54–65.

Sullivan, A.K., Marcus, M., Epstein, M.P., Allen, E.G.,
Anido, A.E., Paquin, J.J. et al. (2005).
Association of FMR1 repeat size with ovarian dys-
function. *Human Reproduction, 20*(2), 402–412.

Sun, H.T., Cohen, S., & Kaufmann, W.E. (2001).
Annexin-1 is abnormally expressed in fragile
X syndrome: Two-dimensional electrophoresis
study in lymphocytes. *American Journal of Medical Genetics, 103*(1), 81–90.

Tassone, F., Hagerman, R.J., Ikle, D.N., Dyer, P.N.,
Lampe, M., Willemsen, R., et al. (1999). FMRP
expression as a potential prognostic indicator
in fragile X syndrome. *American Journal of Medical Genetics, 84*(3), 250–261.

Tassone, F., Hagerman, R.J., Taylor, A.K., Gane,
L.W., Godfrey, T.E., & Hagerman, P.J. (2000).
Elevated levels of FMR1 mRNA in carrier
males: A new mechanism of involvement in
the fragile-X syndrome. *American Journal of
Human Genetics, 66*(1), 6–15.

Tassone, F., Hagerman, R.J., Taylor, A.K., Mills,
J.B., Harris, S.W., Gane, L.W., et al. (2000).
Clinical involvement and protein expression in
individuals with FMR1 premutation. *American
Journal of Medical Genetics, 91,* 144–152.

Van Dam, D., Errijgers, V., Kooy, R.F., Willemsen,
R., Meintjes, E., Oostra, B.A., et al. (2005).
Cognitive decline, neuromotor and behavioral
disturbances in a mouse model for fragile
X associated tremor/ataxia syndrome (FXTAS).
Behavior Brain Research, 162(2), 233–239.

Verkerk, A.J., Piereti, M., Sutcliffe, J.S., Fu, Y.H.,
Kuhl, D.P., Pizutti, A., et al. (1991). Identification of a gene (FMR-1) containing a CGG
repeat coincident with a breakpoint cluster
region exhibiting length variation in fragile X
syndrome. *Cell, 65,* 905–941.

Vianna-Morgante, A.M., & Costa, S.S. (2000).
Premature ovarian failure is associated with maternally and paternally inherited premutation
in Brazilian families with fragile X. *American
Journal of Human Genetics, 67*(1), 254–255.

Watson, C., Hoeft, F., Garrett, A.S., Hall, S.S., &
Reiss, A.L. (2008) Aberrant brain activation
during gaze processing in boys with fragile X
syndrome. *Archives of General Psychiatry, 65*(11),
1315–1323.

Wittenberger, M.D., Hagerman, R.J., Sherman,
S.L., McConkie-Rosell, A., Welt, C.K., Rebar,
R.W., et al. (2007). The FMR1 premutation
and reproduction. *Fertility and Sterility, 87*(3),
456–465.

Zhao, M.G., Toyoda, H. Ko, S.W., Ding, H.K., Wu,
L.J., & Zhuo, M. (2005). Deficits in trace fear
memory and long-term potentiation in a
mouse model for fragile X syndrome. *Journal of
Neuroscience, 25*(32) 7385–7392.

4

Behavioral Phenotypes in Down Syndrome

A Probabilistic Model

George T. Capone

We are beginning to acquire a detailed understanding of the neurobiological factors accounting for phenotypic variation within the general constructs of intelligence, personality, emotion, and behavior. In a postgenomic era, one of the unstated research goals is to establish a detailed description of the neurogenetics of individual difference and the underpinnings of neuropsychiatric conditions and to demystify individuals who make up special populations with neurodevelopmental disability (Karmiloff-Smith, 2006; Scerif & Karmiloff-Smith, 2005). One of the central scientific challenges posed by trisomy 21 is elucidating how complex phenotypes emerge in conjunction with triplication of the > 300 genes mapping to chromosome 21 (Pritchard, Reeves, Dierssen, Patterson, & Gardner, 2008), and from a clinician's perspective, how to account for the expression of uncommon or complex phenotypes that are only occasionally observed.

BEHAVIORAL PHENOTYPES IN NEUROGENETIC SYNDROMES

The study of behavioral phenotypes in individuals with genetic conditions is necessarily concerned with the measurement of observable behavioral phenomena. A behavioral phenotype describes a specific behavior or trait that is more commonly observed in individuals with a particular condition than it is in those without it. Frequency, however, is relative, determined by the statistical power of the observation and whether a comparison is made with individuals who are typically developing or have intellectual disabilities or have a different neurogenetic syndrome (Flint, 1998). Typically, the salient aspects of motor, speech, language, cognitive, social, or maladaptive behavior become the domain of study. The probability that any group of individuals with condition X will manifest trait X is likely to reveal qualitative differences between syndromes, as well as variable expression within syndromes (Dykens, 1995). When the emergence of these attributes in the developing brain is emphasized, their study takes on unparalleled complexity. In children with trisomy 21, psychosocial and developmental dynamics, physiologic variables, and associated medical comorbidities further complicate the measurement and interpretation of behavior (Capone, Goyal, Ares, & Lannigan, 2006).

CHALLENGES FOR BEHAVIORAL PHENOTYPING OF DOWN SYNDROME

When the behavioral phenotype of any neurogenetic syndrome is being considered, etiology clearly matters (Dykens & Hodapp,

The author acknowledges the many children and their families who make this work possible.

2001; Karmiloff-Smith, 1998). Despite shared impairment in cognitive, speech, language, and behavioral domains, the neurobiological differences between fragile X, Down, and Rett syndromes are well documented (Kaufmann, Capone, Carter, & Lieberman, 2008). Compared with chromosomal trisomy, a defect in the *FMR1* gene results in a fairly circumscribed set of neurobiological consequences leading to the pathogenesis of fragile X syndrome (Kaufmann, 2002). In Down syndrome, the genetic contribution of greatest significance results from dosage imbalance for the entire 21st chromosome, which includes in excess of 300 genes (Epstein, 1986b; Nikolaienko, Ngyuen, Crinc, Cios, & Gardiner, 2005). Thus, the opportunity for novel and complex gene-gene or protein-protein interactions is more likely for trisomy than for single-gene or microdeletion disorders (Meechan, Maynard, Gopalakrishna, Wu, & LaMantia, 2007; Roper & Reeves, 2006), making the prospect of finding a "genetic solution" to any phenotype more difficult (Gottesman & Gould, 2003).

The common belief that Down syndrome already enjoys a well-known cognitive-behavioral phenotype, and that there is simply not much left to discover, may be one reason that research into this aspect of trisomy 21 has been difficult to establish. Another set of potential challenges faced by all researchers trying to determine whether any behavioral trait is truly associated with a particular genetic condition (i.e., phenotypic) includes questions about the science itself, the precision of behavioral definitions, nosological constructs, methods of detection, threshold minima, boundary criteria, construct validity, measurement reliability, test floor effects, and age, gender, and developmental effects (Dykens & Hodapp, 2007; Hodapp & Dykens, 2005; Kendell & Jablensky, 2003). These constraints, and the tendency toward academic entrenchment within favored paradigms, limit the prospect for multidisciplinary research and the advancement of theoretical models necessary to stimulate hypothesis testing. There are few shared paradigms or research traditions comprehensive enough to account for the complexity encountered at the genetic, neurobiological, developmental, and behavioral levels of inquiry. Perhaps because of this complexity, genotype-behavioral phenotype research lacks an agreed-upon theoretical model from which to generate predictions or direct hypothesis testing. Compared with more recently described neurogenetic syndromes, Down syndrome languishes in a state of preparadigmatic limbo. The absence of a contemporary research construct becomes painfully evident whenever atypical behavioral traits observed in lower functioning individuals are "explained" with simple developmental metaphors, or in fatalistic terms, resulting in a kind of therapeutic nihilism. Although the contribution of underlying genetic, neurobiological, and other developmental processes has been difficult to determine, recent consensus in articulating a core agenda for Down syndrome research is reason for encouragement (Lott, Patterson, & Seltzer, 2007; NIH Down Syndrome Working Group, 2007).

TOWARD A GENOTYPE—BEHAVIORAL PHENOTYPE RESEARCH AGENDA

What, then, are some questions about trisomy 21 that are potentially informative regarding the genotype–behavioral phenotype connection?

1. Through careful, detailed description can we define patterns of developmental-behavioral-physiological impairment among individuals such that specific subgroups emerge (i.e., clustering)?

2. If such clustering exists, can this result in behavioral phenotypes that are unique to Down syndrome (i.e., syndrome specificity)?

3. Do early-presenting behavior clusters reflect neurobiological impairment in

brain development and organization? What are the earliest indicators signaling the emergence of various behavioral clusters?

4. Are there specific environmental or medical risk factors that contribute to the emergence of certain behavioral clusters?

5. How does cognitive impairment influence the expression of behavior? How does maladaptive behavior influence cognitive function and performance? What are the implications for psychiatric-based diagnosis, treatment, and therapy?

BRAIN DEVELOPMENT IN DOWN SYNDROME

Converging evidence from neuropathological studies indicates that trisomy 21 results in a rather profound disruption of cellular proliferation during early embryogenesis and abnormal neuronal differentiation and connectivity (Coyle, Oster-Granite, & Gearhart, 1986; Guidi et al., 2008). Neuronal impairment results from gene activity that affects cellular function directly, or indirectly as a downstream consequence of gene action. In young subjects with Down syndrome the greatest reduction in brain tissue occurs in the neocortical regions despite preservation of deeper subcortical structures (Jernigan & Bellugi, 1990; Pinter, Eliez, Schmitt, Capone, & Reiss, 2001). A diminution of cortical neurons (Colon, 1972; Davidoff, 1928; Ross, Galaburda, & Kemper, 1984), reduced complexity of dendritic arborizations (Armstrong, Dunn, & Antalffy, 1998; Colon, 1972), and subsequent delays in postnatal myelination (Wisniewski & Schmidt-Sidor, 1989) are frequently cited correlates of smaller brain volume. The temporal and frontal lobes are two regions most consistently affected by these disturbances (Golden & Hyman, 1994; Wang, Koherty, Hesselink, & Bellugi, 1992).

MALADAPTIVE BEHAVIOR IN DOWN SYNDROME

Maladaptive behavior is often a presenting concern of parents on behalf of children with Down syndrome who are recognized to be developing atypically. Measures of developmental progress, cognition, and behavioral function can be particularly informative for understanding brain organization and the integrity of its functions (Capone & Kaufmann, 2007). Although early onset maladaptive behaviors are in no way unique to young children with trisomy 21 or intellectual disability, both the type and severity of certain behaviors could be considered a surrogate biomarker, heralding the occurrence of some neurobiological disturbance. Such behavioral markers could prove highly informative when identified, measured, and monitored throughout the course of early development. Repetitive motor activity, unusual sensory responding, or visual gaze abnormalities, which can manifest before 24 months, are of particular interest. Such behaviors are readily observed and quantified with the use of validated rating scales such as the Aberrant Behavior Checklist (ABC; Aman, Singh, Stewart, & Field, 1985), observation scales such as the Autism Diagnostic Observation Schedule (ADOS) (Lord et al., 2000), and structured Functional Behavioral Analysis (FBA) models.

Maladaptive Behavior in Down Syndrome: Phenotypic?

Despite significant delays across all domains, many young children with Down syndrome enjoy a reputation for their social affection and stubborn persistence or compulsive-like behavior (Evans & Gray, 2000), and fewer serious problem behaviors compared with children with other types of intellectual disability (Chapman & Hesketh, 2000; Fidler, 2005). However, as many as 20%–30% of children with Down syndrome manifest maladaptive behaviors significant enough to warrant consideration as a comorbid condition (Coe et al., 1999; Dykens, Shah, Sagun, Beck, & King, 2002). According to the current nosology of the *Diagnostic and Statistical Manual of Mental Disorders, Fourth Edition, Text Revision (DSM-IV-TR;* American Psychiatric Association, 2000), criteria for anxiety disorders, attention-deficit/hyperactivity disorder (ADHD), disruptive behavior disorders, stereotypy movement disorder, and autism spectrum disorders are sometimes observed in prepubertal children with Down syndrome (Capone et al., 2006), which raises a number of questions. How should we be using a *DSM-IV* criteria–based approach to classification of psychiatric or behavioral disorders in this population, and is this approach useful? How should we interpret a behavioral symptom complex (i.e., autism's core triad of symptoms or hyperactivity-impulsivity) when it occurs in the setting of significantly impaired intellectual and adaptive function? Because maladaptive behaviors are seen in children with low function regardless of etiology, the question of syndrome-specific behavior "as phenotype" can be problematic.

Many individual acts of misbehavior exhibited by children with Down syndrome are not specific to their genetic condition or to children with intellectual disability generally. For example, most externally directed disruptive acts have little to do with genetics or neurobiology and more to do with environmental setting, task demands, and social dynamics in real time. Although the probability of engaging in certain types of disruptive behavior may have a genetically influenced neurobiological basis (i.e., inhibitory control or temperament) (Rubia et al., 2008; Sanson & Prior, 1999), clearly, some types of internally directed behavior, such as stereotypy, may warrant a different consideration (Cunningham & Schreibman, 2008). Internally directed behaviors like repetitive or patterned movements, gaze preference, and guttural vocalizations may occur not in response to specific task de-

mands, but during periods of self-imposed isolation, sensory seeking, and social withdrawal. This restricted range of interests and preferences is typical of children with pervasive developmental disorders (Bishop, Richler, & Lord, 2006) and some children with low intellectual-adaptive function (Vig & Jedrysek, 1999).

Stereotypy as a Behavioral Phenotype in Down Syndrome?

Stereotypy rarely occurs as an isolated motor phenomenon and is often associated with other types of repetition, such as obsessive-compulsive behavior, perseveration, dyskinesia, or vocal and motor tics (Lewis & Bodfish, 1998). They are common to individuals with autism or intellectual disability and can interfere significantly with learning and social adaptive function (Cunningham & Schreibman, 2008). Stereotypic behaviors frequently involve simple or complex patterns of fine or gross motor movements. Common stereotypies include repetitive movements such as body rocking, hand flapping, or using the body to generate object movement, such as paper waving or dangling strings. Sometimes repetitive vocalizations accompany these movements. The movements are typically rhythmic without much variability, persist over time, and are difficult to interrupt despite changes in environmental setting (Rapp & Vollmer, 2005). Transient reduction of stereotypic movements may be observed following antecedent manipulation (environmental enrichment) or consequent interventions (differential reinforcement of alternative behavior), but the urge to engage in stereotypy remains. Because stereotypy appears to confer little social advantage or external reward, it is said to be automatically reinforcing.

Our interest in stereotypic behavior has evolved in an effort to highlight one clinical dimension of the within-syndrome behavioral variation seen in children with Down syndrome. Among the maladaptive behaviors, stereotypy is easy to observe, occurs across a range of functional levels, is associated with other maladaptive behaviors, and is one component of the core triad of autism symptoms. Although the frequency of stereotypy in children with Down syndrome has not been determined, based upon observation and theoretical considerations the incidence and severity would be expected to vary with developmental-cognitive level (Evans & Gray, 2000; Haw, Barnes, Clark, Crichton, & Kohen, 1996).

Stereotypy across *DSM-IV* Group Assignment

Using a clinically referred sample of 305 children with Down syndrome, we administered the ABC rating scale to parents of children aged 2–13 years. Individual ABC scales were analyzed across the entire study population and according to *DSM-IV-TR* diagnostic groupings (APA, 2000). Children with stereotypy often fulfill *DSM-IV-TR* criteria for Autism, Pervasive Developmental Disorder (PDD), or Stereotypic Movement Disorder (SMD) with or without Self-Injurious Behavior (SIB) (Capone, Grados, Kaufmann, Bernad-Ripoll, & Jewell, 2005). Those with anxiety disorders, disruptive behavior disorders (DBDs), and/or ADHD may also exhibit some tendency to engage in stereotypy in stressful situations or under certain environmental conditions (Capone, unpublished).

We observed an ordered decrease in stereotypy score in individuals meeting *DSM-IV-TR* criteria for autism spectrum disorder-stereotypic movement disorder-disruptive behavior disorder (ASD-SMD-DBD) and Down syndrome controls (Table 4.1)

Table 4.1. Intelligence and maladaptive behavior by *DSM-IV-TR group*

	ASD	SMD	DBD	Typical	ANOVA
Subjects (*N*)	112	48	101	44	
Male (%)	82/112(73%)	43/48(89%)	73/101(72%)	31/44(70%)	
Irritability	12.9 ± 9.6	8.9 ± 7.7	13.7 ± 7.5	2.5 ± 2.5	< .00001
Hyperactivity	20.1 ± 9.8	15.9 ± 7.6	24.2 ± 9.5	5.9 ± 5.2	< .00001
Lethargy	18.2 ± 8.9	8.4 ± 6.9	5.1 ± 5.3	1.5 ± 3.0	< .00001
Stereotypy	11.9 ± 4.7	7.4 ± 4.3	3.0 ± 4.0	0.9 ± 1.8	< .00001
Self-injury	2.1 ± 3.2	1.3 ± 2.4	0.8 ± 1.7	0.1 ± 0.7	< .00001
IQ/DQ	25.8 ±12.9	39.1 ± 11.8	47.3 ± 12.5	49.2 ± 13.0	< .00001

Key: *DSM-IV-TR, Diagnostic and Statistical Manual of Mental Disorders, Fourth Edition, Text Revision* (American Psychiatric Association, 2000); ASD, autism spectrum disorder; SMD, stereotypic movement disorder; DBD, disruptive behavior disorder; ANOVA, analysis of variance; *N*, total number in sample; IQ/DQ, intelligence quotient/development quotient.

See text for results of post hoc *t* tests.

(ANOVA < .00001, post hoc t test $p < .00001$ for all comparisons, except DBD versus typical $p < .01$). No interaction was found between Age at evaluation and ABC Stereotypy score ($R^2 = 0.003$). Recurrent hand, extremity, and body movements are the most common stereotypies observed; odd, bizarre, or complex motor stereotypies are less frequent (Carter, Capone, Gray, Cox, & Kaufmann, 2007). The severity of SIB, as determined by three items from the ABC Irritability scale, follows the same gradient as Stereotypy across *DSM-IV-TR* types, suggesting an overlapping or shared etiology between these related phenomena (Table 4.1) (ASD-SMD-DBD-typical) (ANOVA $p < .00001$, post hoc t test $p < .0001$ for ASD versus Typical and ASD versus DBD comparisons; Typical versus SMD and Typical versus DBD $p < .05$).

Stereotypy Clusters with Internalizing Behaviors

In our experience, motor stereotypy is more common in males with Down syndrome than it is in females, and in some individuals it manifests prior to 24 months (Capone et al., 2005). Stereotypy is usually not an isolated finding and may be associated with

functional impairments in attention, sensory responding, and social function, as well as abnormal patterns of feeding and sleep. We have observed stereotypy to intensify during the course of early development, often preceding a general deterioration in developmental skills (cognitive, speech/language, social) and the emergence of other maladaptive behaviors (Capone et al., 2006). Contrawise, stereotypic motor behaviors may become less frequent as developmental progress and maturation proceed.

Internalizing behaviors such as apathy, low motivation, social indifference, and affective blunting as captured on the ABC Lethargy scale are often observed in individuals with stereotypy. We observe an ordered decrease in Lethargy score in individuals meeting *DSM-IV-TR* criteria for ASD-SMD-DBD and Down syndrome controls (Table 4.1) (ANOVA < .00001, post hoc t test $p < .00001$ for all comparisons). No interaction was found between Age at evaluation and ABC Lethargy score ($R^2 = 0.00$). Lethargy, however, is strongly and positively associated with Stereotypy ($R^2 = 0.4612$) (Figure 4.1), whereas a much weaker association is seen between Stereotypy and Hyperactivity score ($R^2 = 0.104$) (Figure 4.2). This raises an interesting

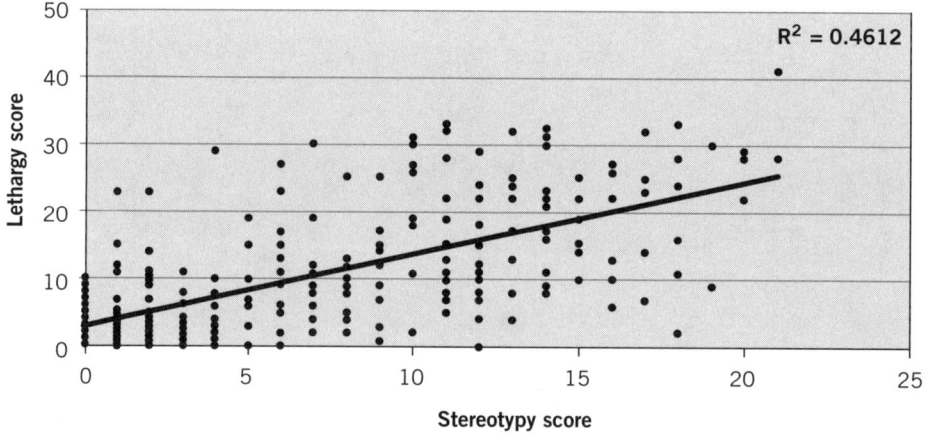

Figure 4.1. ABC Lethargy x Stereotypy. (*Key:* ABC, Aberrant Behavior Checklist; Aman, Singh, Stewart, & Field, 1985.)

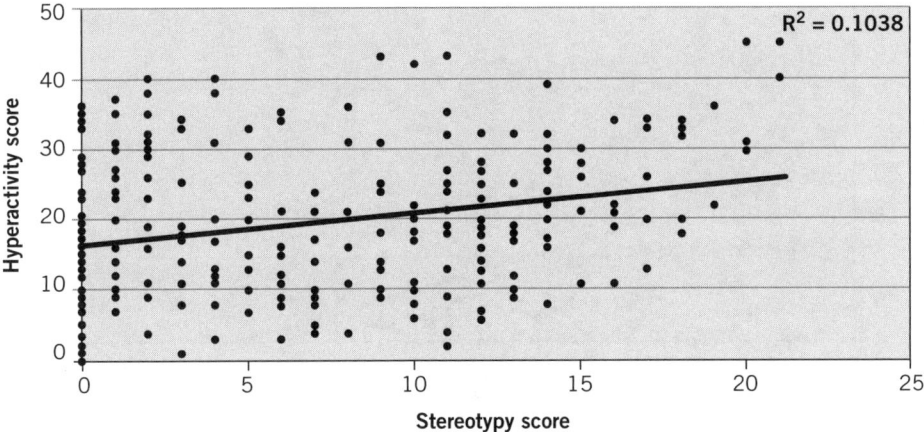

Figure 4.2. ABC Hyperactivity x Stereotypy. (*Key:* ABC, Aberrant Behavior Checklist; Aman, Singh, Stewart, & Field, 1985.)

query. Why should repetitive motor actions and perseveration be so highly correlated with motivation, arousal, affect, and social interest, and apparently less so with motor hyperactivity itself, and what is the relationship of stereotypy to general intelligence?

Maladaptive Behaviors and General Intelligence

Measures of general intelligence (intelligence quotient/development quotient [IQ/DQ]) were not associated with Age (R^2 = 0.007) or gender in our study sample.

However IQ/DQ was negatively associated with both Stereotypy score ($R^2 = 0.2689$) (Figure 4.3) and Lethargy score ($R^2 = 0.222$) (Figure 4.4) on the ABC, but not for measures of Hyperactivity ($R^2 = 0.01$) or Irritability ($R^2 = 0.023$). An ordered decrease in IQ/DQ is seen for ASD-SMD-DBD and Down syndrome controls (Table 4.1) (ANOVA < .00001, post hoc t test $p < .00001$ for all comparisons, except DBD versus SMD $p < .01$, DBD versus Down syndrome control, not significant). This dissociation between IQ/DQ and maladaptive behavior in the DBD group is noteworthy because it underscores differences between the internalizing

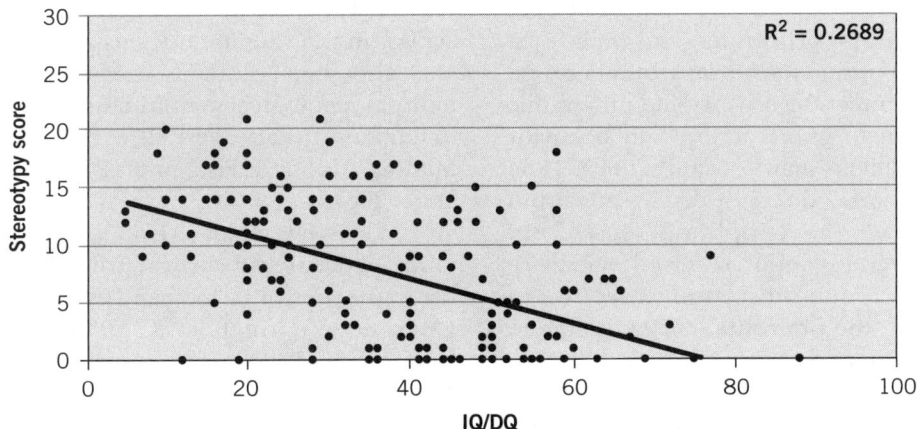

Figure 4.3. ABC Stereotypy x intelligence quotient/development quotient. (*Key:* ABC, Aberrant Behavior Checklist; Aman, Singh, Stewart, & Field, 1985.)

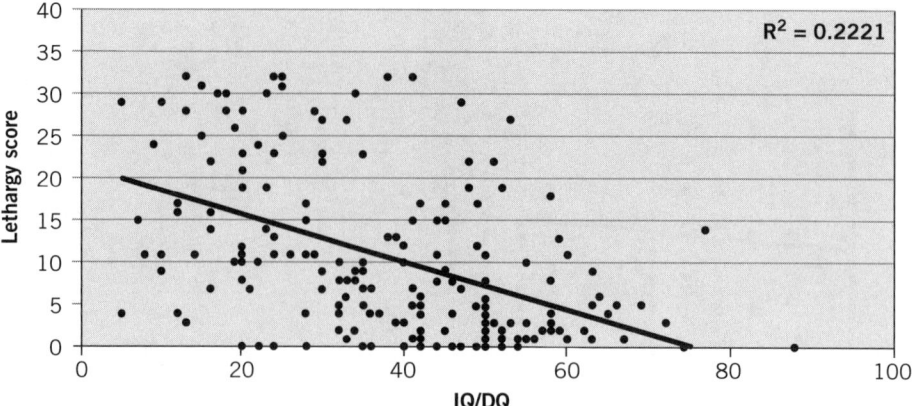

Figure 4.4. ABC Lethargy x intelligence quotient/development quotient. (*Key:* ABC, Aberrant Behavior Checklist; Aman, Singh, Stewart, & Field, 1985.)

and externalizing dimension of maladaptive behavior, and the confounding role of intelligence in their expression. Hyperactive, impulsive, disruptive children with Down syndrome appear to function as well cognitively as their peers without behavior disorders, suggesting the role of other variables.

The robust interaction between Stereotypy and Lethargy suggests that the same circuits regulating arousal, attention, motivation, social interest, and motor control also function in the expression of intelligence and adaptive behavior. Anatomically, the cognitive system is distributed diffusely throughout the fronto-temporo-parietal cortices (Jung & Haier, 2007). The frontal lobes and their connections play a central role in cognitive processing, and their most anterior component, the prefrontal cortex, selects and integrates sensory and mnemonic information to guide action and behavior (Barbas, 2000; Shaw et al., 2006). The prefrontal cortex does not receive any input directly from the peripheral senses, but rather integrates input received via distributed cortical and subcortical structures. In this manner the prefrontal cortex is able to analyze information about the immediate environment and integrate this information with experiential memory and emotive states to direct behavior and motor control in real time (Barsalou, Breazeal, & Smith, 2007).

Frontal-Subcortical Circuits and Behavior

Perseveration and stereotypy are two related repetitive phenomenon that have been modeled in animals that have undergone frontal lobe lesioning (Ridley, 1994). Increased dopaminergic activity in the caudate nucleus results in both repetitive and excessive motor activity, and task perseveration is associated with disruption of frontal cortex due to the loss of inhibitory signal to the striatum and thalamus, which normally prohibits repetitive motor responding. Frontal lesions further exacerbate the stereotypy-inducing effects of elevated dopaminergic activity in the caudate (Ridley, 1994). Hypo- or hyperdopaminergic activity in the caudate nucleus and putamen has also been linked to many of the neuropsychological manifestations of ADHD in humans (Jacaite, Fernell, & Halldin, 2005; Madras, Miller, & Fischman, 2005; Spencer et al., 2007).

At least five major circuits link the frontal lobe and deeper subcortical structures, giving rise to the prefrontal loop and the motor loop (Alexander, Crutcher, & DeLong, 1990; Cummings, 1993) (Figures 4.5 and 4.6). Damage to these regions of the prefrontal cortex or corresponding regions in the caudate nucleus, globus pallidus, or thalamus is associated with signature behavioral

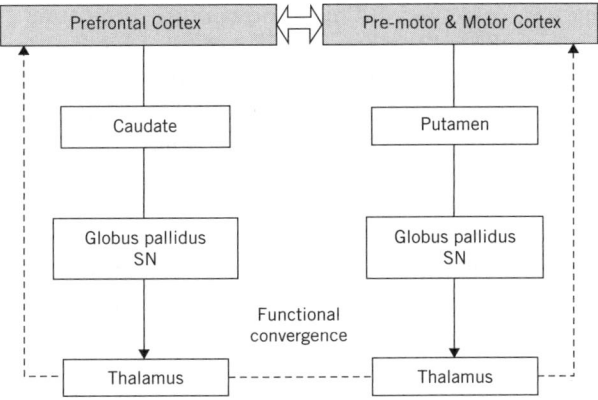

Figure 4.5. Prefrontal and motor loops.

syndromes: notably, irritability and disinhibition with injury to the orbitofrontal cortex; apathy and mutism with injury to the cingulate circuit; and working memory and executive dysfunction with disruption of the dorsolateral prefrontal cortex (Cummings, 1993). Interestingly, the same cognitive, behavioral, and emotional symptoms observed with damage to the prefrontal cortex are reminiscent of the complex symptomatology observed in several neuropsychiatric disorders with a developmental onset (i.e., ADHD, obsessive-compulsive disorder, autism, stereotypy/SIB).

Frontal subcortical circuits are intimately involved in the pathophysiology of ADHD (Sagvolden, Aase, Johansen, & Russell, 2005), OCD (Kang et al., 2004), and self-injury (SIB) in people with intellectual disability (King, 1993) and autism (Rojas et al., 2006; Takarae, Minshew, Luna, & Sweeny, 2007). Like stereotypy, self-injurious and obsessive-compulsive behaviors are exacerbated by stress or anxiety and can be difficult to control voluntarily (Lewis & Bodfish, 1998). The restrictive and repetitive symptom deficits characteristic of children with autism are also related to aspects of cognitive flexibility, working memory, and response inhibition (Lopez, Lincoln, Ozonoff, & Lai, 2005). These symptoms are often treated with medications acting at serotonin and/or dopamine receptors (Grados &

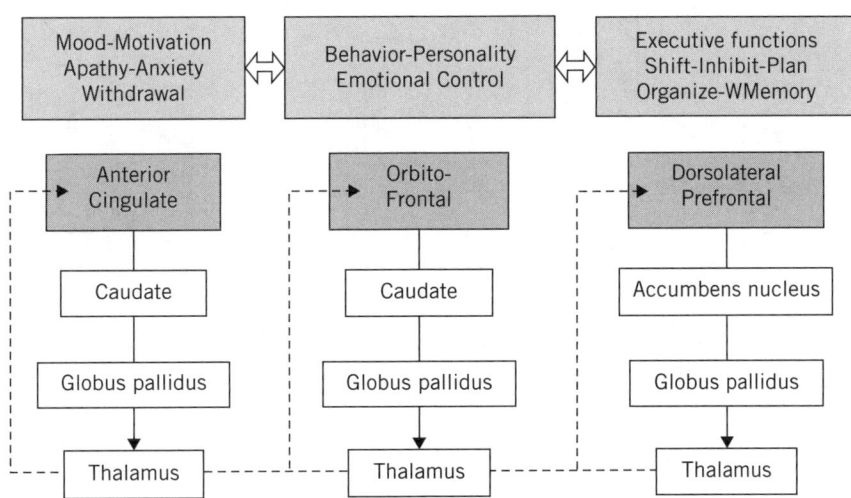

Figure 4.6. Prefrontal-subcortical circuits. (*Key:* WMemory, working memory.)

Riddle, 2001; Grossman & Verobyev, 1998; King, 1993). Treatment with the second-generation antipsychotic risperidone, which exhibits high affinity for serotonin $5HT_{2a}$ and dopamine D_2 receptors, results in diminished stereotypy, diminished hyperactivity, and apparent improvement in lethargy in some children with Down syndrome, severe intellectual impairment, and autism spectrum phenotype (Capone, Grados, Goyal, Smith, & Kammann, 2008). Apparently, children with Down syndrome + ASD phenotype share some of the same characteristics and responses as children with autism, stereotypy, and intellectual disability of undetermined etiology (RUPP Autism Network, 2002).

Frontal-Subcortical Circuits and Stereotypy

Why should symptoms of stereotypy be associated with impairment of prefrontal function? Consistent with the first models established by Ridley (1994) and Cummings (1993) and expanded upon by Sagvolden et al. (2005), we propose that the diminished influence of descending prefrontal glutamatergic efferents to striatum reduces the GABAergic signaling within the globus pallidus and thalamus, resulting in diminished activity of glutamatergic thalmocortical afferents. Attenuation of prefrontal circuits relative to the more robust motor circuits leads to a functional imbalance, allowing activity within the motor loop to emerge as dominant. Further dysregulation of tonic/phasic dopamine control in the caudate-putamen, nucleus accumbens, and prefrontal cortex also contributes to the emergence of fluctuating inattention and stereotypic motor behavior (Prince, 2008; Sagvolden et al., 2005). Simple, repetitive patterns of motor activity that are easily performed become automatically reinforced and established as activity arising within the motor circuit cannot be overcome by a diminished prefrontal network (Figure 4.7). Consistent with a poorly organized prefrontal network and faulty dopaminergic modulation, impairment in higher cortical and executive function is observed. Young children with Down syndrome who display stereotypy are thus more likely to manifest associated deficits in attention, arousal, motivation, mood, behavioral regulation,

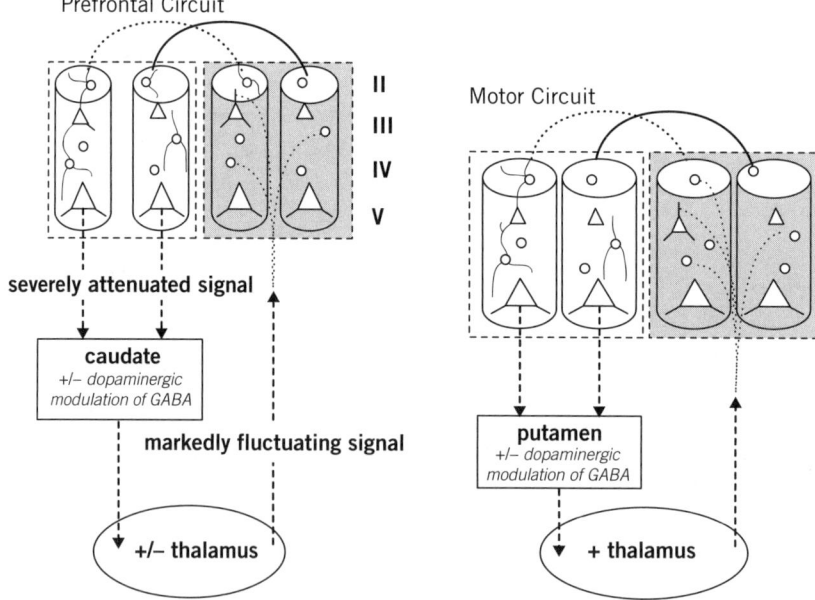

Figure 4.7. Prefrontal-motor circuit imbalance in Down Syndrome with stereotypy.

impulse control, and working memory (Nieoullon, 2002). These same features are often described in individuals with autism (Lopez et al., 2005; Rojas et al., 2006). In children with Down syndrome the location and extent of prefrontal involvement probably determines the specific pattern of deficits observed. For instance, severe neuronal depletion or abnormal connectivity in dorsolateral prefrontal cortex (DLPFC) may have very different cognitive-behavioral consequences, depending on whether anterior cingulate, orbitofrontal, temporal, or limbic structures are also affected (Barbas, 2000) and whether dopaminergic dysfunction reaches threshold significance. Developmentally, the frontal cortex and its connections constitute a most critical landscape, where the neurobiological events played out in utero continue to reverberate with lifelong repercussions for cognitive, behavioral, and emotional function.

Neurogenesis and Cortical Development

Genetic disruption of cell-cycle control during embryogenesis may set into motion critical events that lead to the later emergence of complex behavioral phenotypes in trisomy 21 (Epstein, 1986a). Paramount to the structural integrity of the evolutionarily conserved cerebral cortex is the establishment of sufficient numbers of preneurogenetic proliferative (progenitor) cells in the embryonic brain (Caviness, Takahashi, & Nowakowski, 1995). These progenitor cells comprise the founder population within the germinal matrix from which all neurons and glia of the cortical and subcortical plate are generated. Initially, each progenitor divides symmetrically during mitosis to produce two additional progenitor cells, thereby amplifying its histiogenic potential. After a period of time, asymmetric cell division ensues, producing one neuroblast and only a single progenitor cell with each round of cell division. Events that interfere with the timing or switching from symmetric to asymmetric

cell division would have a profound impact on the development of cortical structure (Rakic, 1995). Prolongation of the cell cycle, for instance, would result in fewer total mitotic events; and fewer progenitors would result in diminished generation of both neuronal and glial cell populations. Any reduction in progenitor cells would result in fewer radially oriented "physiological units" or cortical modules in neocortex (Levitt, Eagleson, & Powell, 2004; Mountcastle, 1979) and incomplete specification of cortical function. In this scenario, prefrontal function as determined by the number and integrity of intact cortical modules dedicated to multimodal information processing could become severely compromised. Thus, prolongation of the cell cycle in progenitor cells is one likely mechanism for neuronal depletion in trisomy 21 (Contestabile et al., 2007), as previously suggested by the Ts16 and TsDn65 mouse models (Chakrabarti, Galdzicki, & Haydar, 2007; Haydar, Nowakowski, Yarowsky, & Krueger, 2000).

Development of Prefrontal Subcortical Connections

Reduced neurogenesis has important downstream consequences in trisomy 21, as it contributes to diminished buffering capacity throughout the extended period of brain development. For example, ingrowth of thalamic and extrathalamic axons is required for the establishment of functional specificity in the cerebral cortex (Molnar & Blakemore, 1995). During the second half of prenatal life, corticocortical and thalamic axons enter the subplate to form transient but functional synapses (Kanold, 2004; Kostovic & Judas, 2007). These axons then disconnect and segregate into the upper cortical plate to form stable connections in layer I, II, III, and V neurons (Rodriguez, Whitson, & Granger, 2004). The timing of these events is critical to the regional specification and functionality of the neocortex (Rakic, 1988). We predict that any reduction in subplate neuron number, or their ability to form transient synapses with thalamic axons, destabilizes this otherwise highly buffered sequence, resulting in imprecise cortical specification and aberrant circuitry (Kanold, 2004; Kostovic & Judas, 2007), which is already compromised by deficient cortical modules (Buxhoeveden, Fobbs, Roy, & Casanova, 2002).

Similarly, reduction in neurotransmitter-mediated trophic support could have additional detrimental effects on the survival, arborization pattern, and biochemical phenotype of developing cortical neurons (Lauder, 1993; Luo, Perscio, & Lauder, 2003). Frontal cortex from a fetal Down syndrome brain reveals a marked reduction in the levels of serotonin and dopamine metabolites, suggesting decreased axonal penetration/survival or decreased synthesis/release of synaptic neurotransmitter from projection neurons in the dorsal raphe nucleus (serotonin) and ventral tegmental area (dopamine) (Whittle, Sartori, Dierssen, Lubec, & Singewald, 2007). It appears, then, that early deficits in cortical neurogenesis compounded by imprecise axonal ingrowth and concomitant loss of serotonin-mediated trophic support is one possible sequence by which atypical patterns of cortical connectivity could emerge in trisomy 21.

Canalization and Brain Development

In typically developing euploid individuals, natural selection has resulted in the canalization of genetic, biochemical, and cellular events effecting brain structure and organization. That is, myriad developmental pathways have evolved in humans to bring about a largely predestined end result (i.e., highly organized, six-layered neocortex) regardless of minor perturbations that may occur in the cellular and physiological environment. Developmental pathways become increasingly constrained over time as cellular development proceeds from undifferentiated and pluripotent toward a highly differ-

entiated end state. The degree to which canalization is maintained in a species through genetic mechanisms is referred to as buffering (Waddington, 1942). It has been proposed that aneuploidy itself results in disruption to the buffering capacity or homeostatic balance inherent to the process of development (Shapiro, 1983). An extension of Waddington's concept introduces the notion of "stable attractors," which pull and shift developmental pathways away from their canalized end point and toward some alternative outcome (Grossman et al., 2003). Despite a myriad of etiological triggers, the developing brain is probably vulnerable to only a finite number of stable attractors. If this is the case, then we might expect to observe similar outcomes (behavior phenotypes) among etiologically distinct neurogenetic syndromes, and we do.

Genotype-Phenotype Predictions Are Probabilistic

By some estimates $> 50\%$ of the human genome plays a role in brain development or function (Rosenberg, 1997), so accounting for all of the interacting molecular and biochemical mechanisms becomes rather daunting. From the perspective of clinicians attempting to deconstruct behavioral phenotypes in Down syndrome, it is necessary and comforting to conceptualize genetics *in toto* as orchestrating molecular neurobiological events in a quasi-predictable rather than strictly deterministic fashion. The vagaries of gene expression, the stochastic nature of the developmental processes, and threshold effects required for the expression and detection of behavioral phenotypes make this so. Specific genes at dosage imbalance may well alter expression proportionately via predictable molecular mechanisms (Epstein, 1990), but the overall developmental complexity of biological networks precludes any genetically informed prediction of cognitive-behavioral phenotype *a priori*. One day it may be possible to determine the specific patterns of gene expression that predict an increased

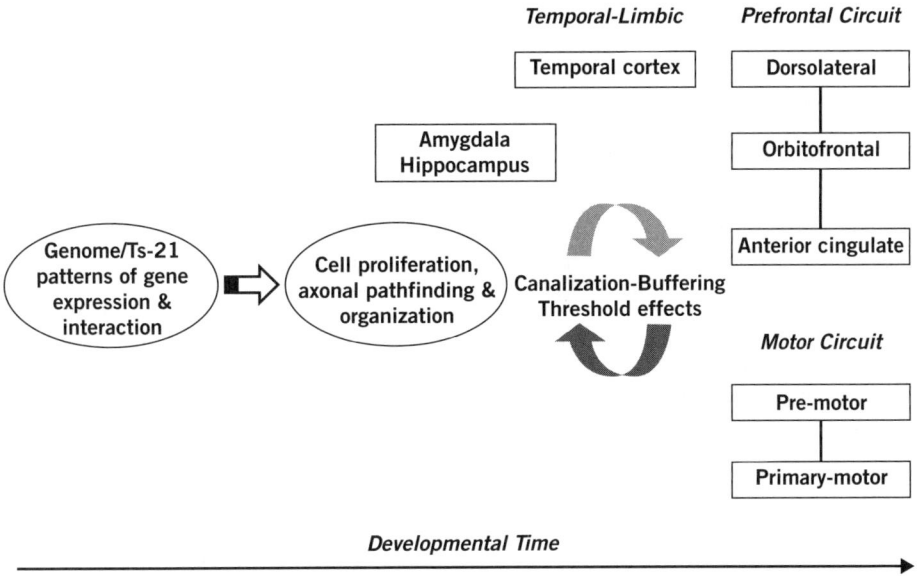

Figure 4.8. Probabilistic model of genotype-behavioral phenotype.

probability of deficits in cortical neurogenesis or the cellular architecture of prefrontal cortex and caudate (Figure 4.8), but currently we are unable to do so for individuals with trisomy 21.

NEW CHALLENGES FOR BEHAVIORAL PHENOTYPING

Our findings provide preliminary support for the existence of specific clusters of behavioral-physiological impairment consistent with the notion of behavioral phenotypes within Down syndrome, which by their very nature appear to reflect underlying differences in brain development and organization. Diffusion tensor imaging tractography and functional magnetic resonance imaging now permit detailed examination and analysis of prefrontal cortex (Shaw et al., 2007), fronto-subcortical (Leh, Ptito, Chakravarty, & Strafella, 2007), and thalamo-cortical circuits in humans (Counsell et al., 2006). Thus, predictions about genetic or neurobiological influences on behavior phenotype based on a prefrontal-subcortical model are becoming testable (Brocki, Fan, & Fossella, 2008; de Geus, Goldberg, Boomsma, & Posthuma, 2008).

In partial response to the questions posed earlier in this chapter, it appears there are many questions worth asking about behavioral phenotype(s) in children with Down syndrome. Their biological uniqueness and humanity alone should be enough to warrant further investigation with contemporary methods and paradigms. Clearly such knowledge would have important implications for early intervention and research. The complexity of this undertaking and implications for behavioral phenotype research are discussed by Karmiloff-Smith under the neuroconstructivist paradigm, in which she famously articulates that "development itself is the key to understanding developmental disorders" (Karmiloff-Smith, 1998). We endorse the call for a more comprehensive understanding of relevant biological, developmental, and behavioral factors in order to establish a meaningful tradition of genotype-behavioral phenotype research. It will be fascinating to determine the degree to which behavioral phenotypes reflect specific patterns of strength-weakness in neocortical functional domains, physiological functions, and medical comorbidities that are unique to trisomy 21. How we choose to frame these questions is at least as important as the answers themselves.

REFERENCES

Alexander, G.E., Crutcher, M.D., & DeLong, M.R. (1990). Basal ganglia-thalamocortical circuits: Parallel substrates for motor, oculomotor, prefrontal and limbic functions. *Progress in Brain Research, 85*, 119–146.

Aman, M., Singh, N., Stewart, A., & Field, C. (1985). The Aberrant Behavior Checklist: A behavior rating scale for the assessment of treatment effects. *American Journal of Mental Deficiency, 89*(5), 485–491.

American Psychiatric Association. (2000). *Diagnostic and statistical manual of mental disorders* (4th ed., text rev.). Washington, DC: Author.

Armstrong, D., Dunn, K., & Antalffy, B. (1998). Decreased dendritic branching in frontal, motor and limbic cortex in Rett syndrome compared with Trisomy 21. *Journal of Neuropathology and Experimental Neurology, 57*(11), 1013–1017.

Barbas, H. (2000). Connections underlying the synthesis of cognition, memory, and emotion in primate prefrontal cortices. *Brain Research Bulletin, 53*(5), 319–330.

Barsalou, L., Breazeal, C., & Smith, L. (2007). Cognition as coordinated non-cognition. *Cognitive Process, 8*, 79–91.

Bishop, S.L., Richler, J., & Lord, C. (2006). Association between restricted and repetitive behaviors and nonverbal IQ in children with autism spectrum disorders. *Child Neuropsychology, 12*(4–5), 247–267.

Brocki, K., Fan, J., & Fossella, J. (2008). Placing neuroanatomical models of executive function in a developmental context: Imaging and imaging-genetic strategies. *Annals of the New York Academy of Sciences, 1129*, 246–255.

Buxhoeveden, D., Fobbs, A., Roy, E., & Casanova, M. (2002). Quantitative comparison of radial cell columns in children with Down syndrome and controls. *Journal of Intellectual Disability Research, 46*(1), 76–81.

Capone, G., Goyal, P., Ares, W., & Lannigan, E. (2006). Neurobehavioral disorders in children, adolescents and young adults with Down syndrome. *American Journal of Medical Genetics, 142C*, 158–178.

Capone, G., Grados, M., Goyal, P., Smith, B., & Kammann, H. (2008). Risperidone use in children with Down syndrome, severe intellectual disability and co-morbid autistic spectrum disorder. *Journal of Developmental and Behavioral Pediatrics, 29*, 106–116.

Capone, G., Grados, M., Kaufmann, W., Bernad-Ripoll, S., & Jewell, A. (2005). Down syndrome and co-morbid autism-spectrum disorder:

Characterization using the Aberrant Behavior Checklist. *American Journal of Medical Genetics, 134A*(4), 373–380.

Capone, G., & Kaufmann, W.E. (2007). Human brain development. In P. Accardo (Ed.), *Capute and Accardo's neurodevelopmental disabilities in infancy and childhood* (3rd ed., Vol. I, pp. 27–57). Baltimore: Paul H. Brookes Publishing Co.

Carter, J., Capone, G., Gray, R., Cox, C., & Kaufmann, W.E. (2007). Autistic-spectrum disorders in Down syndrome: Further delineation and distinction from other behavioral abnormalities. *American Journal of Medical Genetics, Part B (Neuropsychiatric Genetics), 144B*, 87–94.

Caviness, V.S., Takahashi, T., & Nowakowski, R.S. (1995). Numbers, time and neocortical neurogenesis: A general developmental and evolutionary model. *Trends in Neurosciences, 18*(9), 379–383.

Chakrabarti, L., Galdzicki, Z., & Haydar, T.F. (2007). Defects in embryonic neurogenesis and initial synapse formation in the forebrain of the Ts65Dn mouse model of Down syndrome. *Journal of Neuroscience, 27*(43), 11483–11495.

Chapman, R.S., & Hesketh, L.J. (2000). Behavioral phenotype of individuals with Down syndrome. *Mental Retardation and Developmental Disabilities Research Reviews, 6*(2), 84–95.

Coe, D., Matson, J., Russell, D., Slifer, K., Capone, G., Baglio, C., et al. (1999). Behavior problems of children with Down syndrome and life events. *Journal of Autism and Developmental Disorders, 29*(2), 149–156.

Colon, E.J. (1972). The structure of the cerebral cortex in Down's syndrome. *Neuropadiatrie, 3*(4), 362–376.

Contestabile, A., Fila, T., Ceccarelli, C., Bonasoni, P., Bonapace, L., Santini, D., et al. (2007). Cell cycle alteration and decreased cell proliferation in the hippocampal dentate gyrus and the neocortical germinal matrix of fetuses with Down syndrome and in Ts65Dn Mice. *Hippocampus, 17*, 665–678.

Counsell, S., Dyet, L., Larkman, D., Nunes, R., Boardman, J., Allsop, J., et al. (2006). Thalamo-cortical connectivity in children born preterm mapped using probabilistic magnetic resonance tractography. *NeuroImage, 34*, 896–904.

Coyle, J., Oster-Granite, M., & Gearhart, J. (1986). The neurobiologic consequences of Down syndrome. *Brain Research Bulletin, 16*, 773–787.

Cummings, J. (1993). Frontal-subcortical circuits and human behavior. *Archives of Neurology, 50*, 873–880.

Cunningham, A., & Schreibman, L. (2008). Stereotypy in autism: The importance of function. *Research in Autism Spectrum Disorders, 2*(3), 469–479.

Davidoff, L. (1928). The brain in mongolian idiocy. *Archives of Neurology and Psychiatry, 20,* 1229–1257.

de Geus, E., Goldberg, T., Boomsma, D., & Posthuma, D. (2008). Imaging the genetics of brain structure and function. *Biological Psychology, 79,* 1–8.

Dykens, E.M. (1995). Measuring behavioral phenotypes: Provocations from the "new genetics." *American Journal on Mental Retardation, 99*(5), 522–532.

Dykens, E.M., & Hodapp, R.M. (2001). Research in mental retardation: Toward an etiologic approach. *Journal of Child Psychology and Psychiatry, 42*(1), 49–71.

Dykens, E., & Hodapp, R. (2007). Three steps toward improving the measurement of behavior and behavioral phenotype research. *Child and Adolescent Psychiatric Clinics of North America, 16,* 617–630.

Dykens, E.M., Shah, B., Sagun, J., Beck, T., & King, B.H. (2002). Maladaptive behaviour in children and adolescents with Down's syndrome. *Journal of Intellectual Disability Research, 46,* 484–492.

Epstein, C.J. (1986a). *The consequences of chromosome imbalance.* New York: Cambridge University Press.

Epstein, C.J. (1986b). *Trisomy 21 and the nervous system: From cause to cure.* New York: Raven Press.

Epstein, C.J. (1990). The consequences of chromosome imbalance. *American Journal of Medical Genetics, 7,* 31–37.

Evans, D., & Gray, F. (2000). Compulsive-like behavior in individuals with Down syndrome: Its relation to mental age level, adaptive and maladaptive behavior. *Child Development, 71*(2), 288–300.

Fidler, D.J. (2005). The emerging Down syndrome behavioral phenotype in early childhood. *Infants & Young Children, 18*(2), 86–103.

Flint, J. (1998). Behavioral phenotypes: Conceptual and methodological issues. *American Journal of Medical Genetics (Neuropsychiatric Genetics), 81,* 235–240.

Golden, J., & Hyman, B. (1994). Development of the superior temporal neocortex is anomalous in trisomy 21. *Journal of Neuropathology and Experimental Neurology, 53*(5), 513–520.

Gottesman, I.I., & Gould, T.D. (2003). The endophenotype concept in psychiatry: Etymology and strategic intentions. *American Journal of Psychiatry, 160*(4), 636–645.

Grados, M.A., & Riddle, M.A. (2001). Pharmacological treatment of childhood obsessive-compulsive disorder: From theory to practice. *Journal of Clinical Child Psychology, 30*(1), 67–79.

Grossman, A., Churchill, J., McKinney, B., Kodish, I., Otte, S., & Greenough, W. (2003). Experience effects on brain development: Possible contributions to psychopathology. *Journal of Child Psychology and Psychiatry, 44*(1), 33–63.

Grossman, R., & Verobyev, L. (1998). The neurobiology of stereotypic behaviors and stereotypic movement disorders. *Psychiatric Annals, 28*(6), 317–323.

Guidi, S., Bonasoni, P., Ceccarelli, C., Santini, D., Gualtieri, F., & Ciani, E. (2008). Neurogenesis impairment and increased cell death reduce total neuron number in the hippocampal region of fetuses with Down syndrome. *Brain Pathology, 18,* 180–197.

Haw, C., Barnes, T., Clark, K., Crichton, P., & Kohen, D. (1996). Movement disorder in Down's syndrome: A possible marker of the severity of mental handicap. *Movement Disorders, 11*(4), 395–403.

Haydar, T.F., Nowakowski, R.S., Yarowsky, P.J., & Krueger, B.K. (2000). Role of founder cell deficit and delayed neurogenesis in microencephaly of the trisomy 16 mouse. *Journal of Neuroscience, 20*(11), 4156–4164.

Hodapp, R., & Dykens, E. (2005). Measuring behavior in genetic disorders of mental retardation. *Mental Retardation and Developmental Disabilities Research Reviews, 11,* 340–346.

Jacaite, A., Fernell, E., & Halldin, C. (2005). Reduced midbrain dopamine transporter binding in male adolescents with attention-deficit/hyperactivity disorder: Association between striatal dopamine markers and motor hyperactivity. *Biological Psychiatry, 57*(3), 229–238.

Jernigan, T., & Bellugi, U. (1990). Anomalous brain morphology on magnetic resonance images in Williams syndrome and Downs syndrome. *Archives of Neurology, 47,* 529–533.

Jung, R., & Haier, R. (2007). The parieto-frontal integration theory (P-FIT) of intelligence: Converging neuroimaging evidence. *Behavioral and Brain Sciences, 30,* 135–154.

Kang, D.H., Kim, J.J., Choi, J.S., Kim, C.W., Youn, T., Han, M.H., et al. (2004). Volumetric investigation of the frontal-subcortical circuitry in patients with obsessive-compulsive disorder. *Journal of Neuropsychiatry and Clinical Neuroscience, 16*(3), 342–349.

Kanold, P.O. (2004). Transient microcircuits formed by subplate neurons and their role in functional development of thalamocortical connections. *NeuroReport, 15*(14), 2149–2153.

Karmiloff-Smith, A. (1998). Development itself is the key to understanding developmental disorders. *Trends in Cognitive Sciences, 2*(10), 389–398

Karmiloff-Smith, A. (2006). The tortuous route from genes to behavior: A neuroconstructivist approach. *Cognitive, Affective & Behavioral Neuroscience, 6*(1), 9–17.

Kaufmann, W.E. (2002). Neurobiology of fragile X syndrome: From molecular genetics to neurobehavioral phenotype. *Microscopy Research and Techniques, 57*, 131–134.

Kaufmann, W.E., Capone, G., Carter, J., & Lieberman, D. (2008). Genetic intellectual disability. In P. Accardo (Ed.), *Capute and Accardo's neurodevelopmental disabilities in infancy and childhood* (3rd ed., Vol. I, pp. 155–174). Baltimore: Paul H. Brookes Publishing Co.

Kendell, R., & Jablensky, A. (2003). Distinguishing between the validity and utility of psychiatric diagnoses. *American Journal of Psychiatry, 160*, 4–12.

King, B. (1993). Self injury by people with mental retardation: A compulsive behaviour hypothesis. *American Journal on Mental Retardation, 98*(1), 93–112.

Kostovic, I., & Judas, M. (2007). Transient patterns of cortical lamination during prenatal life: Do they have implications for treatment? *Neuroscience and Biobehavioral Reviews, 31*, 1157–1168.

Lauder, J. (1993). Neurotransmitters as growth regulatory signals: Role of receptors and second messengers. *Trends in Neurosciences, 16*(6), 233–240.

Leh, S., Ptito, A., Chakravarty, M., & Strafella, A. (2007). Fronto-striatal connections in the human brain: A probabilistic diffusion tractography study. *Neuroscience Letters, 419*, 113–118.

Levitt, P., Eagleson, K.L., & Powell, E.M. (2004). Regulation of neocortical interneuron development and the implications for neurodevelopmental disorders. *Trends in Neurosciences, 27*(7), 400–406.

Lewis, M., & Bodfish, J. (1998). Repetitive behavior disorders in autism. *Mental Retardation and Developmental Disabilities Research Reviews, 4*, 80–89.

Lopez, B., Lincoln, A., Ozonoff, S., & Lai, Z. (2005). Examining the relationship between executive functions and restricted, repetitive symptoms of autistic disorder. *Journal of Autism and Developmental Disorders, 35*(4), 445–460.

Lord, C., Risi, S., Lambrecht, L., Cook, E., Jr., Leventhal, B., DiLavore, P., et al. (2000). The Autism Diagnostic Observation Schedule–Generic: A standard measure of social and communication deficits associated with the spectrum of autism. *Journal of Autism and Developmental Disorders, 30*, 205–223.

Lott, I., Patterson, D., & Seltzer, M. (2007). Toward a research agenda for Down syndrome. *Mental Retardation and Developmental Disabilities Research Reviews, 13*, 288–289.

Luo, X., Perscio, A., & Lauder, J. (2003). Serotonergic regulation of somatosensory cortical development: Lessons from genetic mouse models. *Developmental Neuroscience, 25*, 173–183.

Madras, B.K., Miller, G.M., & Fischman, A.J. (2005). The dopamine transporter and attention deficit/hyperactivity disorder. *Biological Psychiatry, 57*, 1397–1409.

Meechan, D.W., Maynard, T.M., Gopalakrishna, D., Wu, Y., & LaMantia, A.S. (2007). When half is not enough: Gene expression and dosage in the 22q11 deletion syndrome. *Gene Expression, 13*(6), 299–310.

Molnar, Z., & Blakemore, C. (1995). How do thalamic neurons find their way to the cortex. *Trends in Neurosciences, 18*(9), 389–397.

Mountcastle, V.B. (1979). An organizing principle for cerebral function: The unit module and the distributed system. In F.O. Schmitt & F.G. Worden (Eds.), *The neurosciences: 4th study program* (pp. 21–42). Cambridge, MA: MIT Press.

Nieoullon, A. (2002). Dopamine and the regulation of cognition and attention. *Progress in Neurobiology, 67*, 53–83.

NIH Down Syndrome Working Group. (2007). *National Institutes of Health (NIH) research plan on Down syndrome.* Washington, DC: U.S. Government Printing Office.

Nikolaienko, O., Ngyuen, C., Crinc, L., Cios, K., & Gardiner, K. (2005). Human chromosome 21: Down syndrome gene function and pathway database. *Gene Section Functional Genomics, 364*, 90–98.

Pinter, J., Eliez, S., Schmitt, E., Capone, G., & Reiss, A. (2001). Neuroanatomy of Down's syndrome: A high-resolution MRI study. *American Journal of Psychiatry, 158*(10), 1659–1665.

Prince, J. (2008). Catecholamine dysfunction in attention-deficit/hyperactivity disorder. *Journal of Clinical Psychopharmacology, 28*(3), S39–S45.

Pritchard, M., Reeves, R.H., Dierssen, M., Patterson, D., & Gardner, K.J. (2008). Down syndrome and the genes of human chromosome 21: Current knowledge and future potentials. *Cytogenetic and Genome Research, 121*, 67–77.

Rakic, P. (1988). Specification of cerebral cortical areas. *Science, 241*, 170–176.

Rakic, P. (1995). A small step for the cell, a giant leap for mankind: A hypothesis of neocortical expansion during evolution. *Trends in Neurosciences, 18*(9), 383–388.

Rapp, J., & Vollmer, T. (2005). Stereotypy I: A review of behavioral assessment and treatment. *Research in Developmental Disabilities, 26*, 527–547.

Ridley, R. (1994). The psychology of perservative and stereotyped behavior. *Progress in Neurobiology, 44*(2), 221–231.

Rodriguez, A., Whitson, J., & Granger, R. (2004). Derivation and analysis of basic computational operations of thalamocortical circuits. *Journal of Cognitive Neuroscience, 16*(5), 856–877.

Rojas, D., Peterson, E., Winterrowd, E., Reite, M., Rogers, S., & Tregellas, J. (2006). Regional gray matter volumetric changes in autism associated with social and repetitive behavior symptoms. *BMC Psychiatry, 6*, 56.

Roper, R., & Reeves, R.H. (2006). Understanding the basis for Down syndrome phenotypes. *PLoS Genetics, 2*(3).

Rosenberg, R. (1997). Molecular neurogenetics: The genome is settling the issue. *Journal of the American Medical Association, 278*(15), 1282–1283.

Ross, M., Galaburda, A., & Kemper, T. (1984). Down's syndrome: Is there a decreased population of neurons? *Neurology, 34*, 909–916.

Rubia, K., Halari, R., Smith, A., Mohammed, M., Scott, S., Giampietro, V., et al. (2008). Dissociated functional brain abnormalities of inhibition in boys with pure conduct disorder and in boys with pure attention deficit hyperactivity disorder. *American Journal of Psychiatry, 165*(7), 889–897.

RUPP Autism Network. (2002). Risperidone in children with autism and serious behavioral problems. *New England Journal of Medicine, 347*(5), 314–321.

Sagvolden, T., Aase, H., Johansen, E., & Russell, V. (2005). A dynamic developmental theory of attention-deficit/hyperactivity disorder (ADHD) predominantly hyperactive/impulsive and combined types. *Behavioral and Brain Sciences, 28*, 397–468.

Sanson, A., & Prior, M. (1999). Temperament and behavioral precursors to oppositional defiant disorder and conduct disorder. In H. Quay & A. Hogan (Eds.), *Handbook of disruptive behavior disorders* (pp. 397–417). New York: Kluwer Academic/Plenum.

Scerif, G., & Karmiloff-Smith, A. (2005). The dawn of cognitive genetics? Crucial developmental caveats. *Trends in Cognitive Sciences, 9*(3), 126–135.

Shapiro, B. (1983). Down syndrome—a disruption of homeostasis. *American Journal of Medical Genetics, 14*, 241–269.

Shaw, P., Eckstrand, K., Sharp, W., Blumenthal, J., Lerch, J., Greenstein, D., et al. (2007). Attention-deficit/hyperactivity disorder is characterized by a delay in cortical maturation. *Proceedings of the National Academy of Sciences USA, 104*(49), 19649–19654.

Shaw, P., Greenstein, D., Lerch, J., Clasen, L., Lenroot, R., Gogtay, N., et al. (2006). Intellectual ability and cortical development in children and adolescents. *Nature, 440*, 676–679.

Spencer, T.J., Biederman, J., Madras, B.K., Dougherty, D.D., Bonab, A.A., Livni, E., et al. (2007). Further evidence of the dopamine transporter dysregulation in ADHD: A controlled PET imaging study using altropane. *Biological Psychiatry, 62*(9), 1059–1061.

Takarae, Y., Minshew, N., Luna, B., & Sweeny, J. (2007). Atypical involvement of frontostriatal systems during sensorimotor control in autism. *Psychiatry Research: Neuroimaging, 156*, 117–127.

Vig, S., & Jedrysek, E. (1999). Autistic features in young children with significant cognitive impairment: Autism or mental retardation? *Journal of Autism and Developmental Disorders, 29*(3), 235–248.

Waddington, C.H. (1942). Canalization of development and the inheritance of acquired characters. *Nature, 150*(3811), 563–565.

Wang, P., Koherty, S., Hesselink, J., & Bellugi, U. (1992). Callosal morphology concurs with neurobehavioral and neuropathological findings in two neurodevelopmental disorders. *Archives of Neurology, 49*, 407–411.

Whittle, N., Sartori, S., Dierssen, M., Lubec, G., & Singewald, N. (2007). Fetal Down syndrome brains exhibit aberrant levels of neurotransmitters critical for normal brain development. *Pediatrics, 120*(6).

Wisniewski, K.E., & Schmidt-Sidor, B. (1989). Postnatal delay of myelin formation in brains from Down syndrome infants and children. *Clinical Neuropathology, 8*(2), 55–62.

5

Autism and Prader-Willi Syndrome

Searching for Shared Endophenotypes

Travis Thompson

Achieving a more thorough understanding of complex developmental syndromes and their relations with one another has scientific and practical implications. Researchers as well as clinicians are often puzzled about the reasons two developmental disabilities that are very different nonetheless share important features in common. Parents, educators, and therapists faced with the responsibility for improving the lives of individuals with these disabilities may be even more confused. In this chapter, my attention is focused on two syndromes, autism spectrum disorders (ASDs) and Prader-Willi syndrome (PWS), which have distinctively different features. Individuals with ASDs may have little use of language, often lack social interest, and engage in repetitive stereotyped behavior, such as hand flapping, twirling in circles, or repeatedly turning light switches on and off. In contrast, most people with PWS are socially curious, many are quite talkative, and their ritualistic behavior is often surreptitious, such as skin picking and hiding items. PWS is distinguished by the severe hyperphagia associated with the syndrome, which is unlike eating among most people with ASDs.

There are several phenotypic features the two syndromes share, and, perhaps most importantly, a subset of people with ASDs has a genetic error in the same area of chromosome 15 as individuals with PWS. In this chapter I suggest that discovering these phenotype-genotype relations that are shared among syndromes permits us to identify subtypes of more complex disorders or disabilities and their underlying mechanisms. This approach is similar to that proposed by Dykens, Sutcliffe, and Levitt (2004). They noted that pathophysiological studies implicate altered development of specific neuron types and circuits in the cerebral cortex as part of the processes associated with autism and Prader-Willi syndrome. Gottesman and Shields (1967) originally applied the term *endophenotype* to refer to such subtypes. They described endophenotypes as internal phenotypes discoverable by a biochemical test or microscopic examination, though the concept has expanded considerably since then. An understanding of endophenotypes may lay the foundation for more effective treatments and educational interventions.

Gottesman and Gould (2003) proposed the following criteria for an endophenotype:

1. Associated with disorder in the population

2. Heritable

3. Manifests in an individual whether or not illness is active

4. Endophenotype and illness cosegregate in families

5. Endophenotype identified in probands is found in unaffected relatives at a higher rate than in the general population

PWS and ASD appear to meet the foregoing criteria for a common endophenotype, which will be the focus of the remainder of this chapter: common neurocognitive and behavioral phenotypic features, including a social processing deficit, compulsive behavior, and self-injury. In addition, they share common genetic lesions, deletion of the GABRB3 receptor subunit gene, and deletion of NIPA1, NIPA2, and UBE3A.

Human chromosome 15q11-13 is a complex locus. This region includes a cluster of three GABA(A) receptor subunit (GABR) genes: GABRB3, GABRA5, and GABRG3. Deletion or duplication of 15q11-13 GABR genes occurs in several neurodevelopmental disorders, including PWS, Angelman syndrome (AS), and autism. There is evidence autism can be caused by maternal dysregulation of UBE3A (ubiquitin protein ligase 3A gene), located in an imprinted region on chromosome 15q11-q13 (Nurmi et al., 2001; Veenstra-VanderWeele, Gonen, Leventhal, & Cook, 1999). Although there are several molecular types of autism, deletions of the 15q11-q13 region account for a substantial proportion of the AS patients (Sahoo et al., 2006).

PRADER-WILLI SYNDROME ENDOPHENOTYPE

Prader-Willi syndrome is a contiguous gene syndrome caused by an error on the proximal long arm of chromosome 15 (Thompson & Butler, 2004). The main phenotypic features of PWS include hyperphagia, reduced metabolic rate, and short stature with small hands and feet. The mean full scale IQ is 65. Among the more striking behavioral and cognitive features of PWS are intense preoccupation with food, including stealing and hoarding, tantrums when preferred routines are disrupted, skin picking, other compulsive behavior (e.g., counting, stealing, and hiding objects), social deficits, and facial processing deficits.

PWS is caused by a chromosomal abnormality involving a paternal deletion of a section of the proximal long arm of

chromosome 15 (15q11-q13) in approximately 70% of individuals, two maternal copies of chromosome 15 in approximately 25% of individuals, and an imprinting-center defect in the 15q11-q13 region for the remaining individuals (Bittel & Butler, 2005; Thompson & Butler, 2004). The PWS critical region on chromosome 15 contains three breakpoints (BPs) implicated in the distinct typical deletion patterns. The longer Type I (TI) deletion spans BP1 and BP3 (see Figure 5.1), and the shorter Type II (TII) deletion involves only BP2 and BP3 (Bittel, Kibiryeva, & Butler, 2006). As a result, people with the TI deletion have approximately 500 kb less genetic material than people with the TII deletion, which may result in more significant clinical symptoms (Zarcone et al., 2007) (Figure 5.1). NIPA1 and NIPA2 are located between TI and TII. SNURF-SNRPN, NECDIN, and three GABA receptor genes, including GABRB3, are located between TII and TIII, a region which, if deleted, is associated with more severe PWS symptoms.

AUTISM SPECTRUM DISORDER ENDOPHENOTYPE

Autism spectrum disorders are characterized by several behavioral, cognitive, communication, and emotional phenotypic features (American Psychiatric Association, 2000):

1. Reduced or absent social awareness or empathy

2. Communicative dysfunction, ranging from lack of speech to limited-use of pragmatically appropriate speech

3. Repetitive, fixed, and/or nonfunctional compulsive behavioral routines

Children and youth with autism spectrum disorders express their disability across the three domains to varying degrees from mild

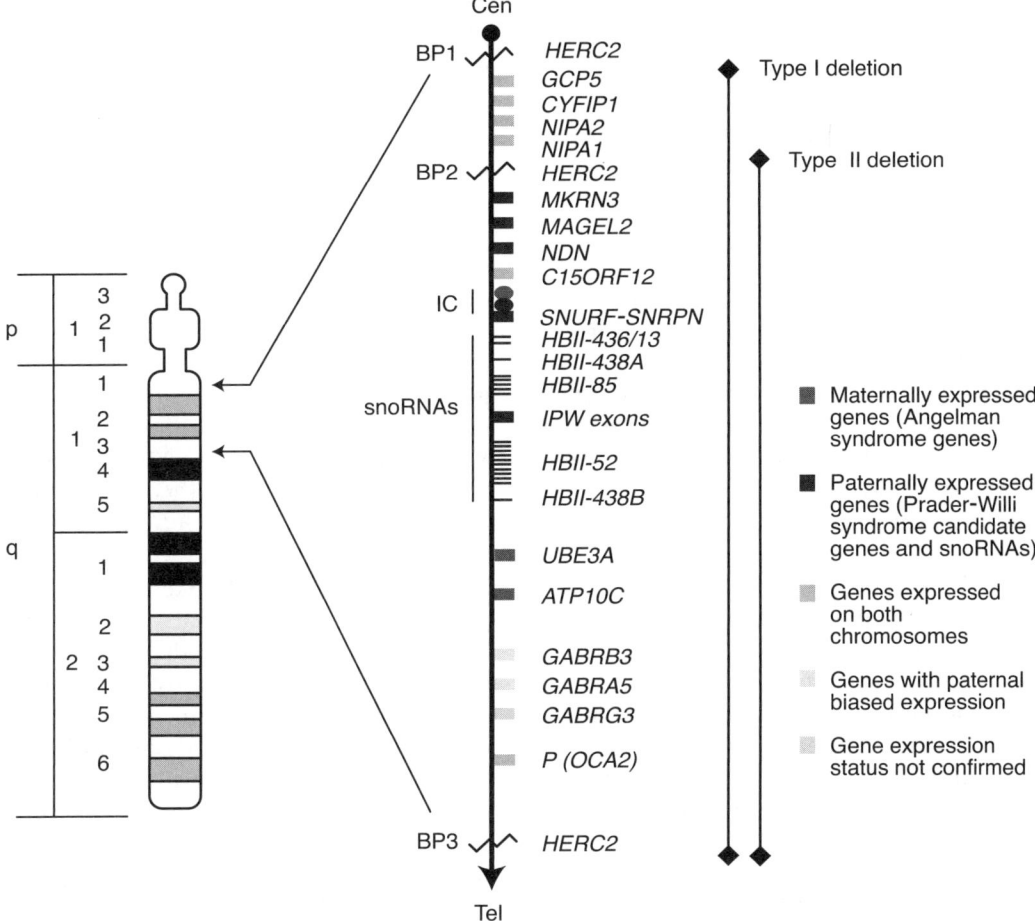

Ideogram of chromosome 15, showing genes located in the typical deletion region of Prader-Willi syndrome

Expert Reviews in Molecular Medicine © 2005 Cambridge University Press

Figure 5.1. Ideogram of chromosome 15, showing the genes located in the typical deletion region of Prader-Willi syndrome. (From Bittel, D.C., & Butler, M.G. [2005]. Prader-Willi syndrome: Clinical genetics, cytogenetics and molecular biology. *Expert Reviews in Molecular Medicine, 7,* 4. Copyright © Cambridge University Press 2005. Reprinted with permission.)

to severe. Generally intellectual ability varies directly with language ability. Lack of social interest appears to be related to dysfunction in the amygdala, cingulate cortex, and orbitofrontal cortex (Baron-Cohen et al., 1999). Children with more severe autism seldom establish eye contact, which appears to

result in lack of activation of the fusiform gyrus, which usually is activated when people look at faces, especially the eyes (Schultz et al., 2000). Nearly all individuals with autism engage in highly preferred nonfunctional routines, such as insisting tasks be done in a very specific order, lining up objects, or turning light switches on and off repeatedly. Among lower functioning individuals with ASDs, stereotyped movements, such as finger flicking, hand flapping, twirling in circles, and repetitive self-injury, may occur.

Autism Heterogeneity

Several lines of evidence indicate that autism is not a single condition, and likely occurs via several genetic mechanisms, which share one or more final common pathways. Miles et al. (2005) reported that Autistic Disorder is of two main types: 1) complex autism, involving dysmorphic features and/or small head size, and 2) essential autism, lacking such dysmoprhic features or reduced head size. The latter individuals have fewer magnetic resonance imaging and electroencephalographic brain abnormalities, higher intellectual functioning, better language and social skills, and better prognosis during intervention. They also appear to be from families with a higher recurrence rate among family members (20% versus 9%) and higher male-to-female ratios.

Increasing evidence suggests that multiple genes may be involved in the etiology of autism, with different genes contributing to specific features of the syndrome. In their recent review, Abrahams and Geschwind (2008, p. 341) pointed out that "several biological themes, including defective synaptic function and abnormal brain connectivity," have emerged from genotype-phenotype research of the past few years. Though in the past it was often believed autism severity was expressed roughly equally in all three domains (communication, socialization, and repetitive, compulsive behavior), more recent research suggests that the three domains often vary independently. Linkage studies have revealed different linkages between specific autism features and specific loci (Alarcon &

Cantor, 2001). This is consistent with the idea that identifying endophenotypes may be at the core of understanding the etiology of autism.

Autism and Chromosome 15

Although genes on several chromosomes have been shown to be associated with autism or autism features, this chapter focuses specifically on chromosome 15. Inherited duplication of the 15q11-q13 chromosomal region has been associated with autism spectrum disorders in numerous studies. Within this region, the ubiquitin protein ligase E3A (UBE3A) and gamma-aminobutyric acid A receptor beta 3 (GABRB3) genes are thought to be central to the etiology of autism (Abrahams & Geschwind, 2008, p. 344). This chromosomal region is causally linked to Prader-Willi syndrome and Angelman syndrome. Levels of UBE3A and GABRB3 are reduced in Angelman syndrome, Rett syndrome, and idiopathic autism (Samaco, Hogart, & LaSalle, 2005). Gene expression studies have pointed specifically to genes in the 15q interval as etiologically linked to autism (e.g., NIPA2 and UBE3A) (Baron, Liu, Hicks, & Gregg, 2006; Gregg et al., 2008). Morrow et al. (2008) used homozygosity mapping in consanguineous autism pedigrees to identify several genetic loci, with limited overlap between pedigrees. Using the Affymetrix Gene-Chip Human Mapping SNP arrays, Murrow et al. (2008) produced a pattern of findings consistent with autosomal recessive inheritance with high penetrance. The homozygous deletions in autism patients appeared to preferentially involve activity-regulated genes. The regulation of expression of some autism candidate genes by neuronal membrane depolarization suggests that neural activity-dependent regulation of synapse development may be a mechanism common to several heterogeneous autism mutations.

PRADER-WILLI SYNDROME AND AUTISM: SHARED PHENOTYPIC FEATURES

Several lines of evidence suggest that individuals with PWS share behavioral, communication, and emotional features in common with autism: 1) social withdrawal (Dykens, Cassidy, & King, 1999); 2) social isolation and poor peer relationships (Dykens & Rosner, 1999); 3) lack of empathy; and 4) may not respond when spoken to, demands must be met immediately (Clarke, Chung, Sturmey, & Webb, 1996). Dimitropoulos and Schultz (2007) point out that these features are more common in the maternal uniparental disomy form of PWS, suggesting that multiple maternal copies of the 15q11-q13 region contain extra genetic material linked to these phenotypes.

Self-Injury in Prader-Willi Syndrome and Autism

Self-injury is found in both Prader-Willi syndrome and autism. Dimitropoulos, Feurer, Butler, and Thompson (2001) found that half of 5-year-olds with PWS engaged in compulsive behavior, including skin picking, and over 80% of adults did so. Among adults with autism in a residential setting, self-injury was common (45%) (Bodfish et al., 1995), but good prevalence figures for younger children with pervasive developmental disorders are not readily available. Symons and Thompson (1997) and Symons, Butler, Sanders, Feurer, and Thompson (1999) studied self-injury in school-age children with autism and related intellectual disabilities, and children and adults with PWS. Skin picking was the most prevalent form of self-injury among PWS participants, with the front of the legs and top of head being disproportionately targeted as most likely self-injury body sites. Individuals with the 15q11-q13 deletion injured significantly more body sites than did individuals

with maternal disomy 15q11-q13, which was positively correlated with Yale Brown Obsessive Compulsive Scale scores (Goodman et al., 1989). Among school-age youth with autism and related intellectual disabilities, 80% of self-injury occurred on only 5% of the body surface: primarily hitting the sides of the head and forehead with fists or against hard surfaces, or biting the back of the hand, the medial side of the first (index) finger, the first web space between the thumb and finger, and the medial aspect of the wrist. Skin picking in PWS is often surreptitious, whereas head hitting and self-biting in autism are overt and often associated with tantrums.

Skin picking among individuals with PWS has been difficult to treat successfully. Shapira et al. (2004) studied 8 adults with PWS during an 8-week open-label trial with the anticonvulsant topiramate. Treatment resulted in a clinically significant improvement in self-injury (i.e., skin picking). In their review of the PWS treatment literature, Dykens and Shah (2003) concluded that selective serotonin reuptake inhibitors have been effective in reducing skin picking, compulsivity, and aggressive episodes in some individuals with Prader-Willi syndrome. Atypical antipsychotics have also proved helpful in individuals with psychotic features or extreme aggression and impulsivity.

Self-injury among individuals with autism and other related developmental disorders has been successfully treated with the opiate antagonist naltrexone (Thompson, Hackenberg, Cerutti, Baker, & Axtell, 1994). In a review of all published naltrexone reports, Symons, Thompson, and Rodriguez (2004) concluded that 80% of subjects treated with naltrexone were reported to improve relative to baseline (i.e., self-injurious behavior [SIB] reduced) during naltrexone administration, and SIB of 47% of subjects was reduced by 50% or more. Atypical antipsychotics, such as risperidone, and selective serotonin reuptake inhibitors are often used to treat irritability in autism, which includes self-injury (McDougle, Stigler, Erickson, & Posey, 2008; Scott & Dhillon, 2008). The differential effects of pharmacological treatments for self-injury in AS and PWS suggest that different mechanisms underlie the two types of self-destructive behavior patterns, which in turn leads us to consider neurochemical processes.

THE POSSIBLE ROLE OF GABA

Genes in the 15q11.2-q12 region responsible for GABRB3 receptors in the frontal cortex are missing or inactive in Prader-Willi syndrome. GABA plays an important inhibitory role for monoamines, such as dopamine in the frontal cortex and basal ganglia. We found higher levels of peripheral GABA in PWS, which presumably is compensatory for dysfunctional GABA receptors. GABA plasma levels were negatively correlated with skin picking in PWS (Compulsive Behavior Checklist [Gedye 1992, 1996]; Pearson $r = -.42$; $p = .032$) (Ebert et al., 1997). This suggests that the missing or imprinted GABRA3 receptor gene may lead to insufficient regulation of dopamine cells. Overactive dopamine cells in frontal cortex may in turn lead to increased compulsive behavior. Holsen and Thompson (2004) studied compulsive behavior in a group of 16 people with Prader-Willi syndrome and a comparison group of 19 people with intellectual disability. Using eye-blink rate as an indirect measure of central nervous system dopamine, we found a higher eye-blink rate in the Prader-Willi syndrome group and a relationship between compulsive behavior and eye-blink rate across both groups. These findings suggest a gene-brain-behavior link between dopamine levels and GABAergic mechanisms in individuals with Prader-Willi syndrome.

GABRB3 has been found to be decreased among individuals with autism (Samaco, Hogart, & LaSalle, 2005). Nurmi et al. (2001) examined the role of the 15q region in savant symptoms in autism. When their sample was divided on the basis of mean score for the "savant skills" cluster, the logarithm of the odds under a recessive model at D15S511, within the GABRB3 gene, increased from 0.6 to 2.6 in the families in which target individuals had greater savant skills. These data are consistent with the genetic contribution of a 15q locus to autism susceptibility in a subset of affected individuals exhibiting savant skills. Similar types of skills have been noted in individuals with Prader-Willi syndrome, which results from deletions of this chromosomal region.

HERITABILITY OF PRADER-WILLI SYNDROME AND AUTISM

The vast majority of cases of PWS are sporadic, but there has been evidence since the 1960s that familial PWS occurs as well. Gabilan (1962) reported a family with affected brother and sister, and another in which the parents of the affected individual were first cousins. Jancar (1971), Hall and Smith (1972), and Clarren and Smith (1977) reported familial recurrence as well. One mechanism involves balanced rearrangement with a breakpoint in 15q13 in related male carriers.

Evidence for heritability is very strong for autism spectrum disorders. The first clear evidence that autism has a genetic component came when Folstein and Rutter (1977) reported that 36% of children (from a very small sample) who were identical twins also had autism, whereas none of the reported dizygotic twins had autism. Ritvo and his colleagues (1989) reported the likelihood that a second child would have autism depended on the sex of the first child (with autism). The risk was estimated to be 7% if the first child with autism was a boy and 14.5% if it was a girl. Numerous studies since then have revealed generally consistent findings. The risk for identical twins (if one has autism, the other will as well) is in the range of 30%–50%, the risk ratio for nonidentical twins, brothers and sisters, and parents is in the 4%–10% range, and the risk is higher if the first child (with autism) is a girl. However, if the family already has two children with an ASD diagnosis, the likelihood that a third will also have autism jumps to approximately 35%.

Other evidence consistent with a familial form of autism comes from studies of the Broad Autism Phenotype, that refers to siblings or parents who exhibit some but not all of the features of autism in milder degrees (Piven, 2001). In a test of this notion, Losh and Piven (2007) studied parents of individuals with autism (13 of whom were identified as "aloof"), and 22 control parents, who were administered the "Eyes Test," a social-cognitive measure of ability to read complex psychological states from viewing only the eye region of faces. The subgroup of parents defined as "aloof" displayed significant social-cognitive deficits on the "Eyes Test." Impaired social-cognitive ability was associated with low quality of friendships and problems with pragmatic language use, which mirror those documented in autism. Fifteen percent of the autism and Specific Language Impairment (SLI) parents showed severe impairments. This suggests impaired communication is part of the Broader Autism Phenotype and a broader SLI phenotype, especially among male family members.

Constantino and colleagues (2006) measured social impairment in three groups: 1) children with ASDs from families with more than one child with an ASD, and their closest-in-age brothers without autism; 2) children with any pervasive developmental disorder, including autism, and their closest-in-age brothers; and 3) children with psychopathology unrelated to autism, and their closest-in-age brothers. Greatest social impairment was seen among siblings of children with autism from families with more than one child with an ASD, followed by siblings whose target child had any pervasive developmental disorder, as compared with siblings of children who had some type of psychopathology unrelated to autism. In a related study, Adolphs, Piven, and colleagues (2008) investigated face processing to measure how viewers make use of information from specific facial features to judge emo-

tions. Parents of children with autism who were socially aloof showed a reduction in processing the eye region in faces, together with enhanced processing of the mouth, compared to a control group of parents of neurotypical children, as well as to non-aloof parents of children with autism whose pattern of face processing was intermediate.

A recent collaborative study of a group of families from the Autism Genetic Resource Exchange (AGRE), an international autism database, reveals strong evidence for dominant transmission to male offspring (Zhao et al., 2007). They propose that among a small minority of families, the risk of autism in male offspring is near 50%, the vast majority for whom male offspring have a low risk. They suggest so-called Sporadic Autism occurs in the low-risk families due to spontaneous mutation with high penetrance in males and relatively poor penetrance in females. High-risk families are from those children, most often females who carry a new causative mutation but are unaffected themselves. They, in turn, transmit the mutation in dominant fashion to their offspring.

DISCUSSION AND CONCLUSIONS

Though there are important phenotypic differences between autism and Prader-Willi syndrome, there are also striking similarities, notably compulsive ritualistic behavior, intolerance of changes in routines, and social aloofness and lack of social skills. While there is evidence that genes related to cell adhesion molecules may play a role in the etiology of autism, consistent and compelling evidence indicates autism and PWS share one or more genetic errors in the 15q11q13 region. While PWS and AS phenotypes both include self-injurious behavior, they appear to involve different mechanisms. The most consistent findings point to GABA receptor genes and UBE3A as common lesions in the two syndromes. Though PWS is nearly always sporadic, inherited cases have been recognized for many years. It is increasingly clear some forms of autism run in families, and milder forms (Broad Autism Phenotype) are relatively common among other family members. This suggests the two disorders share a compulsive socially deficient endophenotype causally linked to GABRA3, GABRB3, and UBE3AQ.

REFERENCES

Abrahams, B.S., & Geschwind, D.H. (2008). Advances in autism genetics: On the threshold of a new neurobiology. *Nature Reviews Genetics, 9,* 341–355.

Adolphs, R., Spezio, M.L., Parlier, M., & Piven, J. (2008). Distinct face-processing strategies in parents of autistic children. *Current Biology, 18,* 1090–1093.

Alarcon, M., & Cantor, R.M. (2001). Quantitative trait loci mapping of serum IgE in an isolated Hutterite population. *Genetic Epidemiology, 21*(Suppl. 1), S224–S229.

American Psychiatric Association. (2000). *Diagnostic and Statistical Manual of Mental Disorders.* (4th ed., text rev.). Washington, DC: Author.

Baron, C.A., Liu, S.Y., Hicks, C., & Gregg, J.P. (2006). Utilization of lymphoblastoid cell lines as a system for the molecular modeling of autism. *Journal of Autism and Developmental Disorders, 36,* 973–982.

Baron-Cohen, S., Ring, H.A., Wheelwright, S., Bullmore, E.T., Brammer, M.J., Simmons, A., et al. (1999). Social intelligence in the normal and autistic brain: An fMRI study. *European Journal of Neuroscience, 11,* 1891–1898.

Bittel, D.C., & Butler, M.G. (2005). Prader-Willi syndrome: Clinical genetics, cytogenetics and molecular biology. *Expert Reviews in Molecular Medicine, 7,* 1–20.

Bittel, D.C., Kibiryeva, N., & Butler, M.G. (2006). Expression of 4 genes between chromosome 15 breakpoints 1 and 2 and behavioral outcomes in Prader-Willi syndrome. *Pediatrics, 118,* e1276–e1283.

Bodfish, J.W., Crawford, T.W., Powell, S.B., Parker, D.E., Golden, R.N., & Lewis, M.H. (1995). Compulsions in adults with mental retardation: Prevalence, phenomenology, and comorbidity with stereotypy and self-injury. *American Journal on Mental Retardation, 100,* 183–192.

Clarke, D.J., Boer, H., Chung, M.C., Sturmey, P., & Webb, T. (1996). Maladaptive behaviour in Prader-Willi syndrome in adult life. *Journal of Intellectual Disabilities Research, 40*(Pt. 2), 159–165.

Clarren, S.K., & Smith, D.W. (1977). Prader-Willi syndrome: Variable severity and recurrence risk. *American Journal of Disabled Child, 131,* 798–800.

Constantino, J.N., Lajonchere, C., Lutz, M., Gray, T., Abbacchi, A., McKenna, K., et al. (2006). Autistic social impairment in the siblings of children with pervasive developmental disorders. *American Journal of Psychiatry, 163,* 294–296.

DeFraites, E.B., Thurmon, T.F., & Farhadian, H. (1975). Familial Prader-Willi syndrome. In Bergsma, D. (Ed.), *Genetic forms of hypogonadism* (pp. 123–126). New York: National Foundation March of Dimes.

Dimitropoulos, A., Feurer, I.D., Butler, M.G., & Thompson, T. (2001). Emergence of compulsive behavior and tantrums in children with Prader-Willi syndrome. *American Journal on Mental Retardation, 106,* 39–51.

Dimitropoulos, A., & Schultz, R.T. (2007). Autistic-like symptomatology in Prader-Willi syndrome: A review of recent findings. *Current Psychiatry Report, 9,* 159–164.

Dykens, E.M., Cassidy, S.B., & King, B.H. (1999). Maladaptive behavior differences in Prader-Willi syndrome due to paternal deletion versus maternal uniparental disomy. *American Journal on Mental Retardation, 104,* 67–77.

Dykens, E.M., & Rosner, B.A. (1999). Refining behavioral phenotypes: Personality-motivation in Williams and Prader-Willi syndromes. *American Journal on Mental Retardation, 104*(2), 158–169.

Dykens, E., & Shah, B. (2003). Psychiatric disorders in Prader-Willi syndrome: Epidemiology and management. *CNS Drugs, 17,* 167–178.

Dykens, E., Sutcliffe, J.S., & Levitt, P. (2004). Contrasting autism and 15q11-q13 disorders: Behavioral, genetic, and pathophysiological issues. *Mental Retardation and Developmental Disability Research Reviews, 10,* 284–291.

Ebert, M.H., Schmidt, D.E., Thompson, T., & Butler, M.G. (1997). Elevated plasma gamma-aminobutyric acid (GABA) levels in individuals with either Prader-Willi syndrome or Angelman syndrome. *Journal of Neuropsychiatry and Clinical Neurosciences, 9,* 75–80.

Fernandez, F., Berry, C., & Mutton, D. (1987). Prader-Willi syndrome in siblings, due to unbalanced translocation between chromosomes 15 and 22. *Archives of the Disabled Child, 62,* 841–843.

Folstein, S., & Rutter, M. (1977). Infantile autism: A genetic study of 21 twin pairs. *Journal of Child Psychology and Psychiatry, 18,* 297–321.

Gabilan, J.C. (1962). Syndrome de Prader, Labhardt et Willi. *Journal of Pediatrics (Paris), 1,* 179.

Gedye, A. (1992). Recognizing obsessive-compulsive disorder in clients with developmental disabilities. *Habilitative Mental Healthcare Newsletter, 11,* 73–77.

Gedye, A. (1996). Issues involved in recognizing obsessive-compulsive disorder in developmentally disabled clients. *Seminars in Clinical Neuropsychiatry, 1,* 142–147.

Goodman, W.K., Price, L.H., Rasmussen, S.A., Mazure, C., Fleischmann, R.L., Hill, C.L., et al. (1989). The Yale-Brown Obsessive Compulsive Scale. I. Development, use, and reliability. *Archives of General Psychiatry, 46*(11), 1006–1011.

Gottesman, I.I., & Gould, T.D. (2003). The endophenotype concept in psychiatry: Etymology and strategic intentions. *American Journal of Psychiatry, 160,* 636–645.

Gottesman, I.I., & Shields, J. (1967). A polygenic theory of schizophrenia. *Proceedings of the National Academy of Sciences USA, 58,* 199–205.

Gregg, J.P., Lit, L., Baron, C.A., Hertz-Picciotto, I., Walker, W., Davis, R.A., et al. (2008). Gene expression changes in children with autism. *Genomics, 91*(1), 22–29.

Hall, B.D., & Smith, D.W. (1972). Prader-Willi syndrome: A resume of 32 cases including an instance of affected first cousins, one of whom is of normal stature and intelligence. *Journal of Pediatrics, 81,* 286–293.

Holsen, L., & Thompson, T. (2004). Compulsive behavior and eye blink in Prader-Willi syndrome: Neurochemical implications. *American Journal on Mental Retardation, 109*(3), 197–207.

Jancar, J. (1971). Prader-Willi syndrome (hypotonia, obesity, hypogonadism, growth and mental retardation). *Journal of Mental Deficiency Research, 15,* 20–29.

Losh, M., & Piven, J. (2007). Social-cognition and the broad autism phenotype: Identifying genetically meaningful phenotypes. *Journal of Child Psychology and Psychiatry, 48,* 105–112.

McDougle, C.J., Stigler, K.A., Erickson, C.A., & Posey, D.J. (2008). Atypical antipsychotics in children and adolescents with autistic and other pervasive developmental disorders. *Journal of Clinical Psychiatry, 69*(Suppl. 4), 15–20.

Miles, J.H., Takahashi, T.N., Bagby, S., Sahota, P.K., Vaslow, D.F., Wang, C.H., et al. (2005). Essential versus complex autism: Definition of fundamental prognostic subtypes. *American Journal of Medical Genetics A, 135,* 171–180.

Morrow, E.M., Yoo, S.Y., Flavell, S.W., Kim, T.K., Lin, Y., Hill, R.S., et al. (2008). Identifying autism loci and genes by tracing recent shared ancestry. *Science, 321,* 218–223.

Nurmi, E.L., Bradford, Y., Chen, Y., Hall, J., Arnone, B., Gardiner, M.B., et al. (2001). Linkage disequilibrium at the Angelman syndrome gene UBE3A in autism families. *Genomics, 77,* 105–113.

Piven, J. (2001). The broad autism phenotype: A complementary strategy for molecular genetic studies of autism. *American Journal of Medical Genetics, 105,* 34–35.

Ritvo, E., et al. (1989). The UCLA-University of Utah epidemiologic survey of autism: Prevalence. *American Journal of Psychiatry, 146,* 194–199.

Sahoo, T., Peters, S.U., Madduri, N.S., Glaze, D.G., German, J.R., Bird, L.M., et al. (2006). Microarray based comparative genomic hybridization testing in deletion bearing patients with Angelman syndrome: Genotype-phenotype correlations. *Journal of Medical Genetics, 43,* 512–516.

Samaco, R.C., Hogart, A., & LaSalle, J.M. (2005). Epigenetic overlap in autism-spectrum neurodevelopmental disorders: MECP2 deficiency causes reduced expression of UBE3A and GABRB3. *Human Molecular Genetics, 14,* 483–492.

Schultz, R.T., Gauthier, I., Klin, A., Fulbright, R.K., Anderson, A.W., Volkmar, F., et al. (2000). Abnormal ventral temporal cortical activity during face discrimination among individuals with autism and Asperger syndrome. *Archives of General Psychiatry, 57,* 331–340.

Scott, L.J., & Dhillon, S. (2008). Spotlight on risperidone in irritability associated with autistic disorder in children and adolescents. *CNS Drugs, 22,* 259–262.

Shapira, N.A., Lessig, M.C., Lewis, M.H., Goodman, W.K., Driscoll, D.J. (2004). Effects of topiramate in adults With Prader-Willi syndrome. *American Journal on Mental Retardation, 109*(4), 301–309.

Symons, F.J., Butler, M.G., Sanders, M.D., Feurer, I.D., & Thompson, T. (1999). Self-injurious behavior and Prader-Willi syndrome: Behavioral forms and body locations. *American Journal on Mental Retardation, 104,* 260–269.

Symons, F.J., & Thompson, T. (1997). Self-injurious behaviour and body site preference. *Journal of Intellectual Disabilities Research, 41*(Pt. 6), 456–468.

Symons, F.J., Thompson, A., & Rodriguez, M.C. (2004). Self-injurious behavior and the efficacy of naltrexone treatment: A quantitative synthesis. *Mental Retardation and Developmental Disabilities Research Review, 10,* 193–200.

Thompson, T., & Butler, M.G. (2004). Prader Willi Syndrome: Clinical, behavioral and genetic findings. In M.C. Wolraich (Ed.), *Disorders of learning and behavior: A practical guide to assessment and management* (p. 4). Hamilton, Ontario: B.C. Decker.

Thompson, T., Hackenberg, T., Cerutti, D., Baker, D., & Axtell, S. (1994). Opioid antagonist effects on self-injury in adults with mental retardation: Response form and location as determinants of medication effects. *American Journal on Mental Retardation, 99,* 85–102.

Veenstra-VanderWeele, J., Gonen, D., Leventhal, B.L., & Cook, E.H. (1999). Mutation screening of the UBE3A/E6-AP gene in autistic disorder. *Molecular Psychiatry, 4,* 64–67.

Zarcone, J., Napolitano, D., Peterson, C., Breidbord, J., Ferraioli, S., Caruso-Anderson, M., et al. (2007). The relationship between compulsive behaviour and academic achievement across the three genetic subtypes of Prader-Willi syndrome. *Journal of Intellectual Disabilities Research, 51*(Pt. 6), 478–487.

Zhao, X., et al. (2007). A unified genetic theory for sporadic and inherited autism. *Proceedings of the National Academy of Sciences USA, 104,* 12831–12836.

Williams Syndrome

Psychological Characteristics

Carolyn B. Mervis and Angela E. John

Williams syndrome is a complex neurodevelopmental disorder caused by a deletion of ~25 genes on one copy of chromosome 7q11.23 (Hillier et al., 2003; Osborne, 2006). The prevalence of this syndrome is 1 in 7,500 live births (Strømme, Bjørnstad, & Ramstad, 2002). Physical and medical manifestations include a characteristic facial appearance (see Figure 6.1), congenital heart disease (especially supravalvar aortic stenosis [SVAS]), connective tissue abnormalities such as hernias or diverticuli of the bladder or colon, and failure to thrive or growth deficiency (Morris, 2006). Infants and young children with Williams syndrome have developmental delay, and older children typically have intellectual or learning disabilities. Williams syndrome is also associated with characteristic cognitive (Mervis et al., 2000) and personality (Klein-Tasman & Mervis, 2003) profiles. In this chapter, we first provide a brief description of the earliest studies of the psychological characteristics of children and adolescents with Williams syndrome and then summarize findings from our research group's recent studies of the

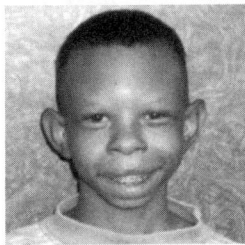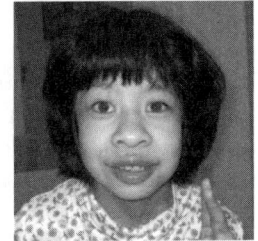

Figure 6.1. Photographs of children with Williams syndrome. All of the children pictured have classic Williams syndrome deletions. Ages (from left to right): 9 months, 5 years, 11 years, and 11 years. (*Source:* Mervis & Morris, 2007.)

psychological characteristics of individuals with this syndrome.

HISTORICAL CONTEXT

The articles that led to the syndrome's name (Williams syndrome in North America and

We are very grateful to the children and families who have participated so enthusiastically in our studies. The authors' research and preparation of this chapter were supported by National Institute of Neurological Disorders and Stroke grant R01NS35102 and National Institute of Child Health and Human Development grant R37 HD29957.

the United Kingdom; Williams-Beuren syndrome in Europe) were published by cardiologists (Beuren, Apitz, & Harmjanz, 1962; Williams, Barratt-Boyes, & Lowe, 1961) reporting on small groups of individuals who had intellectual disability, SVAS, and a characteristic facial appearance. The genetic basis for Williams syndrome was discovered about 30 years later (Ewart et al., 1993), and a blood test for this syndrome became available commercially a few years afterward. The first article focusing on the psychological characteristics was published in 1964 by von Arnim and Engel. The authors, who were unaware of the articles by Williams et al. and Beuren et al., stated that the children being described "show striking resemblances in both physical and mental characteristics. They stand out and are clearly distinguishable from other [children with intellectual disability] . . . and we conclude that they must belong to a well-defined group suffering from a special syndrome" (p. 366). von Arnim and Engel state that the primary psychological characteristics are "an unusual command of language" (p. 367); intellectual ability that initially appears to be much higher than the actual level, but which is invariably in the severe intellectual disability range; difficulty interacting with other children their age but uninhibited, instantly friendly, and polite when interacting with adults even if they are strangers; difficulty with drawing; hypersensitivity to frustration; and unreasonable levels of anxiety and insecurity.

Over the next two decades, very little was reported on the psychological characteristics of Williams syndrome. The articles that were published focused on standardized assessment performance and/or on the language abilities of small samples of children. Bennett, LaVeck, and Sells (1978) reported that on the McCarthy Scales of Children's Abilities (McCarthy, 1972), all seven participants earned their highest score on one of the scales that measures verbal ability and their lowest score on the gross motor scale. Six of the seven earned their next lowest

score on the scale measuring visuospatial construction (including drawing). MacDonald and Roy (1988) compared the performance of children with Williams syndrome to that of children with other forms of intellectual disability matched for chronological age (CA) and performance on the Peabody Picture Vocabulary Test–Revised (PPVT-R; Dunn & Dunn, 1981) and found that the Williams syndrome group performed at the same level as the contrast group on the other language assessments but performed well below the contrast group on tests of visuospatial construction. Kataria, Goldstein, and Kushnick (1984) and Meyerson and Frank (1987) reported that the language abilities of children with Williams syndrome were well below CA expectations.

Another article published in 1988, however, returned to von Arnim and Engel's theme that individuals with Williams syndrome have excellent language despite severe intellectual disability, including great difficulty on tests of drawing. Bellugi, Marks, Bihrle, and Sabo (1988) argued that although adolescents with Williams syndrome were still in Piaget's preoperational period (as evidenced by an inability to conserve number or quantity), they nevertheless were able to comprehend and produce complex linguistic constructions, including passives, conditionals, and tag questions. Furthermore, they had excellent vocabularies, including a variety of unusual words. The combination of the ability to comprehend and produce reversible passives with the inability to conserve led Bellugi et al. to argue that Williams syndrome provided strong evidence of the independence of language from cognition. This argument, taken as support for language modularity, quickly attracted the attention of researchers concerned with the relation between language and cognition and catapulted Williams syndrome to the forefront of the debate on the modularity of language. As a result, additional researchers began to study children with Williams syndrome, and at the same time, pundits concerned with modularity began to write about the syndrome, taking a much more strident position than either von Arnim and Engel or Bellugi et al. Piattelli-Palmarini (2001, p. 887) offered a particularly provocative statement: "For instance, children with Williams syndrome have barely measurable general intelligence and require constant parental care, yet they have an exquisite mastery of syntax and vocabulary."

More recent research offers a more nuanced view of the overall intellectual abilities and language abilities of individuals with Williams syndrome. At the same time, studies of the personality characteristics of individuals with Williams syndrome have offered more support for the characteristics identified by von Arnim and Engel (1964). In the remainder of this chapter, we briefly summarize some of the research conducted in our laboratory addressing the psychological characteristics of individuals with Williams syndrome, focusing on overall intellectual ability, language and sociocommunicative abilities, personality, and psychopathology.

OVERALL INTELLECTUAL ABILITY

The most commonly used assessments of the overall intellectual abilities of individuals with developmental disability are the Wechsler tests (e.g., various versions of the Wechsler Intelligence Scale for Children [WISC] or the Wechsler Adult Intelligence Scale [WAIS]). These assessments have often been administered to individuals with Williams syndrome. At a group level, mean Verbal IQ is typically about 5 points higher than mean Performance IQ (Howlin, Davies, & Udwin, 1998; Searcy, Lincoln, Rose, Klima, & Bavar, 2004), and this difference usually is significant. However, for individual participants, the difference between Verbal and Performance IQ is much less likely to be significant. For example, Searcy et al. (2004) reported that Verbal IQ on the WAIS-R (Wechsler, 1981)

was significantly higher than Performance IQ for only 24% of their sample; for 1% Performance IQ was significantly higher than Verbal IQ. Thus, the pattern of performance on the Wechsler composites does not capture the consistent finding from the early studies of individuals with Williams syndrome that verbal abilities are much stronger than visuospatial construction abilities. Examination of the mix of subtests included in the Verbal and Performance composites suggests that the Wechsler measures are unlikely to be effective in capturing the finding that verbal abilities are considerably more advanced than visuospatial construction abilities because the Performance Composite (or Perceptual Reasoning Composite on the WISC-IV; Wechsler, 2003) includes not only subtests measuring visuospatial construction (e.g., Block Design) but also subtests that measure nonverbal reasoning (e.g., Picture Completion on prior versions of the WISC and Matrix Reasoning and Picture Concepts on the WISC-IV). In addition, prior versions of the WISC and WAIS included the Arithmetic subtest as part of the Verbal Composite. Furthermore, because the subtests are only normed to 3 standard deviations below the general population mean, many individuals with Williams syndrome score at floor on one or more subtests, making significant differences less likely.

To avoid these problems, our research group has used a different measure of overall intellectual ability, the Differential Ability Scales (DAS or DAS–II; Elliott, 1990, 2007), whose cluster structure clearly separates visuospatial construction abilities from nonverbal reasoning abilities. The six core subtests form three clusters: Verbal, Nonverbal Reasoning, and Spatial (visuospatial construction). Furthermore, the DAS–II is normed to 4 standard deviations below the general population mean, making it much less likely that children with Williams syndrome will score at floor on individual

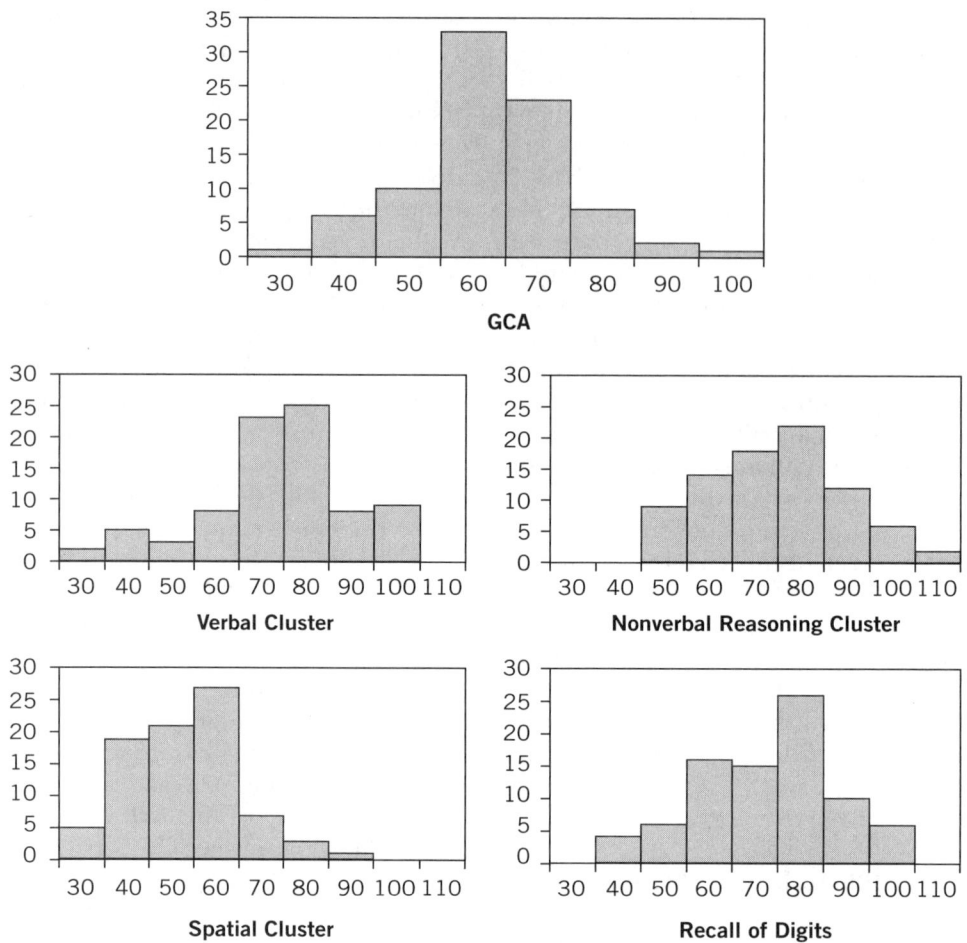

Figure 6.2. Differential Ability Scales–Second Edition (DAS-II; Elliott, 2007) General Conceptual Ability (GCA) and Verbal Cluster, Nonverbal Reasoning Cluster, Spatial Cluster, and Recall of Digits standard scores for 83 children with Williams syndrome ages 4–17 years who completed either the Early Years or School-Age Level of the DAS-II. Recall of Digits *T* scores have been converted to the same scale as GCA and Cluster standard scores. All children have classic deletions and have not been diagnosed with an autism spectrum disorder.

subtests. In Figure 6.2, we present a series of histograms depicting the overall performance (General Conceptual Ability [GCA]; similar to IQ) of 83 4- to 17-year-olds with Williams syndrome on the DAS–II, along with their performance on each of the three core clusters and on a supplemental subtest measuring verbal short-term memory. All of these children had classic-length deletions, and none had been diagnosed with an

autism spectrum disorder (ASD). As can be seen from the figure, performance on the DAS–II more reliably captures the early insight that children with Williams syndrome have considerably stronger verbal than visuospatial construction abilities. It is also clear that overall level of intellectual ability is considerably higher than the severe intellectual disability range suggested by von Arnim and Engel (1964) and Bellugi et al. (1988). In particular, the mean GCA was 62.98 (SD = 11.79, range: 35–98); mean cluster standard scores were 73.88 (SD = 16.20, range: 32–105) for Verbal, 74.99 (SD = 14.07, range: 46–109) for Nonverbal Reasoning, and 53.61 (SD = 12.49, range: 32–94) for Spatial; and the mean Recall of Digits standard score was 73.41 (SD = 15.58, range: 40–101). Thus, the mean Spatial Cluster standard score was ~20 points below the mean standard scores for the remaining clusters and the Recall of Digits-Forward subtest, not only capturing the insight regarding the relation between verbal and visuospatial abilities from the early studies, but also expanding it to suggest that visuospatial construction abilities are considerably more limited than nonverbal reasoning and verbal short-term memory abilities, which are at about the same level as verbal abilities. This general pattern holds not just at the group level but also at the level of the individual child. The performance of 72% of the children was significantly better on the Verbal Cluster than on the Spatial Cluster; similarly, 75% performed significantly better on the Nonverbal Reasoning Cluster than on the Spatial Cluster, and 73% did not evidence a significant difference in performance on the Verbal and Nonverbal Reasoning Clusters. Overall, 86% of the children performed significantly better on the Verbal Cluster, the Nonverbal Reasoning Cluster, or both than on the Spatial Cluster, and 2% performed significantly better on the Spatial Cluster than on the Verbal Cluster.

This pattern of considerably stronger performance on subtests measuring verbal abilities or nonverbal reasoning abilities than on subtests measuring visuospatial construction is apparent even in toddlers and young preschoolers. We have administered the Mullen Scales of Learning (Mullen, 1995) to 117 2- to 4-year-olds with Williams syndrome who had classic-length deletions and had not been diagnosed with an ASD. The Mean

Early Learning Composite (DQ) was 60.84 [SD = 11.07, range: 49 (floor)–96]. Subtest performance is measured in T scores (general population mean = 50, SD = 10, range: 20–80). Mean T scores were 28.50 (SD = 9.56, range: 20–51) for Visual Reception (measuring primarily nonverbal reasoning), 29.04 (SD = 9.72, range: 20–55) for Receptive Language, and 32.43 (SD = 9.32, range: 20–49) for Expressive Language. In sharp contrast, the mean T score for Fine Motor (measuring primarily visuospatial construction) was 21.15 (SD = 3.06, range: 20–41), with 82% of the children scoring at floor (T = 20).

In collaboration with Karen Berman's research group at the National Institutes of Mental Health, our research group has conducted several neuroimaging studies comparing normal-IQ adults with Williams syndrome who have classic deletions and have not been diagnosed with an ASD to gender, CA-, and IQ-matched groups of individuals from the general population. We have identified an area of reduced gray matter and reduced sulcal depth in the intraparietal sulcus (Kippenhan et al., 2005; Meyer-Lindenberg et al., 2004; Meyer-Lindenberg, Mervis, & Berman, 2006) in the Williams syndrome group. The results of a series of functional magnetic resonance imaging (fMRI) studies investigating ventral and dorsal processing indicated that ventral processing was intact. However, dorsal stream function during a task that was a two-dimensional analogue to pattern construction (the signature weakness for individuals with Williams syndrome) differed from that of the control group. In particular, the abnormality in the intraparietal sulcus served as a roadblock to dorsal stream information flow, as indicated by the results of a path analysis which showed that the only difference between the Williams syndrome group and the control group was that the path from the intraparietal sulcus to the later dorsal stream region was significant for the control group but not for the Williams syndrome group. The convergence of the behavioral assessment and neuroimaging results strongly suggests that a gene (or genes) in the Williams syndrome region, in transaction with other genes and the environment, contributes to the development of visuospatial construction skills.

LANGUAGE ABILITIES: VOCABULARY AND GRAMMAR

Many researchers who are currently studying Williams syndrome began their research in this area because of prior claims that individuals with this syndrome have both an excellent command of grammar and an excellent vocabulary, including the use of unusual words, evidencing unusual semantic organization. The argument that individuals with Williams syndrome use unusual words was based primarily on the findings of Bellugi and her colleagues (Bellugi et al., 1988; Bellugi, Lichtenberger, Jones, Lai, & St. George, 2000; Bellugi, Wang, & Jernigan, 1994) that on semantic fluency tasks (which require the participant to name as many members of a researcher-provided category as possible), children and adolescents with Williams syndrome are much more likely than either CA- and IQ-matched individuals with Down syndrome or younger typically developing children to produce low-frequency words. The category typically studied was "animal." Several studies conducted by other laboratories have failed to replicate this finding, however. Instead, results indicated that the animal names produced are similar to those produced either by matched individuals with other forms of intellectual disability or by younger typically developing children matched for mental age (MA) (e.g., Jarrold, Hartley, Phillips, & Baddeley, 2000; Levy & Bechar, 2003; Lukács, 2005; Mervis, Morris, Bertrand, & Robinson, 1999; Temple, Almazan, & Sherwood, 2002; Volterra, Capirci, Pezzini, Sabbadini, & Vicari, 1996).

Bellugi et al. (e.g., 1988, 1994, 2000) also noted that individuals with Williams syndrome often use a relatively complex word in a context that indicates partial but not complete understanding. The most common example given is, "I have to evacuate the glass" (Bellugi et al., 2000, p. 13), said while the participant was emptying a glass of water into a sink. We have also observed this type of partially correct word use of words that seem more advanced than the rest of the participant's vocabulary. However, this phenomenon is not restricted to individuals with Williams syndrome; Miller and Gildea (1987), reporting on a study of typically developing school-age children, noted that these types of partially correct use of a word are common when children use words that they have recently learned.

Despite these indications that the vocabularies of children and adolescents with Williams syndrome do not include unexpected unusual words and are not atypically organized, there is considerable evidence that concrete vocabulary is a relative strength for individuals with this syndrome. Researchers who have included multiple standardized assessments in their studies (e.g., Bellugi et al., 1988; Brock, Jarrold, Farran, Laws, & Riby, 2007; Mervis & Becerra, 2007) have consistently reported that the highest mean standard score was on the PPVT, which measures receptive vocabulary (primarily names for objects, actions, and descriptors). We have administered the most recent version of the PPVT (PPVT-4; Dunn & Dunn, 2007) to 88 4- to 49-year-olds with Williams syndrome. The mean standard score was 81.81 ($SD = 13.55$, range: 42–124), with 83% scoring in the normal range (≥ 70) and 6% scoring at or above the general population mean (≥ 100). We have also administered the Expressive Vocabulary Test-2 (EVT-2; Williams, 2007) to the same individuals. Most of the items on this measure, which was conormed with the PPVT-4, require the participant to name the object, object part, or action indicated in a picture; on a small proportion of the items, the participant is asked to provide a synonym for a word produced by the examiner in the context of a picture. The descriptive statistics for the EVT-2 were almost identical to those for the PPVT-4: the mean standard score was 80.81 (SD = 13.54, range: 47–120), with 82% scoring in the normal range and 10% scoring at or above the general population mean.

In sharp contrast to this relative strength in concrete vocabulary, individuals with Williams syndrome have considerable difficulty with relational/conceptual vocabulary. Relational vocabulary includes both terms for basic relational concepts (e.g., spatial, temporal, quantitative, and dimensional terms) and more advanced relational concepts such as conjunctions and disjunctions (e.g., *and, or, although, nevertheless, neither . . . nor*). We (Mervis & John, 2008) compared the performance of 92 5- to 7-year-olds with Williams syndrome on the Test of Relational Concepts (TRC; Edmonston & Litchfield Thane, 1988), which measures receptive knowledge of simple relational concepts, and the PPVT-III (Dunn & Dunn, 1997). The mean PPVT-III standard score was 86.73 (SD = 13.67, range: 59–118), 30 points higher than the children's mean TRC standard score of 55.79 (SD = 21.37, range: 25–104). The children's performance on the TRC was at about the same level as their performance on the DAS Pattern Construction subtest, the signature weakness for individuals with Williams syndrome. However, the pattern of TRC errors indicated that the children had difficulty with relational concepts in general, rather than specifically with spatial concepts. This pattern of findings is consistent with Walsh's (2003) argument that spatial, temporal, and quantitative processing are all controlled by a common magnitude system that is located in the inferior parietal cortex. This is the area in which Meyer-Lindenberg et al. (2004, 2006) identified a region of reduced gray matter and reduced sulcal depth that served as a roadblock to dorsal stream information flow, suggesting

the possibility of a common basis for the findings of extreme difficulty in both visuospatial construction and relational language.

Although most individuals with Williams syndrome eventually acquire simple relational concepts, older children demonstrate clear difficulties when they are tested on more advanced relational concepts. We tested 29 9- to 11-year-olds on the Clinical Evaluation of Language Fundamentals-IV (CELF-IV; Semel, Wiig, & Secord, 2003) Formulated Sentences subtest, which includes both simple and advanced relational concepts. Twelve participants earned a scaled score of 1 (the lowest possible). Ten children had participated in both the TRC study and, an average of 4 years later, the CELF-IV Formulated Sentences study. The correlation between performances in the two studies was r = .87, indicating strong continuity in relational language ability over the age range of the two studies (Mervis & John, 2008).

The initial argument that the grammatical abilities of individuals with Williams syndrome were considerably more advanced than expected for their overall intellectual abilities was based on the comparison by Bellugi et al. (e.g., 1988) of the grammatical abilities of a small sample of adolescents with Williams syndrome with those of CA- and IQ-matched adolescents with Down syndrome. Bellugi et al. found that the Williams syndrome group spoke in complete, largely grammatical sentences that included complex constructions, and that they also comprehended constructions such as passives, conditionals, and tag questions. In contrast, the Down syndrome group spoke in short, often ungrammatical utterances, and was much less likely than the Williams syndrome group to comprehend complex grammatical constructions. This finding has been replicated several times (e.g., Klein & Mervis, 1999; Mervis, Robinson, Rowe, Becerra, & Klein-Tasman, 2003; Vicari et al., 2004). However, these differences most likely reflect the inordinate difficulty that individuals with Down syndrome have with the grammatical aspects of language, rather than that individuals with Williams syndrome have unusually good grammar. Studies comparing the grammatical abilities of individuals with Williams syndrome with those of either CA- and IQ-matched individuals with intellectual disability due to causes other than Down syndrome or to MA-matched typically developing children have consistently indicated that the grammatical abilities of the Williams syndrome group are at or slightly below that of the comparison group (see review in Mervis & Becerra, 2007).

The most common measure of receptive grammatical ability used in studies of individuals with Williams syndrome is the Test for Reception of Grammar (TROG or TROG-2; Bishop, 1989, 2003a) or a translation of this test. This test measures understanding of constructions ranging from simple positive statements to sentences containing center-embedded relative clauses. Our research group has administered the TROG-2 to 159 5- to 46-year-olds with Williams syndrome. Mean standard score was 71.48 ($SD = 16.24$, range: 55 [floor]–116), with 31% earning the lowest possible standard score. The relative difficulty of different grammatical constructions for individuals with Williams syndrome follows the same pattern as for MA-matched typically developing children (e.g., Lukács, 2005; Volterra et al., 1996). The performance of individuals with Williams syndrome on the TROG is strongly correlated with their verbal working memory ability, and this relation is significantly stronger for children and adolescents with Williams syndrome than for MA-matched typically developing children (Robinson, Mervis, & Robinson, 2003).

LANGUAGE ABILITIES: SOCIO-COMMUNICATIVE ABILITIES

von Arnim and Engel's (1964) observation that children and adolescents with Williams syndrome have difficulty interacting with CA peers, even if the peers also have intellectual disability, but are uninhibited in interacting with adults suggests that individuals with Williams syndrome may have significant difficulties with social communication. Laws and Bishop (2004) reported that 15 of 19 children and adults with Williams syndrome met the criterion for pragmatic language impairment on the Children's Communication Checklist (CCC; Bishop, 1998), evidencing significant difficulties in all areas of pragmatics measured, with particular difficulty in the use of stereotyped conversation, inappropriate initiation, and overdependence on context. These types of pragmatic difficulties overlap with those associated with the autism spectrum, and Philofsky, Fidler, and Hepburn (2007) found no difference between the performance of a group of school-age children with Williams syndrome and a CA-matched group of children with autism on two of the CCC-2 (Bishop, 2003b) pragmatics scales.

Socio-communicative difficulties have also been reported in toddlers and young preschoolers with Williams syndrome (Laing et al., 2002; Rowe, Peregrine, & Mervis, 2005), with the types of difficulties identified again overlapping with those on the autism spectrum. To provide a more in-depth examination of the socio-communicative abilities of young children with Williams syndrome, our research group, in collaboration with Bonnie Klein-Tasman and Cathy Lord, administered the Autism Diagnostic Observation Schedule (ADOS; Lord et al., 2000) Module 1 to 29 30- to 63-month-olds with Williams syndrome who had limited or no expressive language (Klein-Tasman, Mervis, Lord, & Phillips, 2007). The results were consistent with and extended those of previous studies. More than half of the children evidenced difficulties with pointing, giving, showing, and appropriate use of eye contact, and many children also had difficulty with initiation of joint attention, response to the researcher's bids for joint attention, and integration of eye gaze with other behaviors. As indicated in Table 6.1, the socio-communicative difficulties identified were significant enough that more than half of the children met or exceeded the ADOS cutoff for autism spectrum for the Communication domain and/or the Reciprocal Social Interaction domain, and almost half the children met or exceeded the autism spectrum cutoff for the Total Score. By itself, performance on the ADOS is not sufficient to make a diagnosis of autism or an ASD. However, the three children who met the ADOS algorithm for "autism" were later clinically diagnosed with autism. More recently we compared the ADOS performance of this group of children with Williams syndrome to that of CA- and MA-matched groups of children with clinical diagnoses of autism, clinical diagnoses of pervasive developmental disorder-not otherwise specified (PDD-NOS), or developmental delay of mixed etiologies (Klein-Tasman, Phillips, Lord, Mervis, & Gallo, in press). The children with Williams syndrome showed significantly less difficulty with socio-communication than the children with autism but about the same level of difficulty as the children with PDD-NOS. Furthermore, the Williams syndrome group showed significantly more difficulty with reciprocal social interaction than the mixed-etiology group, indicating that the problems identified were not due simply to developmental delay. Based on these findings, we (Klein-Tasman et al., 2007, in press) have argued that children with Williams syndrome who evidence socio-communicative difficulties that overlap with the autism spectrum should be evaluated for the possibility that they have PDD-NOS (or, in rare cases, autism), and if they meet clinical criteria, they should be given the diagnosis. That is, the prior diagnosis of Williams syndrome should not lead to

Table 6.1. Autism Diagnostic Observation Schedule (ADOS; Lord et al., 2000) classification as a function of domain

Domain	Classification		
	Nonspectrum	Autism Spectrum	Autism
Communication	8	15	6
Reciprocal Social Interaction	13	8	8
Total	15	11	3

diagnostic overshadowing if the child meets clinical criteria for an ASD.

PERSONALITY AND PSYCHOPATHOLOGY

By pointing out that individuals with Williams syndrome "have a great ability to establish interpersonal contacts" but that this "stands against a background of insecurity and anxiety," von Arnim and Engel (1964, p. 376) highlighted the paradoxical nature of social behavior, personality, and psychopathology in Williams syndrome. These patterns are well captured in the Williams Syndrome Personality Profile (WSPP) we proposed (Klein-Tasman & Mervis, 2003) based on the Multidimensional Personality Questionnaire (MPQ; Tellegen, 1985). This profile was based on parental ratings of 22 8- to 10-year-olds with Williams syndrome and 20 CA- and IQ-matched children with intellectual disability of mixed etiologies. The Williams syndrome group was rated significantly higher than the mixed-etiology group on five items: gregarious, people-oriented, visible, tense, and sensitive. Using a combination of these items, we correctly classified 21 of 22 children with Williams syndrome and 17 of 20 children in the mixed-etiology group, yielding a sensitivity of .96 and a specificity of .85. We have recently analyzed more than 100 additional MPQs for individuals with Williams syndrome; sensitivity was >.90 for ages 5–20 years but then decreased considerably. We did not have enough additional participants with mixed etiologies to compute specificity.

Although von Arnim and Engel (1964) did not mention distractibility as a major problem for individuals with Williams syndrome, all of the studies that have relied on parental responses to standardized questionnaires have identified difficulty focusing as a major concern (e.g., Dilts, Morris, & Leonard, 1990; Pagon, Bennett, LaVeck, Stewart, & Johnson, 1987). These difficulties, as well as problems with frustration and emotional outbursts noted by von Arnim and Engel and the sensitivity and tenseness identified in the WSPP (Klein-Tasman & Mervis, 2003), are often reported in children who have sensory integration problems (e.g., Ayres & Robbins, 2005). Successful integration of sensory information gives meaning to environmental experiences and allows the child to pick out the key information from the endless amount of sensations picked up by his or her body. Children who do not acquire developmentally mature perceptual and sensory integrative abilities often display maladaptive emotional and physical responses to environmental stimuli. Such difficulties are found in children with an ASD (Ben-Sasson et al., 2009) and children with attention-deficit/hyperactivity disorder (ADHD; Dunn & Bennett, 2002). To determine whether children with Williams syndrome demonstrate difficulties with sensory integration, we administered the Short Sensory Profile (SSP; McIntosh, Miller, Shyu, & Dunn, 1999) to the parents of 72 4- to 10-year-olds who had classic deletions and had not been diagnosed with an ASD (John & Mervis, 2008). As shown in Figure 6.3, results indicated that based on the Overall classification, only 10%

of the children were considered to have normal sensory processing, with most of the children classified as having definite abnormalities. The SSP also provides classifications for seven specific types of sensory-processing abilities. More than 50% of our sample scored in the definitely abnormal range for four of the seven subscales: Auditory Filtering (ability to use and screen out sounds), Underresponsiveness/Seeks Sensation (level of noticing sensory events), Low Energy/Weak (ability to use muscles to move), and Visual/Auditory Sensitivity (response to sights and sounds); the children's classifications on these subscales are shown in Figure 6.3. This pattern suggests that children with Williams syndrome are likely to have hypersensitivity to sound, which may lead to clinically significant problems with specific phobia, and to be highly distractible, which may lead to clinically significant levels of inattention.

To address the possibility of clinically significant anxiety and ADHD difficulties in children with Williams syndrome, our research group (Leyfer, Woodruff-Borden, Klein-Tasman, Fricke, & Mervis, 2006) conducted an interview study of the parents of

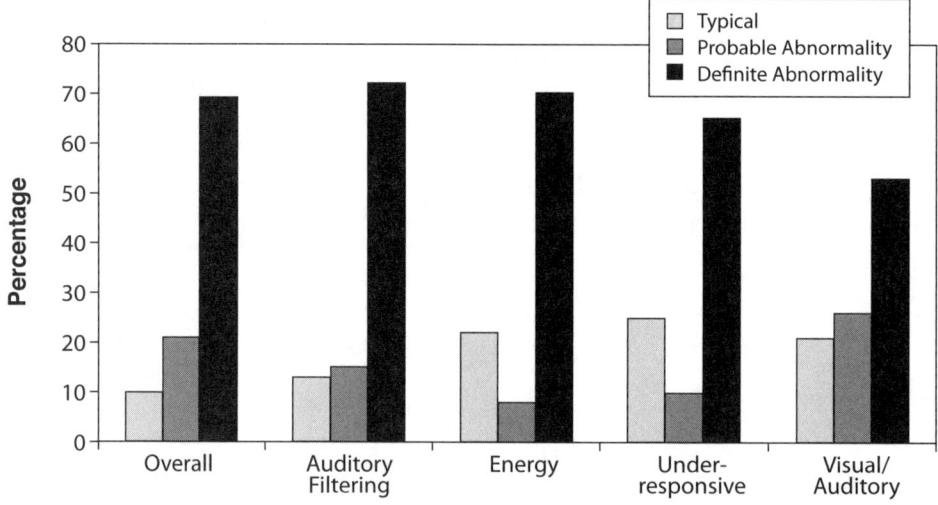

Figure 6.3. Classification of 72 children with Williams syndrome ages 4–10 years on the Short Sensory Profile (McIntosh, Miller, Shyu, & Dunn, 1999). All children have classic deletions and have not been diagnosed with an autism spectrum disorder.

119 4- to 16-year-olds with Williams syndrome using the Anxiety Disorders Interview Schedule-Parent (ADIS-P; Silverman & Albano, 1996). The ADIS-P is a structured interview that provides diagnoses from the *Diagnostic and Statistical Manual of Mental Disorders, Fourth Edition,* (*DSM-IV;* American Psychiatric Association [APA], 1994) for all anxiety and related disorders, including ADHD. The majority of children (78.2%) received at least one *Diagnostic and Statistical Manual of Mental Disorders, Fourth Edition, Text Revision* (*DSM-IV-TR;* APA, 2000) diagnosis, with 64.7% diagnosed with ADHD and 58.8% diagnosed with at least one anxiety disorder. These percentages are considerably higher than either those reported for children in the general population or those reported in the largest study of psychiatric disorders in children with intellectual disability that used a structured diagnostic interview that yields *DSM-IV-TR* diagnoses (Dekker & Koot, 2003). In these studies, the percentage of children with *DSM-IV-TR* diagnoses of ADHD were 4%–7% for children in the general population (APA, 2000) and 6.1% for children with intellectual disability (Dekker & Koot, 2003). The preva-

lence of any anxiety disorder was 9.8% for children in the general population (Shaffer et al., 1996) and 10.5% for children with intellectual disability (Dekker & Koot, 2003).

The most common *DSM-IV-TR* diagnosis for children with Williams syndrome was ADHD (64.7%). Unlike in the general population, there was no gender difference in the prevalence of ADHD for children with Williams syndrome (Leyfer et al., 2006). These findings are not surprising, given the very high proportion of 4- to 10-year-olds with Williams syndrome who have difficulties with auditory filtering as indicated by parental responses on the SSP (72% definitely abnormal, 15% probably abnormal; see Figure 6.3) and the fact that the mean score on the SSP Auditory Filtering subscale was almost identical for boys and girls. In cases in which the parents considered these abnormalities to be impairing, the child would be likely to receive a diagnosis of ADHD. ADHD diagnoses were significantly more frequent for the 7- to 10-year-olds (79.5%) than for either the 4- to 6-year-olds (55.0%) or the 11- to 16-year-olds (57.1%). In Figure 6.4, the distribution of ADHD diagnoses by type (predominantly inattentive,

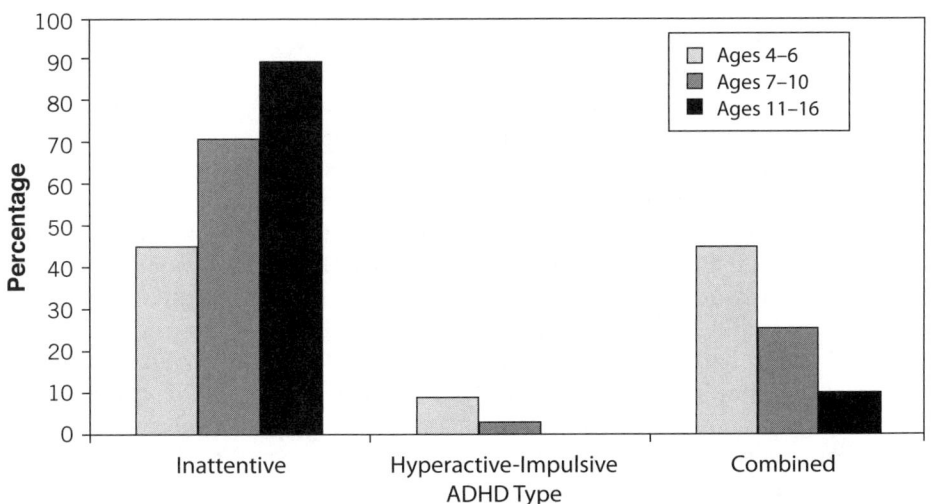

Figure 6.4. Attention-deficit/hyperactivity disorder (ADHD) type as a function of chronological age for children with Williams syndrome who had been diagnosed with ADHD. Only children who had an ADHD diagnosis are included in the figure. Overall, 55% of children between 4 and 6 years, 79.5% of children between 7 and 10 years, and 57.1% of children between 11 and 16 years were diagnosed with ADHD.

Table 6.2. Prevalence of anxiety disorders in children with Williams syndrome and in the general population[a]

	Group	
	Williams syndrome	General population
DSM-IV diagnosis	%	%
Separation Anxiety[b]	6.7	2.3
Social Phobia	1.7	4.5
Specific Phobia[c]	53.8	1.3
Panic Disorder	0.8	1.4
General Anxiety Disorder[c]	11.8	3.1
Obsessive-Compulsive Disorder	2.8	—

[a]General population prevalence based on Shaffer et al. (1996).

[b]$p < .01$.

[c]$p < .001$.

Key: *DSM-IV, Diagnostic and Statistical Manual of Mental Disorders* (American Psychiatric Association, 1994).

predominantly hyperactive-impulsive, combined) is shown as a function of CA. Note that children who do not have an ADHD diagnosis are not included in Figure 6.4; the percentages reported are for the subgroup of children who do have an ADHD diagnosis. As indicated in the figure, as CA increases, children with ADHD are increasingly likely to be diagnosed with ADHD–predominantly inattentive and increasingly less likely to be diagnosed with ADHD–combined. ADHD–predominantly hyperactive is rare.

The percentage of children with Williams syndrome who met *DSM-IV-TR* criteria for a variety of anxiety disorders is shown in Table 6.2, along with the percentages reported by Shaffer et al. (1996), the largest epidemiological study of anxiety disorders in children in the general population. The most common anxiety disorder for children with Williams syndrome was specific phobia; 91% of the children who had at least one anxiety disorder had specific phobia (53.8% of the entire sample). The most common specific phobia was for loud noises (28% of the entire sample). This finding is consistent with both the findings from our sensory processing study (John & Mervis, 2008) and prior studies in which parents were asked to report on hypersensitivity to

sound. For example, Levitin, Cole, Lincoln, and Bellugi (2005) found that 80% of parents of individuals with Williams syndrome reported that their child had hypersensitivity to sound. If this hypersensitivity leads to both fear and adaptive impairment, the child will be diagnosed with specific phobia. Hypersensitivity to sound and the sometimes resulting specific phobia may be related to sensorineural hearing loss, which Marler, Elfenbein, Ryals, Urban, and Netzloff (2005) identified in 14 of 18 school-age children studied. The majority of children who had specific phobia of loud noises also had at least one other specific phobia, usually medically related.

The next most common anxiety disorder was generalized anxiety disorder (GAD), which increased significantly with age (consistent with the pattern found in the general population; e.g., Strauss, Lease, Last, & Francis, 1988). GAD was diagnosed in none of the children with Williams syndrome ages 4–6 years, 13.6% of the children ages 7–10 years, and 22.9% of the children ages 11–16 years. Leyfer et al. (2006) argued that based on clinical experience, the *DSM-IV-TR* criteria for GAD often do not capture the nature of worrying in Williams syndrome. Most parents report that their child worried in anticipation of both events they expected to dislike and events that they expected to enjoy. This worry, which is perseverative, is typically manifested as repeated questioning about the upcoming event, leading most parents to avoid telling their child about the event for as long as possible. This anxiety about upcoming events typically occupies a significant amount of time and is consistently reported as impairing, but it is not captured by the *DSM-IV-TR* criteria for GAD.

Based on consistent findings that individuals with Williams syndrome reliably evidence disinhibited approach behavior and overfriendliness, Bellugi and her colleagues (Bellugi, Adolphs, Cassady, & Chiles, 1999) hypothesized that amygdala function in Williams syndrome is abnormally low. The additional consistent finding of increased nonsocial anxiety (especially specific phobia) suggested to us, however, that other neural mechanisms are likely also involved. As part of our original structural MRI study of adults with Williams syndrome and typical IQs, we identified a region in the orbitofrontal cortex that had reduced gray matter (Meyer-Lindenberg et al., 2004). Results of our fMRI studies of these participants and matched controls from the general population (Meyer-Lindenberg et al., 2005, 2006) indicated that for the Williams syndrome group, the amygdala reacted significantly less than expected to socially relevant threatening visual stimuli (angry and fearful faces) and significantly more than expected to socially irrelevant threatening visual stimuli (scenes not including people). Underreaction of the amygdala has been associated with inappropriate approach and overreaction to specific phobia in the general population (see Meyer-Lindenberg et al., 2006). Further analyses indicated that in contrast to the control group, the orbitofrontal cortex region that had been found to have reduced gray matter was not activated at all in the Williams syndrome group in response to either type of threatening stimulus, and so did not regulate the

amygdala. This type of functional abnormality would lead to the behavioral patterns of social disinhibition and increased nonsocial-specific phobia that is characteristic of individuals with Williams syndrome (Meyer-Lindenberg et al., 2005, 2006).

SUMMARY AND CONCLUSION

Williams syndrome is associated with a wide range of intellectual ability, from severe intellectual disability to average intelligence. The mean IQ is in the mild disability range. However, this mean masks large and significant differences between visuospatial construction abilities, which are a significant weakness for most individuals, and verbal ability and/or nonverbal reasoning ability, one or both of which are typically significant relative strengths. There is also a profile of relative strengths and weaknesses within the language domain, in which concrete vocabulary (both receptive and expressive) is the strongest ability and is within the typical range for most individuals with Williams syndrome; grammatical ability is at about the level expected for overall level of functioning; and conceptual/relational vocabulary is a clear weakness, with performance at a level similar to that for visuospatial construction. This latter finding is consistent with neuroimaging results indicating a region of reduced gray matter and reduced sulcal depth in the intraparietal sulcus, leading to impaired dorsal stream functioning beyond this region. This region has been implicated in spatial, quantitative, and temporal processing. A second clear weakness is in the socio-communicative (pragmatic) aspects of language. Difficulties that are greater than expected for overall level of ability have been identified for coordinating attention between a communicative partner and an object, comprehension and use of communicative gestures, and eye gaze. These difficulties overlap those on the autism spectrum, and in some cases, the possibility of a comorbid ASD should be considered. The personality profile for Williams syndrome is characterized by gregariousness, overfriendliness, high visibility, tenseness, and oversensitivity. Most children and adolescents are socially disinhibited, demonstrating inappropriate approach to strangers, and many have difficulties with self-regulation that lead to outbursts of temper. Sensory modulation abnormalities are very common and likely contribute to the high prevalence of both ADHD and specific phobia. A region of reduced gray matter in the orbitofrontal cortex results in this region's failure to regulate the amygdala, leading to both social disinhibition and non-social specific phobia. Further research that more specifically describes the psychological characteristics associated with Williams syndrome and the associated similarities and differences in brain structure and functioning relative to the general population would both expand our understanding of this syndrome and allow for the development of interventions to help individuals achieve their maximum potential.

Williams syndrome is caused by a hemizygous contiguous deletion of ~25 genes on chromosome 7q11.23. Some of these genes have been implicated in particular medical or psychological characteristics associated with Williams syndrome (see review in Mervis & Morris, 2007). Future research linking specific genes, in transaction with other genes and the environment, to both structural and functional characteristics of the brains of individuals with Williams syndrome and to psychological characteristics associated with this syndrome would further our understanding of both Williams syndrome and typical processes of development.

REFERENCES

American Psychiatric Association. (1994). *Diagnostic and Statistical Manual of Mental Disorders* (4th ed.). Washington, DC: Author.

American Psychiatric Association. (2000). *Diagnostic and Statistical Manual of Mental Disorders* (4th ed., text rev.). Washington, DC: Author.

Ayres, A.J., & Robbins, J. (2005). *Sensory integration and the child: Understanding hidden sensory challenges.* Los Angeles: Western Psychological Services.

Bellugi, U., Adolphs, R., Cassady, C., & Chiles, M. (1999). Towards the neural basis for hypersociability in a genetic syndrome. *NeuroReport, 10,* 1653–1657.

Bellugi, U., Lichtenberger, L., Jones, W., Lai, Z., & St. George, M. (2000). The neurocognitive profile of Williams syndrome: A complex pattern of strengths and weaknesses. *Journal of Cognitive Neuroscience, 12*(Suppl. 1), 7–29.

Bellugi, U., Marks, S., Bihrle, A., & Sabo, H. (1988). Dissociation between language and cognitive functions in Williams syndrome. In D. Bishop & K. Mogford (Eds.), *Language development in exceptional circumstances* (pp. 177–189). London: Churchill Livingstone.

Bellugi, U., Wang, P., & Jernigan, T.L. (1994). Williams syndrome: An unusual neuropsychological profile. In S.H. Broman & J. Grafman (Eds.), *Atypical cognitive deficits in developmental disorders: Implications for brain function* (pp. 23–56). Hillsdale, NJ: Erlbaum.

Bennett, C., LaVeck, B., & Sells, C.J. (1978). The Williams elfin facies syndrome: The psychological profile as an aid in syndrome identification. *Pediatrics, 61,* 303–306.

Ben-Sasson, A., Hen, L., Fluss, R., Cermak, S., Engel-Yeger, B., & Gal, E. (2009). A meta-analysis of sensory modulation symptoms in individuals with autism spectrum disorders. *Journal of Autism and Developmental Disorders, 39,* 1–11.

Beuren, A.J., Apitz, J., & Harmjanz, D. (1962). Supravalvular aortic stenosis in association with mental retardation and a certain facial appearance. *Circulation, 27,* 1235–1240.

Bishop, D.V.M. (1989). *Test for Reception of Grammar.* Manchester, United Kingdom: Chapel Press.

Bishop, D.V.M. (1998). Development of the Children's Communication Checklist (CCC): A method for assessing qualitative aspects of communicative impairment in children. *Journal of Child Psychology and Psychiatry, 39,* 879–891.

Bishop, D.V.M. (2003a). *Test for Reception of Grammar* (version 2). London: Psychological Corporation.

Bishop, D.V.M. (2003b). *The Children's Communication Checklist* (2nd ed.). London: Psychological Corporation.

Brock, J., Jarrold, C., Farran, E.K., Laws, G., & Riby, D.M. (2007). Do children with Williams syndrome really have good vocabulary knowledge? Methods for comparing cognitive and linguistic abilities in developmental disorders. *Clinical Linguistics & Phonetics, 21,* 673–688.

Dekker, M.C., & Koot, H.M. (2003). DSM-IV disorders in children with borderline to moderate intellectual disability. I: prevalence and impact. *Journal of the American Academy of Child and Adolescent Psychiatry, 42,* 915–922.

Dilts, C., Morris, C.A., & Leonard, C.O. (1990). A hypothesis for the development of a behavioral phenotype in Williams syndrome. *American Journal of Medical Genetics Supplement 6,* 126–131.

Dunn, W., & Bennett, D. (2002). Patterns of sensory processing in children with attention deficit hyperactivity disorder. *Occupational Therapy Journal of Research, 22,* 4–15.

Dunn, L.E., & Dunn, L.E. (1981). *Peabody Picture Vocabulary Test–Revised.* Circle Pines, MN: American Guidance Service.

Dunn, L.E., & Dunn, L.E. (1997). *Peabody Picture Vocabulary Test* (3rd ed.). Circle Pines, MN: American Guidance Service.

Dunn, L.E., & Dunn, D.M. (2007). *Peabody Picture Vocabulary Test* (4th ed.). Minneapolis, MN: Pearson Assessments.

Edmonston, N.K., & Litchfield Thane, N. (1988). *TRC: Test of Relational Concepts.* Austin, TX: PRO-ED.

Elliott, C.D. (1990). *Differential Ability Scales.* San Antonio, TX: Psychological Corporation.

Elliott, C.D. (2007). *Differential Ability Scales* (2nd ed.). San Antonio, TX: Psychological Corporation.

Ewart, A.K., Morris, C.A., Atkinson, D., Jin, W., Sternes, K., Spallone, P., et al. (1993). Hemizygosity at the *elastin* locus in a developmental disorder, Williams syndrome. *Nature Genetics, 5,* 11–16.

Hillier, L.W., Fulton, R.S., Fulton, L.A., Graves, T.A., Pepin, K.H., Wagner-McPherson, C., et al. (2003). The DNA sequence of chromosome 7. *Nature, 424,* 157–164.

Howlin, P., Davies, M., & Udwin, O. (1998). Cognitive functioning in adults with Williams syndrome. *Journal of Child Psychology and Psychiatry, 39,* 183–189.

Jarrold, C., Hartley, S.J., Phillips, C., & Baddeley, A.D. (2000). Word fluency in Williams syndrome: Evidence for unusual semantic organization? *Cognitive Neuropsychiatry, 5,* 293–319.

John, A.E., & Mervis, C.B. (2008, July). *Sensory modulation in children with Williams syndrome.* Presented at the International Professional Meeting of the Williams Syndrome Association, Anaheim, CA.

Kataria, S., Goldstein, D.J., & Kushnick, T. (1984). Developmental delays in Williams ("elfin facies") syndrome. *Applied Research in Mental Retardation, 5,* 419–423.

Kippenhan, J.S., Olsen, R.K., Mervis, C.B., Morris, C.A., Kohn, P., Meyer-Lindenberg, A., & Berman, K.F. (2005). Genetic contributions to human gyrification: Sulcal morphometry in Williams syndrome. *Journal of Neuroscience, 25,* 7840–7846.

Klein, B.P., & Mervis, C.B. (1999). Cognitive strengths and weaknesses of 9- and 10-year-olds with Williams syndrome or Down syndrome. *Developmental Neuropsychology, 16,* 177–196.

Klein-Tasman, B.P., & Mervis, C.B. (2003). Distinctive personality characteristics of 8-, 9-, and 10-year-old children with Williams syndrome. *Developmental Neuropsychology, 23,* 271–292.

Klein-Tasman, B.P., Mervis, C.B., Lord, C., & Phillips, K.D. (2007). Socio-communicative deficits in young children with Williams syndrome: Performance on the Autism Diagnostic Observation Schedule. *Child Neuropsychology, 13,* 444–467.

Klein-Tasman, B.P., Phillips, K.D., Lord, C.E., Mervis, C.B., & Gallo, F. (in press). More than meets the eye: Overlap with the autism spectrum in young children with Williams syndrome. *Journal of Developmental and Behavioral Pediatrics.*

Laing, E., Butterworth, G., Ansari, D., Gsödl, M., Longhi, E., Panagiotaki, G., et al. (2002). Atypical development of language and social communication in toddlers with Williams syndrome. *Developmental Science, 5,* 233–246.

Laws, G., & Bishop, D. (2004). Pragmatic language impairment and social deficits in Williams syndrome: A comparison with Down's syndrome and specific language impairment. *International Journal of Language and Communication Disorders, 39,* 45–64.

Levitin, D.J., Cole, K., Lincoln, A., & Bellugi, U. (2005). Aversion, awareness, and attraction: Investigating claims of hyperacusis in the Williams syndrome phenotype. *Journal of Child Psychology and Psychiatry, 46,* 514–523.

Levy, Y., & Bechar, T. (2003). Cognitive, lexical, and morpho-syntactic profiles of Israeli children with Williams syndrome. *Cortex, 29,* 255–271.

Leyfer, O.T., Woodruff-Borden, J., Klein-Tasman, B.P., Fricke, J.S., & Mervis, C.B. (2006). Prevalence of psychiatric disorders in 4 to 16-year-olds with Williams syndrome. *American Journal of Medical Genetics Part B, 141B,* 615–622.

Lord, C., Risi, S., Lambrecht, L., Cook, E.H., Jr., Leventhal, B.L., DiLavore, P.C., et al. (2000). The Autism Diagnostic Observation Schedule–Generic: A standard measure of social and communication deficits associated with the spectrum of autism. *Journal of Autism and Developmental Disorders, 30,* 205–223.

Lukács, A. (2005). *Language abilities in Williams syndrome.* Budapest, Hungary: Akadémiai Kiadó.

MacDonald, G.W., & Roy, D.L. (1988). Williams syndrome: A neuropsychological profile. *Journal of Clinical and Experimental Neuropsychology, 10,* 125–131.

Marler, J.A., Elfenbein, J.L., Ryals, B.M., Urban, Z., & Netzloff, M.L. (2005). Sensorineural hearing loss in children and adults with Williams syndrome. *American Journal of Medical Genetics Part A, 138A,* 318–327.

McCarthy, D. (1972). *McCarthy Scales of Children's Abilities.* New York: Psychological Corporation.

McIntosh, D.N., Miller, L.J., Shyu, V., & Dunn, W. (1999). *Short Sensory Profile.* New York: Psychological Corporation.

Mervis, C.B., & Becerra, A.M. (2007). Language and communicative development in Williams syndrome. *Mental Retardation and Developmental Disabilities Research Reviews, 13,* 3–15.

Mervis, C.B., & John, A.E. (2008). Vocabulary abilities of children with Williams syndrome: Strengths, weaknesses, and relation to visuospatial construction ability. *Journal of Speech, Language, and Hearing Research, 51,* 967–982.

Mervis, C.B., & Morris, C.A. (2007). Williams syndrome. In M.M.M. Mazzocco & J.L. Ross (Eds.), *Neurogenetic developmental disorders: Variation of manifestation in childhood* (pp. 199–262). Cambridge, MA: MIT Press.

Mervis, C.B., Morris, C.A., Bertrand, J., & Robinson, B.F. (1999). Williams syndrome: Findings from an integrated program of research. In H. Tager-Flusberg (Ed.), *Neurodevelopmental disorders* (pp. 65–110). Cambridge, MA: MIT Press.

Mervis, C.B., Robinson, B.F., Bertrand, J., Morris, C.A., Klein-Tasman, B.P., & Armstrong, S.C. (2000). The Williams syndrome cognitive profile. *Brain and Cognition, 44,* 604–628.

Mervis, C.B., Robinson, B.F., Rowe, M.L., Becerra, A.M., & Klein-Tasman, B.P. (2003). Language abilities of individuals who have Williams syndrome. In L. Abbeduto (Ed.), *International Review of Research in Mental Retardation* (vol. 27, pp. 35–81). Orlando, FL: Academic Press.

Meyer-Lindenberg, A., Hariri, A.R., Munoz, K.E., Mervis, C.B., Mattay, V.S., Morris, C.A., & Berman, K.F. (2005). Neural correlates of genetically abnormal social cognition in Williams syndrome. *Nature Neuroscience, 8,* 991–993.

Meyer-Lindenberg, A., Kohn, P., Mervis, C.B., Kippenhan, J.S., Olsen, R., Morris, C.A., & Berman, K.F. (2004). Neural basis of genetically determined visuospatial construction deficit in Williams syndrome. *Neuron, 43,* 623–631.

Meyer-Lindenberg, A., Mervis, C.B., & Berman, K.F. (2006). Neural mechanisms in Williams syndrome: A unique window to genetic influences on cognition and behavior. *Nature Reviews: Neuroscience, 7,* 380–393.

Meyerson, M.D., & Frank, R.A. (1987). Language, speech, and hearing in Williams syndrome: Intervention approaches and research needs. *Developmental Medicine and Child Neurology, 29,* 258–262.

Miller, G.A., & Gildea, P.M. (1987). How children learn words. *Scientific American, 257*(3), 94–99.

Morris, C.A. (2006). The dysmorphology, genetics, and natural history of Williams-Beuren syndrome. In C.A. Morris, H.M. Lenhoff, & P.P. Wang (Eds.), *Williams-Beuren syndrome: Research, evaluation, and treatment* (pp. 3–17). Baltimore: Johns Hopkins University Press.

Mullen, E.M. (1995). *Mullen Scales of Early Learning.* Circle Pines, MN: American Guidance Service.

Osborne, L.R. (2006). The molecular basis of a multisystem disorder. In C.A. Morris, H.M. Lenhoff, & P.P. Wang (Eds.), *Williams-Beuren syndrome: Research, evaluation, and treatment* (pp. 18–58). Baltimore: Johns Hopkins University Press.

Pagon, R.A., Bennett, F.C., LaVeck, B., Stewart, K.B., & Johnson, J. (1987). Williams syndrome: Features in late childhood and adolescence. *Pediatrics, 80,* 85–91.

Philofsky, A., Fidler, D.J., & Hepburn, S. (2007). Pragmatic language profiles of school-age children with autism spectrum disorders and Williams syndrome. *American Journal of Speech-Language Pathology, 16,* 368–380.

Piattelli-Palmarini, M. (2001). Speaking of learning: How do we acquire our marvellous facility for expressing ourselves in words? *Nature, 411,* 887–888.

Robinson, B.F., Mervis, C.B., & Robinson, B.W. (2003). Roles of verbal short-term memory and working memory in the acquisition of grammar by children with Williams syndrome. *Developmental Neuropsychology, 23,* 13–31.

Rowe, M.L., Peregrine, E., & Mervis, C.B. (2005, April). *Communicative development in toddlers with Williams syndrome.* Presented at the Society for Research in Child Development, Atlanta, GA.

Searcy, Y.M., Lincoln, A.J., Rose, F.E., Klima, E.S., & Bavar, N. (2004). The relationship between age and IQ in adults with Williams syndrome. *American Journal on Mental Retardation, 109,* 231–236.

Semel, E., Wiig, E.H., & Secord, W.A. (2003). *Clinical evaluation of language fundamentals* (4th ed.). San Antonio, TX: Harcourt Assessment.

Shaffer, D., Fisher, P., Dulcan, M.K., Davies, M., Piacentini, J., Schwab-Stone, M.E., et al. (1996). The NIMH Diagnostic Interview Schedule for Children Version 2.3 (DISC-2.3): Description, acceptability, prevalence rates, and performance in the MECA Study. Methods for the Epidemiology of Child and Adolescent Mental Disorders Study. *Journal of the American Academy of Child and Adolescent Psychiatry, 35,* 865–877.

Silverman, W.K., & Albano, A.M. (1996). *The Anxiety Disorders Interview Schedule for DSM-IV: Parent Interview Schedule.* San Antonio, TX: Graywind Publications, a Division of the Psychological Corporation.

Strauss, C.C., Lease, C.A., Last, C.G., & Francis, G. (1988). Overanxious disorder: An examination of developmental differences. *Journal of Abnormal Child Psychology, 16,* 433–443.

Strømme, P., Bjørnstad, P.G., & Ramstad, K. (2002). Prevalence estimation of Williams syndrome. *Journal of Child Neurology, 17,* 269–271.

Tellegen, A. (1985). Structures of mood and personality and their relevance to assessing anxiety, with an emphasis on self-report. In A.H. Tum & J.D. Maser (Eds.), *Anxiety and the anxiety disorders* (pp. 681–716). Hillsdale, NJ: Lawrence Erlbaum Associates.

Temple, C.M., Almazan, M., & Sherwood, S. (2002). Lexical skills in Williams syndrome: A cognitive neuropsychological analysis. *Journal of Neurolinguistics, 15,* 463–495.

Vicari, S., Bates, E., Caselli, M.C., Pasqualetti, P., Gagliardi, C., Tonucci, F., & Volterra, V. (2004). Neuropsychological profile of Italians with Williams syndrome: An example of a dissociation between language and cognition? *Journal of the International Neuropsychological Society, 10,* 862–876.

Volterra, V., Capirci, O., Pezzini, G., Sabbadini, L., & Vicari, S. (1996). Linguistic abilities in Italian children with Williams syndrome. *Cortex, 32,* 663–677.

von Arnim, G., & Engel, P. (1964). Mental retardation related to hypercalcaemia. *Developmental Medicine and Child Neurology, 6,* 366–377.

Walsh, V. (2003). A theory of magnitude: Common cortical metrics of time, space and quantity. *Trends in Cognitive Sciences, 7,* 483–488.

Wechsler, D. (1981). *Wechsler Adult Intelligence Scale–Revised.* New York: Psychological Corporation.

Wechsler, D. (2003). *Wechsler Intelligence Scale for Children* (4th ed.). San Antonio, TX: Psychological Corporation.

Williams, J.C.P., Barratt-Boyes, B.G., & Lowe, J.B. (1961). Supravalvular aortic stenosis. *Circulation, 24,* 1311–1318.

Williams, K.T. (2007). *Expressive Vocabulary Test* (2nd ed.). Minneapolis, MN: Pearson Assessments.

7

Developmental Influences on Psychological Phenotypes

Gene S. Fisch

The causes of developmental delay and/or intellectual disability (ID) are generally dichotomized into familial (or cultural-familial) or genetic categories. There is, however, an overlap between the two categories, and the extent to which each factor contributes to the disorder has never been properly delineated and remains controversial. Currently, researchers have identified about 1,000 genetic causes of ID (Sarimski, 1997), and their phenotypes are many and varied. It now appears that genetic abnormalities which produce learning disabilities (LDs) and ID occur in excess of 1% of the general population.

Genetic abnormalities that produce ID may engender a cognitive-behavioral profile of strengths and weaknesses that evolve as children age. In children with Down syndrome, for example, strengths and weaknesses are age-related. Children with Down syndrome continue to develop cognitive skills as they age, but their rate of development is lower than that of other children in the general population. As a result, IQ scores decline with increasing age (Carr, 1992; Wishart, 1993).

Cognitive-behavioral profiles and age-related features of deficits permit us to draw inferences about how brain development may be related to cognitive ability as children age. Consequently, the aim of this chapter is to compare and contrast cognitive profiles and adaptive and maladaptive behavior profiles, as related to age, in children diagnosed with one of three genetic disorders: the fragile X mutation (FRAXA), neurofibromatosis type 1 (NF1), and Williams syndrome.

FRAGILE X SYNDROME

FRAXA produces the fragile X syndrome (FXS), which was first identified cytogenetically by Lubs (1969). FXS results from a mutation in the *FMR1* gene (FRAXA) located on the long arm of the X chromosome, Xq27.3, which arises primarily from a marked increase in the number of CGG repeats in the 5' end of the FMR1 promotor region (see Chapter 3). In addition to ID and macro-orchidism, affected males appear clinically to have an unusually large head, long face, large jaw, anteverted ears, and a predisposition to seizures, autism, and connective tissue dysplasia (Opitz & Sutherland, 1984). FRAXA females may be affected, but their features tend to be milder, owing to

We thank Nancy Carpenter, Ph.D., and Birhan Say, M.D. (retired) of the Chapman Institute in Tulsa, OK; Patricia Howard-Peebles, Ph.D., Anne Maddalena, Ph.D., and Susan Black, M.D., formerly of GIVF, Fairfax, VA, and Joseph Shulman, M.D., GIVF, Fairfax, VA; Walter Nance, M.D., Medical College of Virginia, Richmond, VA; Roger Stevenson, M.D., and Richard Simensen, Ph.D., Greenwood Genetics Center, Greenwood, SC; Jeanette Holden, Ph.D., Ongwanada Resource Centre, Kingston, Ontario; Terri Monkaba of the Williams Syndrome Association; and Mary-Ann Wilson of the NF1 Mid-Atlantic Chapter, Hampden, MD, for their support of and contributions to this project.

the presence of an additional, normal X chromosome.

Developmental delay is detected early in infants with the FRAXA mutation (Bailey, Hatton, & Skinner, 1998). Daily living skills initially appear significantly better than socialization and communication skills, but those differences diminish by mid-adolescence (Dykens, Hodapp, Ort, & Leckman, 1993). Children with FXS are more withdrawn socially compared with age-matched Down syndrome controls (Cohen et al., 1988).

Children with FRAXA often exhibit autism-like features, such as gaze aversion, hand flapping, and body rocking, and often manifest self-injurious behavior. An association between FRAXA and autism in males with the FRAXA full mutation was first noted by Brown et al. (1982), but later it was reported to be confounded with ID (Fisch, 1992). However, the association between autism and FRAXA remains contentious.

Early reports of declines in IQs were reported by Hagerman, Smith, and Mariner (1983). Other researchers also noted declines (e.g., Borghgraef, Fryns, Dielkens, Pyck, & Van den Berghe, 1987; Lachiewicz, Gullion, Spiridigliozzi, & Aylsworth, 1987). In their prospective, longitudinal study, Fisch and colleagues examined FRAXA children and adolescents and found that IQs declined in nearly all males (Fisch et al., 1996) and most females (Fisch, Carpenter, et al., 1999) they examined. These results are compared here with those obtained from children with Williams Syndrome or NF1.

WILLIAMS SYNDROME

Williams syndrome is a microdeletion disorder produced by a ~1.5-Mb loss of genomic material on chromosome 7q11.23 (see Chapter 6 of this book). About 25 genes have been identified in the deleted region, the haploinsufficiency of which affects cardiac, renal, and pulmonary functions. Individuals with Williams syndrome also manifest unusual craniofacial features: an "elf-like" appearance associated with a small, upturned nose,

everted ears, apple cheeks, stellate irises, wide lips, and a narrow philtrim. Neurological dysfunction is manifested by hypotonia and reduced cerebellum, among other features.

Intellectual function as reflected by IQ ranges from mild to moderate ID (Greer, Brown, Pai, Choudry, & Klein, 1997). The profile of cognitive deficits is typified by strengths in verbal abilities, although language acquisition (vocabulary and grammar) is usually delayed (Mervis & Robinson, 2000). Strengths in quantitative abilities have also been noted (Fisch et al., 2001), but weaknesses in short-term memory were observed as well. Fisch et al. also found that overall adaptive behavior was delayed in children with Williams syndrome, with particular weakness in daily living skills, as discussed here.

Establishing genotype-phenotype relationships in children and adolescents with Williams syndrome has been problematic. Frangiskakis et al. (1996) found that the LIM-kinase gene was deleted in children with Williams syndrome who exhibited visual-spatial problems. However, Tassabehji et al. (1999) and Wang et al. (1999) were unable to confirm these findings.

Attempts at establishing a trajectory of cognitive-behavioral development have proved equally vexing. Early reports of test-retest IQs showed stability over time, but Gosch and Pankau (1996) noted that, although Draw-a-Person tests were stable, IQ scores on the Columbia Mental Maturity Scale (Burgemeister, Blum, & Lorge, 1972) decreased significantly after 2 years.

Difficulties in visual-spatial processing affect various other psychomotor functions, such as drawing or copying block designs (Bellugi, Wang, & Jernigan, 1994), as well as activities involving visuomotor coordination (Withers, 1996). Some visual-spatial problems have been attributable to visual difficulties (Winters, Pankau, Amm, Gosch, & Wessel, 1996) and the relatively high frequency of strabismus found in Williams syndrome (Kapp, von Noorden, & Jenkins, 1995). However, Atkinson and colleagues (2001) compared Williams syndrome individuals with and without visual stereo deficits to determine whether these affect visual-spatial ability and found no significant differences between these two groups.

NEUROFIBROMATOSIS TYPE 1

NF1 is the most common autosomal single-gene disorder, the gene for which was identified at the locus on 17q11 by Wallace and colleagues (1990) and Viskochil and colleagues (1990). LD has been observed in 40%–60% of individuals with NF1, and ID is found in another 5%–8% (North et al., 1997). Neurofibromin, the protein produced by the NF1 gene, has been implicated in learning and memory (Brambilla et al., 1997) and may be the mechanism for learning deficits in NF1 (Costa et al., 2002).

LD occurs more frequently in children with NF1 than in their siblings (Stine and Adams, 1989). The most common problems they observed in children with NF1 were visual-perceptual-motor delay and spelling and arithmetic disabilities, similar to those found in children with attention deficit disorder.

Studies of features of cognitive-behavioral development correlated with age in children and adolescents with NF1 are few in number. Riccardi and Eichner (1986) report an increase in cognitive function from childhood to adulthood. However, Ferner, Hughes, and Weinman (1996) found no significant differences in IQ between adults and children with NF1. Hyman et al. (2003) noted that although children with NF1 had significantly lower IQ scores than their sibling controls, the two groups showed comparable increases in IQ scores after retesting 8 years later.

Maladaptive behavior has been observed in children and adolescents with NF1. Using the Child Behavior Checklist (CBCL; Achenbach & Edelbrock, 1991), Johnson, Saal, Lovell, and Schorry (1999) examined children with NF1 and their unaffected siblings and found that children with NF1 had significantly more emotional problems than unaffected siblings. Barton and North (2004)

confirmed these findings but also found that social problems were associated with attention deficits in NF1.

OUR MULTICENTER STUDY

Method

Subjects

We enrolled 108 children and adolescents, 4–15 years of age, who were diagnosed with the FRAXA full mutation, Williams syndrome, or NF1. Forty-four children were diagnosed with the FRAXA mutation, 34 were diagnosed with Williams syndrome, and 30 were diagnosed with NF1. In this report, all subjects with FRAXA were males. Of those with Williams syndrome, 15 were males and 19 were females. Of those with NF1, 21 were males and 9 were females. Subjects were recruited from and tested at one of eight sites in the United States and Canada and were participants in our longitudinal study of children and adolescents with a known genetic abnormality. They were tested for the first time between 1991 and 2004. As part of our longitudinal study, 2 years after their initial evaluation families of participants were contacted again for a follow-up assessment (retest).

Materials

Cognitive-Behavioral Measures

Cognitive abilities were obtained with the *Stanford-Binet Intelligence Scale, Fourth Edition* (SBFE; Thorndike, Hagen, & Sattler, 1986). The SBFE is a well-standardized instrument, with excellent reliability and validity. The SBFE was used because it can be used with both children and adolescents, from ages 2 to 23 years. The SBFE is composed of four major cognitive areas: Verbal Reasoning, Quantitative Reasoning, Abstract/Visual Reasoning, and Short-term Memory. Standard

Area Scores (SAS) for these cognitive areas are used to develop a profile of an individual's strengths and weaknesses. In addition to cognitive abilities, adaptive behavior was also assessed, with the Vineland Adaptive Behavior Scales (VABS; Sparrow, Balla, & Cicchetti, 1984). Like the SBFE, the VABS is a reliable, well-standardized instrument that consists of three primary domains: Communication, Socialization, and Daily Living Skills. There are two additional scales that can be used to identify maladaptive behavior.

DNA and Clinical Testing

All participants had been diagnosed previously by molecular DNA techniques to determine FRAXA status or Williams syndrome status. FRAXA full-mutation status was established by standard Southern blotting techniques (see Fisch et al., 1996). Williams syndrome status was determined by the ELN probe. NF1 status was ascertained clinically, according to the standard clinical criteria published by Stumpf et al. (1988).

Procedure

All participants were individually administered the SBFE at home or in a clinical setting. Parents and/or caregivers of the participants were interviewed afterward with the VABS. Test scores were tallied by this author, after which they were entered into an Excel database. Statistical results were obtained from SAS 9.1 statistical computing software.

Results

Summary descriptive statistics for children assessed can be found in Table 7.1. The range and mean ages for children tested were approximately the same; the greatest difference observed was between children with FRAXA and children with Williams syndrome. Mean IQ scores of children with FRAXA and Williams syndrome were not significantly different from one another. On the other hand, mean IQs for children with NF1 were significantly higher. Mean adaptive behavior scores (or developmental quotient [DQ]) of children with FRAXA and Williams syndrome were comparable, and significantly lower than the mean DQ for children with NF1.

Age Correlated with Cognitive Abilities

Composite IQ scores related to age are presented in Figures 7.1a–c. There is a significant negative correlation of IQ with age in children and adolescents with FRAXA, as

Table 7.1. Summary descriptive statistics for children and adolescents assessed

Genetic disorder	Males: females	Mean age at first test (*SD*)	Mean initial IQ (*SD*)	Mean initial DQ (*SD*)
FRAXA (*N* = 44)	44:0	7.9 (3.0)	51.3 (11.1)	60.7 (13.4)
WBS (*N* = 34)	15:19	9.4 (3.7)	52.1 (10.7)*	57.3 (12.9)
NF1 (*N* = 30)	21:9	8.2 (3.2)	88.3 (13.2)	81.9 (21.0)

From Fisch, G.S., Carpenter, N., Howard-Peebles, P.N., Holden, J.J.A., Tarleton, J., Simensen, R., et al. (2007). Studies of age-correlated features of cognitive-behavioral development in children and adolescents with genetic disorders. *American Journal of Medical Genetics: Part A, 143,* 2481; reprinted by permission.

Key: SD, standard deviation; IQ, intelligence quotient; DQ, developmental quotient/mean adaptive behavior scores; FRAXA, fragile X mutation; WBS, Williams-Beuren syndrome [Williams syndrome]; NF1, neurofibromatosis type 1; *N,* total number in sample; *One child with Williams syndrome had a DQ score only. Therefore, mean IQ for this group is for *N* = 33.

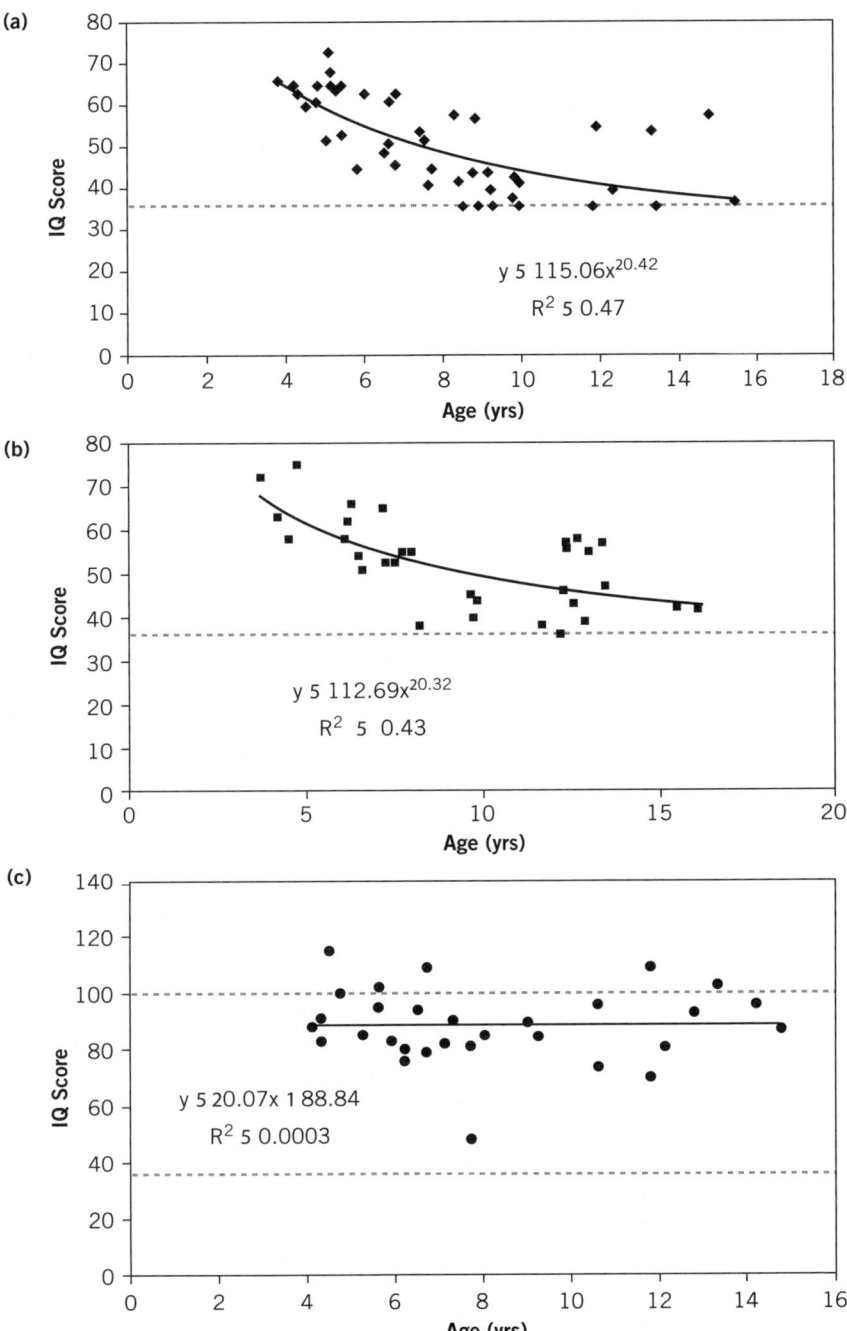

Figure 7.1.   a) Age-correlated IQ scores for children with fragile X mutation (FRAXA). (The dashed line denotes the floor value for the test.) b) Age-correlated intelligence quotient (IQ) scores for children with Williams syndrome. (The dashed line denotes the floor value for the test.) c) Age-correlated IQ scores for children with neurofibromatosis type 1 (NF1). (The lower dashed line denotes the floor value for the test.) (From Fisch, G.S., Carpenter, N., Howard-Peebles, P.N., Holden, J.J.A., Tarleton, J., Simensen, R., et al. [2007]. Studies of age-correlated features of cognitive-behavioral development in children and adolescents with genetic disorders. *American Journal of Medical Genetics: Part A, 143,* 2483; reprinted by permission.)

had been noted previously (Fisch et al., 1996). There is a floor value for the IQ test administered (36), and many older children tested at or near the floor value. As a result, we chose to represent the decline by using an exponential nonlinear regression line. For FRAXA males, the Pearson correlation coefficient (square root of the coefficient of determination) shows a strong, significant inverse relationship between composite IQ and age ($r = -0.68$; $p < .01$), indicating that 47% of the variance in IQ was due to age (see Figure 7.1a). Likewise, among children and adolescents with Williams syndrome, IQ scores are inversely and significantly correlated to age ($r = -0.65$; $p < .01$), indicating that 43% of the variance in IQ score was due to age (see Figure 7.1b). The inverse relationship between age and IQ in Williams syndrome is about the same as that which was observed among same-age males with FRAXA. On the other hand, among children and adolescents with NF1, there is no relationship between composite IQ score and age ($r = -.01$; $p = $ NS; see Figure 7.1c).

Age Correlated with Adaptive Behavior

Adaptive behavior composite scores (DQs) related to age are shown in Figures 7.2a–c. As we also found previously, there is a significant negative relationship between age and DQ ($r = -.43$; $p < .01$) among children and adolescents with FRAXA. In a like fashion, there is a significant negative correlation between age and DQ ($r = -.65$; $p < .01$) among children and adolescents with Williams syndrome. As with IQ scores, children and adolescents with NF1 show no significant relationship between age and DQ ($r = -.18$; $p = $ NS).

Age Correlated with Maladaptive Behavior

We examined maladaptive behavior with Scale I from the VABS Maladaptive Behavior Domain. Maladaptive behavior related to age is presented in Figures 7.3a–c. Dashed lines in each figure represent the boundary between acceptable and unacceptable levels of maladaptive behavior, according to the VABS technical manual. As Figure 7.3a shows, of the 24 children with FRAXA for whom data were available, 16 (67%) demonstrated unacceptably high levels of maladaptive behavior; and, although levels of maladaptive behavior decline with age, the relationship is not statistically significant ($r = -0.39$; $p > .05$). Maladaptive scores from 10 of 30 (33%) children and adolescents with Williams syndrome (Figure 7.3b) exhibit unusually high levels of maladaptive behavior. Unlike FRAXA, however, there appears to be no relationship between maladaptive behavior and age ($r = .11$, $p = $ NS). Children with NF1 show a modestly increasing but nonsignificant relationship between maladaptive behavior and age in ($r = .25$, $p = $ NS) (see Figure 7.3c). Ten of 30 children with NF1 (33%) also show unusually high levels of maladaptive behavior. The proportion of children with FRAXA with unusually high maladaptive behavior scores is significantly greater than the proportion of children with NF1 ($\chi^2 = 5.30$; $p < .02$). The proportion of children with FRAXA with unusually high maladaptive behavior scores is also significantly greater that those with Williams syndrome ($\chi^2 = 5.30$; $p < .02$). There was no difference in the proportion of children with unusually high maladaptive scores between those with Williams syndrome and those with NF1 ($\chi^2 = 1.0$; $p = 1.0$).

Cognitive Profiles

Cognitive and adaptive behavior profiles in each of the three groups were examined (see Figures 7.4 and 7.5). Interestingly, roughly similar cognitive profiles were noted in all three groups. That is, all three groups appear to exhibit relative cognitive strengths in Verbal Reasoning and Quantitative Reasoning, and relative weaknesses in Visual/Spatial Reasoning and Short-term Memory. Using a mixed effects model

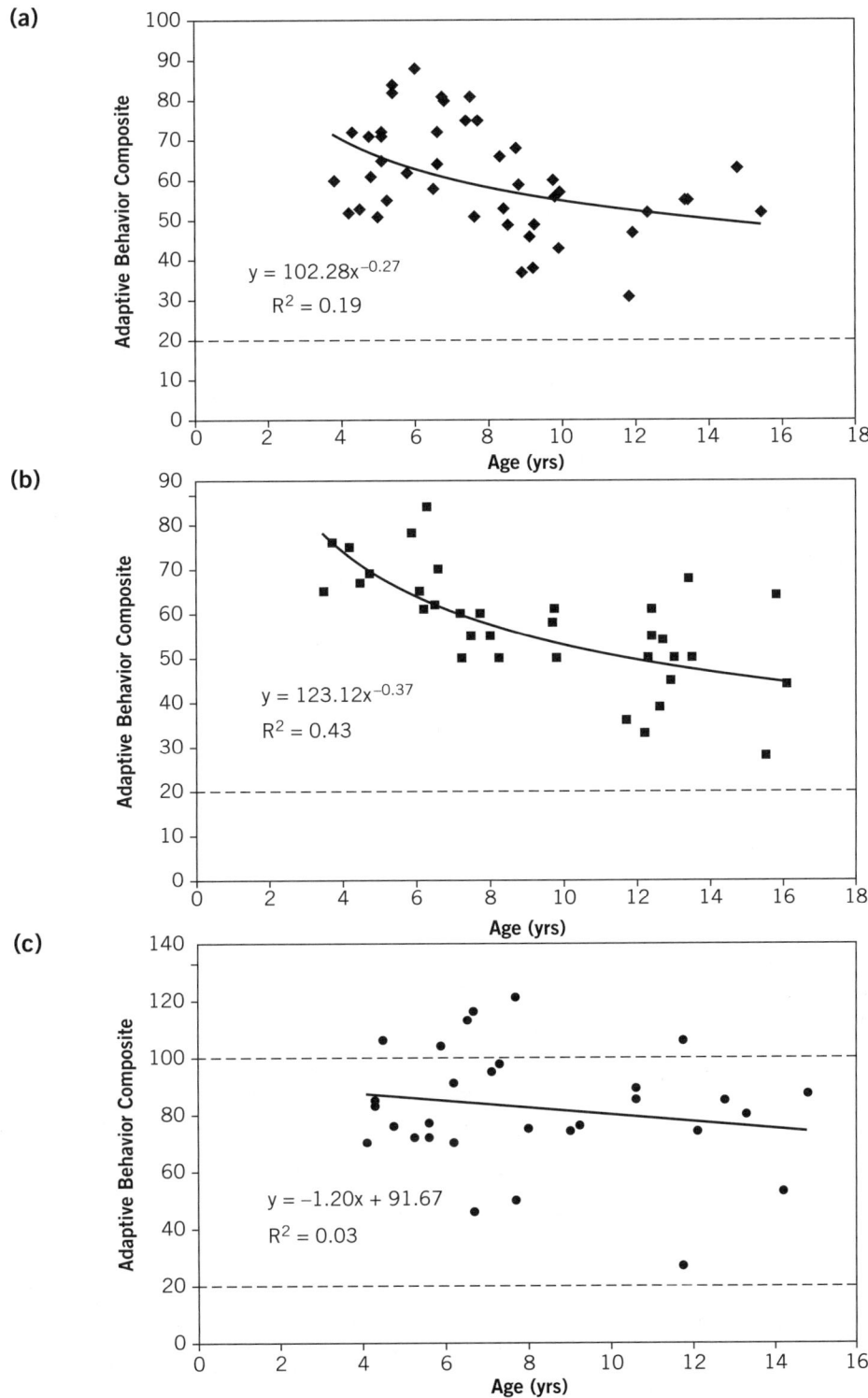

Figure 7.2. a) Age-correlated adaptive behavior scores (or developmental quotient [DQ]) scores for children with fragile X mutation (FRAXA). (The dashed line denotes the floor value for the test.) b) Age-correlated DQ scores for children with Williams syndrome. (The dashed line denotes the floor value for the test.) c) Age-correlated DQ scores for children with neurofibromatosis type 1 (NF1). (The lower dashed line denotes the floor value for the test.) (From Fisch, G.S., Carpenter, N., Howard-Peebles, P.N., Holden, J.J.A., Tarleton, J., Simensen, R., et al. [2007]. Studies of age-correlated features of cognitive-behavioral development in children and adolescents with genetic disorders. *American Journal of Medical Genetics: Part A, 143,* 2484; reprinted by permission.)

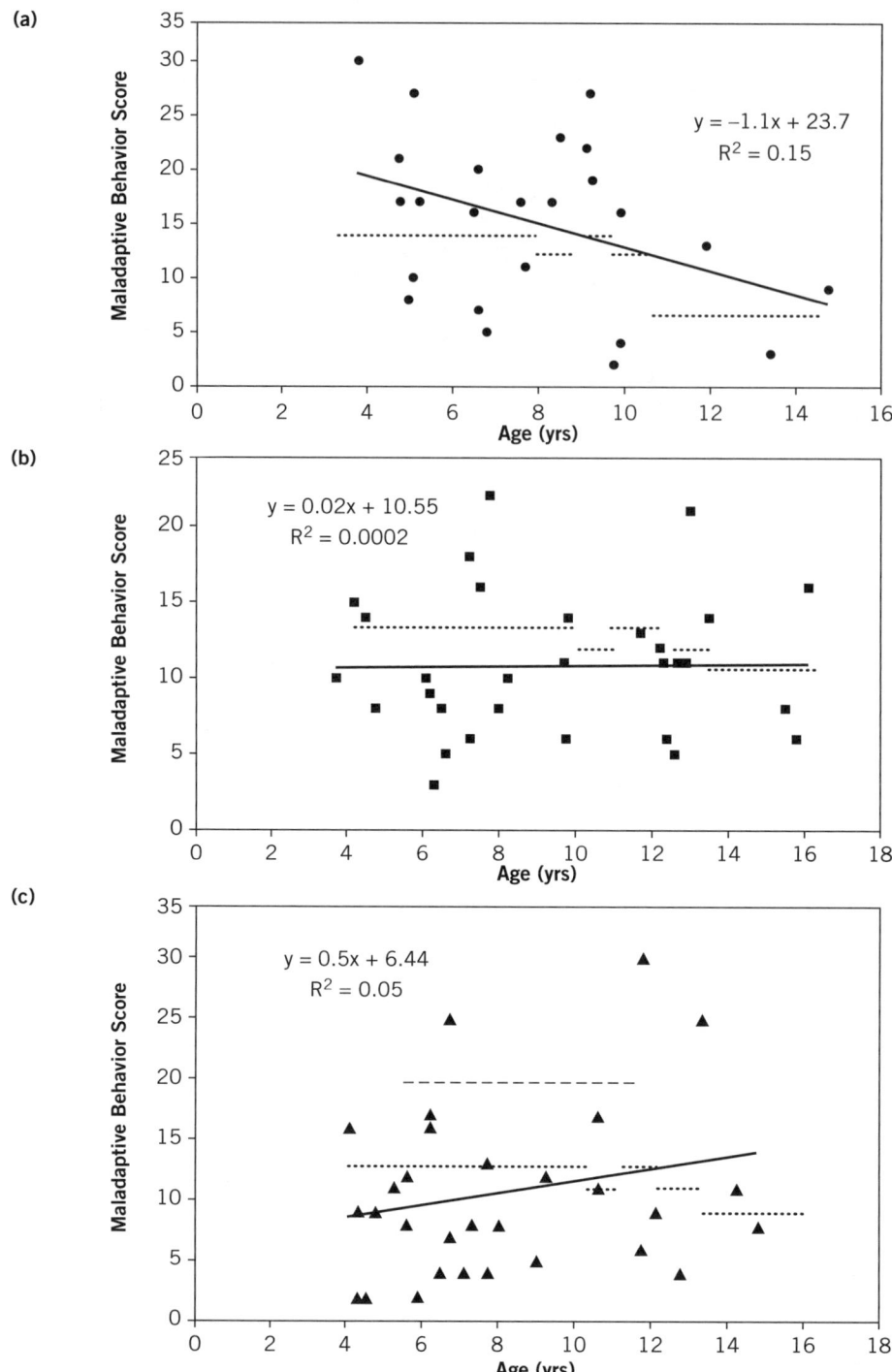

Figure 7.3. a) Age correlated with maladaptive behavior scores for male children and adolescents with fragile X mutation (FRAXA) (total number in sample [*N*] = 24). The horizontal dashed lines represent the upper boundaries between acceptable and unacceptable levels of maladaptive behavior. b) Age correlated with maladaptive behavior scores for children and adolescents with Williams syndrome (*N* = 30). The horizontal dashed lines represent the upper boundaries between acceptable and unacceptable levels of maladaptive behavior. c) Age correlated with maladaptive behavior scores for children and adolescents with neurofibromatosis type 1 (NF1) (*N* = 30). The horizontal dashed lines represent the upper boundaries between acceptable and unacceptable levels of maladaptive behavior. (From Fisch, G.S., Carpenter, N., Howard-Peebles, P.N., Holden, J.J.A., Tarleton, J., Simensen, R., et al. [2007]. Studies of age-correlated features of cognitive-behavioral development in children and adolescents with genetic disorders. *American Journal of Medical Genetics: Part A, 143,* 2484; reprinted by permission.)

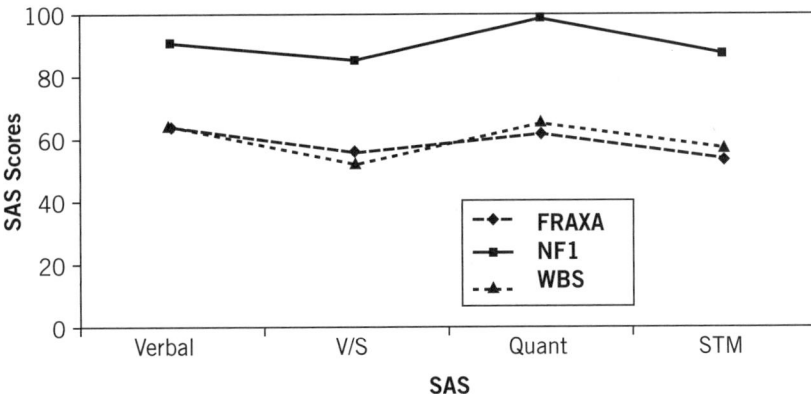

Figure 7.4. Cognitive profiles of Standard Area Scores (SAS) from *Stanford-Binet Intelligence Scale, Fourth Edition* (Thorndike, Hagen, & Sattler, 1986) for children and adolescents with fragile X mutation (FRAXA), Williams syndrome (Williams-Beuren Syndrome [WBS]), or neurofibromatosis type 1 (NF1). (*Key:* V/S, Visual-Spatial Processing; Quant, Quantitative Reasoning; STM, Short-Term Memory.) (From Fisch, G.S., Carpenter, N., Howard-Peebles, P.N., Holden, J.J.A., Tarleton, J., Simensen, R., et al. [2007]. Studies of age-correlated features of cognitive-behavioral development in children and adolescents with genetic disorders. *American Journal of Medical Genetics: Part A, 143,* 2485; adapted by permission.)

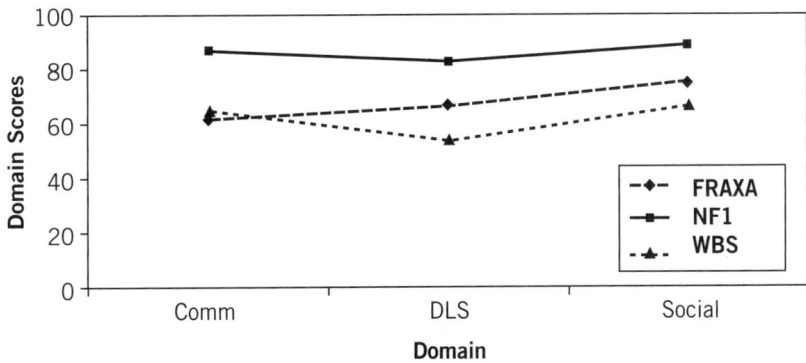

Figure 7.5. Adaptive behavior profiles of domain scores from the Vineland Adaptive Behavior Scales (VABS; Sparrow, Balla, & Cicchetti, 1984) for children and adolescents with fragile X mutation (FRAXA), Williams syndrome (Williams-Beuren Syndrome [WBS]), or neurofibromatosis type 1 (NF1). (*Key:* Comm, Communication; DLS, Daily Living Skills; Social, Socialization.) (From Fisch, G.S., Carpenter, N., Howard-Peebles, P.N., Holden, J.J.A., Tarleton, J., Simensen, R., et al. [2007]. Studies of age-correlated features of cognitive-behavioral development in children and adolescents with genetic disorders. *American Journal of Medical Genetics: Part A, 143,* 2485; adapted by permission.)

(SAS 9.1), we found statistically significant differences among the three genetic disorders ($F = 302.8$; $p < .0001$) and across SAS area scores (17.98; $p < .0001$). Using a one-way analysis of variance (ANOVA) procedure to examine each genetic disorder, we found statistically significant differences among SASs for children with FRAXA ($F = 7.97$; $p < .001$). Post hoc Bonferroni-adjusted tests found there were significant differences between Verbal Reasoning scores and Visual/Spatial scores, and Verbal Reasoning and Short-term Memory scores. There were also significant differences between Quantitative Reasoning scores and Visual/Spatial scores, and Quantitative Reasoning and Short-term Memory scores. Among children with Williams syndrome, there were also statistically significant differences among the four SAS area scores ($F = 9.26$; $p < .0001$). Although the cognitive profiles of children with Williams syndrome are similar to those children with FRAXA, statistically significant differences were noted between Verbal Reasoning and Visual/Spatial Reasoning, but not between Verbal Reasoning and Short-term Memory. However, there were significant differences between Quantitative Reasoning and both Visual/Spatial Reasoning and Short-term Memory. Among children with NF1, there were also statistically significant differences among the four SASs ($F = 5.05$; $p < .003$). No significant differences were noted between Verbal Reasoning and the other three SAS scores. However, significant differences were noted between Quantitative Reasoning and both Visual/Spatial Reasoning and Short-term Memory.

Adaptive Behavior Profiles

Unlike cognitive ability, adaptive behavior profiles among the three groups were not similar to one another. Using a mixed effects model, we found statistically significant differences among the three genetic disorders ($F = 4.56$; $p < 0.01$) and across Domain scores (10.14; $p < 0.0001$). Univariate ANOVA found significant differences among the three adaptive domain scores for children with FRAXA ($F = 6.92$; $p < 0.002$). Post hoc Bonferroni-adjusted tests found mean scores in the Socialization Domain significantly higher than either Daily Living Skills or Communication Domain scores. Among children with Williams syndrome, there were significant differences in Domain scores also ($F = 4.87$; $p < .01$). Post hoc Bonferroni-adjusted tests show that Socialization and Communication skills were significantly higher than Daily Living Skills.

There were also significant differences among adaptive domain scores in children with NF1 ($F = 3.96$; $p < 0.02$). Post hoc Bonferroni-adjusted tests found Daily Living Skills to be significantly lower than either Communication Domain or Socialization Domain.

DISCUSSION

Retrospective cross-sectional studies previously reported a negative correlation between age and IQ in males with the FRAXA full mutation (e.g., Fisch et al., 1991; Sutherland and Hecht, 1985). More recently, we found that children and adolescents with the FRAXA full mutation exhibited longitudinal declines in retest IQ and DQ in nearly all males (Fisch et al., 1996) and most females (Fisch, Carpenter, et al., 1999). In this cross-sectional phase of our study, we found age to be negatively correlated with IQ and DQ in children and adolescents with Williams syndrome, which is comparable to what we observed in children with FRAXA in the same age range and similar to the longitudinal results obtained by Gosch and Pankau (1996). These results, however, are at variance with those obtained by Udwin et al. (1996), who examined test-retest IQs of 23 young adults with Williams syndrome. Udwin et al. initially tested children with the Wechsler Intelligence Scale for Children–Revised (WISC-R; Wechsler, 1974), then retested the young adults 8 years later with the Wechsler Adult Intelligence Scale–Revised (WAIS-R; Wechsler, 1981). They found that Full Scale IQ, Verbal IQ, and Performance IQ all increased significantly during that interval. However, testing with the WISC-R and retesting with the WAIS-R results in markedly increased IQs, especially among individuals with ID (Spitz, 1988; Vance, Brown, Hankins, & Furgerson, 1987).

We should note again that the data presented here concerned IQ and DQ cross-sectional analyses between subjects and within genetic groups. Preliminary results from our test-retest within-subject studies have revealed that although longitudinal declines in IQ and DQ are statistically significant among children with FRAXA, declines in IQ and DQ observed in children with Williams syndrome are not statistically significant; nor are there any statistically significant longitudinal changes in IQ and DQ in children with NF1 (data not shown). For children with Williams syndrome, the lack of statistical significance may likely stem from the relatively small test-retest sample size ($N = 17$). However, our longitudinal results should also impart a cautionary note regarding attempts to draw prospective inferences from cross-sectional data.

Despite the similarity in the negative correlation between age and IQ or DQ in FRAXA and Williams syndrome, their phenotypes are markedly different. Shyness and gaze aversion are common behaviors in children and adolescents with FRAXA (Wolff, Gardner, Paccla, & Lappen, 1989), whereas children with Williams syndrome are quite outgoing and sociable (Morris & Mervis, 2000). Children with FRAXA have severely limited speech and expressive language (Fisch, Holden, et al., 1999). On the other hand, children with Williams syndrome are noted for their expressive verbal skills, which are superior to their nonverbal abilities, particularly among those with higher IQs (Jarrold, Baddeley, Hewes, & Phillips, 2001). We also found strengths in Verbal Reasoning and in the Communication Domain, and weakness in Visual/Spatial Reasoning (cf. Figures 7.4 and 7.5). Interestingly, and despite differences in neurobiological and neurophysiological development produced by their respective genetic abnormalities, declines in IQs have been observed in children with Down syndrome (Carr, 1992; Wishart, 1993).

Inverse correlations of age with IQ or DQ are not necessarily associated with decline in cognitive function, as we and others have noted. Dykens et al. (1989) reported that although age-equivalent cognitive ability was directly related to increasing chronological age in children with FRAXA, the increases gradually slowed, forming a "plateau." Other researchers have also found that the developmental trajectories among young males with FRAXA were lower than those of typically developing children (Bailey et al., 1998, 2002). That is, with regard to cognitive ability, as children with FRAXA age, they fall farther and farther behind typically developing children in the same age range. From a mathematical perspective, age equivalence and IQ (or DQ) are inversely related. Thus, declining IQ and DQ scores that bottom out represent the mirror image of age equivalences that plateau. The plateau effect of age equivalence may simply be a "ceiling effect," the opposite feature of the "floor effect" found in declining IQ and DQ, rather than any specific neurocognitive factor in development.

In addition to overall cognitive-behavioral abilities, Atkinson et al. (2001) found plateaus in both visuospatial and verbal ability in children with Williams syndrome and that verbal skills improve more rapidly than visuospatial ability. Jarrold et al. (2001) also found that adolescents and young adults exhibited significantly higher age-equivalence verbal scores compared with their nonverbal scores, but that preschoolers performed equally well on verbal and nonverbal tasks administered from the Differential Ability Scales (Elliott, 1990). Jarrold et al. (2001) suggested that neurodevelopmental trajectories for verbal and visuospatial skills differ in both rate of development (slope) and initial level of ability (intercept). That is, the rate of development of verbal abilities is greater than that of visuospatial skills; and, at some point between childhood and adolescence, their trajectories intersect, after which verbal skills significantly exceed nonverbal skills.

If we compare the rate of cognitive development (IQ) with that of adaptive behavior (DQ) in our study, we find that, among children with Williams syndrome, the regression functions for composite IQ and DQ, as related to age, exhibit similar slopes. Thus, overall cognitive and adaptive behaviors develop at about the same rate, unlike the specific trajectories of verbal and visuospatial ability.

Earlier, Greer et al. (1997) also employed the SBFE and VABS in their study of children with Williams syndrome, whose ages were similar to those in our sample. Like Jarrold et al. (2001), they found no significant differences in mean SASs in Verbal Reasoning compared with Visual/Spatial Reasoning. Their results support the hypothesis that the developmental trajectory of verbal skills is different from that of visual/spatial skills in late childhood or early adolescence.

As for maladaptive behavior, the proportion of children with FRAXA with unusually high maladaptive behavior scores is significantly greater than among children with Williams syndrome or NF1. Greer et al. (1997) examined maladaptive behavior in children with Williams syndrome with the CBCL. *T* scores from the CBCL indicate that about two-thirds of their sample exhibited significantly high levels of maladaptive behavior. Using the Developmental Behaviour Checklist (Einfeld & Tonge, 1992), Einfeld, Tonge, and Florio (1997) also found similar levels of maladaptive behavior in children with Williams syndrome. Our study showed a significantly lower proportion than either Greer et al. or Einfeld et al., which may have been produced by the different instruments used.

Unlike children with FRAXA or Williams syndrome, we found that IQs and DQs among children with NF1 show no age-related changes. Hyman et al. (2003) examined test-retest IQs in children and adolescents with NF1 and controls, and found a small increase in IQs in both groups. However, both groups were tested initially with the WISC-R, after which they were retested with the WAIS-III. As we noted earlier, retesting with the WAIS typically results in an artificial increase in IQ (Spitz, 1988; Vance et al., 1987). Therefore, the modest increase observed by Hyman et al. may have been an artifact resulting from the change in IQ tests or, possibly, the consequence of the so-called Flynn effect.

Negative correlations of age with IQ or DQ found in children and adolescents with certain genetic abnormalities (e.g., Down syndrome, velocardiofacial syndrome, FRAXA, or Williams syndrome) but not in others (e.g., Prader-Willi syndrome or NF1) requires some explanation. One possibility is that there may be a threshold of impairment associated with a criterion IQ or DQ, above which cognitive-behavioral skills develop at the same rate as those of other individuals in the general population, whereas children and adolescents who are more severely affected probably sustain more extensive or critically important neurodevelopmental damage and acquire new behaviors at a slower rate than other children their age. Unlike the lack of change in test-retest scores we observed in children with NF1, however, we found longitudinal declines in IQs in most females with the FRAXA full mutation (Fisch, Carpenter, et al., 1999) whose IQs also fell in the same borderline-deficit to low normal range. This issue may be resolved by neuroimaging studies.

Declining IQ or DQ and differences in cognitive-behavioral profiles in children with FRAXA and Williams syndrome have several implications for researchers and parents. Not unexpectedly for researchers, our results suggest that neurobiological processes associated with cognitive functions and ID likely vary from one genetic disorder to another. Furthermore, genetic abnormalities that produce declines in IQs and/or DQs will likely have a different pathophysiology from those genetic disorders that do not. Previously, Thompson and Kim (1996) noted that certain features of memory, such as declarative or explicit memory, are typically stored in the medial-temporal lobe of the hippocampus, whereas basic associative processes are also stored in the amygdala and cerebellum. Slower rates of cognitive development that present phenotypically as decline in IQ may be associated with subtle differences within specific structures (e.g., hippocampus and amygdala) that would, in turn, produce different cognitive-behavioral profiles as well as decreases in IQ and DQ. Animal models have shown that disturbances in mineralocorticoid and glucocorticoid processes affect the limbic system, and loss of limbic mineralocorticoid produces distinct cognitive deficits (Berger et al., 2006). Therefore, neurobiological and neurophysiological features of genetic disorders that produce ID need to be examined in greater detail.

Planning educational and behavioral programs based upon what is known about rates of cognitive-behavioral development for these disorders will likely be essential. That is, children with FRAXA and Williams syndrome will not necessarily develop cognitively as rapidly as other children with ID (cf. Dykens et al., 1989, 1993; Fisch et al, 1996). Therefore, parents, caregivers, teachers, and others involved with their care and management need to be alerted to the extent to which these children are able to acquire new behaviors.

CONCLUSIONS

We examined 108 children, ages 4–15 years, with one of three genetic disorders: FRAXA, Williams syndrome, or NF1. Results of our cross-sectional analysis show significant negative correlations between age and IQ, and age and DQ, in children with FRAXA and Williams syndrome, but not in children with NF1. Cognitive and adaptive behavior profiles also note similar patterns—but not levels—of strengths and weaknesses in all three groups of genetic disorders. All three groups of children have unusually high proportions of maladaptive behavior, ranging from 17% of children with NF1 to 67% of children with FRAXA. Preliminary data from our longitudinal study indicate that there are significant declines in IQ and DQ in children with FRAXA, but declines in IQ and DQ in children with Williams syndrome were not statistically significant. These results also show the continued need to examine age-

related cognitive-behavioral changes in children and adolescents with all genetic disorders that produce LD or ID.

REFERENCES

Achenbach, T.M., & Edelbrock C. (1991). *The Child Behavior Checklist and Revised Child Behavior Profile.* Burlington, VT: University Associates in Psychiatry.

Atkinson, J., Anker, J., Braddick, O., Nokes, L., Mason, A., & Braddick, F. (2001). Visual and visuospatial development in young children with Williams syndrome. *Developmental Medicine and Child Neurology, 43,* 330–337.

Bailey, D.B., Jr., Hatton, D.D., & Skinner, M. (1998). Early developmental trajectories of males with fragile X syndrome. *American Journal of Mental Retardation, 103,* 29–39.

Bailey, D.B., Jr., Hatton, D.D., Skinner M., & Mesibov, G. (2002). Autistic behavior, FMR1 protein, and developmental trajectories in young males with fragile X syndrome. *Journal of Autism and Developmental Disorders, 31,* 165–174.Barton, B., & North, K. (2004). Social skills of children with neurofibromatosis type 1. *Developmental Medicine and Child Neurology, 46,* 553–563.

Bellugi, U., Wang, P.P., & Jernigan, T.L. (1994). Williams syndrome: An unusual neuropsychological profile. In S.H. Broman & J. Grafman (Eds.), *Atypical cognitive deficits in developmental disorders: Implication for brain function* (pp. 23–56). Hillsdale, NJ: Lawrence Erlbaum Associates.

Berger, S., Wolfer, D.P., Selbach, O., Alter, H., Erdmann, G., Reichardt, H.M., et al. (2006). Loss of the limbic mineralocorticoid receptor impairs behavioral plasticity. *Proceedings of the National Academy of Sciences USA, 103,* 195–200.

Borghgraef, M., Fryns, J.P., Dielkens, A., Pyck, K., & Van den Berghe, H. (1987). Fragile (X) syndrome: A study of the psychological profile in 23 prepubertal patients. *Clinical Genetics, 32,* 179–186.

Brambilla, R., Gnesutta, N., Minichiello, L., White, G., Roylance, A.J., Herron, C.E., et al. (1997). A role for the Ras signalling pathway in synaptic transmission and long-term memory. *Nature, 390,* 281–286.

Brown, W.T., Jenkins, E.C., Friedman, E., Brooks, J., Wisniewski, K., Raguthu, S., et al. (1982). Autism is associated with the fragile-X syndrome. *Journal of Autism and Developmental Disorders, 12,* 303–308.

Burgemeister, B., Blum, L., & Lorge, J. (1972). *Columbia Mental Maturity Scale.* New York: Harcourt, Brace, Jovanovich.

Carr, J. (1992). Longitudinal research in Down syndrome. *International Review of Research in Mental Retardation, 18,* 197–223

Cohen, I.L., Fisch, G.S., Sudhalter, V., Wolf-Schein, E.G., Hanson, D., Hagerman, R., et al. (1988). Social gaze, social avoidance, and repetitive behavior in fragile X males: A controlled study. *American Journal of Mental Retardation, 92,* 436–446.

Costa, R.M., Federov, N.B., Kogan, J.H., Murphy, G.G., Stern, J., Ohno, M., et al. (2002). Mechanism for the learning deficits in a mouse model of neurofibromatosis type 1. *Nature, 415,* 526–530.

Dykens, E.M., Hodapp, R.M., Ort, S., Finucane, B., Shapiro, L.R., & Leckman, J.F. (1989). The trajectory of cognitive development in males with fragile X syndrome. *Journal of the American Academy of Child and Adolescent Psychiatry, 28,* 422–426.

Dykens, E.M., Hodapp, R.M., Ort, S.I., & Leckman, J.F. (1993). Trajectory of adaptive behavior in males with fragile X syndrome. *Journal of Autism and Developmental Disorders, 23,* 135–145.

Einfeld, S.L., & Tonge, B.J. (1992). *Manual for the Developmental Behaviour Checklist.* Clayton, Melbourne, and Sydney: Monash University Centre for Developmental Psychiatry and School of Psychiatry, University of New South Wales.

Einfeld, S.L., Tonge, B.J., & Florio, T. (1997). Behavioral and emotional disturbance in individuals with Williams syndrome. *American Journal of Mental Retardation, 102,* 45–53.

Elliott, C.D. (1990). *Differential Ability Scales.* San Antonio, TX: Psychological Corporation.

Ferner, R.E., Hughes, R.A.C., & Weinman, J. (1996). Intellectual impairment in neurofibromatosis 1. *Journal of Neurological Science, 138,* 25–33.

Fisch, G.S. (1992). Is autism associated with the fragile X syndrome? *American Journal of Medical Genetics, 43,* 47–55.

Fisch, G.S., Arinami, T., Froster-Iskenius, U., Fryns, J.P., Curfs, L.M., Borghgraef, M., et al. (1991). Relationship between age and IQ among fragile X males: A multicenter study. *American Journal of Medical Genetics, 38,* 481–487.

Fisch, G.S., Carpenter, N., Holden, J.J., Howard-Peebles, P.N., Maddalena, A., Borghgraef, M., et al. (1999). Longitudinal changes in cognitive and adaptive behavior in fragile X females: A prospective multicenter analysis. *American Journal of Medical Genetics, 83,* 308–312.

Fisch, G.S., Carpenter, N., Howard-Peebles, P.N., Holden, J.J.A., Tarleton, J., Simensen, R., et al. (2007). Studies of age-correlated features of cognitive-behavioral development in children and adolescents with genetic disorders. *American Journal of Medical Genetics: Part A, 143,* 2478–2489.

Fisch, G.S., Carpenter, N., Howard-Peebles, P.N., Tarleton, J., Holden, J.J.A., & Simensen, R.J. (2001). *Longitudinal changes in cognitive ability and adaptive behavior in children and adolescents with fragile X or Williams syndrome.* Presented at the 10th International Workshop on Fragile X and XLMR, September 19–22, Frascati, Italy.

Fisch, G.S., Holden, J.J., Carpenter, N.J., Howard-Peebles, P.N., Maddalena, A., Pandya, A., et al. (1999). Age-related language characteristics of children and adolescents with fragile X syndrome. *American Journal of Medical Genetics, 83,* 253–256.

Fisch, G.S., Simensen, R., Tarleton, J., Chalifoux, M., Holden, J.J., Carpenter, N., et al. (1996). Longitudinal study of cognitive abilities and adaptive behavior levels in fragile X males: A prospective multicenter analysis. *American Journal of Medical Genetics, 64,* 356–361.

Frangiskakis, J.M., Ewart, A.K., Morris, C.A., Mervis, C.B., Bertrand, J., Robinson, B.F., et al. (1996). LIM-kinase1 hemizygosity implicated in impaired visuospatial constructive cognition. *Cell, 86,* 59–69.

Gosch, A., & Pankau, R. (1996). Longitudinal study of the cognitive development in children with Williams-Beuren syndrome. *American Journal of Medical Genetics, 61,* 26–29.

Greer, M.K., Brown, F.R., III, Pai, G.S., Choudry, S.H., & Klein, A.J. (1997). Cognitive, adaptive, and behavioral characteristics of Williams syndrome. *American Journal of Medical Genetics (Neuropsychiatric Genetics), 74,* 521–525.

Hagerman, R., Smith, A.C.M., & Mariner, R. (1983). Clinical features of the fragile X syndrome. In R.J. Hagerman & P.M. McBogg (Eds.), *The fragile X syndrome: Diagnosis, biochemistry, intervention* (pp. 17–54). Dillon, CO: Spectra.

Hyman, S.L., Gill, D.S., Shores, E.A., Steinberg, A., Joy, P., Gibikote, S.V., et al. (2003). Natural history of cognitive deficits and their relationship to MRI T2-hyperintensities in NF1. *Neurology, 60,* 1139–1145.

Jarrold, C., Baddeley, A.D., Hewes, A.K., & Phillips, C. (2001). A longitudinal assessment of diverging verbal and non-verbal abilities in the Williams syndrome phenotype. *Cortex, 37,* 423–431.

Johnson, N.S., Saal, H.M., Lovell, A.M., & Schorry, E.K. (1999). Social and emotional problems in children with neurofibromatosis type 1: Evidence and proposed interventions. *Journal of Pediatrics, 134,* 767–772.

Kapp, M.E., von Noorden, G.K., & Jenkins, R. (1995). Strabismus in Williams syndrome. *American Journal of Ophthamology, 119*, 355–360.

Lachiewicz, A.M., Gullion, C.M., Spiridigliozzi, G.A., & Aylsworth, A.S. (1987). Declining IQs of young males with the fragile X syndrome. *American Journal of Mental Retardation, 92*, 272–278.

Lubs, H.A. (1969). A marker X chromosome. *American Journal of Human Genetics, 21*, 231–244.

Mervis, C.B., & Robinson, B.F. (2000). Expressive vocabulary ability of toddlers with Williams syndrome or Down syndrome: a comparison. *Developmental Neuropsychology, 17*, 111–126.

Morris, C.A., & Mervis, C.B. (2000). Williams syndrome and related disorders. *Annual Review of Genomics and Human Genetics, 1*, 461–484.

North, K.N., Riccardi, V., Samango-Sprouse, C., Ferner, R., Moore, B., Legius, E., et al. (1997). Cognitive function and academic performance in neurofibromatosis. 1: Consensus statement from the NF1 Cognitive Disorders Task Force. *Neurology, 48*, 1121–1127.

Opitz, J.M., & Sutherland, G.R. (1984). Conference report: International Workshop on the fragile X and X-linked mental retardation. *American Journal of Medical Genetics, 17*, 5–94.

Riccardi, V.M., & Eichner, J.E. (1986). *Neurofibromatosis: Phenotype, natural history and pathogenesis.* Baltimore: Johns Hopkins University Press.

Sarimski, K. (1997). Behavioural phenotypes and family stress in three mental retardation syndromes. *European Child and Adolescent Psychiatry, 6*, 26–31.

Sparrow, S.S., Balla, D.A., & Cicchetti, D.V. (1984). *Vineland Adaptive Behavior Scales.* Circle Pines, MN: American Guidance Service.

Spitz, H.H. (1988). Inverse relationship between the WISC-R/WAIS-R score disparity and IQ level in the lower range of intelligence. *American Journal of Mental Retardation, 92*, 376–378.

Stine, S.B., & Adams, W.V. (1989). Learning problems in neurofibromatosis patients. *Clinical Orthopaedics and Related Research, 245*, 43–48.

Stumpf, D.A., Alksne, J.F., Annegers, J.F., Brown, S.S., Conneally, P.M., Housman, D., et al. (1988). Neurofibromatosis: Conference statement. *Archives of Neurology, 45*, 575–578.

Sutherland, G.R., & Hecht, F. (1985). *Fragile sites on human chromosomes.* (pp. 113–131). New York: Oxford University Press.

Tassabehji, M., Metcalfe, K., Karmiloff-Smith, A., Carette, M.J., Grant, J., Dennis, N., et al. (1999). Williams syndrome: Use of chromosomal microdeletions as a tool to dissect cognitive and physical phenotypes. *American Journal of Human Genetics, 64*, 118–125.

Thompson, R.F., & Kim, J.J. (1996). Memory systems in the brain and localization of a memory. *Proceedings of the National Academy of Sciences USA, 93*, 13438–13444.

Thorndike, R.L., Hagen, E.P., & Sattler, J.M. (1986). *Stanford-Binet Intelligence Scale* (4th ed.). Chicago: Riverside.

Udwin, O., Davies, M., & Howlin, P. (1996). A longitudinal study of cognitive abilities and educational attainment in Williams syndrome. *Developmental Medicine and Child Neurology, 38*, 1020–1029.

Vance, H.R., Brown, W., Hankins, N., & Furgerson, S.C. (1987). A comparison of the WISC-R and the WAIS-R with special education students. *Journal of Clinical Psychology, 43*, 377–380.

Viskochil, D., Buchberg, A.M., Xu, G., Cawthon, R.M., Stevens, J., Wolff, R.K., et al. (1990). Deletions and a translocation interrupt a cloned gene at the neurofibromatosis type 1 locus. *Cell, 62*, 187–192.

Wallace, M.R., Marchuk, D.A., Andersen, L.B., Letcher, R., Odeh, H.M., Saulino, A.M., et al. (1990). Type 1 neurofibromatosis gene: identification of a large transcript disrupted in three NF1 patients. *Science, 249*, 181–186.

Wang, M.S., Schinzel, A., Kotzot, D., Balmer, D., Casey, R., Chodirker, B.N., et al. (1999). Molecular and clinical correlation study of Williams-Beuren syndrome: No evidence of molecular factors in the deletion region or imprinting affecting clinical outcome. *American Journal of Medical Genetics, 86*, 34–43.

Wechsler, D. (1974). *Wechsler Intelligence Scale for Children–Revised.* San Antonio, TX: Harcourt Assessment.

Wechsler, D. (1981). *Wechsler Adult Intelligence Scale–Revised.* San Antonio, TX: Harcourt Assessment.

Winters, M., Pankau, R., Amm, M., Gosch, A., & Wessel, A. (1996). The spectrum of ocular features in the Williams syndrome. *Clinical Genetics, 49*, 28–31.

Wishart, J.G. (1993). The development of learning difficulties in children with Down's syndrome. *Journal of Intellectual Disabilities Research, 37*, 389–403.

Withers, S. (1996). A new clinical sign in Williams syndrome. *Archives of Disease in Childhood, 75*, 1.

Wolff, P.H., Gardner, J., Paccla, J., & Lappen, J. (1989). The greeting behavior of fragile X males. *American Journal of Mental Retardation, 93*, 406–411.

8

Is There a Behavioral Phenotype in Children with Fetal Alcohol Spectrum Disorders?

Piyadasa W. Kodituwakku

Evidence converging from animal and human research over the past 35 years has established that prenatal alcohol exposure produces a range of morphological and behavioral outcomes, often referred to as fetal alcohol spectrum disorders (FASDs), in offspring (Riley & McGee, 2005). At the severe end of this spectrum is fetal alcohol syndrome (FAS), which is characterized by prenatal and postnatal growth retardation, a unique pattern of facial anomalies, and central nervous system anomalies (Jones & Smith, 1973). The majority of alcohol-exposed children do not meet the strict criteria of fetal alcohol syndrome, displaying only some or even none of the physical dysmorphia (Sampson et al., 1997). The Institute of Medicine report on fetal alcohol syndrome (Stratton, Howe, & Battaglia, 1996) recommended the terms *partial fetal alcohol syndrome, alcohol-related birth defects,* and *alcohol-related neurodevelopmental disorders* to label different clusters of birth defects in those who do not meet the criteria of FAS. Common to all children along the spectrum, and contributing to most negative life outcomes in them, are cognitive and behavioral

difficulties that are associated with central nervous system dysfunction.

There exists a large body of literature documenting that children with FASDs exhibit a wide range of cognitive impairments, including diminished intellectual functioning (Streissguth, Barr, & Sampson, 1990), impairments in attention (Coles et al., 1997), executive control dysfunction (Kodituwakku, Handmaker, Cutler, Weathersby, & Handmaker, 1995; Kodituwakku, Kalberg, & May, 2001; Rasmussen, 2005), language difficulties (Adnams et al., 2001), impairments in learning and memory (Mattson & Roebuck, 2002), impairments of social cognition (Monnot, Lovallo, Nixon, & Ross, 2002), motor problems (Connor, Sampson, Streissguth, Bookstein, & Barr, 2006), limitations of adaptive behavior (Whaley, O'Connor, & Gunderson, 2001), and multiple secondary disabilities (Streissguth et al., 2004). Despite the advances in delineating cognitive and behavioral problems in children with FASDs, it remains unresolved whether there is a unique pattern of cognitive-behavioral functioning or a behavioral phenotype associated with FASDs. The

Preparation of this chapter was supported by National Institute on Alcohol Abuse and Alcoholism, National Institutes of Health grant 1P20 AA017068, and the Center for Development and Disability, The University of New Mexico School of Medicine.

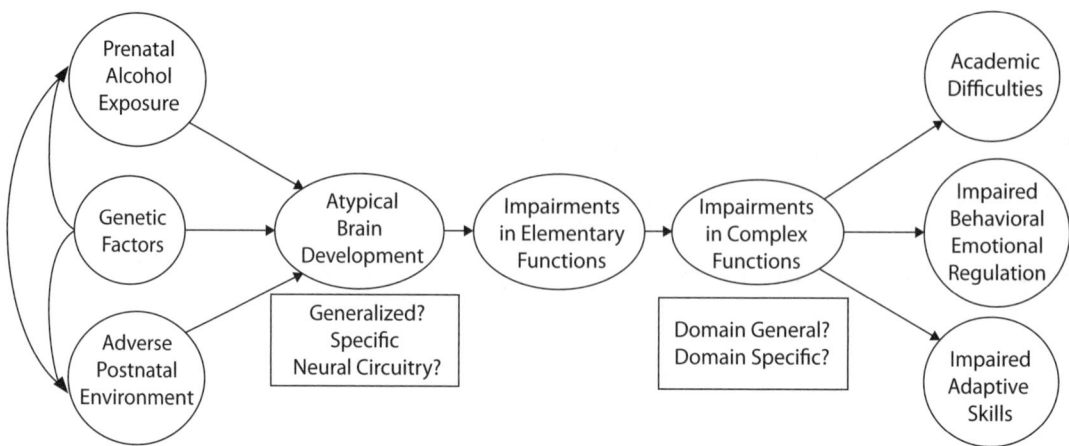

Figure 8.1. A neurodevelopmental model of the behavioral phenotype in fetal alcohol spectrum disorder. (Reprinted from *Neuroscience and Biobehavioral Reviews, 31*[2], Kodituwakku, P.W. Defining the behavioral phenotype in children with fetal alcohol spectrum disorders: A review [p. 194], Copyright 2007, with permission from Elsevier.)

identification of a cognitive-behavioral phenotype will aid the diagnosis of alcohol-exposed children without dysmorphia and planning of appropriate interventions for all children affected by alcohol.

O'Brien and Yule define a behavioral phenotype as "a characteristic pattern of motor, cognitive, linguistic and social observations that is consistently associated with a biological disorder" (2000, p. 2). Adopting this broad definition, we use the term *behavioral* in this chapter to encompass cognitive test data as well as observations of social and adaptive behaviors. It is important, however, to make a distinction between intermediate phenotypes, known as endophenotypes, and end-state phenotypes. This distinction has proved to be very useful in the study of unique behavioral profiles associated with numerous neurodevelopmental syndromes of a genetic origin (Gottesman & Gould, 2003; Viding & Blakemore, 2007). As Gottesman and Gould point out, endophenotypes represent relatively elementary processes (e.g., neurochemical, neuroanatomical, or cognitive) that provide clues to genetic underpinnings of a disease rather than the disease syndrome itself. For example, impairments in sensory motor gating and eye tracking have proved to be useful endophenotypes in explaining the behavioral syndrome of schizophrenia (Gottesman & Gould, 2003).

Although FASD is not a genetic syndrome, the concepts of phenotype and endophenotypes can fruitfully be applied to relating anomalous brain development and behavioral dysfunction in children with FASDs. As shown in Figure 8.1, prenatal alcohol exposure contributes to atypical brain development, which leads to dysfunction of elementary processes such as deficient eye blink conditioning. Animal models of FASD have provided important clues to the identification of candidate elementary processes reflecting alcohol-related brain damage (Johnson, Stanton, Goodlett, & Cudd, 2008). Just as endophenotypes provide clues to genetic underpinnings of a disease, so do elementary functions reflect anomalous brain development in FASD. We propose that deficits in specific elementary functions lead to disabilities in downstream, end-state neuropsychological functions through the process of ontogenetic development. As shown in Figure 8.1, impaired neuropsychological functions contribute to negative life outcomes, which have been referred to as secondary disabilities. In agreement with the Vygotskian notion of lower and higher cognitive functions (Vygotsky, 1978), we use the terms *primary disability* to denote impairments in the elementary functions and *secondary disability* to denote impairments in higher-order functions and observable behaviors.

Accordingly, the task of defining a behavioral phenotype involves answering three primary questions: Do individuals prenatally exposed to alcohol display a unique pattern of elementary functions? Do individuals prenatally exposed to alcohol display a unique pattern of higher-order functions? Do individuals prenatally exposed to alcohol display a unique pattern of observable behavior? It is evident from Figure 8.1 that the establishment of a causal link between prenatal alcohol exposure and primary and secondary disabilities is methodologically challenging, as environmental and genetic factors interactively introduce variability into cognitive-behavioral functioning. The severity of the outcomes is also known to vary as a function of exposure (e.g., quantity and frequency) and maternal (e.g., age, body weight) variables (Jacobson, 1998; May, 1995).

Notwithstanding the variability of outcomes, there are emerging patterns of cognitive functioning and parent- and teacher-reported behavioral problems in children with FASDs. The main aim of this chapter is to delineate these patterns after presenting an overview of human research addressing cognitive functioning and behavior patterns of children affected by alcohol. Following the paths depicted in Figure 8.1, the chapter first summarizes findings from neuroimaging studies of FASDs by way of setting the stage for identifying candidate intermediate phenotypes. The next section of the chapter presents a summary of the literature addressing elementary or lower-level functions in individuals with FASDs. The third section reviews the literature on higher-order cognitive functions, including intellectual functioning, attention and information processing, executive function, language, visual perception, number processing, memory, and social cognition. The fourth section

summarizes findings on adaptive behavior and parent/teacher-rated behavioral problems in individuals with FASDs. The last section attempts to integrate the information from the other sections and proposes a hypothesis on the behavioral phenotype of FASD.

NEUROIMAGING STUDIES

Neuroimaging studies and animal models of FASD have demonstrated that some regions of the brain are more vulnerable to the effects of alcohol than other regions (Riley & McGee, 2005). Volumetric analyses have revealed evidence of hypoplasia in the parietal, temporal, and frontal lobes, with volume reductions being most pronounced in the parietal region (Archibald et al., 2001). Using voxel-based morphometry, Sowell et al. (2001) found significant anomalies in the perisylvian cortices of the parietal and temporal lobes of children with FASDs. Compared with controls, alcohol-exposed participants had relative increases in the gray matter and decreases in the white matter in the perisylvian cortices of the temporal and parietal lobes. Furthermore, Sowell, Thompson, Mattson, et al. (2002) found shape abnormalities in the same regions of the brain in alcohol-exposed adolescents. These investigators also observed significantly reduced brain growth in the ventral regions of the frontal lobes, particularly in the left hemisphere. Sowell, Thompson, Peterson, et al. (2002) observed a gray matter surface asymmetry in the temporal lobes of control subjects, where the right hemisphere surface was larger than the left. This asymmetry was reduced in alcohol-exposed adolescents.

In view of the finding that the mid-line structures in the developing brain are particularly vulnerable to the deleterious effects of ethanol, the integrity of the corpus callosum has been the focus of number of studies. Bookstein et al. (Bookstein, Sampson, Streissguth, & Connor, 2001; Bookstein,

Streissguth, Sampson, Connor, & Barr, 2002) have documented that adults with prenatal alcohol exposure (FAS and fetal alcohol effects, or FAE) significantly differed from controls on morphometric indices of the corpus callosum. Using diffusion tensor imaging (DTI), which provides a measure of tissue microstructural integrity, researchers have found abnormalities in the posterior corpus callosum of alcohol-exposed individuals (Ma et al., 2005; Wozniak et al., 2006). A recent DTI study of children and adolescents with FASDs has also shown microstructural abnormalities in the posterior corpus callosum and in the right temporal lobes (Sowell et al., 2008).

Some studies have suggested that the basal ganglia structures are vulnerable to the effects of prenatal alcohol exposure. Mattson, Riley, Sowell, et al. (1996) found that, compared with controls, basal ganglia volume was reduced in children with FASDs, particularly in the caudate. Cortese et al. (2006) also reported reduced volume of the caudate in alcohol-exposed children. Additionally, these investigators observed that, compared with the control group, the FASD group showed an elevated metabolite ratio of N-acetyl-asparate to creatine (NAA/cr), which is an index of neural function, in the left caudate. The authors speculated that the increased NAA/cr ratio may indicate a lack of normal cell death, pruning, or myelination during development. Reductions in cerebellar volume in individuals with FASDs have also been reported (Archibald et al., 2001). Examination of regional volumetric differences in the cerebellum has revealed greater reduction in size of the anterior than of the posterior vermis (Sowell et al., 1996).

Researchers have recently employed functional magnetic resonance imaging to investigate neural activation patterns in children with FASDs during the performance of cognitive tasks assessing executive control functioning (Fryer et al., 2007; Malisza et al., 2005). Fryer et al. (2007) found that the FASD group differed from controls in blood oxygen level–dependent (BOLD) response patterns during performance of a response

inhibition task (go/no-go), although the two groups did not differ in response latency and performance accuracy. Results of this study were interpreted as showing that the frontal-striatal circuitry was vulnerable to the effects of alcohol exposure. Malisza et al. (2005) also observed group differences in BOLD responses during the performance of a spatial working memory task.

Thus, structural and functional brain imaging studies show that specific brain regions or specific neural circuitries are selectively affected in children and adults with FASDs. In view of these findings, it is reasonable to ask the question: Do functions at elementary or higher-order levels show selective impairments corresponding to selective anomalies in the brain? We suggested previously that elementary functions may better reflect regional anomalies of the brain than higher-order functions, as the former functions are "close to the scene," so to speak.

ELEMENTARY FUNCTIONS

The importance of distinguishing between elementary and higher-order functions has long been recognized in clinical neurology (Hughlings Jackson, 1958; Luria, 1966) and developmental psychology (Vygotsky, 1978). Vygotsky believed that elementary functions, such as sensation, reactive attention, and sensory motor intelligence, are essentially innate and are less influenced by experience and culture than higher-order functions. Given that animals and humans share these lower-level functions, the study of teratogenic effects on them allows the generalization of animal research to humans. In this chapter we examine the effects of prenatal alcohol exposure on three main elementary functions: eye-blink conditioning (EBC), sensory processing, and elementary perception.

Eye-Blink Conditioning

Eye-blink conditioning (EBC) is a well-established paradigm of associative learning in which the eye-blink reflex is conditioned to a neutral stimulus (tone) that predicts an unconditioned stimulus (puff of air). It is known that EBC engages a well-defined neural circuitry including the cerebellum and the brainstem. Given that the cerebellar region is sensitive to the effects of ethanol, researchers have used EBC as a biomarker of the effects of prenatal alcohol exposure in animal models (Brown, Calizo, & Stanton, 2008; Green, Rogers, Goodlett, & Steinmetz, 2000). A consistent finding in these studies is that alcohol-exposed animals exhibit later-onset or later-peaked conditioned responses, suggesting slow and impaired conditioning. There is also evidence that deficient EBC persists into adulthood in alcohol-exposed animals (Green et al., 2000). Coffin et al. compared children with FASDs, dyslexia, attention-deficit/hyperactivity disorder, and typically developing controls on EBC and found that those with FASDs and dyslexia compared with the other two groups showed impairments in the acquisition of conditioned responses (Coffin, Baroody, Schneider, & O'Neill, 2005). Children with FASDs and dyslexia showed longer latencies and more timing errors in conditioned responses than the other two groups. Recently Jacobson et al. (2008) reported a study in which a group of 5-year-old children with FAS were compared with age-matched controls on EBC. These investigators found that the FAS group was significantly impaired in the acquisition of conditioned responses.

Sensory Processing

Researchers have documented that infants with prenatal alcohol exposure show poor habituation to sensory stimulation and low arousal level as assessed by the Brazelton Scale (Streissguth, Barr, & Martin, 1983). There is also evidence that infants with prenatal alcohol exposure display sleep disturbances (Rosett et al., 1979). Researchers found that exposure to alcohol throughout gestation was associated with a decrease in the total time spent sleeping and restlessness

with major body movements in infants. Although it is known that sensory issues persist into adulthood as indicated by caregiver reports (Streissguth, Bookstein, Barr, Press, & Sampson, 1998), little is known about the extent and nature of these problems. In a recent study, Franklin et al. reported that caregivers rated children with FASDs as having sensory and behavioral problems (Franklin, Deitz, Jirikowic, & Astley, 2008). Given that the source of the data in this study was caregivers, one cannot draw causal conclusions about sensory problems in FASDs because of confounding variables. Animal models of sensory processing in FASDs, however, allow systematic control of confounding variables (Schneider et al., 2008). Schneider et al. (2008) compared four groups of monkeys, with the use of positron emission tomography: prenatally alcohol-exposed, prenatal stress, prenatal alcohol + prenatal stress, and sucrose controls. Results showed that the patterns of habituation and the magnitude of aversion to repetitive tactile stimulation were associated with prenatal stress and prenatal alcohol exposure. These investigators also found that increased D_2 receptor binding in the striatum was related to increased withdrawal and reduced habituation.

Elementary Perceptual Functions

It has been documented that children with FASDs display impaired vision and a wide range of ocular anomalies, including microcornea, retinal dysplasia, and optic nerve hypoplasia (Stromland, 2004). Church and Kaltenbach (1997) found that hearing disorders are commonly associated with prenatal alcohol exposure. Hearing problems are often associated with vestibular disturbances. It is unknown how these impairments of vision and hearing lead to perceptual problems at elementary and complex levels. It is reasonable to expect that sensory losses limit sensory input during critical periods of brain development, thus creating lasting changes in the brain. As noted previously, Sowell et al. (2001) reported anomalies in the posterior parietal and temporal lobes, including increased gray matter and reduced white matter. It is possible, therefore, that limited sensory inputs affect critical events in brain development, such as diminished pruning and myelination.

No published data are available on elementary visual processes such as detection of form and motion coherence thresholds (Gunn et al., 2002) in children with FASDs. In a recent study we found that children with FASDs performed less efficiently than controls on a visual backward masking task, suggesting difficulty in processing visual information (Verney, unpublished data).

In summary, studies of elementary functions show that children with FASDs display impairments in the acquisition of conditioned responses in classical conditioning paradigms. Alcohol-affected infants show slow habituation to repeated sensory stimulation and sleep disturbances. Compared with complex functions, relatively little is known about elementary functions in children with FASDs.

Higher-Order Functions

Several questions regarding higher-order functions are relevant for delineating the behavioral phenotype: 1) Do children with FASDs show an even or uneven profile of intellectual functioning? 2) On tests of information processing, do children with FASDs show difficulty with specific types of materials (e.g., verbal, visual)? 3) At a complex level, do children with FASDs show difficulty in specific domains or across the board? Previously published reviews (Kodituwakku, 2007; Mattson & Riley, 1998; Riley & McGee, 2005) have partially addressed these questions.

Intellectual Ability

Given that intellectual tests form an essential part of cognitive test batteries, intellectual functioning of children with FASDs has been extensively studied (see Mattson & Riley, 1998, for a review). Researchers have consistently found diminished intellectual performance in children with FASDs (Mattson, Riley, Gramling, Delis, & Jones, 1997; Streissguth et al., 1990), with average IQs ranging from mild intellectual disability to the borderline range. Mattson et al. (1997) found that children with heavy prenatal alcohol exposure displayed diminished intellectual functioning, irrespective of the presence or absence of physical features of fetal alcohol syndrome. Some researchers have obtained evidence for a dose-dependent decrement of intellectual ability, finding exposure to 1 ounce of absolute alcohol a day to be related to a decrease in nearly 5 full-scale IQ points (Streissguth, Barr, Sampson, Darby, & Martin, 1989).

Some studies have sought to answer the question of whether the intellectual profile in children with FASDs is even or uneven. Studies that contrasted Verbal and Performance IQs have yielded equivocal results (Mattson and Riley, 1998). Variability in subject backgrounds (e.g., social-ethnic) and amounts of prenatal alcohol exposure probably would account for these equivocal results. Studies involving children with heavy prenatal alcohol exposure show comparable decrement in both Verbal and Performance IQs (Mattson et al., 1997). Adnams et al. (2001) also reported that children with FASDs were impaired at more intellectually demanding tests of the Griffiths Scales (e.g., Speech-Hearing, Practical Reasoning, Performance) (Griffiths, 1954), but were unimpaired at less demanding tests such as Gross Motor. Therefore, it is reasonable to conclude that children with FASDs display impaired performance on complex intellectual tasks across the board.

Attention and Speed of Information Processing

Attentional skills of children with FASDs have been the focus of numerous studies, because attentional deficits are considered to be a pronounced characteristic of this group (Nanson & Hiscock, 1990; Streissguth et al., 1986). Streissguth et al. (1986) reported that children with FASDs showed impaired performance in vigilance. Nanson and Hiscock (1990) found that children with FASDs, like those with attention deficit disorder (ADD), had difficulty with the investment, organization, and maintenance of attention over time and had difficulty in response inhibition. In view of the observed similarities in children with ADD and FASDs, Coles et al. (1997) compared these two groups on a battery of tests assessing different components of attention: alert, sustain, encode, and shift. Results of this study revealed distinct patterns of performance in the two groups, with the FASD group showing greater impairments in encode and shift components and the ADD group displaying greater impairments in alert and sustain components.

Thus, the finding that the FASD group was unimpaired in sustained attention compared with typically developing controls in the study by Coles et al. (1997) is at variance with the findings reported by Streissguth et al. (1986) and Nanson and Hiscock (1990). A number of variables, including differences in subject selection (clinic-referred versus nonreferred) and differential sensitivity of the tests that were employed, would account for the previously noted inconsistencies in results. In a subsequent study, Coles, Platzman, Lynch, and Freides (2002) found an interaction between groups and the modality of sustained attention, with the FASD group performing worse on visual than on auditory sustained attention. Lee, Mattson, and Riley (2004) also found that children with FASDs performed less well than controls on a visual test of sustained attention (Test of Variable Attention; Greenberg & Waldman, 1993). A logistic regression analysis revealed, however, that auditory tests assessing freedom from distractibility (Digit Span and Arithmetic) distinguished alcohol-exposed children from controls better than the Test of Visual Sustained Attention. Inasmuch as the Digit Span and the Arithmetic subtests of the Wechsler Intelligence Scale for Children, Third Edition (Wechsler, 1991) involve intellectual ability, it is possible that the complexity of the task, instead of its modality, discriminated between the alcohol-exposed and typically developing groups.

Specific effects of prenatal alcohol exposure on information processing have been the subject of a number of studies. Jacobson (1998) found that infants with prenatal alcohol exposure showed slower processing speed than comparison groups on the Fagan Test of Infant Intelligence (Fagan, Singer, Montie, & Shepherd, 1986). Consistent with this finding, Burden, Jacobson, and Jacobson (2005) found that prenatal alcohol exposure was associated with impaired processing efficiency in school-age children. These investigators also noted that tasks involving effortful processing rather than those involving automatic processing discriminated between the FASD and control groups. Roebuck, Mattson, and Riley (2002) found that children with FASDs made more errors than typically developing controls on a test of finger localization, with the group effect being greater on relatively demanding trials that involved interhemispheric transfer of information. Thus, the pattern emerging from the previously discussed studies of information processing is that the FASD group has difficulty in rapidly processing relatively complex information, irrespective of the type of material being processed.

Executive Functioning

The term *executive functioning* refers to a range of abilities involved in the attainment of goals efficiently in nonroutine situations. These include planning, monitoring of goal-directed behavior in the face of interference, shifting response sets in response to the changes in task demands, and error correction. Given that tasks assessing these skills require conscious effort or supervisory attention, it is reasonable to hypothesize that children with FASDs have impairments in executive control (Kodituwakku et al., 1995; Mattson, Goodman, Caine, Delis, & Riley, 1999).

There is a growing body of literature supporting the previously presented hypothesis (Kodituwakku, Kalberg, & May, 2001; Rasmussen, 2005). Researchers have found that the FASD group performs less well than controls on tests assessing cognitive planning, such as the Progressive Planning Test (Kodituwakku et al., 1995) and the California Tower Task (Mattson et al., 1999). A process analysis of performance on the Progressive Planning Test has revealed, however, that the FASD group had difficulty only with complex problems that involved the management of subgoals in working memory (Kodituwakku, unpublished data).

Researchers have obtained evidence that prenatal alcohol exposure is associated with impaired cognitive set shifting, reflected by increased perseverative errors on tests such as the Wisconsin Card Sorting Test (WCST) (Carmichael Olson, Feldman, Streissguth, Sampson, & Bookstein, 1998; Coles et al., 1997; Grant & Berg, 1993). The type of set shifting involved in the WCST is also called extradimensional set shifting, as the subject is required to sort cards by a specific dimension and then to shift attention to sort by a different dimension, using the examiner's feedback. Extradimensional set shifting is often contrasted with intradimensional set shifting, in which the examinee is required to shift attention within the same dimension (e.g., from one color to another color or one shape to another shape). The visual discrimination reversal learning test exemplifies a measure of intradimensional set shifting (Rolls, Hornak, Wade, & McGrath, 1994) because this test requires the examinee to respond to one of two stimuli (rewarding) and withhold responding to the other one. After reaching a learning criterion, the reinforcement contingencies are reversed, requiring the subject to alter responses. Thus, the reversal learning test involves shifting the response-reward associations or affective set shifting and hence predominantly activates the orbito-frontal cortex. It is known that the dorsolateral-striatal circuitry subserves the performance of extradimensional set shifting.

Kodituwakku, May, Clericuzio, and Weers (2001) investigated intradimensional set shifting in children with FASDs, with the use of a visual discrimination reversal learning task that was adapted for use with humans. Results showed that the FASD group was impaired in the ability to make reward-response reversals and that impaired reversal learning was associated with parent-rated behavioral problems in these children. Thus, children with FASD display impaired performance in both cognition-based executive function and emotion-related self-regulation. A comparison of effect sizes has revealed, however, that children affected by alcohol are relatively more impaired at extradimensional than at intradimensional set shifting.

There is evidence that children with FASDs employ ineffective strategies during problem solving (Mattson et al., 1999) and show defective skills in concept formation (McGee, Schonfeld, Roebuck-Spencer, Riley, & Mattson, 2008). Children with FASDs have also been found to have impairments in nonverbal and verbal fluency (Schonfeld, Mattson, Lang, Delis, & Riley, 2001), particularly in letter fluency (generation of words beginning with certain letters under specific constraints). Noland et al. (2003) reported that children exposed to alcohol performed less well than controls on a tapping inhibition task. There is also evidence that children with FASDs have deficiencies in inhibiting reflexive eye movements (antisaccades), a function that involves executive control (Green, Munoz, Nikkel, & Reynolds, 2007).

In summary, children and adults with FASDs show impaired performance on tests assessing executive control skills. On closer examination, however, one finds that children with FASDs display greater difficulty with more complex tasks of executive functioning than with less complex ones. For example, the FASD group has greater difficulty with solving planning problems using two rules (highly constrained condition) than it does with one rule (minimally constrained). The FASD group displays relatively more difficulty with extradimensional set shifting than with intradimensional set shifting. Furthermore, there is evidence that the FASD group has greater difficulty with letter fluency than with category fluency.

Language

Compared with the areas of attention and executive control, relatively limited information is available on language skills in children with FASDs. One reason for this is that some researchers have failed to find notable language impairments in children affected by alcohol (Fried, O'Connell, & Watkinson, 1992; Greene, Ernhart, Martier, Sokol, & Ager, 1990). In contrast, researchers have found significant language impairments in clinic-referred samples of children with FASDs. Using standardized tests, researchers have documented that prenatal alcohol exposure is associated with impairments in naming (Mattson, Riley, Gramling, Delis, & Jones, 1998), grammatical and semantic abilities (Becker, Warr-Leeper, & Leeper, 1990), and pragmatics (Abkarian, 1992). Streissguth et al. (1994) reported a relationship between prenatal alcohol exposure and decreased performance on the Word Attack subtest of the Woodcock Reading Mastery Tests (Woodcock, 1973), which require subjects to pronounce unfamiliar words. A number of population-based studies from South Africa and Italy revealed that children with FAS were markedly impaired in grammar comprehension (Adnams et al., 2001; Kodituwakku, Adnams, et al., 2006; Kodituwakku, Coriale, et al., 2006). As mentioned previously, a number of investigators have demonstrated that the FASD group shows greater difficulty than controls in verbal fluency, particularly in letter fluency (Schonfeld et al., 2001).

The previously described discrepancies in the findings related to language could be attributed partly to differences in participant characteristics. It appears that the studies reporting negative results included children, on average, with lower levels of exposure than those studies reporting positive results. Whereas the majority of children in the study by Greene et al. did not have significant morphological anomalies consistent with FAS, almost all children in studies reported by Adnams et al. (2001), Kodituwakku, Adnams, et al. (2006), and Kodituwakku, Coriale, et al. (2006) had a diagnosis of FAS. It is also possible that the studies reporting positive findings employed relatively complex tests of language involving working memory. For example, the Word Attack Test (Streissguth et al., 1994), the Grammar Comprehension Test (Kodituwakku, Adnams, et al., 2006), and verbal fluency tasks all involved phonological working memory, in addition to the processing of linguistic information per se. Coggins, Timler, and Olswang (2007) reported that children with FASDs also display impairments in social communication. Thus it is reasonable to conclude that children with FASDs show impairments specifically in those language skills involving phonological working memory and social pragmatics.

Visual Perception and Visual Construction

As noted previously, little is known about elementary perceptual processes in children with FASD. Uecker and Nadel (1996) reported that children with FASD were relatively unimpaired in visual perception as assessed by a test of facial recognition, but were markedly impaired at tests assessing visual construction, such as the Beery Visual Motor Integration and Clock Drawing tests (Beery, 1982). Mattson, Gramling, Delis, Jones, and Riley (1996) investigated whether children with FASDs differentially processed global and local features of visual stimuli.

Results showed that the FASD group was impaired in copying and recalling local features of hierarchical stimuli, such as a large letter *D* (global) made up of small letters *y* (local features).

The previously presented findings suggest that children with FASDs are impaired at visual perceptual tasks that demand integration of information. The results of Uecker and Nadel (1996) demonstrate that children with FASDs performed as competently as controls on a task of visual matching, but experienced significant difficulty with drawing designs that required planning and visual-motor integration. The basis of the differential effect in processing global and local features needs to be further investigated.

Learning and Memory

Animal research has provided consistent evidence that the hippocampus, a region subserving memory, is specifically vulnerable to the toxic effects of alcohol during brain development (Berman & Hannigan, 2000; Savage, Becher, de la Torre, & Sutherland, 2002). Using various learning paradigms (e.g., spatial navigation), researchers have now established the association between hippocampal damage resulting from prenatal alcohol exposure and learning and memory deficits in animals (Savage et al., 2002). A number of researchers have extended the findings in animal research to humans by using tests that have been found to be sensitive to hippocampal functioning. Uecker and Nadel (1996) utilized the Smith and Milner Memory for 16 Objects Task to assess learning and memory in children with FASDs and found that the alcohol-affected group had difficulty with delayed, but not with immediate, object recall. Uecker and Nadel (1996) also found that the FASD group also had difficulty with nonhippocampal visual tasks, suggesting that children affected by alcohol had generalized deficits in the visual domain. Using a computerized (virtual) version of the Water Maze, another test that indexes hip-

pocampal functioning, Hamilton, Kodituwakku, Sutherland, and Savage (2003) found that children with FASDs demonstrated difficulty in place learning.

Using standardized tests, numerous investigators have found that children with FASDs are impaired at learning and memory tasks compared with demographically matched controls (Mattson & Riley, 1999; Mattson, Riley, Delis, Stern, & Jones, 1996; Willford, Richardson, Leech, & Day, 2004). The results of these studies make it possible to draw a number of conclusions. First, children with FASDs display greater difficulty than controls learning both verbal and visual materials, a pattern commensurate with IQ. Second, alcohol-exposed children are impaired at explicit memory, but unimpaired at implicit memory. Third, free recall of information is challenging for children with FASDs, whereas their recognition memory is relatively intact. Therefore, it appears that children with prenatal alcohol exposure show impairments in memory processes that involve conscious effort, such as free recall and organization.

Social Cognition

There is a growing body of literature showing that children with prenatal alcohol exposure display social impairments (O'Connor et al., 2006; Schonfeld, Paley, Frankel, & O'Connor, 2008). Given that impairments have been demonstrated in animal models of FASD (Kelly, Day, & Streissguth, 2000; Lugo, Marino, Cronise, & Kelly, 2003), it is reasonable to hypothesize that alcohol-induced brain damage contributes to social difficulties in alcohol-exposed children.

A number of investigators have sought to identify the cognitive mechanisms underlying social problems in children affected by alcohol and then to relate these mechanisms to central nervous system dysfunction (Schonfeld, Paley, Frankel, & O'Connor, 2006; Schonfeld et al., 2008). Bishop, Gahagan, and Lord (2007) compared children with autism and FASD on the

Autism Diagnostic Observation Scale (Lord et al., 1989) and found that the two groups differed in social interaction and communication. The autism group, but not the FASD group, showed difficulties in initiating social interaction, sharing affect, and using nonverbal communication. Socially inappropriate behaviors and difficulty with peers were found, however, in both groups. This suggests that social difficulties in children with FASDs may therefore be related to their impaired self-regulation, rather than a lack of empathy. Schonfeld et al. (2006, 2008) reported evidence supporting this hypothesis. These investigators found that executive functions accounted for a significant proportion of variance in parent- and teacher-rated social problems in children with FASDs.

Number Processing

On standardized test batteries of academic achievement, alcohol-exposed children obtain relatively lower arithmetic scores (Streissguth et al., 1994). This finding has led researchers to ask whether children with FASDs display a specific impairment in number processing. Kopera-Frye, Dehaene, and Streissguth (1996) investigated number processing in children with FASDs, with the use of a comprehensive test battery based on recent developments in cognitive neuroscience (Dehaene, 1997). This study revealed that children with prenatal alcohol exposure performed without difficulty on relatively simple tasks of number processing, such as number reading and number writing, but displayed impairments on relatively complex tasks, such as calculation and cognitive estimation.

Burden et al. (2005) reported specific impairments in number comparison in children with prenatal alcohol exposure. The demonstration of a differential effect requires, however, comparing number-processing tasks with non-number-processing tasks of equal complexity. It is reasonable to hypothesize that only complex number-processing tasks discriminate alcohol-exposed children and controls.

Summary

Children with FASDs seem to perform worse than controls on complex tasks of cognitive functioning across a wide range of domains. On tests of intellectual functioning and memory, the FASD group displays difficulties in both verbal and visual tasks, suggesting a generalized pattern of impairments. Children with FASDs perform more slowly and make more errors than controls on tests of information processing. On tests consisting of simple and complex items from different domains of cognitive functioning, the FASD group tends to perform worse than controls on complex, but not simple, items.

Behavioral Dysfunction

As Figure 8.1 shows, the previously noted cognitive impairments contribute to a range of behavioral outcomes, including academic difficulties, social skills impairments, and regulation of emotions. As mentioned previously, impairments in executive control functioning leads to difficulties in social interactions (Schonfeld et al., 2006) in children with FASDs. Evidence from longitudinal studies of FASD shows that alcohol-affected individuals experience a range of adverse outcomes, including academic failures, trouble with the law, confinement, psychiatric problems, dependent living, and substance use disorders (Spohr, Willms, & Steinhausen, 2007; Streissguth et al., 2004). Through questionnaires and caregiver interviews, researchers have also obtained evidence that children or adults with FASDs display impairments in adaptive behavior (Whaley et al., 2001), attentional difficulties and conduct problems (Nash et al., 2006), and psychiatric illness (O'Connor et al., 2002). An important question that one can ask about these reported behavioral problems is whether a unique pattern associated with FASD exists.

In an attempt to answer this question, Streissguth et al. (1998) administered a checklist to parents and caregivers of 472 individuals with prenatal alcohol exposure. Based on descriptors used by caregivers to describe their children with FASD, this checklist comprised items tapping a number of areas: communication and speech, personal manner, emotions, motor skills and activities, social skills and interactions, academic performance, and bodily and physiological functions. Results showed that 36 items emerged as associated with FASD, including being unaware of consequences, attention problems, poor judgment, inability to take a hint, overreaction, and being overly friendly (the checklist comprising these 36 items is known as the Personal Behavior Checklist-36). Most of these items can be characterized as indexing "dysexecutive behaviors" seen in a number of disorders, including attention-deficit/hyperactivity disorder. Kodituwakku et al. (2001) found that neuropsychological measures of executive functioning, particularly of orbito-frontal functioning, were associated with scores of this checklist. As noted previously, Schonfeld et al. (2006, 2008) reported that executive control dysfunction in children with FASDs predicted their social impairments and response to a social skills training program. These results make it possible to draw the conclusion that executive control problems account for a significant proportion of behavioral problems in children with FASDs.

Dysexecutive behavior is a hallmark of a number of disorders, including attention-deficit/hyperactivity disorder, conduct disorder, Tourette syndrome, obsessive-compulsive disorder, and schizophrenia, which are collectively called disorders of action regulation (Pennington, 2002). To determine if the FASD group differed from children with ADHD, Nash et al. (2006) administered the Child Behavior Checklist (Achenbach & Edelbrock, 1991) and found that a number of items discriminated between the two groups: acting young, cruelty, absence of guilt, lying or cheating, stealing from home, and stealing outside the home. Thus, predominantly delinquent

behaviors in FASD account for the group difference. It remains unresolved, however, whether delinquent behaviors in FASD are a product of environmental and genetic factors or if they are an inherent characteristic of the disorder. Lynch, Coles, Corley, and Falek (2003) failed to find an association between delinquency and prenatal alcohol exposure in a community sample, suggesting that these traits may not be an essential characteristic of the FASD phenotype.

With respect to adaptive behaviors, the FASD group usually scores lower than controls across domains (Kalberg et al., 2006; Olson, Feldman, Streissguth, Sampson, & Bookstein, 1998). A number of investigators, however, have documented that children with FASDs show relatively greater social impairments, particularly when they reach adolescence (Whaley et al., 2001). As noted previously, social skills involve executive skills, and as children grow older the social demands increase, probably taxing executive capacities.

Turning to academic skills, diminished intellectual functioning and difficulties in information processing and attention seem to contribute to poor school performance in children with FASDs (Carmichael Olson et al., 1998). In a recent population-based study conducted in Italy, teachers rated children with FASDs as being hyperactive and inattentive. However, inattentive behaviors, but not hyperactivity, distinguished the FASD group from typical controls (Kodituwakku, Coriale, et al., 2006). Inattentiveness can be considered to be associated with slow information processing and diminished intellectual functioning.

SUMMARY AND CONCLUSIONS

Using a broad definition of behavioral phenotype, we included under the term *behavior* both higher-level cognitive functions as well as parent/caregiver-rated behaviors. Higher-level cognitive functions were distinguished from elementary functions, which correspond to endophenotypes in genetic syndromes. The previously discussed review reveals a pattern emerging from both the lower-level and higher-level functions in children with FASDs. Results from various conditioning paradigms and studies of sensory processing show that alcohol-exposed children are slow at the acquisition of responses and demonstrate evidence of slow habituation. The following observations were made with regard to the higher-level cognitive functions: 1) children with FASDs show lower intellectual functioning than controls, and their verbal and nonverbal skills are diminished; 2) children with FASDs are slower than controls at information processing; and 3) when task demands (complexity) increase, the performance of the FASD group declines at a greater rate compared with controls, irrespective of the domain of functioning. Behavioral and academic problems in children with FASDs reported by teachers and parents are associated with the cognitive problems noted previously.

In a recent review of the literature on end-state cognitive functions and behaviors, the author concluded that the essence of the behavioral phenotype associated with FASD could be defined as a generalized impairment in processing complex information (Kodituwakku, 2007). Given that the current review finds impaired information processing even at a lower level, the previously presented conclusion should be modified as a generalized impairment in processing and integration of information. This generalized impairment is indexed by slowness in the acquisition of conditioned responses at the elementary level and by inefficiency in processing complex information at the higher-order level. One is justified in arguing that children with FASDs employ inefficient strategies and activate neural circuitries different from those typically used when they succeed in completing a higher-order task.

Given the previously presented conclusion, the challenging theoretical problem that one has to tackle concerns how regional anomalies in the brain lead to a generalized impairment in information processing. We propose that the answer lies in the ontogenetic developmental processes. As Karmiloff-Smith and colleagues (Karmiloff-Smith, 2006; Thomas & Karmiloff-Smith, 2002) have demonstrated, the assumption that atypical development can produce selective impairments while the rest of the system develops typically, which they dubbed the assumption of "residual normality," is untenable in developmental neuroscience. As noted previously, Fryer et al. (2007) have demonstrated that the FASD group displayed a different pattern of BOLD activation compared with controls during performance of a go/no-go task. Therefore, the brain regions without anomalies may participate in the execution of functions of the damaged regions and thus undergo alterations through the processes of plasticity during development. Understanding these developmental processes may be the new frontier in research aimed at uncovering the mechanisms that produce the behavioral phenotype in FASD.

REFERENCES

Abkarian, G.G. (1992). Communication effects of prenatal alcohol exposure. *Journal of Communication Disorders, 25*(4), 221–240.

Achenbach, T.M., & Edelbrock C. (1991). The Child Behavior Checklist and Revised Child Behavior Profile. Burlington, VT: University Associates in Psychiatry.

Adnams, C.M., Kodituwakku, P.W., Hay, A., Molteno, C.D., Viljoen, D., & May, P.A. (2001). Patterns of cognitive-motor development in children with fetal alcohol syndrome from a community in South Africa. *Alcoholism, Clinical and Experimental Research, 25*(4), 557–562.

Archibald, S.L., Fennema-Notestine, C., Gamst, A., Riley, E.P., Mattson, S.N., & Jernigan, T.L. (2001). Brain dysmorphology in individuals with severe prenatal alcohol exposure. *Developmental Medicine and Child Neurology, 43*(3), 148–154.

Becker, M., Warr-Leeper, G.A., & Leeper, H.A., Jr. (1990). Fetal alcohol syndrome: A description of oral motor, articulatory, short-term memory, grammatical, and semantic abilities. *Journal of Communication Disorders, 23*(2), 97–124.

Beery, K.E. (1982). *Developmental Test of Visual-Motor Integration* (2nd rev. ed.). Chicago: Follett.

Berman, R.F., & Hannigan, J.H. (2000). Effects of prenatal alcohol exposure on the hippocampus: Spatial behavior, electrophysiology, and neuroanatomy. *Hippocampus, 10*(1), 94–110.

Bishop, S., Gahagan, S., & Lord, C. (2007). Re-examining the core features of autism: A comparison of autism spectrum disorder and fetal alcohol spectrum disorder. *Journal of Child Psychology and Psychiatry, and Allied Disciplines, 48*(11), 1111–1121.

Bookstein, F.L., Sampson, P.D., Streissguth, A.P., & Connor, P.D. (2001). Geometric morphometrics of corpus callosum and subcortical structures in the fetal-alcohol-affected brain. *Teratology, 64*(1), 4–32.

Bookstein, F.L., Streissguth, A.P., Sampson, P.D., Connor, P.D., & Barr, H.M. (2002). Corpus callosum shape and neuropsychological deficits in adult males with heavy fetal alcohol exposure. *Neuroimage, 15*(1), 233–251.

Brown, K.L., Calizo, L.H., & Stanton, M.E. (2008). Dose-dependent deficits in dual interstimulus interval classical eyeblink conditioning tasks following neonatal binge alcohol exposure in rats. *Alcoholism, Clinical and Experimental Research, 32*(2), 277–293.

Burden, M.J., Jacobson, S.W., & Jacobson, J.L. (2005). Relation of prenatal alcohol exposure to cognitive processing speed and efficiency in childhood. *Alcoholism, Clinical and Experimental Research, 29*(8), 1473–1483.

Carmichael Olson, H., Feldman, J.J., Streissguth, A.P., Sampson, P.D., & Bookstein, F.L. (1998). Neuropsychological deficits in adolescents with fetal alcohol syndrome: Clinical findings. *Alcoholism, Clinical and Experimental Research, 22*(9), 1998–2012.

Church, M.W., & Kaltenbach, J.A. (1997). Hearing, speech, language, and vestibular disorders in the fetal alcohol syndrome: A literature review. *Alcoholism, Clinical and Experimental Research, 21*(3), 495–512.

Coffin, J.M., Baroody, S., Schneider, K., & O'Neill, J. (2005). Impaired cerebellar learning in children with prenatal alcohol exposure: A comparative study of eyeblink conditioning in children with ADHD and dyslexia. *Cortex, 41*(3), 389–398.

Coggins, T.E., Timler, G.R., & Olswang, L.B. (2007). A state of double jeopardy: Impact of prenatal alcohol exposure and adverse environments on the social communicative abilities of school-age children with fetal alcohol spectrum disorder.

Language, Speech, and Hearing Services in School, 38(2), 117–127.

Coles, C.D., Platzman, K.A., Lynch, M.E., & Freides, D. (2002). Auditory and visual sustained attention in adolescents prenatally exposed to alcohol. *Alcoholism, Clinical and Experimental Research, 26*(2), 263–271.

Coles, C.D., Platzman, K.A., Raskind-Hood, C.L., Brown, R.T., Falek, A., & Smith, I.E. (1997). A comparison of children affected by prenatal alcohol exposure and attention deficit, hyperactivity disorder. *Alcoholism, Clinical and Experimental Research, 21*(1), 150–161.

Connor, P.D., Sampson, P.D., Streissguth, A.P., Bookstein, F.L., & Barr, H.M. (2006). Effects of prenatal alcohol exposure on fine motor coordination and balance: A study of two adult samples. *Neuropsychologia, 44*(5), 744–751.

Cortese, B.M., Moore, G.J., Bailey, B.A., Jacobson, S.W., Delaney-Black, V., & Hannigan, J.H. (2006). Magnetic resonance and spectroscopic imaging in prenatal alcohol-exposed children: Preliminary findings in the caudate nucleus. *Neurotoxicology and Teratology, 28*(5), 597–606.

Dehaene, S. (1997). *The number sense: How the mind creates mathematics.* New York: Oxford University Press.

Fagan, J., Singer, L., Montie, J., & Shepherd, P. (1986). Selective screening device for the early detection of normal or delayed cognitive development in infants at risk for later mental retardation. *Pediatrics, 78,* 1021–1026.

Franklin, L., Deitz, J., Jirikowic, T., & Astley, S. (2008). Children with fetal alcohol spectrum disorders: Problem behaviors and sensory processing. *The American Journal of Occupational Therapy, 62*(3), 265–273.

Fried, P.A., O'Connell, C.M., & Watkinson, B. (1992). 60- and 72-month follow-up of children prenatally exposed to marijuana, cigarettes, and alcohol: Cognitive and language assessment. *Journal of Developmental and Behavioral Pediatrics 13*(6), 383–391.

Fryer, S.L., Tapert, S.F., Mattson, S.N., Paulus, M.P., Spadoni, A.D., & Riley, E.P. (2007). Prenatal alcohol exposure affects frontal-striatal BOLD response during inhibitory control. *Alcoholism, Clinical and Experimental Research, 31*(8), 1415–1424.

Gottesman, I.I., & Gould, T.D. (2003). The endophenotype concept in psychiatry: Etymology and strategic intentions. *American Journal of Psychiatry, 160,* 636–645.

Grant, D., & Berg, F. (1993). *Wisconsin Card Sorting Test.* Odessa, FL: Psychological Assessment Resources.

Green, C.R., Munoz, D.P., Nikkel, S.M., & Reynolds, J.N. (2007). Deficits in eye movement control in children with fetal alcohol spectrum disorders. *Alcoholism, Clinical and Experimental Research, 31*(3), 500–511.

Green, J.T., Rogers, R.F., Goodlett, C.R., & Steinmetz, J.E. (2000). Impairment in eyeblink classical conditioning in adult rats exposed to ethanol as neonates. *Alcoholism, Clinical and Experimental Research, 24*(4), 438–447.

Greenberg, L.M., & Waldman, I.D. (1993). Developmental normative data on the Test of Variables of Attention (T.O.V.A.). *Journal of Child and Adolescent Psychiatry, 34*(6), 1019–1030.

Greene, T., Ernhart, C.B., Martier, S., Sokol, R., & Ager, J. (1990). Prenatal alcohol exposure and language development. *Alcoholism, Clinical and Experimental Research, 14*(6), 937–945.

Griffiths, R. (1954). *The abilities of babies: A study of mental measurements.* London: University of London Press.

Gunn, A., Cory, E., Atkinson, J., Braddick, O., Wattam-Bell, J., Guzzetta, A., et al. (2002). Dorsal and ventral stream sensitivity in normal development and hemiplegia. *Neuroreport, 13*(6), 843–847.

Hamilton, D.A., Kodituwakku, P., Sutherland, R.J., & Savage, D.D. (2003). Children with fetal alcohol syndrome are impaired at place learning but not cued-navigation in a virtual Morris water task. *Behavioral Brain Research, 143*(1), 85–94.

Hughlings Jackson, J. (1958). On the anatomical and physiological localization of the movements in the brain. In J. Taylor (Ed.), *Selected writings of John Hughlings Jackson* (Vol. 1). New York: Basic Books.

Jacobson, S.W. (1998). Specificity of neurobehavioral outcomes associated with prenatal alcohol exposure. *Alcoholism, Clinical and Experimental Research, 22*(2), 313–320.

Jacobson, S.W., Stanton, M.E., Molteno, C.D., Burden, M.J., Fuller, D.S., Hoyme, H.E., et al. (2008). Impaired eyeblink conditioning in children with fetal alcohol syndrome. *Alcoholism, Clinical and Experimental Research, 32*(2), 365–372.

Johnson, T.B., Stanton, M.E., Goodlett, C.R., & Cudd, T.A. (2008). Eyeblink classical conditioning in the preweanling lamb. *Behavioral Neuroscience, 122*(3), 722–729.

Jones, K.L., & Smith, D.W. (1973). Recognition of the fetal alcohol syndrome in early infancy. *Lancet, 2*(7836), 999–1001.

Kalberg, W.O., Provost, B., Tollison, S.J., Tabachnick, B.G., Robinson, L.K., Eugene Hoyme, H., et al. (2006). Comparison of motor delays in young children with fetal alcohol syndrome to those with prenatal alcohol exposure and with no prenatal alcohol exposure. *Alcoholism,*

Clinical and Experimental Research, 30(12), 2037–2045.

Karmiloff-Smith, A. (2006). The tortuous route from genes to behavior: A neuroconstructivist approach. *Cognitive, Affective & Behavioral Neuroscience, 6*(1), 9–17.

Kelly, S.J., Day, N., & Streissguth, A.P. (2000). Effects of prenatal alcohol exposure on social behavior in humans and other species. *Neurotoxicology and Teratology, 22*(2), 143–149.

Kodituwakku, P.W. (2007). Defining the behavioral phenotype in children with fetal alcohol spectrum disorders: A review. *Neuroscience and Biobehavioral Reviews, 31*(2), 192–201.

Kodituwakku, P.W., Adnams, C.M., Hay, A., Kitching, A.E., Burger, E., Kalberg, W.O., et al. (2006). Letter and category fluency in children with fetal alcohol syndrome from a community in South Africa. *Journal of Studies on Alcohol, 67*(4), 502–509.

Kodituwakku, P., Coriale, G., Fiorentino, D., Aragon, A.S., Kalberg, W.O., Buckley, D., et al. (2006). Neurobehavioral characteristics of children with fetal alcohol spectrum disorders in communities from Italy: Preliminary results. *Alcoholism, Clinical and Experimental Research, 30*(9), 1551–1561.

Kodituwakku, P.W., Handmaker, N.S., Cutler, S.K., Weathersby, E.K., & Handmaker, S.D. (1995). Specific impairments in self-regulation in children exposed to alcohol prenatally. *Alcoholism, Clinical and Experimental Research, 19*(6), 1558–1564.

Kodituwakku, P.W., Kalberg, W., & May, P.A. (2001). The effects of prenatal alcohol exposure on executive functioning. *Alcohol Research and Health, 25*(3), 192–198.

Kodituwakku, P.W., May, P.A., Clericuzio, C.L., & Weers, D. (2001). Emotion-related learning in individuals prenatally exposed to alcohol: An investigation of the relation between set shifting, extinction of responses, and behavior. *Neuropsychologia, 39*(7), 699–708.

Kopera-Frye, K., Dehaene, S., & Streissguth, A.P. (1996). Impairments of number processing induced by prenatal alcohol exposure. *Neuropsychologia, 34*(12), 1187–1196.

Lee, K.T., Mattson, S.N., & Riley, E.P. (2004). Classifying children with heavy prenatal alcohol exposure using measures of attention. *Journal of the International Neuropsychological Society, 10*(2), 271–277.

Lord, C., Rutter, M., Goode, S., Heemsbergen, J., Jordan, H., Mawhood, L., et al. (1989). Autism Diagnostic Observation Schedule: A standardized observation of communicative and social behavior. *Journal of Autism and Developmental Disorders, 19*, 185–212.

Lugo, J.N., Jr., Marino, M.D., Cronise, K., & Kelly, S.J. (2003). Effects of alcohol exposure during development on social behavior in rats. *Physiology and Behavior, 78*(2), 185–194.

Luria, A.R. (1966). *Higher cortical functions in man.* New York: Basic Books.

Lynch, M.E., Coles, C.D., Corley, T., & Falek, A. (2003). Examining delinquency in adolescents differentially prenatally exposed to alcohol: The role of proximal and distal risk factors. *Journal of Studies on Alcohol, 64*(5), 678–686.

Ma, X., Coles, C.D., Lynch, M.E., Laconte, S.M., Zurkiya, O., Wang, D., et al. (2005). Evaluation of corpus callosum anisotropy in young adults with fetal alcohol syndrome according to diffusion tensor imaging. *Alcoholism, Clinical and Experimental Research, 29*(7), 1214–1222.

Malisza, K.L., Allman, A.A., Shiloff, D., Jakobson, L., Longstaffe, S., & Chudley, A.E. (2005). Evaluation of spatial working memory function in children and adults with fetal alcohol spectrum disorders: A functional magnetic resonance imaging study. *Pediatric Research, 58*(6), 1150–1157.

Mattson, S.N., Goodman, A.M., Caine, C., Delis, D.C., & Riley, E.P. (1999). Executive functioning in children with heavy prenatal alcohol exposure. *Alcoholism, Clinical and Experimental Research, 23*(11), 1808–1815.

Mattson, S.N., Gramling, L., Delis, D.C., Jones, K.L., & Riley, E.P. (1996). Global-local processing in children prenatally exposed to alcohol. *Child Neuropsychology, 2,* 165–175.

Mattson, S.N., & Riley, E.P. (1998). A review of the neurobehavioral deficits in children with fetal alcohol syndrome or prenatal exposure to alcohol. *Alcoholism, Clinical and Experimental Research, 22*(2), 279–294.

Mattson, S.N., & Riley, E.P. (1999). Implicit and explicit memory functioning in children with heavy prenatal alcohol exposure. *Journal of the International Neuropsychological Society, 5*(5), 462–471.

Mattson, S.N., Riley, E.P., Delis, D.C., Stern, C., & Jones, K.L. (1996). Verbal learning and memory in children with fetal alcohol syndrome. *Alcoholism, Clinical and Experimental Research, 20*(5), 810–816.

Mattson, S.N., Riley, E.P., Gramling, L., Delis, D.C., & Jones, K.L. (1997). Heavy prenatal alcohol exposure with or without physical features of fetal alcohol syndrome leads to IQ deficits. *Journal of Pediatrics, 131*(5), 718–721.

Mattson, S.N., Riley, E.P., Gramling, L., Delis, D.C., & Jones, K.L. (1998). Neuropsychological comparison of alcohol-exposed children with or without physical features of fetal alcohol syndrome. *Neuropsychology, 12*(1), 146–153.

Mattson, S.N., Riley, E.P., Sowell, E.R., Jernigan, T.L., Sobel, D.F., & Jones, K.L. (1996). A decrease in the size of the basal ganglia in children with fetal alcohol syndrome. *Alcoholism, Clinical and Experimental Research, 20*(6), 1088–1093.

Mattson, S.N., & Roebuck, T.M. (2002). Acquisition and retention of verbal and nonverbal information in children with heavy prenatal alcohol exposure. *Alcoholism, Clinical and Experimental Research, 26*(6), 875–882.

May, P.A. (1995). A multi-level comprehensive approach to the prevention of fetal alcohol syndrome (FAS) and other alcohol related birth defects (ARBD). *International Journal of Addiction, 30,* 549–602.

McGee, C.L., Schonfeld, A.M., Roebuck-Spencer, T.M., Riley, E.P., & Mattson, S.N. (2008). Children with heavy prenatal alcohol exposure demonstrate deficits on multiple measures of concept formation. *Alcoholism, Clinical and Experimental Research, 32*(8), 1388–1397.

Monnot, M., Lovallo, W.R., Nixon, S.J., & Ross, E. (2002). Neurological basis of deficits in affective prosody comprehension among alcoholics and fetal alcohol-exposed adults. *Journal of Neuropsychiatry and Clinical Neurosciences, 14*(3), 321–328.

Nanson, J.L., & Hiscock, M. (1990). Attention deficits in children exposed to alcohol prenatally. *Alcoholism, Clinical and Experimental Research, 14*(5), 656–661.

Nash, K., Rovet, J., Greenbaum, R., Fantus, E., Nulman, I., & Koren, G. (2006). Identifying the behavioural phenotype in fetal alcohol spectrum disorder: Sensitivity, specificity and screening potential. *Archives of Women's Mental Health, 9*(4), 181–186.

Noland, J.S., Singer, L.T., Arendt, R.E., Minnes, S., Short, E.J., & Bearer, C.F. (2003). Executive functioning in preschool-age children prenatally exposed to alcohol, cocaine, and marijuana. *Alcoholism, Clinical and Experimental Research, 27*(4), 647–656.

O'Brien, G., & Yule, W. (2000). *Behavioural phenotypes: Clinics in developmental medicine.* (p. 138). London: MacKeith Press.

O'Connor, M.J., Frankel, F., Paley, B., Schonfeld, A.M., Carpenter, E., Laugeson, E.A., et al. (2006). A controlled social skills training for children with fetal alcohol spectrum disorders. *Journal of Consulting and Clinical Psychology, 74*(4), 639–648.

O'Connor, M.J., Shah, B., Whaley, S., Cronin, P., Gunderson, B., & Graham, J. (2002). Psychiatric illness in a clinical sample of children with prenatal alcohol exposure. *The American Journal of Drug and Alcohol Abuse, 28*(4), 743–754.

Olson, H.C., Feldman, J.J., Streissguth, A.P., Sampson, P.D., & Bookstein, F.L. (1998). Neuropsychological deficits in adolescents with fetal alcohol syndrome: Clinical findings. *Alcoholism, Clinical and Experimental Research, 22*(9), 1998–2012.

Pennington, B.F. (2002). *The development of psychopathology: Nature and nurture.* New York: Guilford Press.

Rasmussen, C. (2005). Executive functioning and working memory in fetal alcohol spectrum disorder. *Alcoholism, Clinical and Experimental Research, 29*(8), 1359–1367.

Riley, E.P., & McGee, C.L. (2005). Fetal alcohol spectrum disorders: An overview with emphasis on changes in brain and behavior. *Experimental Biology and Medicine, 230*(6), 357–365.

Roebuck, T.M., Mattson, S.N., & Riley, E.P. (2002). Interhemispheric transfer in children with heavy prenatal alcohol exposure. *Alcoholism, Clinical and Experimental Research, 26*(12), 1863–1871.

Rolls, E.T., Hornak, J., Wade, D., & McGrath, J. (1994). Emotion-related learning in patients with social and emotional changes associated with frontal lobe damage. *Journal of Neurology, Neurosurgery, and Psychiatry, 57*(12), 1518–1524.

Rosett, H.L., Snyder, P., Sander, L.W., Lee, A., Cook, P., Weiner, L., et al. (1979). Effects of maternal drinking on neonate state regulation. *Developmental Medicine and Child Neurology, 21*(4), 464–473.

Sampson, P.D., Streissguth, A.P., Bookstein, F.L., Little, R.E., Clarren, S.K., Dehaene, P., et al. (1997). Incidence of fetal alcohol syndrome and prevalence of alcohol-related neurodevelopmental disorder. *Teratology, 56*(5), 317–326.

Savage, D.D., Becher, M., de la Torre, A.J., & Sutherland, R.J. (2002). Dose-dependent effects of prenatal ethanol exposure on synaptic plasticity and learning in mature offspring. *Alcoholism, Clinical and Experimental Research, 26*(11), 1752–1758.

Schneider, M.L., Moore, C.F., Gajewski, L.L., Larson, J.A., Roberts, A.D., Converse, A.K., et al. (2008). Sensory processing disorder in a primate model: Evidence from a longitudinal study of prenatal alcohol and prenatal stress effects. *Child Development, 79*(1), 100–113.

Schonfeld, A.M., Mattson, S.N., Lang, A.R., Delis, D.C., & Riley, E.P. (2001). Verbal and nonverbal fluency in children with heavy prenatal alcohol exposure. *Journal of Studies on Alcohol, 62*(2), 239–246.

Schonfeld, A.M., Paley, B., Frankel, F., & O'Connor, M.J. (2006). Executive functioning predicts social skills following prenatal alcohol exposure. *Child Neuropsychology, 12*(6), 439–452.

Schonfeld, A.M., Paley, B., Frankel, F., & O'Connor, M.J. (2008). Behavioral regulation as a predictor of response to children's friendship training in children with fetal alcohol spectrum disorders. *The Clinical Neuropsychologist,* 1–18.

Sowell, E.R., Jernigan, T.L., Mattson, S.N., Riley, E.P., Sobel, D.F., & Jones, K.L. (1996). Abnormal development of the cerebellar vermis in children prenatally exposed to alcohol: Size reduction in lobules I–V. *Alcoholism, Clinical and Experimental Research, 20*(1), 31–34.

Sowell, E.R., Johnson, A., Kan, E., Lu, L.H., Van Horn, J.D., Toga, A.W., et al. (2008). Mapping white matter integrity and neurobehavioral correlates in children with fetal alcohol spectrum disorders. *Journal of Neuroscience, 28*(6), 1313–1319.

Sowell, E.R., Thompson, P.M., Mattson, S.N., Tessner, K.D., Jernigan, T.L., Riley, E.P., et al. (2001). Voxel-based morphometric analyses of the brain in children and adolescents prenatally exposed to alcohol. *Neuroreport, 12*(3), 515–523.

Sowell, E.R., Thompson, P.M., Mattson, S.N., Tessner, K.D., Jernigan, T.L., Riley, E.P., et al. (2002). Regional brain shape abnormalities persist into adolescence after heavy prenatal alcohol exposure. *Cerebral Cortex, 12,* 856–865.

Sowell, E.R., Thompson, P.M., Peterson, B.S., Mattson, S.N., Welcome, S.E., Henkenius, A.L., et al. (2002). Mapping cortical gray matter asymmetry patterns in adolescents with heavy prenatal alcohol exposure. *Neuroimage, 17*(4), 1807–1819.

Spohr, H.L., Willms, J., & Steinhausen, H.C. (2007). Fetal alcohol spectrum disorders in young adulthood. *Journal of Pediatrics, 150*(2), 175–179, 179 e171.

Stratton, K., Howe, C., & Battaglia, F. (Eds.). (1996). *Fetal alcohol syndrome: Diagnosis, epidemilogy, prevention, and treatment.* Washington, DC: National Academy Press.

Streissguth, A.P., Barr, H.M., & Martin, D.C. (1983). Maternal alcohol use and neonatal habituation assessed with the Brazelton scale. *Child Development, 54*(5), 1109–1118.

Streissguth, A.P., Barr, H.M., Olson, H.C., Sampson, P.D., Bookstein, F.L., & Burgess, D.M. (1994). Drinking during pregnancy decreases word attack and arithmetic scores on standardized tests: Adolescent data from a population-based prospective study. *Alcoholism, Clinical and Experimental Research, 18*(2), 248–254.

Streissguth, A.P., Barr, H.M., & Sampson, P.D. (1990). Moderate prenatal alcohol exposure: Effects on child IQ and learning problems at age 7 1/2 years. *Alcoholism, Clinical and Experimental Research, 14*(5), 662–669.

Streissguth, A.P., Barr, H.M., Sampson, P.D., Darby, B.L., & Martin, D.C. (1989). IQ at age 4 in relation to maternal alcohol use and smoking during pregnancy. *Developmental Psychology, 25*, 3–11.

Streissguth, A.P., Barr, H.M., Sampson, P.D., Parrish-Johnson, J.C., Kirchner, G.L., & Martin, D.C. (1986). Attention, distraction and reaction time at age 7 years and prenatal alcohol exposure. *Neurobehavioral Toxicology and Teratology, 8*(6), 717–725.

Streissguth, A.P., Bookstein, F.L., Barr, H.M., Press, S., & Sampson, P.D. (1998). A fetal alcohol behavior scale. *Alcoholism, Clinical and Experimental Research, 22*(2), 325–333.

Streissguth, A.P., Bookstein, F.L., Barr, H.M., Sampson, P.D., O'Malley, K., & Young, J.K. (2004). Risk factors for adverse life outcomes in fetal alcohol syndrome and fetal alcohol effects. *Journal of Developmental and Behavioral Pediatrics, 25*(4), 228–238.

Stromland, K. (2004). Visual impairment and ocular abnormalities in children with fetal alcohol syndrome. *Addiction Biology, 9*(2), 153–157; discussion 159–160.

Sulik, K.K. (2005). Genesis of alcohol-induced craniofacial dysmorphism. *Experimental Biology and Medicine, 230*(6), 366–375.

Thomas, M., & Karmiloff-Smith, A. (2002). Are developmental disorders like cases of adult brain damage? Implications from connectionist modelling. *The Behavioral and Brain Sciences, 25*(6), 727–750; discussion 750–787.

Uecker, A., & Nadel, L. (1996). Spatial locations gone awry: Object and spatial memory deficits in children with fetal alcohol syndrome. *Neuropsychologia, 34*(3), 209–223.

Viding, E., & Blakemore, S.-J. (2007). Endophenotype approach to developmental psychopathology. *Behavior Genetics, 37,* 51–60.

Vygotsky, L.S. (1978). *Mind and society: The development of higher mental process.* Cambridge, MA: Harvard University Press.

Wechsler, D. (1991). *Wechsler Intelligence Scale for Children* (3rd ed.). San Antonio, TX: Harcourt Assessment.

Whaley, S.E., O'Connor, M.J., & Gunderson, B. (2001). Comparison of the adaptive functioning of children prenatally exposed to alcohol to a nonexposed clinical sample. *Alcoholism, Clinical and Experimental Research, 25*(7), 1018–1024.

Willford, J.A., Richardson, G.A., Leech, S.L., & Day, N.L. (2004). Verbal and visuospatial learning and memory function in children with moderate prenatal alcohol exposure. *Alcoholism, Clinical and Experimental Research, 28*(3), 497–507.

Woodcock, R.W. (1973). *Woodcock Reading Mastery Test Manual.* Circle Pines, MN: American Guidance Service.

Wozniak, J.R., Mueller, B.A., Chang, P.N., Muetzel, R.L., Caros, L., & Lim, K.O. (2006). Diffusion tensor imaging in children with fetal alcohol spectrum disorders. *Alcoholism, Clinical and Experimental Research, 30*(10), 1799–1806.

Treatment of Neurogenetic Syndromes

Functional Behavioral Assessment

Its Value in the Treatment of Maladaptive Behaviors in Individuals with Neurogenetic Syndromes

Theodosia R. Paclawskyj

In the literature on people with neurodevelopmental disabilities, maladaptive behaviors such as self-injury, aggression, and property destruction are defined as responses that result in physical harm or imminent risk of harm to the individual, others, or the environment (Lowry & Sovner, 1991). Although physical injury and/or property destruction are the immediate outcomes of the behaviors, the longer-term limitations on daily activities and social interaction increase the risk of emotional disorders for the individual as well as his or her family (Hastings & Brown, 2000; Matson & Coe, 1992). Difficulty managing the individual in the school or vocational setting generally leads to more restrictive placement, with minimal to no contact with typically functioning peers (Borthwick-Duffy, 1994; Jacobson, 1982; Joyce, Ditchfield, & Harris, 2001). Finally, the costs of the above levels of care can be overwhelming for families, and significant state and federal funding is required for chronic care (National Institutes of Health, Consensus Development Panel on Destructive Behavior in Persons with Developmental Disabilities, 1989).

Risk factors for maladaptive behaviors include the diagnosis of severe or profound intellectual disability, difficulties with expressive language, the presence of medical or psychiatric disorders, and diagnosis of specific genetic disorders (Borthwick-Duffy, 1994; Dura, 1997; Emerson et al., 2001). Such disorders include Lesch-Nyhan syndrome, Cornelia de Lange syndrome, and Smith-Magenis syndrome, among others (Basile, Villa, Selicorni, & Molteni, 2007; Dykens & Smith, 1998; Nyhan, 1976).

Unfortunately, research in genetic disorders and research in the assessment and treatment of maladaptive behaviors in individuals with developmental disabilities have followed disparate paths for many years. The investigation of behavioral phenotypes has focused on the identification of overt behaviors and characteristics that have a high probability of manifesting in people with a particular syndrome (Dykens, 1995). Many behavioral phenotypes describe a range of maladaptive behaviors and/or psychiatric symptoms. Maladaptive behaviors are often described topographically (e.g., self-injury may be defined by the forms in which it manifests, such as hand biting, head banging, or hair pulling; Table 9.1) and are reported by frequency of occurrence within the population of interest. Similarly, more global descriptors of behavioral states (e.g., hyperactivity) tend to be documented via standardized behavioral rating scales, with the end goal of defining a behavioral profile of specific genetic disorders.

The numbers of syndrome-specific research studies have grown at a rapid rate over the past decade (Dykens & Hodapp, 2007). However, the majority have been structural rather than functional in focus; that is, the phenotype and/or known neurophysiological etiology is described, but the manifestation of

Table 9.1. Examples of common maladaptive behaviors

Self-injury: Head banging, hand-to-head hitting, face slapping, pulling own hair, biting self, eye gouging, shin kicking, knee-to-head hitting, skin picking, self-scratching, self-pinching, pulling off fingernails or toenails, inserting foreign objects into body orifices, stabbing self with objects, and so forth

Aggression: Hitting, kicking, punching, slapping, head butting, biting, scratching, pinching, choking, hair pulling, grabbing and pulling, pushing/tackling others to the floor, throwing objects at others, spitting at others, and so forth

Property destruction: Throwing objects, breaking objects, ripping items, knocking over furniture, punching/kicking holes in the wall, breaking glass, pulling pictures off the walls, sweeping items off tables, forcefully banging on objects, kicking objects, and so forth

Screaming

Yelling

Pica: Ingestion of inedible items

Elopement: Darting off from adults to another room, outside of the house or classroom, in public, and so forth

Fecal smearing: Smearing fecal matter over self, walls, or objects

the behavior(s) in a dynamic social environment is rarely investigated. The current research challenge is to better investigate the interrelationship between the social environment and the expression of behaviors ascribed to a particular phenotype, especially as such behaviors are mediated by reactions and interpretations of caregivers (Chertkoff-Walz & Benson, 2002; Hodapp & Dykens, 2005).

PART I: APPLIED BEHAVIOR ANALYSIS AND FUNCTIONAL ASSESSMENT

The hallmark of the field of applied behavior analysis (ABA; formerly referred to as "behavior modification") is assessing and understanding functional relationships between behavior and the environment. The term *applied behavior analysis* was coined by Baer, Wolf, and Risley (1968) to describe the generalization of principles of learning derived from the laboratory to problems of social significance (therefore the term *applied*). *Behavior* defined the unit of analysis as observable and measurable responses of humans, whereas *analysis* emphasized a reliance on methods to demonstrate rather than hypothesize conclusions, with the use of experimental control. ABA research studies on maladaptive behaviors have been con-

ducted for many decades and constitute the most comprehensive body of research that the field has produced (Didden, Duker, & Korzilius, 1997; Matson & Coe, 1992).

Within ABA, the methods of functional assessment of maladaptive behaviors have been described as the most effective tool for the development of successful individualized behavioral interventions in people with intellectual disabilities (Laties & Mace, 1993; Repp, Felce, & Barton, 1988). By identifying the relationships between a maladaptive behavior and events or stimuli in the environment, functional assessment can predict which interventions will be effective prior to the actual implementation of a behavioral treatment (Iwata, Vollmer, & Zarcone, 1990). Current best practice requires a functional assessment to be completed prior to the start of treatment (Arndorfer & Miltenberger, 1993; Association for Behavior Analysis, Task Force on the Right to Effective Behavioral Treatment, 1988; National Institutes of Health, Consensus Development Panel on Destructive Behaviors in Persons with Developmental Disabilities, 1989).

Didden, Duker, and Korzilius (1997) conducted a meta-analytic study on the effectiveness of treatments for maladaptive behaviors in people with intellectual disabilities. Examining variables such as functional level of the client, etiology of the behavior, intervention setting, duration of assessment and treatment, secondary handicaps, and pretreatment functional assessment, they determined that use of functional assessment was the only significant variable in the prediction of treatment success (defined as a clinically significant reduction in the target behavior), replicating similar findings by Scotti, Evans, Meyer, and Walker (1991). From a behavioral standpoint, any intervention for maladaptive behaviors must be preceded by a functional assessment for successful treatment to occur, regardless of diagnosis.

Therefore, the techniques of functional assessment hold promise for the development of effective treatments for maladaptive behaviors in individuals with neurodevelopmental disabilities of varying etiologies. At minimum, they can identify which, if any, factors in the social environment may have an impact on maladaptive behaviors; at best, they can prescribe an effective individualized treatment approach.

Concepts in Applied Behavior Analysis

Reinforcement

Operant conditioning is the process of learning any voluntary behavior. When speaking of the "function" of a maladaptive behavior, an applied behavior analyst literally is referring to the reason(s) it persists, even if an individual has the cognitive ability to understand that such behavior is not socially acceptable. The persistence of a behavior across time occurs because of the naturally occurring phenomenon of reinforcement, defined as the mechanism through which the probability of recurrence of a behavior increases based on the consequence that follows that behavior. If the consequence is a stimulus that is *added* to the environment ("contingent presentation"), the process is referred to as positive reinforcement. In contrast, if the consequence is that a stimulus is *removed* from the environment ("contingent removal"), the process is negative reinforcement. A common misconception is that *positive* and *negative* refer to consequences that are seen as either socially good or bad or that *negative reinforcement* refers to punishment; however, this is not the case. An added stimulus is simply something that the person "receives" following the behavior (positive reinforcement), whereas a stimulus that is removed refers to something being "avoided or delayed" following the behavior (negative reinforcement). Both processes result in behavior persistence or increase over time. Reinforcement is in place for learning and maintenance of any operant behavior, and awareness of reinforcement is not necessary for the behavior to occur (Alberto & Troutman, 2006; Cooper, Heron, & Heward, 2007). See Table 9.2 for examples of positive and negative reinforcement.

Table 9.2. Examples of positive reinforcement and negative reinforcement

Positive reinforcement
1. Employee continues to offer suggestions during staff meetings after receiving compliments from the director
2. Adult continues to purchase lottery tickets after winning $50.00
3. Child continues to raise hand in class after being praised for good participation
4. Student continues to insult teacher after other students laugh

Negative reinforcement
1. Adult continues to purchase a new brand of pain reliever when headaches are reduced
2. Driver more consistently parks in appropriate spots after paying a hefty fine
3. Child continues to walk a different way home to avoid a loud dog
4. Student continues to self-injure during class when fine motor tasks are removed as a result

Reinforcement of Maladaptive Behavior

Although a behavior may be socially unacceptable in nature, if it persists over time then reinforcement is in place. Given that reinforcement is subcategorized into positive and negative, the common reinforcers for maladaptive behaviors are likewise subdivided. A further division lies in the physical domain of the reinforcer. If it is extrinsic to the individual, it is classified as a *social reinforcer;* if intrinsic, it is termed *nonsocial.* Therefore, as seen in Table 9.3, four subdivisions are possible (Alberto & Troutman, 2006; Carr, 1977; Cooper, Heron, & Heward, 2007; Matson & Vollmer, 1995; O'Neill, Horner, Albin, Storey, & Sprague, 1990).

Within the category of social positive reinforcement, two consequences commonly occur. The first is attention or some type of reaction from others, whether caregivers or peers. The reaction may be commonly perceived as socially undesirable (e.g., a reprimand); however, if a child's behavior increases following a reprimand, then that reaction serves to reinforce the behavior. The other social positive reinforcer is access to a tangible item; the common scenario is that in which a child has to wait for or is denied access to a preferred item such as a toy or food, and when the child engages in maladaptive behavior, this sometimes results in access to the item in an attempt by the caregiver to calm the child. What inadvertently develops is a pattern of responding in which the child's maladaptive behavior persists in response to periodically receiving the desired item.

Within the domain of social negative reinforcement, the issue becomes one of escape or avoidance. Most often this occurs in educational or vocational settings during which tasks are presented; if maladaptive behaviors ensue and are followed by a delay in or removal of the task, a reinforcement pattern develops. An identical pattern may exist if the nonpreferred scenario is one of an undesired activity or interaction with others when the child wants to be alone.

The variables responsible for nonsocial positive reinforcement are less clear. As it is not possible within ABA to observe the physiological or emotional responses occurring in the body, especially in the case of an individual who is functionally nonverbal and cannot report on internal states, the specific reinforcement mechanism can only be hypothesized rather than confirmed. In some cases, sensory stimulation from a behavior such as self-injury may be responsible for the persistence and maintenance of this behavior. Another possible nonsocial positive reinforcer may be the release of endorphins in cases of severe self-injury. Some individuals whose self-injury diminishes following treatment with an opiate antagonist such as naltrexone are assumed to be reinforced by the release of endorphins (Symons, Thompson, & Rodriguez, 2004).

Finally, nonsocial negative reinforcement is equally challenging to discern. Response to medical treatment may be the only method of determining that amelioration of discomfort may be the reinforcer for maladaptive behaviors. For example, some female adolescents with cyclical increases in maladaptive behavior during menses may

Table 9.3. Subcategories of reinforcement for maladaptive behaviors

	Positive	Negative
Social	Attention/reaction from others Access to preferred tangible items	Escape/avoidance of tasks Escape from nonpreferred activities/interactions
Nonsocial	Sensory stimulation Endorphin release	Amelioration of physical/emotional discomfort

respond to prophylactic treatment with analgesics during the menstrual cycle (Carr, Smith, Giacin, Whelan, & Pancari, 2003).

The essence of the functional assessment is to identify which of the above reinforcement categories play a role in the persistence of maladaptive behaviors for an individual. In many cases, multiple reinforcers may be present, especially when multiple topographies of maladaptive behaviors exist. The challenge comes in selecting and implementing sufficient assessment strategies to be able to demonstrate via data collection that functional relationships exist.

Functional Relationships

As mentioned previously, the interaction between two environmental events that change in a predictable manner—in this case, a behavior and a reinforcer—is defined as a *functional relationship*. To identify a functional relationship, ABA relies on data collection and data graphing to determine if a behavior varies in the manner expected when a particular variable is manipulated. Therefore, if data collected on a specific behavior show persistence or increase in the behavior, the behavior analyst proceeds with assessment to determine what reinforcement may be present for that particular behavior. This process of data collection is fundamental to functional assessment (Iwata, Vollmer, & Zarcone, 1990; Linscheid, Iwata, & Foxx, 1996; Neef & Iwata, 1994; Sprague & Horner, 1995; Vollmer & Smith, 1996).

Functional Assessment

A functional assessment defines the topography of the behavior, its frequency and duration, the antecedent events that may occasion the behavior, and the reinforcers that may maintain the behavior (Lennox & Miltenberger, 1989; Sprague & Horner, 1995). In a majority of circumstances, the use of functional assessment may improve the effectiveness and efficiency of behavioral treatment (Horner, 1994). Failure to use a functional assessment, however, may lead to delay in implementing effective treatment procedures (Lennox & Miltenberger, 1989), countertherapeutic effects on the target behavior from an arbitrarily selected treatment (Solnick, Rincover, & Peterson, 1977), or unnecessary exposure to aversive procedures (Iwata, Dorsey, Slifer, Bauman, & Richman, 1982).

Terminology

The term *functional assessment* refers to the set of nonexperimental procedures used to identify possible functions of specific behaviors (Arndorfer & Miltenberger, 1993; Sturmey, 1994; Vollmer & Smith, 1996). In contrast, *functional analysis* refers to systematic procedures that can be used to experimentally manipulate environmental conditions in a controlled setting (Horner, 1994; Lennox & Miltenberger, 1989). This distinction is observed throughout the following literature review.

Methods of Assessment: Caregiver Interview

A complete behavioral interview should elicit information about the topography of the behavior setting events, antecedent and consequent events, and rate of the behavior. Setting events (now referred to as Motivating Operations by most behavior analysts; Cooper, Heron, & Heward, 2007) may include a multitude of variables such as physiological states of hunger, thirst, fatigue, physical illness, medication side effects, or others; social stressors such as exposure to aversive events (e.g. family arguments or even domestic violence); recent pleasant events; deprivation from or satiation to individual reinforcers; ambient environmental conditions such as lighting, temperature, noise level, number of people present, and others; specific teaching strategies (e.g., one-to-one, group instruction, independent work); or scheduling issues such as changes in daily routine, cancellation of preferred activities, or a long period of unstructured time.

The Functional Analysis Interview Form (FAIF) is the only structured interview available to assess function of behavior. The FAIF is a component of the manual *Functional Assessment of Problem Behavior: A Practical Assessment Guide,* published by O'Neill, Horner, Albin, Storey, and Sprague (1990). The FAIF consists of nine sections of open-ended questions and short forms in areas of 1) the problem behaviors, 2) potential ecological events, 3) events and situations that predict occurrences of the problem behavior, 4) identifying the functions of the problem behavior, 5) the efficiency of the problem behavior, 6) the person's primary mode of communication, 7) functional alternative behaviors, 8) history of the problem behaviors, and 9) previous attempts at treatment. The interview requires 45–90 minutes for completion. O'Neill et al. emphasize that their manual is intended to be used with flexibility by clinicians with training in applied behavior analysis.

The Functional Analysis Screening Tool (FAST; Iwata & DeLeon, 1996) is a more recently developed instrument that consists of 4 questions on the informant-client relationship, 8 open-ended questions on the topography of behavior, and 16 yes/no questions covering the four domains of social and nonsocial functions. Research on the FAST still is emerging, although it is commonly used in applied settings (Ellingson, Miltenberger, & Long, 1999).

Methods of Assessment: Behavioral Checklists

Although highly desirable as the simplest forms of functional assessment, behavioral checklists alone cannot constitute a reliable and valid method of assessment. Despite the lack of required training for administration, brevity of administration time, and possibility of assessing low-rate maladaptive behaviors, research has not identified an instrument with good psychometric properties and predictive value for treatment. However, the utility of such checklists lies in providing corroborating data for other methods of interview and direct observation. Two of the most commonly used scales are the Motivation Assessment Scale (MAS; Durand & Crimmins, 1988) and the Questions About Behavioral Function (QABF; Matson & Vollmer, 1995).

The MAS consists of prespecified descriptions of situations in which maladaptive behavior may occur. Using a 7-point Likert-type scale, the informant must rate the likelihood that problem behavior will occur in each circumstance. The psychometric properties of the MAS have been extensively studied; however, most studies have reported inadequate internal reliability and validity (Singh et al., 1993; Spreat & Connelly, 1996; Zarcone, Rodgers, Iwata, Rourke, & Dorsey, 1991). Solid psychometric data tend to be found only in those studies using participants with high-frequency self-injury and informants with training in behavior analysis.

The QABF has had some initial demonstrations of better psychometric properties (Matson & Vollmer, 1995; Paclawskyj, Matson, Rush, Smalls, & Vollmer, 2000). The QABF is a 25-item questionnaire scored on a 4-point Likert-type scale that is used to identify behavioral function. In addition to assessing escape, attention, tangible, and nonsocial reinforcement, the QABF examines other variables, such as physical distress and social avoidance. The QABF was found to identify functions for 84% of participants in a recent study. In addition, use of the QABF significantly resulted in more success with behavioral treatment (Matson, Bamburg, Cherry, & Paclawskyj, 1999). Comparison with analog assessments yielded only moderate correlations (Paclawskyj, Matson, Smalls, & Vollmer, 2001).

With both interviews and checklists, the examiner does not observe the behavior directly, but must rely on the informant's recollection, which may not always be accurate (Kazdin, 1980; Linscheid et al., 1996). Informants must be able to accurately discriminate and report a client's behavior in behavior analytic terms (Sturmey, 1996). Rating scales in general may be subject to errors of leniency or severity, central tendency, and the halo effect (Pedhazur & Pedhazur-Schmelkin, 1991). Errors of leniency or severity imply a tendency to give ratings that are consistently too high or too low, whereas errors of central tendency suggest avoidance of extreme categories. With the halo effect, a rater's general impressions bias ratings on more specific and distinct aspects of the issue being assessed.

Methods of Assessment: Direct Observation

Direct observation methods are a foundation of behavior analysis (Baer et al., 1968), and the technology for conducting direct observations has become more refined over the years. Two commonly used formats for direct observations include the scatterplot and the antecedent-behavior-consequence (ABC) assessment (Lennox & Miltenberger, 1989).

Touchette, MacDonald, and Langer (1985) introduced the scatterplot assessment as a means to compare levels of the target behavior with time of day/activity variables. Each occurrence of the target behavior is plotted on a grid with time of day in rows and consecutive days in columns. Differentially high rates of the behavior can be observed by patterns on the temporal grid. When behaviors occur at particular times of the day, a clinician can further examine those activities for further variables to assess, including type of activity, effort involved, people present, physical location, time relationships to meals and medications, and so forth. Although the scatterplot does not provide information on

all of the maintaining variables of the behavior, it provides useful preliminary information that guides the direction of future assessment. See Figure 9.1 for an example of scatterplot recording.

The ABC assessment involves more comprehensive recording in that the observer notes each instance of the target behavior and the events that immediately preceded and followed it (Bijou, Peterson, & Ault, 1968; Kazdin, 1980). To note antecedents, the observer must record what happened in the environment immediately prior to the episode of maladaptive behavior: the activity, any specific instruction given or comment made to the individual, any items given to or removed from the individual, any noises or sudden changes in the environment, the specific people in the environment, and so forth, in as much detail as possible. Similarly, the consequences are described with equal precision: what stopped, changed, and began after the behavior occurred; how caregivers and peers responded; what was physically altered in the environment; and so forth. Antecedents and consequences are documented for each instance of target behavior. When sufficient data are collected over time, the clinician evaluates the data for suspected patterns—that is, consistent precursors to or consequences of the maladaptive behaviors.

The primary advantage of direct observation is that one can observe the range of antecedent and consequent events in a manner more objective than verbal report (Linscheid et al., 1996). The method has low cost relative to experimental techniques and can pinpoint time of day or other specific variables (Sprague & Horner, 1995). Mace (1994) noted that direct observation may often be used to identify idiosyncratic reinforcement conditions and provide an estimate of reinforcer schedules established in the natural environment.

The major limitation of direct observation is that it yields correlational data that do

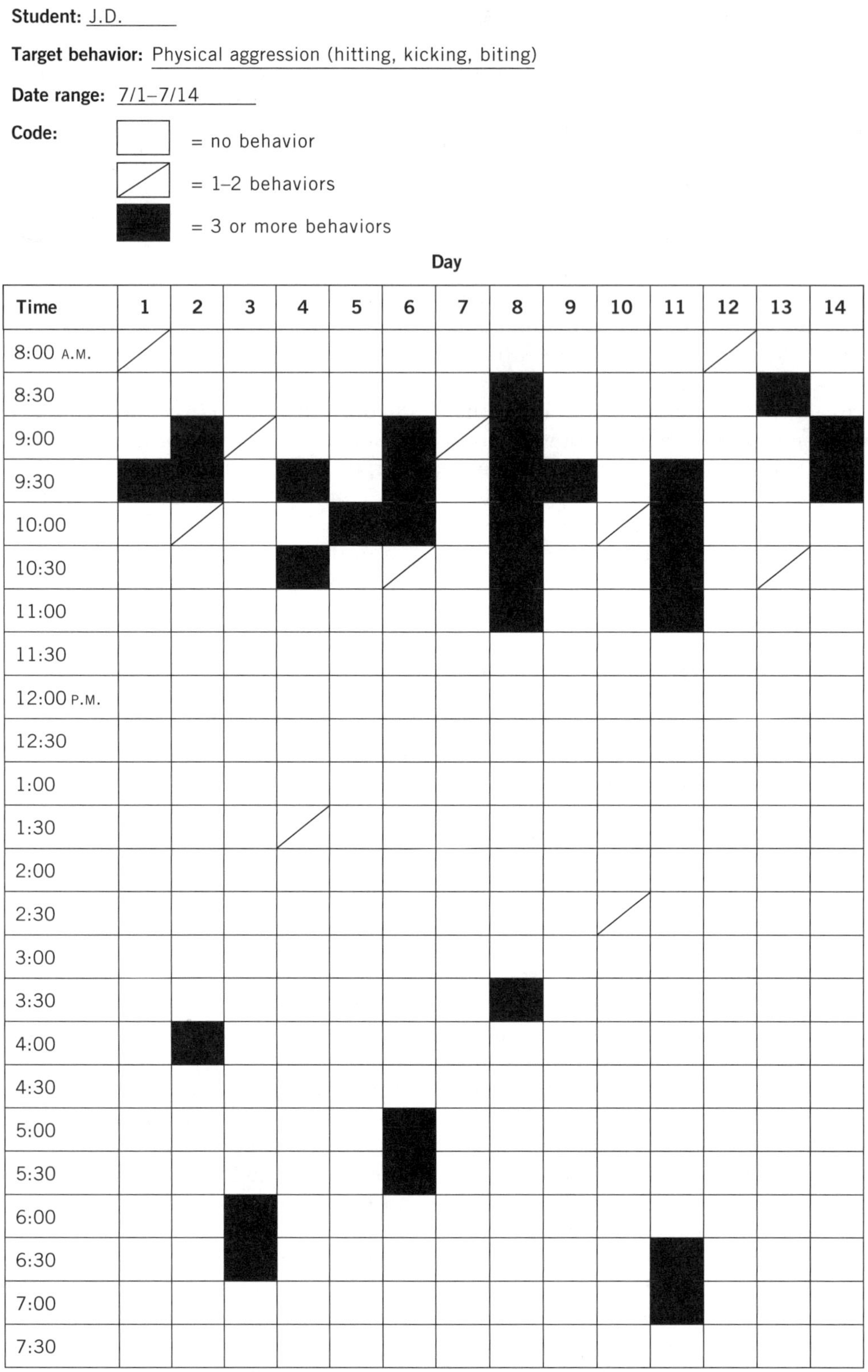

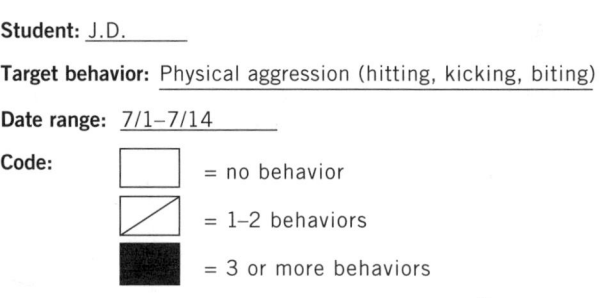

Student: J.D.

Target behavior: Physical aggression (hitting, kicking, biting)

Date range: 7/1–7/14

Code:
☐ = no behavior
◩ = 1–2 behaviors
■ = 3 or more behaviors

Day

Time	1	2	3	4	5	6	7	8	9	10	11	12	13	14
8:00 A.M.	◩											◩		
8:30								■					■	
9:00		■	◩	■		■	◩	■						■
9:30	■	■		■		■		■			■			■
10:00		◩		■	■	■		■		◩	■			
10:30				■		◩		■			■		◩	
11:00								■			■			
11:30														
12:00 P.M.														
12:30														
1:00														
1:30				◩										
2:00														
2:30										◩				
3:00														
3:30								■						
4:00		■												
4:30														
5:00						■								
5:30						■								
6:00			■											
6:30			■								■			
7:00											■			
7:30														

Figure 9.1. Sample scatterplot assessment data sheet.

142

not always correspond to experimental results (Mace, 1994; Sprague & Horner, 1995). An example of correlational data would be the child who bangs her head and receives teacher attention each time. The hypothesis from this observation may be that attention maintains the behavior. However, the child may actually have a sinus headache, and a treatment of teaching request for attention may not reduce the behavior. Such correlational errors may be due to observation of caregivers who are inconsistent in their response to the target behavior (Lerman & Iwata, 1993).

Functional Analyses

In contrast to methods of functional assessment that yield hypothesized functions for behavior, functional analysis techniques involve a demonstration of behavioral function utilizing the test-control methodology of single-case research design (Alberto & Troutman, 2006; Cooper, Heron, & Heward, 2007). Analog conditions that replicate common scenarios in an individual's environment are simulated in a clinic setting; typically such conditions are simulated for 10- to 20-minute sessions. A now-commonly used set of procedures was first described by Iwata and colleagues (1982). Three test conditions and one control condition were developed to assess the following functions: attention from others, escape from task demands, and potential nonsocial reinforcement. Each condition was simulated for a 10-minute period of time; all conditions were repeated until differentiation across conditions was seen in the graph of rates of behavior across conditions. In the attention condition, the therapist pretended to be engaged in work and ignored the child unless the child displayed a maladaptive behavior, at which point the therapist provided a brief reprimand. Children with a history of displaying maladaptive behavior reinforced by attention tended to display high rates of behavior in this condition. In the demand condition, academic demands were systematically presented to the child and removed for 30 seconds contingent on maladaptive behaviors to assess a potential escape/avoidance function. In the alone condition, the child was observed through a one-way mirror to determine if the maladaptive behavior persisted in the absence of environmental and social stimulation to evaluate the possibility of a nonsocial function. These three test conditions were compared with a control condition of play in which attention was delivered on a frequent schedule, demands were absent, and preferred items were present. If a child displayed high rates of behavior in one or more test conditions when compared with the control condition, that test condition then served to demonstrate the function of the maladaptive behavior.

Functional analyses provide the only empirical demonstration of functional relationships and the most direct and reliable match between function and treatment (Mace, 1994; Sprague & Horner, 1995). The validity of this approach has been established for self-injurious behavior (Day, Rea, Schussler, Larsen, & Johnson, 1988), aggression (Slifer, Ivancic, Parrish, Page, & Burgio, 1986), stereotypy (Sturmey, Carlsen, Crisp, & Newton, 1988), and disruptive behavior (Carr & Durand, 1985). However, this technique is the most costly to administer and requires the most extensive training; therefore, it is not easily replicated by teachers or other caregivers (Lennox & Miltenberger, 1989; Sprague & Horner, 1995). Administration may require from 1 to 2 weeks of several sessions per day (Iwata et al., 1994). When behaviors are also of high intensity and risk (e.g., sexual assault), a functional analysis may not be ethically appropriate (Vollmer & Smith, 1996).

Comprehensive Assessment

There are a number of issues that clinicians face when selecting which of these method-

Student: J.D.

Target behavior: Physical aggression (hitting, kicking, biting)

Date: 7/1

Time	Antecedent	Behavior	Consequence
8:35 A.M.	J. seated at desk, staff member presented five-piece puzzle.	J. hit staff member and pushed puzzle to the floor.	Staff member told J., "I'm going to wait until you calm down before I work with you."
9:45 A.M.	J. standing in line behind M. waiting for the bathroom; M. stumbled backward and fell into J.	J. screamed and kicked M. twice before staff member came; J. then bit staff member on the arm.	Staff member put J. in time-out chair for 2 minutes.

Figure 9.2.   Sample antecedent-behavior-consequence (ABC) assessment data collection sheet.

ologies to use. Surveys have found that clinicians in applied settings used interviews, checklists, and ABC data collection (Figure 9.2) and typically were not able to conduct functional analyses because of the level of resources required (Desrochers, Hile, & Williams-Moseley, 1997; Ellingson, Miltenberger, & Long, 1999). Several authors have called for the use of multiple assessment measures, reasoning that concordance in results would provide the most valid conclusions on functional relationships. Lennox and Miltenberger (1989) and Mace, Lalli, and Pinter-Lalli (1991) stressed that although experimental methods provide the most rigorous methodology, direct observation and rating scales are still essential in that they contribute to a multimodal assessment and minimally provide more benefit than treatment selection without any method of functional assessment. Even if these indirect methods identify the function in fewer cases than analog assessments, they would still reduce the

amount of time before treatment is implemented (Vollmer, Marcus, Ringdahl, & Roane, 1995).

PART II: FUNCTIONAL ASSESSMENT AND GENETIC DISORDERS

Research in ABA has tended to de-emphasize diagnosis and stress the topography and function of maladaptive behaviors. The unfortunate outcome is that even with the small sample sizes used in single-case research design, participant grouping may be too heterogeneous and result in significant sources of error. That is, if a participant with a genetic disorder and a behavioral phenotype including maladaptive behavior is included in a sample with others who may not have a biological predisposition to such behavior, the results may be clouded, albeit to an uncertain degree. Both individuals may develop a learned pattern of responding if their self-injury is consistently followed by a reinforcing consequence, such as attention from caregivers. However, the long-term prognosis of maintained behavior reduction following ABA-based treatment and the risk of recurrence is unknown if different neurophysiological mechanisms mediate the predisposition to display the behavior. The existing research in functional assessment of maladaptive behaviors in specific populations of individuals with genetic disorders is a promising start and may ultimately serve to answer these long-term questions.

Lesch-Nyhan Syndrome

The syndrome that perhaps most typifies an idiosyncratic biological predisposition to self-injury is Lesch-Nyhan syndrome. Individuals with this syndrome have a deficiency in the enzyme HGPRT (Nyhan, 1976), which results in a complex neurobiological profile of excessive production of uric acid. Treatment with allopurinol, however, does not result in amelioration of symptoms. The majority of individuals display severe intellectual disability and are nonambulatory, requiring full physical support to remain upright. The onset of self-injury in these individuals tends to occur by age 2 when teeth erupt, and is rapid and severe (Bergen, Holborn, & Scott-Huyghebaert, 2002). Lip, cheek, and finger biting are the common initial topographies of self-injury, whereas many individuals display progressively more topographies that can include picking, scratching, and catching fingers and body parts in wheelchairs or furniture. Many individuals express a preference for mechanical restraint devices and become upset if such devices are removed. In addition, complete extraction of all teeth also is a common strategy to reduce the persistent tissue damage from self-biting.

Yet despite the intense and damaging forms of self-injury that are an integral feature of this disorder, initial functional assessment research across eight single-case design studies has demonstrated that some individuals tend to display rates of self-injury that vary depending on the level of social attention available in the environment, and that contingent withdrawal of attention via ignoring or time-out can yield decreases in rates of self-injury in some individuals (e.g., Bergen, Holborn, & Scott-Huyghebaert, 2002; Hall, Oliver, & Murphy, 2001). In the former study, the authors conducted a functional analysis to specifically assess access to attention as a potential function of self-injury. Four conditions were conducted in which attention was provided for all attempts at self-injury (eye gouging; attempts were defined as lifting arms above the shoulders), attention was delivered contingent on behavior incompatible with self-injury, noncontingent attention was delivered throughout the session, and no attention was provided throughout the session. Ironically, they found the lowest rates of attempts in the attention for attempts and no attention conditions. Bergen and colleagues concluded that the higher rates of self-injury attempts occurred in the other two sessions as attention served as a trigger (discriminative stimulus) for self-injury that would be followed by additional attention. This argument could be plausible but was not confirmed by the data; in fact, an alternative hypothesis of high rates of attention creating an overly stimulating environment was not evaluated. This would have necessitated adding a condition in which attention was removed contingent on self-injury; in this manner, escape from overstimulation could have served as a negative reinforcer for the self-injury. Ironically, Obi (1997) described a case study in which the hypothesized function was nonsocial negative reinforcement in the form of alleviating anxiety from the removal of restraints. However, this hypothesis was not confirmed via functional analysis, and observable behaviors that could indicate anxiety were not assessed.

In contrast, the study by Hall and colleagues (2001) examined the display of self-injury in three young boys across an 18-month period. Caregiver interviews and direct observations were conducted at 3- to 4-month intervals. In some circumstances, the boys' self-injury was noted to increase when they were in an environment with low caregiver attention; however, no analog sessions were conducted to replicate this situation.

Cornelia de Lange Syndrome

Cornelia de Lange syndrome is another genetic disorder in which self-injurious behavior is highly probable, with estimated proportions of 36%–63% of the population; however, it tends to develop as children grow older (Basile, Villa, Selicorne, & Volteni, 2007). Oliver, Arron, Hall, Sloneem, Forman, and McClintock (2006) assessed 16 individuals with Cornelia de Lange syndrome in a functional analysis focused on the continuous presence or absence of social attention. Nine participants displayed some level of self-injury during the assessment. For three of these nine participants, self-injury was significantly higher in the no-attention condition than in the continuous-attention condition, suggesting that self-injury may occur to access adult interaction. However, the remaining six participants with self-injury did not show consistent patterns, indicating that the function(s) of their self-injury did not include a social positive component. As with Lesch-Nyhan syndrome, these findings demonstrate that baseline levels of self-injury originally mediated by neurophysiological variables may also come under environmental control in some individuals, necessitating both an awareness of the potential development of social functions for the behavior and the need for behavioral intervention to reduce self-injury to more manageable levels.

Smith-Magenis Syndrome

Smith-Magenis syndrome (see Chapter 2, this volume) also has a behavioral phenotype that describes a high probability of self-injury, especially self-biting, self-hitting, and, in approximately 25% of individuals, peeling off finger/toenails and/or inserting objects into body orifices. Additional maladaptive behaviors include aggression, tantrums, property destruction, and stereotypies, among others. Significant sleep disturbance with an inverted circadian rhythm cycle and diminished rapid eye movement sleep characterizes approximately 75% of individuals with this syndrome; sleep disturbance can significantly increase the probability of maladaptive behaviors (Dykens & Smith, 1998; Finucane, Dirrigl, & Simon, 2001). One trend noted in this population is that as children get older, the number of specific topographies of self-injury increases. Nail yanking in particular does not manifest until the early teens. Finucane et al. (2001) noted that there may be complex factors present in this pattern. From the neurophysiological standpoint, it is possible that as sensory neuropathy progresses as a function of the syndrome, more intense forms of self-injury appear. However, nail yanking is seen by caregivers as very high risk, thereby resulting in a high probability of reinforcement from the social environment. These authors merely posed this hypothesis; clearly, this area requires experimental analysis.

Prader-Willi Syndrome

Although the disorder is associated with primary issues of hyperphagia, lack of satiety, and behavior problems associated with denied access to food (a clear functional relationship), an associated feature of Prader-Willi syndrome is self-injurious skin picking, seen in 58%–86% of individuals with this syndrome (see Chapter 5, this volume). Skin picking is positively correlated with compulsive behaviors in this population and therefore has been conceptualized and treated as such using pharmacological (selective serotonin reuptake inhibitors) and behavioral (response prevention)

patterns of functions warrants continued investigation, as such information could drive both treatment of maladaptive behaviors and prevention of such in young children who do not as yet manifest them.

SUMMARY

Functional assessment procedures allow for an in-depth, individualized assessment of maladaptive behaviors that can lead to a high probability of successful reduction of such behaviors with treatment. Despite the potential neurophysiological origins of maladaptive behaviors in individuals with genetic disorders, such behaviors may nonetheless come under control of environmental contingencies and persist or exacerbate without adequate behavioral assessment and treatment. Multiple potential responses by caregivers can inadvertently reinforce the occurrence of maladaptive behaviors and bring them under operant control. In some instances, common reactions of reprimanding a child or removing him or her from the environment may result in increased behaviors when adult reactions or avoidance of nonpreferred activities is reinforcing for the child.

Clinicians in any discipline should be aware of such issues so as to better prescribe a comprehensive assessment and treatment mechanism for a patient presenting with maladaptive behaviors that may correspond to a behavioral phenotype. The presence of such behaviors does not necessarily imply that they will persist across the life span. In contrast, early behavioral treatment may prevent a further increase in frequency and severity as the child ages. Further research is warranted to better understand syndrome-specific patterns in terms of behavioral function.

REFERENCES

Alberto, P.A., & Troutman, A.C. (2006). *Applied behavior analysis for teachers* (7th ed.). Upper Saddle River, NJ: Pearson Merrill Prentice Hall.

Arndorfer, R.E., & Miltenberger, R.G. (1993). Functional assessment and treatment of challenging behavior: A review with implications for early childhood. *Topics in Early Childhood Special Education, 13,* 82–105.

Association for Behavior Analysis, Task Force on the Right to Effective Behavioral Treatment. (1988). The right to effective behavioral treatment. *Journal of Applied Behavior Analysis, 21,* 381–384.

Baer, D.M., Wolf, M.M., & Risley, T.R. (1968). Some current dimensions of applied behavior analysis. *Journal of Applied Behavior Analysis, 1,* 91–97.

Basile, E., Villa, L., Selicorni, A., & Molteni, M. (2007). The behavioural phenotype of Cornelia de Lange syndrome: A study of 56 individuals. *Journal of Intellectual Disability Research, 51,* 671–681.

Bergen, A.E., Holborn, S.W., & Scott-Huyghebaert, V.C. (2002). Functional analysis of self-injurious behavior in an adult with Lesch-Nyhan syndrome. *Behavior Modification, 26,* 187.

Bijou, S.W., Peterson, R.F., & Ault, M.H. (1968). A method to integrate descriptive and field studies at the level of data and empirical concepts. *Journal of Applied Behavior Analysis, 1,* 175–191.

Borthwick-Duffy, S. (1994). Prevalence of destructive behaviors: A study of aggression, self-injury, and property destruction. In T. Thompson & D.B. Gray (Eds.), *Destructive behavior in developmental disabilities: Diagnosis and treatment* (pp. 3–23). Thousand Oaks, CA: Sage.

Carr, E.G. (1977). The motivation of self-injurious behavior: A review of some hypotheses. *Psychological Bulletin, 84,* 800–816.

Carr, E.G., & Durand, V.M. (1985). Reducing problem behaviors through functional communication training. *Journal of Applied Behavior Analysis, 18,* 11–126.

Carr, E.G., Smith, C.E., Giacin, T.A., Whelan, B.M., & Pancari, J. (2003). Menstrual discomfort as a biological setting event for severe problem behavior: Assessment and intervention. *American Journal on Mental Retardation, 108,* 117–133.

Chertkoff-Walz, N., & Benson, B.A. (2002). Behavioral phenotypes in children with Down syndrome, Prader-Willi syndrome, or Angelman syndrome. *Journal of Developmental and Physical Disabilities, 14,* 307–321.

Cooper, J.O., Heron, T.E., & Heward, W.L. (2007). *Applied behavior analysis* (2nd ed.). New York: Prentice Hall.

Day, R.M., Rea, J.A., Schussler, N.G., Larsen, S.E., & Johnson, W.L. (1988). A functionally based approach to the treatment of self-injurious behavior. *Behavior Modification, 12,* 565–589.

Desrochers, M.N., Hile, M.G., & Williams-Moseley, T.L. (1997). Survey of functional assessment procedures used with individuals who display mental retardation and severe problem behaviors. *American Journal on Mental Retardation, 101,* 535–546.

Didden, R., Duker, P.C., & Korzilius, H. (1997). Meta-analytic study on treatment effectiveness for problem behaviors with individuals who have mental retardation. *American Journal on Mental Retardation, 101,* 387–399.

Didden, R., Korzilius, H., & Curfs, L.M.G. (2007). Skin-picking in individuals with Prader-Willi syndrome: Prevalence, functional assessment, and its comorbidity with compulsive and self-injurious behaviors. *Journal of Applied Research in Intellectual Disabilities, 20,* 409–419.

Dura, J. (1997). Expressive communicative ability, symptoms of mental illness, and aggressive behavior. *Journal of Clinical Psychology, 53,* 307–318.

Durand, V.M., & Crimmins, D.B. (1988). Identifying the variables maintaining self-injurious behavior. *Journal of Autism and Developmental Disorders, 18,* 99–117.

Dykens, E.M. (1995). Measuring behavioral phenotypes: Provocations from the "new genetics." *American Journal on Mental Retardation, 99,* 522–532.

Dykens, E.M., & Hodapp, R.M. (2007). Three steps toward improving the measurement of behavior in behavioral phenotype research. *Child and Adolescent Psychiatric Clinics of North America, 16,* 617–630.

Dykens, E.M., & Smith, A.C.M. (1998). Distinctiveness and correlates of maladaptive behaviour in children and adolescents with Smith-Magenis syndrome. *Journal of Intellectual Disability Research, 42,* 481–489.

Ellingson, S.A., Miltenberger, R.G., & Long, E.S. (1999). A survey of the use of functional assessment procedures in agencies serving individuals with developmental disabilities. *Behavioral Interventions, 14,* 187–198.

Emerson, E., Kiernan, C., Alborz, A., Reeves, D., Mason, H., Swarbrick, R., et al. (2001). The prevalence of challenging behaviors: A total population study. *Research in Developmental Disabilities, 22,* 77–93.

Finucane, B., Dirrigl, K.H., & Simon, E.W. (2001). Characterization of self-injurious behaviors in children and adolescents with Smith-Magenis syndrome. *American Journal on Mental Retardation, 106,* 52–58.

Hall, S., Oliver, C., & Murphy, G. (2001). Self-injurious behaviour in young children with Lesch-Nyhan syndrome. *Developmental Medicine & Child Neurology, 43,* 745–749.

Hastings, R.P., & Brown, T. (2000). Functional assessment and challenging behaviors: Some future directions. *Journal of the Association for Persons with Severe Handicaps, 25,* 229–240.

Hodapp, R.M., & Dykens, E.M. (2005). Measuring behavior in genetic disorders of mental retardation. *Mental Retardation and Developmental Disabilities Research Reviews, 11,* 340–346.

Horner, R.H. (1994). Functional assessment: Contributions and future directions. *Journal of Applied Behavior Analysis, 27,* 401–404.

Iwata, B.A., & DeLeon, I. (1996). *The Functional Analysis Screening Tool.* Gainesville, FL: University of Florida, Florida Center on Self-Injury.

Iwata, B.A., Dorsey, M.F., Slifer, K.E., Bauman, K.E., & Richman, G.S. (1982). Towards a functional analysis of self-injury. *Analysis and Intervention in Developmental Disabilities, 2,* 3–20.

Iwata, B.A., Pace, G.M., Dorsey, M.F., Zarcone, J.R., Vollmer, T.R., Smith, R.G., et al. (1994). The functions of self-injurious behavior: An experimental epidemiological analysis. *Journal of Applied Behavior Analysis, 27,* 215–240.

Iwata, B.A., Vollmer, T.R., & Zarcone, J.R. (1990). The experimental (functional) analysis of behavior disorders: Methodology, applications, and limitations. In A.C. Repp & N.N. Singh (Eds.), *Perspectives on the use of nonaversive and aversive interventions for persons with developmental disabilities* (pp. 301–330). Sycamore, IL: Sycamore.

Jacobson, J.W. (1982). Problem behavior and psychiatric impairment within a developmentally disabled population: Behavior frequency. *Applied Research in Mental Retardation, 3,* 121–139.

Joyce, T., Ditchfield, H., & Harris, P. (2001). Challenging behavior in community services. *Journal of Intellectual Disability Research, 45,* 130–138.

Kazdin, A.E. (1980). *Behavior modification in applied settings.* Homewood, IL: Dorsey.

Laties, V.G., & Mace, F.C. (1993). Taking stock: The first 25 years of the Journal of Applied Behavior Analysis. *Journal of Applied Behavior Analysis, 26,* 513–525.

Lennox, D.B., & Miltenberger, R.G. (1989). Conducting a functional assessment of problem behavior in applied settings. *Journal of the Association for Persons with Severe Handicaps, 14,* 304–311.

Lerman, D.C., & Iwata, B.A. (1993). Descriptive and experimental analyses of variables maintaining self-injurious behavior. *Journal of Applied Behavior Analysis, 26,* 293–319.

Linscheid, T.R., Iwata, B.A., & Foxx, R.M. (1996). Behavioral assessment. In J.W. Jacobson

& J.A. Mulick (Eds.), *Manual of diagnosis and professional practice in mental retardation.* Washington, DC: American Psychological Association.

Lowry, M., & Sovner, R. (1991). The functional significance of problem behavior: A key to effective treatment. *The Habilitative Mental Healthcare Newsletter, 10,* 59–62.

Mace, F.C. (1994). The significance and future of functional analysis methodologies. *Journal of Applied Behavior Analysis, 27,* 385–392.

Mace, F.C., Lalli, J.S., & Pinter-Lalli, E. (1991). Functional analysis and treatment of aberrant behavior. *Research in Developmental Disabilities, 12,* 155–180.

Matson, J.L., Bamburg, J.W., Cherry, K.E., & Paclawskyj, T.R. (1999). A validity study on the Questions About Behavioral Function (QABF) scale: Predicting treatment success for self-injury, aggression, and stereotypies. *Research in Developmental Disabilities, 20,* 163–176.

Matson, J.L., & Coe, D.A. (1992). Applied behavior analysis: Its impact on the treatment of mentally retarded emotionally disturbed people. *Research in Developmental Disabilities, 13,* 171–189.

Matson, J.L., & Vollmer, T.R. (1995). *Users guide: Questions about behavior function (QABF).* Baton Rouge, LA: Scientific Publishers.

Millichap, D., Oliver, C., McQuillan, S., Kalsy, S., Lloyd, V., & Hall, S. (2003). Descriptive functional analysis of behavioral excesses shown by adults with Down syndrome and dementia. *International Journal of Geriatric Psychiatry, 18,* 844–854.

National Institutes of Health, Consensus Development Panel on Destructive Behaviors in Persons with Developmental Disabilities. (1989). *Treatment of destructive behaviors in persons with developmental disabilities* (NIH Publication No. 91-2410). Bethesda, MD: U.S. Department of Health and Human Services.

Neef, N.A., & Iwata, B.A. (1994). Current research on functional analysis methodologies: An introduction. *Journal of Applied Behavior Analysis, 27,* 211–214.

Nyhan, W.L. (1976). Behavior in the Lesch-Nyhan syndrome. *Journal of Autism and Childhood Schizophrenia, 6,* 235–252.

Obi, C. (1997). Restraint fading and alternative management strategies to treat a man with Lesch-Nyhan syndrome over a 2-year period. *Behavioral Interventions, 12,* 195–202.

Oliver, C., Arron, K., Hall, S., Sloneem, J., Forman, D., & McClintock, K. (2006). Effects of social context on social interaction and self-injurious behavior in Cornelia de Lange syndrome.

American Journal on Mental Retardation, 111, 184–192.

O'Neill, R.E., Horner, R.H., Albin, R.W., Storey, K., & Sprague, J.R. (1990). *Functional analysis of problem behavior: A practical assessment guide.* Sycamore, IL: Sycamore.

Paclawskyj, T.R., Matson, J.L., Rush, K.S., Smalls, Y., & Vollmer, T.R. (2000). Questions about behavioral function (QABF): A behavioral checklist for functional assessment of aberrant behavior. *Research in Developmental Disabilities, 21,* 223–229.

Paclawskyj, T.R., Matson, J.L., Smalls, Y., & Vollmer, T.R. (2001). Assessment of the convergent validity of the Questions About Behavioral Function scale with analogue functional analysis and the Motivation Assessment Scale. *Journal of Intellectual Disability Research, 45,* 484–494.

Pedhazur, E., & Pedhazur-Schmelkin, L. (1991). *Measurement, design, and analysis: An integrated approach.* Hillsdale, NJ: Lawrence Erlbaum Associates.

Repp, A.C., Felce, D., & Barton, L.E. (1988). Basing the treatment of stereotypic and self-injurious behaviors on hypotheses of their causes. *Journal of Applied Behavior Analysis, 21,* 281–289.

Scotti, J.R., Evans, I.M., Meyer, L.H., & Walker, P. (1991). A meta-analysis of intervention research with problem behavior: Treatment validity and standards of practice. *American Journal on Mental Retardation, 96,* 233–256.

Singh, N.N., Donatelli, L.S., Best, A., Williams, D.E., Barrera, F.J., Lenz, M.W., et al. (1993). Factor structure of the Motivation Assessment Scale. *Journal of Intellectual Disability Research, 37,* 65–74.

Slifer, K.S., Ivancic, M.T., Parrish, J.M., Page, T.J., & Burgio, L.D. (1986). Assessment and treatment of multiple behavior problems exhibited by a profoundly retarded adolescent. *Journal of Behavior Therapy and Experimental Psychiatry, 17,* 203–213.

Solnick, J.V., Rincover, A., & Peterson, C.R. (1977). Some determinants of punishing effects of time-out. *Journal of Applied Behavior Analysis, 10,* 415–424.

Sprague, J.R., & Horner, R.H. (1995). Functional assessment and intervention in community settings. *Mental Retardation and Developmental Disabilities Research Reviews, 1,* 89–93.

Spreat, S., & Connelly, L. (1996). Reliability analysis of the Motivation Assessment Scale. *American Journal on Mental Retardation, 100,* 528–532.

Sturmey, P. (1994). Assessing the functions of aberrant behaviors: A review of psychometric instruments. *Journal of Autism and Developmental Disorders, 24,* 293–304.

Sturmey, P. (1996). *Functional analysis in clinical psychology.* Oxford, England: John Wiley & Sons.

Sturmey, P., Carlsen, A., Crisp, A.G., & Newton, J.T. (1988). A functional analysis of multiple aberrant responses: A refinement and extension of Iwata et al.'s methodology. *Journal of Mental Deficiency Research, 32,* 31–46.

Symons, F.J., Thompson, A., & Rodriguez, M.C. (2004). Self-injurious behavior and the efficacy of naltrexone treatment: A quantitative synthesis. *Mental Retardation and Developmental Disability Research Reviews, 10,* 193–200.

Touchette, P.E., MacDonald, R.F., & Langer, S.N. (1985). A scatter plot for identifying stimulus control of problem behavior. *Journal of Applied Behavior Analysis, 18,* 343–351.

Vollmer, T.R., Marcus, B.A., Ringdahl, J.E., & Roane, H.S. (1995). Progressing from brief assessments to extended experimental analyses in the evaluation of aberrant behavior. *Journal of Applied Behavior Analysis, 28,* 561–576.

Vollmer, T.R., & Smith, R.G. (1996). Some current themes in functional analysis research. *Research in Developmental Disabilities, 17,* 229–249.

Wehmeyer, M., Bourland, G., & Ingram, D. (1993). An analogue assessment of hand stereotypies in two cases of Rett syndrome. *Journal of Intellectual Disability Research, 37,* 95–102.

Zarcone, J.R., Rodgers, T.A., Iwata, B.A., Rourke, D.A., & Dorsey, M.F. (1991). Reliability analysis of the Motivation Assessment Scale: A failure to replicate. *Research in Developmental Disabilities, 12,* 349–360.

10

Psychiatric Diagnosis in Individuals with Neurodevelopmental Disability

Richard B. Ferrell

This chapter begins with a selective historical review and then moves to a discussion of psychiatric diagnosis in general, followed by specific problems in psychiatric diagnosis in people with neurodevelopmental disability (NDD). Little is known about the care of people with disability in antiquity. Psychiatry as a medical specialty is only about 200 years old, as is medical specialization in general, except for surgery (Shorter, 1997).

Jean-Nicolas Corvisart (1755–1821) worked at the Charité Hospital in Paris at the end of the 18th century. He founded the French method of physical examination and clinical thinking. In the French method the student would examine the patient at the bedside—this in itself was new. Then Professor Corvisart would also examine the patient and give a short lecture on the findings and diagnosis. Corvisart was a founder of modern medical diagnostic thinking. He taught that doctors should conduct post-mortem examinations on their patients. He attempted to correlate symptoms with physical and post-mortem findings. The French method anticipated modern clinical-pathological conferences, clinical case conferences, and bedside clinical rounds (Nuland, 2005).

Jean-Etienne-Dominique Esquirol (1772–1840), a student of Philippe Pinel, anticipated modern views in his suggestion that some mental disorders might have a basis in psychosocial stress and might not be exclusively the result of brain disease, which was the view of 19th-century biological psychiatry (Shorter, 1997). Esquirol advocated for diagnostic refinement and humane treatment. He gave the first accurate description of mental disability as separate from insanity, as psychiatric illness was then called. Despite Esquirol's teaching, NDD and psychiatric illness were often lumped together until the late 19th or early 20th century (Brown & Radford, 2007).

Institutional care was rare before the 19th century. With few exceptions, before 1800, the best care was home care. An exception was the hospital of the Priory of St. Mary of Bethlehem, later known as Bedlam, which was founded in the 13th century. Bedlam was acquired by the city of London in 1547 and continued in operation until 1948. As late as 1815 Bedlam housed only 122 patients. It was probably not quite as bad as its reputation as depicted in *The Rake's Progress*, by William Hogarth.

There were psychiatric hospitals in Baghdad and Cairo in the 9th century during the European Dark Ages, when knowledge and enlightened attitudes from classical Western Europe were preserved in Arabian culture (Shorter, 1997).

Knowing how people thought and felt in a time different from our own is a difficult task for historians and for all of us. Care for people with disability has been far from ideal throughout history, as it still is today. Sometimes attitudes are reflected in the art or literature of a culture that are not obvious from written history. A positive example is in the work of the artist Diego Velázquez (1599–1660), who painted for the Spanish Court (Encyclopaedia Britannica, 2008). Velázquez painted the Spanish royal family, but also painted the *bufones*—people with physical and intellectual disability who resided in the Court—and did so with sensitivity and dignity, and without any sense of pity or condescension. It is worth a trip to the Prado in Madrid to see these wonderful paintings.

One of the *bufones* is thought to have had congenital hypothyroidism or cretinism. The word *cretin* is an English word derived from French/Latin *chrétien/christianus,* that is, "incapable of sin," perhaps parallel to the 20th century view that people with disability are incapable of neurosis or psychiatric illness.

Literary examples of this same idea exist, especially in the Russian tradition in the figure of the Holy Fool or fool in Christ described by Dostoevsky and Tolstoy, and others and perhaps most dramatically depicted in the opera *Boris Godunov* by Modest Mussorgsky. In this opera the Holy Fool refuses to pray for the Tsar Boris, whom he regards as a murderer. This act of denial contributes strongly to the psychological disintegration of the Tsar.

In the 19th century care of people with NDD and psychiatric illness moved from jails, almshouses, pits, pens, stalls, and cages to asylums and state school institutions, as they were built.

Dorothea Dix (1802–1887) was a pioneer in bringing humane treatment of people with psychiatric illness to the United States (Shorter, 1997). Dix was a semi-invalid who taught school in Worcester and Boston, Massachusetts. She also taught Sunday school in the Cambridge jail, where she observed the mistreatment of prisoners with mental illness and disabilities. She advocated for legislation in Massachusetts and in fifteen other states to create or improve asylums and was responsible

for the creation of 32 hospitals in all. She was an American counterpart of European figures such as Philippe Pinel, Vincenzo Chiarugi, William Tuke, and others in Europe and America whose names are long lost but who were the reformers of the care of people with psychiatric illness and NDD in the late 18th and 19th centuries. We should be aware that our modern assumptions and stereotypes of asylums might be inaccurate when applied to the early asylums and to the intent of the asylums. The word *asylum* means a place of safety or refuge. In this case, it meant safe from the stressful "exciting factors" that characterized life during the Industrial Revolution. It is important to remember that the intent of the 18th- and 19th-century reformers and of the asylum movement and the development of the state schools was altruistic. No one foresaw that they would become snake pits and hellholes by the end of the 19th century (Shorter, 1997).

DIAGNOSIS IN PSYCHIATRY

Accurate psychiatric diagnosis matters because it guides treatment. As in all of medicine, inaccurate diagnosis can result in ineffective, inappropriate, or dangerous treatment. Overdiagnosis leads to overtreatment, and underdiagnosis leads to failure to recognize and treat treatable disorders, with unnecessary suffering and loss of freedom as results. There is now general agreement that people with NDD can have the same illnesses as people without disability, including psychiatric illnesses. In a sense, we have made a circle from lumping NDD and psychiatric illness together, to separating them as if they were mutually exclusive, to a now more integrated, holistic, and scientific understanding.

Yet there is still resistance to an integrated view of people. Many U.S. states have two systems of care, one for individuals with disability and one for individuals with psychiatric illness. People with disability can easily fall through the cracks between service delivery systems, especially when those systems are grossly underfunded and are therefore unable to take necessary responsibility for treatment, services, and support. People with NDD are two to four times more likely to have psychiatric illness than the general population (Fletcher, Loschen, Stavrakaki, & First, 2007b).

GENERAL PROBLEMS IN PSYCHIATRIC DIAGNOSIS

Psychiatric diagnosis is not easy. It is a part of medicine where it is still entirely appropriate to talk about art and science. Signs and symptoms often seem to overlap from one disorder to the next. Different examiners seeing patients at different times may observe different phenomena, or regard clinical phenomena differently, or draw different diagnostic conclusions. We encounter charts that have accumulated multiple diagnoses over time: schizophrenia, bipolar disorder, schizoaffective disorder, obsessive-compulsive disorder (OCD), an anxiety disorder, an impulse control disorder, or posttraumatic stress disorder (PTSD). An important remedy for this problem is having a longitudinal developmental view of the individual and his or her life history. Modern diagnostic thinking relies on both cross-sectional information, that is, current signs and symptoms, and a view of the history of health and illness over time. In reality both current signs and symptoms and longitudinal history are important, and neither should be discounted. Drawing diagnostic conclusions exclusively from a snapshot approach can cause problems.

Unlike many areas of general medicine, there are not many tests in psychiatry that have diagnostic precision. Psychiatric diagnosis, excepting some disorders that have neurologic or general medical underpinnings, is based on the history, the psychiatric interview, and the mental status examination.

Psychiatric diagnostic criteria are written and periodically revised by a committee. The expert consensus model is still the best method for defining diagnoses that we have, but it is often based more on informed opinion than on hard science. Psychiatric diagnosis depends largely on what patients are able to tell us about thoughts, feelings, and experiences.

The diagnostic criteria for two major diagnoses serve as examples. Diagnostic criteria for major depressive disorder are depressed mood, decreased interest or decreased pleasure, weight loss and loss of appetite, sleep change, psychomotor agitation or retardation, fatigue, loss of energy, feelings of worthlessness or guilt, decreased ability to think or concentrate, and thoughts of death or suicide (American Psychiatric Association, 2000). Five of these signs and symptoms must occur together in a 2-week period to make the diagnosis. At least six of these (thinking about death, feelings of worthlessness or guilt, depressed mood, loss of interest, trouble concentrating, and loss of energy) depend to some extent on the patient's ability to report these things. If the person has an impaired ability to comprehend, appreciate, or report these symptoms, then making this diagnosis is difficult.

It is not possible to know directly the emotional or sensory or cognitive experiences of others, except to the extent that they tell us about them or to the extent that we can draw inferences from behavior or appearance. Irritability, aggression toward others, self-injurious behavior, regression, cognitive changes, somatic complaints or hypochondriasis might be depressive equivalents in people with NDD (Bradley et al., 2007). Yet general psychiatric principles still apply in the diagnostic appraisal of people with NDD. A separate psychiatry is not needed for people with disability but, rather, a refined diagnostic approach that accounts for possible differences in symptom presentation in people with NDD. *The Diagnostic Manual–Intellectual Disability* (*DM-ID*) is an excellent attempt to do just that (Fletcher, Loschen, Stavrakaki, & First, 2007a).

Diagnostic criteria for schizophrenia in the *Diagnostic and Statistical Manual of Mental*

Disorders, Fourth Edition, Text Revision (*DSM-IV-TR;* American Psychiatric Association, 2000) include delusions, hallucinations, disorganized speech, grossly disorganized or catatonic behavior, and negative symptoms, such as affective flattening or avolition. How does one know if a person is experiencing a hallucination? One sees notes in charts such as "patient appears to be responding to internal stimuli." To what internal stimuli, if any, is the person responding? If the person appears distracted, is it because he or she is seeing visions, having sciatic pain, or experiencing constipation?

In the sixth edition of his textbook in 1899, Emil Kraepelin distinguished manic-depressive psychosis and dementia praecox, or schizophrenia as it was later called, a valid distinction today (Encyclopædia Britannica, 2009). Kraepelin filled out little cards about each of his patients and kept a running account of the course of their illnesses (Shorter, 1997). In Kraepelin's view the course of illness and prognosis were what mattered. This way of thinking was a departure from the usual diagnostic practice of 19th-century biological psychiatry, which relied more on the presentation of signs and symptoms at a point in time.

SPECIAL DIAGNOSTIC PROBLEMS IN PEOPLE WITH NEURODEVELOPMENTAL DISABILITY

NDD disorders may have their own particular psychological or behavioral phenotype that overlaps with signs and symptoms of psychiatric illness. If a person has impaired communication skill, making a psychiatric diagnosis can be very difficult.

Often patients are admitted to our care with a long list of psychoactive medications: mood stabilizer drugs, antipsychotic drugs, antidepressant drugs, benzodiazepines, or other tranquilizing, calming, or hypnotic drugs. I understand how this happens and how prescribers are reluctant to alter preexisting regimens for fear that some catastrophe will occur. Yet layered-on polypharmacy increases the risk of iatrogenic symptoms and can make accurate diagnosis even more difficult.

Careful reduction or discontinuance of drugs of dubious merit can reduce unwanted effects, reveal baseline psychopathology, and increase diagnostic accuracy. On our inpatient neurobehavioral unit at New Hampshire Hospital we have the luxury of being able to taper and discontinue medicines in a serial manner to see if they were treating a disorder, were doing no good, or were making things worse. Sometimes, but not always, patients get better when we do this. We ask guardians, patients, and community caregivers if they know what particular drugs are for and whether they know or have an opinion about whether they are helping. A good question to ask is, Are there any of these drugs that you would recommend not stopping? When tapering and discontinuing mood stabilizer drugs, it is important to be certain that they are not prescribed for seizure prophylaxis.

Some common behavioral clusters might be considered "normal" for people with NDD who function at a lower intellectual and emotional level, as they also might be for young children. Examples include soothing or comforting repeated habitual behavior, overvalued ideas, and talking to oneself or to an imaginary friend. Where do you draw the line between a soothing or comforting repeated habitual behavior and a compulsion that you might think needs treatment? At what point does a recurrent or overvalued idea become an obsession? When does an incorrect obsessional idea become a delusion, or talking to an imaginary friend represent psychosis?

Some other ideas are worth considering (Ferrell, Wolinsky, Kauffman, Flashman, & McAllister, 2004). Mixed memory, attention, speech, or language deficits can occur and confound accurate diagnosis. Here emphasis is on the importance of careful history collecting from family members and other caregivers.

Filter effects of central nervous system injury, from traumatic brain injury or other causes, can alter the presentation of psychiatric pathology. Consider, for example, how the following likely would present:

1. Depression in a person who cannot speak

2. Manic hyperactivity in a person with quadriplegia

3. Hallucinations in a person with minimal speech and language function

Seminal work by Robert Sovner, published in 1986, described factors that limit or compromise our ability to make cogent psychiatric diagnoses in people with NDD (Sovner, 1986). These factors include intellectual distortion, psychosocial masking, cognitive disintegration, and baseline exaggeration.

1. Intellectual distortion: "Concrete thinking and impaired communication skills result in inability of patient to label their own experiences and report them." Diminished ability to think abstractly and to observe and describe one's own behavior and feeling states limits diagnosis because most Axis I diagnoses depend on self-report about symptoms. The observations and reports of family members and caregivers are thus ever more important. This phenomenon makes diagnosis of psychosis, for example, very difficult in moderate to severe NDD. *NDD makes it hard for people to report their symptoms.*

2. Psychosocial masking: "Impoverished social skill and life experiences lead to unsophisticated presentation and lack of poise during interview and can result in missed symptoms and misattribution of nervousness and silliness as psychiatric features." This results in a bland clinical presentation without detail based in broad or rich life experiences. Mania might lack its usual grandiosity. Delusions might resemble simple fears.

Psychosocial masking means that impaired social skills can be misinterpreted as psychiatric illness.

3. Cognitive disintegration: "Stress-induced disruption of information processing leads to bizarre presentation and psychotic-like state that may be misdiagnosed as schizophrenia." Sovner describes this as similar to pseudodementia. The stressed brain is overwhelmed and intellectually compromised and produces transient symptoms such as hallucination, withdrawal, and illusory or psychosis-like states. *Stress can cause transient symptoms that can lead to overdiagnosis.*

4. Baseline exaggeration: "Increase in severity of preexisting cognitive deficits and maladaptive behaviors that creates difficulty in establishing illness features, target symptoms, and outcome measures." *That is, stress can cause an increase in baseline behaviors and does not necessarily mean that there is new psychiatric illness.*

Another confounding phenomenon is called *diagnostic overshadowing.* To quote Reiss et al., "This phenomenon is called diagnostic overshadowing because the hypothesis is that intellectual subnormality is such a salient feature of mental retardation that accompanying emotional disturbances are overshadowed in importance by the presence of intellectual retardation" (Reiss, Levitan, & Szysko, 1982). Diagnostic overshadowing leads to under diagnosis rather than over diagnosis.

If one is thinking parsimoniously, behavioral, emotional, or functional change could be regarded as resulting from the primary NDD syndrome, thus discounting the likelihood of an added-on psychiatric disorder. We all tend to think in this parsimonious way to some extent, so we have to be aware of this possible error. One might miss a treatable psychiatric illness because of attributing every thing to NDD. *Diagnostic overshadowing means a tendency to underestimate psychiatric or emotional disturbance and misattribute psychiatric symptoms to NDD.*

Generally, people with mild disability present with more typical (*DSM-IV-TR*) psychiatric presentations than do people with more severe disability. The more severe the level of disability, the more difficult psychiatric diagnosis becomes. This phenomenon is the essence of NDD limiting our ability to make accurate psychiatric diagnoses.

The following cases describe two patients whose stories demonstrate diagnostic difficulties, but which also contain a kernel of hope for helping people with disability and psychiatric illness.

Patient 1

A 21-year-old woman with mild lifelong NDD was conceived in a group home where her mother resided. Her father was also a resident. As an infant she was briefly cared for by her mother, and then subsequently by grandparents. She was sexually abused as a young child over a period of several years. She lived at a youth center from age 11 to age 21. A treatment plan goal at the center was less than two violent outbursts per week. She "aged out" and moved to a supported family caregiver situation, where she assaulted her caregiver. She was admitted to our care. She showed frequent episodes of violence, which occurred two or three times per day. Typically, there was no recognizable antecedent for the violent behavior. There were multiple staff injuries. There was no compelling evidence of psychosis, major affective illness, or intracranial pathology. Treatment with psychoactive drugs, including several antipsychotic drugs, did not help. Treatment with clozapine and a behavioral treatment plan did eventually reduce the frequency and severity of violent outbursts. Violent outbursts now occur once or twice a month. They are less intense and shorter. There is still no clear psychiatric diagnosis, except possible PTSD and mild intellectual disability.

Initially, motor symptoms suggesting catatonia, such as posturing, and hypomotility alternating with extreme hypermotility were evident. These psychomotor symptoms responded to lorazepam.

Violent episodes often appeared to have a volitional component. She would appear to wait for periods of staff vulnerability, such as shift changes, or when staff members were busy with another patient.

In mild to moderate NDD, psychiatric diagnosis does not vary much from the diagnostic process that is used in people without disability. We usually reach the limits of conventional psychiatric diagnosis with people in the low end of the moderate range of intellectual disability and even more so in the severe and profound ranges. The history from the patient, the psychiatric interview, and the mental status examination then begin to give insufficient data for diagnostic clarity. Greater reliance on observation of the patient and on the historical reports of others is then needed. When all sources of data have been exhausted, clinical acuity and suffering will sometimes compel caregivers to conduct empirical treatment trials, because failing to do so is not compassionate or acceptable. Although Patient 1 had only mild NDD, she was not able to tell staff what was wrong. PTSD might be a relevant diagnosis. She certainly had trauma, but there was no evidence that she had flashbacks or nightmares or startled easily. She denied memory of the early trauma. Clozapine is not an indicated treatment for PTSD, and the full syndrome was lacking.

The staff used clozapine because her situation was desperate, and in a sense because the staff was desperate. Unless things got better, she was doomed to an institutional life. There was doubt that she could receive sufficient care even in a relatively well-staffed locked unit. Clozapine was not chosen because the staff thought she had schizophrenia or any kind of psychosis. It was used because it is a last-ditch drug in severe, unexplained behavioral dyscontrol situations. This patient now resides in a residential facility.

Patient 1 is a good example of one kind of limitation of psychiatric diagnosis in NDD. I am still not sure that she has an Axis I diagnosis. One might think that she has impulse control disorder, but some of the dyscontrol episodes did not seem very impulsive. There was also behavioral plan treatment and enormous staff effort, but I believe clozapine was a crucial factor.

Patient 1—Main Points

1. The severity of behavioral dyscontrol required an attempt at therapeutic intervention.
2. The case shows a limitation of psychiatric diagnosis in trying to account for behavioral dyscontrol in some people with NDD.

A brief comment about catatonia in NDD: Catatonia is eminently treatable. Symptoms of catatonia, such as motor disturbances, negativism, and disinhibited behavior, can overlap or mimic symptoms of other neuropsychiatric disorders or conditions, making diagnosis difficult. People with autism spectrum disorders appear to be especially vulnerable to catatonia (Wing & Shah, 2006). A lorazepam challenge and/or a lorazepam treatment trial can be conducted to assist in diagnosis as a diagnostic probe.

Patient 2

A 43-year-old man has lifelong severe NDD. He has never had speech, although he can phonate. A first presentation suggested depression; he improved when treated with an antidepressant medicine. We considered decreased irritability and increased social interaction to be evidence of improvement. A second presentation several years later was more confusing. Symptoms were loud vocalization, pacing, weight loss, and agitation. Staff noticed trouble swallowing, choking, and aspiration of food. A laryngeal diversion procedure solved this problem. On a third presentation, staff treated a urinary tract infection and empirically treated depression. He improved. On a fourth admission, fever, râles, and rhonchi were present. Treatment of pneumonia led to decreased agitation. He was subsequently hospitalized with a Stage IV decubitus and a newly diagnosed malabsorption syndrome. Symptoms are assaultive and self-injurious behavior and marked weight loss. The usual signs and symptoms of affective disorder or psychosis could not be discerned. Improvement occurred with treatment of the decubitus wound, the malabsorption syndrome, and empirical treatment with clozapine.

In this case it was virtually impossible to tell if there was a separate psychiatric diagnosis. Staff treated him as if there was such a diagnosis. This case illustrates the importance of searching carefully for occult physical illness.

Patient 2—Main Points

1. In puzzling situations in people with NDD look for undiagnosed medical illness.
2. There can be more than one problem.
3. Psychiatric diagnosis using the usual *DSM-IV-TR* criteria is difficult to impossible in people with severe and profound NDD.
4. Empirical drug treatment trials can serve as diagnostic probes and may have therapeutic benefit.
5. Clinical monitoring is necessary to recognize positive or negative results and to stop unhelpful treatment trials.

RECOMMENDATIONS

Sometimes the best we can do is to generate a hypothesis by asking the question, What disorder does this symptom cluster look most like? The next step is to initiate a treatment trial and monitor results closely. Sometimes a response to a particular treatment gives us some idea or clue about the original problem when we cannot initially define a diagnosis.

Treatment trials can serve as a useful diagnostic probe, although they must be viewed with circumspection. Again, the procedure is to identify the most prominent symptom cluster and then start a trial of treatment. Close monitoring is necessary to recognize positive or negative results and stop unhelpful treatment trials.

The *DM-ID* is an important step in refining psychiatric diagnostic thinking with regard to people with NDD (Fletcher et al., 2007a). For most psychiatric diagnoses diagnostic criteria need not be adapted for people with NDD, especially when the degree of disability is mild to moderate.

For a diagnosis of a major depressive episode, four not five symptoms, one of which is depressed mood, loss of interest or pleasure, or irritability, are required. People with severe or profound NDD generally do not have the cognitive capacity to experience feelings of guilt or worthlessness.

With regard to schizophrenia, *DM-ID* emphasizes the importance of corroborative history and a longitudinal approach to history taking and diagnosis. Self-talk is common in people with ID and is not necessarily hallucinosis.

DM-ID pays special attention to OCD and PTSD because of their frequent occurrence in people with ID. With regard to OCD, the person may not recognize symptoms as unreasonable or excessive and may not attempt to ignore or suppress them. Ordering, hoarding, telling or asking, or rubbing may replace counting, checking, and hand-washing rituals. An examiner may not be able to elicit or recognize obsessional thoughts. The diagnosis of delusional disorder should not be considered for people with severe ID.

I recommend an approach to psychiatric diagnosis in individuals with NDD based on a careful longitudinal history from multiple sources. It is important to ask about changes in the frequency or intensity of previous or "baseline" signs and symptoms. Careful physical, neurological, and mental status exams are essential. A general laboratory screen, an electroencephalogram, and, in many cases, cerebral imaging are also important. Other significant elements of a comprehensive evaluation are psychological assessment including behavioral analysis, neuropsychological evaluation, and genetic testing, if indicated.

We must look to find whatever data we can get, and then form a diagnostic idea or hypothesis. We develop a rational treatment plan, weighing the risks and benefits of each proposed treatment, and give it a try. We do not have to be right all of the time, and we will not be, but we must avoid sloppy thinking. The fundamental problem is that psychiatric diagnosis is an imprecise business to begin with, and the greater the level of intellectual disability, the murkier and more difficult it becomes. The thing that most limits psychiatric diagnosis in neurodevelopment disability is the disability itself, whereby the individual cannot reflect on his or her inner state, feelings, emotions, thinking, or life experiences and communicate those to us.

We must always deal with some uncertainty in medicine. In treating people with NDD there is often a lot of uncertainty. Suffering compels us to act when knowledge is still incomplete and diagnosis still unclear. We want diagnosis to be based on signs and symptoms and, if possible, on objective, physical, laboratory, and radiographic findings. Yet, we can also hit a diagnostic dead end and be compelled by clinical exigency to try treatment anyway. It is crucial to consider and look for occult or undiagnosed physical illness. Patient 2 illustrates this point.

Some attributes we should cultivate in ourselves as caregivers include intellectual humility in the face of our ignorance, compassion and empathy, open-minded curiosity, and circumspect thinking.

Broad-based community supports plus careful and circumspect neuropsychiatric assessment and treatment make a potent therapeutic combination (Ferrell et al., 2004). We are all still learning. Sometimes the best we can do is to continue learning from our patients, from our reading and study, and from each other, and to have as much compassion as we can muster for our patients. This is difficult but intellectually and emotionally satisfying and important work.

REFERENCES

American Psychiatric Association. (2000). *Diagnostic and statistical manual of mental disorders* (4th ed., text rev.). Washington, DC: Author.

Bradley, E., Summers, J., Brereton, A., Einfeld, S., Havercamp, S., Holt, G., et al. (2007). Intellectual disabilities and behavioral, emotional and psychiatric disturbances. In I. Brown & M. Percy (Eds.), *A comprehensive guide to intellectual and developmental disabilities* (pp. 645–666). Baltimore: Paul H. Brookes Publishing Co.

Brown, I., & Radford, J. (2007). Historical overview of intellectual and developmental disabilities. In I. Brown & M. Percy (Eds.), *A comprehensive guide to intellectual and developmental disabilities* (pp. 17–33). Baltimore: Paul H. Brookes Publishing Co.

Encyclopaedia Britannica. (2008). *Diego Velázquez.* Retrieved September 30, 2008, from http://www.search.eb.com/eb/article-61332.

Encyclopædia Britannica. (2009). *Emil Kraepelin.* Retrieved May 12, 2009, from Encyclopædia Britannica Online: http://www.search.eb.com/eb/article-9046179

Ferrell, R., Wolinsky, E., Kauffman, C., Flashman, L., & McAllister, T. (2004). Neuropsychiatric syndromes in adults with intellectual disability: Issues in assessment and treatment. *Current Psychiatry Reports, 6,* 380–390.

Fletcher, R., Loschen, E., Stavrakaki, C., & First, M. (2007a). *Diagnostic Manual–Intellectual Disability.* Kingston, NY: NADD Press.

Fletcher, R., Loschen, E., Stavrakaki, C., & First, M. (2007b). Introduction. In R. Fletcher, E. Loschen, C. Stavrakaki, & M. First (Eds.), *Diagnostic Manual–Intellectual Disability* (pp. 1–7). Kingston, NY: NADD Press.

Nuland, S. (2005). *Doctors: The history of scientific medicine revealed through biography.* Chantilly, VA: The Teaching Company Limited Partnership.

Reiss, S., Levitan, G., & Szysko, J. (1982). Emotional disturbance and mental retardation: Diagnostic overshadowing. *American Journal of Mental Deficiency, 86,* 567–574.

Shorter, E. (1997). *A history of psychiatry: From the era of the asylum to the age of Prozac.* New York: John Wiley & Sons.

Sovner, R. (1986). Limiting factors in using DSM-III criteria with mentally ill/mentally retarded persons. *Psychopharmacology Bulletin, 22*(4), 1055–1059.

Wing, L., & Shah, A. (2006). A systematic examination of catatonia-like pictures in autism spectrum disorders. *International Review of Neurobiology, 72,* 21–39.

11

Speech-Language Therapy for Children with Social, Emotional, and Behavioral Disorders

Janet E. Turner and Mary K. Boyle

Children who have neurodevelopmental disabilities face many challenges in their attempts to build and use the skills they need for functioning in everyday life. A major complicating factor for these children, their families, and the professionals who help them is the frequent presence of coexisting disorders. Two commonly coexisting disorders are the broadly defined diagnoses of communication disorders (CDs) and social-emotional-behavioral disorders (SEBDs). Serious functional difficulties can result from disorders in either area alone, but the cumulative impact of comorbid disorders takes a much larger toll on function. Detecting the presence of these coexisting disorders and understanding the interplay between them are essential steps in determining children's needs and developing appropriate treatment programs to improve their participation in everyday life.

This chapter describes speech-language treatment designed for children who have CDs in conjunction with SEBDs. A brief review of some studies describing the coexistence of these disorders is presented. Then, a case study of a child with significant impairments in communication and social-emotional-behavioral skills is presented. Review of her story illustrates how speech-language treatment services can be designed to address children's interrelated needs.

Treatment that addresses the "whole child" has better potential to effect change, in one or more areas of need. Finally, comments are offered about future directions that clinical services and research may take to help children with this profile.

The focus of this chapter, on the intersection between CDs and SEBDs, is not meant to imply that comorbidities are limited to these areas. In fact, children with neurodevelopmental disabilities often have cognitive, academic, and motor difficulties, in addition to difficulties in language and social-emotional and behavioral skills (Hart, 2004; Hill, 2001; Raitano, Pennington, Tunick, Boada, & Shriberg, 2004; Rescorla, 2002; Tager-Flusberg, 1999; Webster, Majnemer, Platt, & Shevell, 2005). The case review presented in this chapter illustrates the complexity that additional comorbidities bring to assessment and treatment processes in the presence of documented communication and social-emotional-behavior difficulties.

DIAGNOSTIC TERMINOLOGY: HISTORY AND USE

Many factors affect how diagnostic labels are applied. Examples include purpose (e.g., research versus clinical service, access to services or reimbursement), characteristics of patient populations, advances in knowledge, and changes in practice guidelines. Discipline-specific practices also affect the use of terminology.

Ultimately, however, assessment for treatment planning involves capturing the child's profile. This process, for clinical or research purposes, can be described as behavioral phenotyping, or documenting "observable characteristics of an organism, which are the joint product of both genotypic and environmental influences" (Gottesman & Gould, 2003, p. 636). Specification of behavioral phenotypes involves identification of characteristics that are likely to occur within a certain group of patients and unlikely to occur in other groups of patients. That is, these traits would ideally have a high degree of sensitivity and specificity (Dykens, 1995; Dykens & Hodapp, 2007; Mervis, 2004).

Behavioral phenotyping research has evolved over time, as a review of the study of genetic neurodevelopmental disorders in the field of psychiatry reveals (Feinstein & Singh, 2007). Initial stages of this research involved careful narrative accounts of behavioral traits obtained through observation. These descriptions of salient traits then served as the basis for development of diagnostic categories. A second generation of research, from which prevalence data come, involved assessment of patients within and across diagnostic boundaries, with the use of standardized instruments and behavioral checklists. This categorically bound research is narrower in scope than "true" behavioral phenotyping, which involves the review of a whole array of behaviors, not just traits already incorporated into diagnostic criteria.

Heterogeneity is now acknowledged within psychiatric diagnostic classification systems (Feinstein & Singh, 2007; Gottesman & Gould, 2003). Sources of heterogeneity within diagnostic classifications for complex mental health and communication disorders include their multifactorial and polygenetic origins, susceptibility to environmental influences, and the current absence of strong diagnostic-specific physiological markers (Gottesman & Gould, 2003). Not so long ago, characterizing between-group differences was necessary to establish the presence of diagnoses and differentiate patterns across diagnoses (e.g., intellectual disability from autism). Renewed interest in within-group differences, which "form the heart and soul of phenotypic work," has the potential to capture the range of behaviors and individual differences that may be tied to genetic and neurologic underpinnings of a disorder (Dykens & Hodapp, 2007). Such work may lead to a redrawing of diagnostic boundaries in order to capture unique profiles. Knowledge of between- and within-group differences also serves as the basis for diagnostic assessment and treatment planning and implementation.

Here, the term *communication disorder* refers to speech, language, and communication disorders. (Disorders of hearing, although critical to address, are not discussed here.) In a pediatric population, the general term *speech disorder* refers to atypical speech production (e.g., motor speech disorder) and delays in speech production with respect to chronological age. Language disorder is itself a broad construct. Language disorders include problems with input (receptive) and output (expressive) deficits that may occur in phonology (speech sounds in a language for spoken and written communication), semantics (meaning), syntax (grammar), or pragmatics (language use in context). In clinical populations, language disorders appear as impairments in many combinations of these areas.

Pragmatic language disorders, in particular, frequently coexist with mental health or psychiatric diagnoses because adapting language to use in context requires social and emotional knowledge and skills, as well as at least basic verbal and nonverbal linguistic skills (Gilmour, Hill, Place, & Skuse, 2004; Im-Bolter & Cohen, 2007). Examples of the pragmatic language weaknesses often reported in the presence of social-emotional and behavioral disorders include poor adaptation to listeners, difficulty with conversational or narrative discourse skills, and trouble using discourse to accomplish tasks with others (Gallagher, 1993).

The term *social-emotional-behavioral disorder* refers to a broad group of psychiatric diagnoses. Social-emotional disorders may be distinguished from behavioral disorders. Examples of true disorders and observable traits in this general area that are reported to coexist in children who have language disorders are wide-ranging. They include immaturity, inattention, hyperactivity, impulsivity, frustration, aggression, oppositional defiant disorders, conduct disorders, low self-esteem, low self-confidence, adjustment disorder, avoidant disorder, social withdrawal, depression, and anxiety (Cantwell, Baker, & Mattison, 1981; Egger & Angold, 2006; Gallagher, 1993; Gilmour et al., 2004).

COEXISTENCE OF COMMUNICATION DISORDERS AND SOCIAL-EMOTIONAL-BEHAVIORAL DISORDERS

Documentation of coexisting CDs and SEBDs dates back more than two decades. Several factors indicate strong ties between CDs and SEBDs, such that this connection is "highly relevant for a majority of the children identified as language impaired" (Redmond & Rice, 1998, p. 689). These disorders coexist across levels of severity, albeit at a higher rate in children whose diagnoses are more severe (Beitchman, 1985; Camarata, Hughes, & Ruhl, 1988; Hart, Fujiki, Brinton, & Hart, 2004). Coexistence occurs whether the sample population is ascertained from clinics addressing social-emotional-behavioral populations (Cohen, Davine, Horodessky, Lipsett, & Isaacson, 1993) or from settings where children with communication disorders access care (Cantwell et al., 1981). Relationships exist between these disorder

clusters over time (Beitchman et al., 2001; Brownlie et al., 2004; Glogowska, Roulstone, Peters, & Enderby, 2006). Gallagher acknowledged the variability in subject age, referral source, assessment tools, and criteria for clinical significance present in studies about coexisting communication and behavior disorders completed at that time, then observed that, despite all this variability, "what is most remarkable is the consistent finding that the prevalence of overlap among these populations is substantial" (1999, p. 2).

Rates of Comorbidity

In a frequently cited study, 600 children (ages 2–16 years) from diverse backgrounds who presented to a community speech and hearing clinic over a 3-year period participated in a psychiatric evaluation (Cantwell & Baker, 1987). The children were divided into three groups based on their general CD profile: speech problems alone, language problems alone, or combined speech and language disorders. Thirty-one percent of the children with speech disorders alone had coexisting psychiatric disorders, 58% of children with speech and language disorders did, and 78% of children with language disorders alone had coexisting psychiatric disorders. Rates of coexisting behavioral disorders ranged from a low of 14% (in children with speech disorder alone) to a high of 47% (language disorder alone). Attention disorders ranged from 10% (speech disorders alone) to 33% (language disorders alone), and emotional disorders ranged from 15% (speech disorders alone) to 31% (language disorders alone).

Specific psychiatric disorders had different prevalence rates for the three groups of communication disorders into which the children had been sorted. For example, in children with speech disorders who had psychiatric disorders, emotional disorders were more common (e.g., anxiety-related, adjustment, and affective disorders). Behavioral disorders (e.g., attention, conduct, and oppositional disorders) were more common in children who had language disorders alone. Cantwell and Baker (1987) reported that children who had language disorders (with or without speech disorders) were at greater risk for psychiatric and developmental disorders than were children who had only speech disorders, but they observed that the children with only speech disorders still had a risk for psychiatric disorders that was two to three times greater than that seen in the general population.

Redmond and Rice (1998) assessed the congruence between parent and teacher ratings of behavioral problems in a longitudinal study of children with receptive-expressive language impairment ($n = 17$) and their age-matched peers ($n = 20$). Standardized rating scales included the Child Behavior Checklist (Achenbach, 1991a) and the Teacher Report Form (Achenbach, 1991b), chosen so that results could be compared. Teachers and parents completed the scales when the children completed kindergarten and then again when they finished first grade. Means of both rating scales fell into the typical range for age for children with typical language and language impairment, but ratings for children with language impairment were significantly higher (i.e., worse) than for their peers. Group differences occurred on the following scales: Withdrawn, Social Problems, Attention Problems, and Internalizing Scales. Teachers rated the children with more Social Problems and Internalizing problems than their parents did. The study showed little agreement between parent and teacher rating; some differences between raters may have occurred because of differences in rating scales.

In a follow-up study with the same populations, Redmond and Rice (2002) found more convergence between the formerly disparate parent and teacher ratings for children with language impairment at levels that approached expectations for typically functioning children. An exception to this pattern of parent and teacher ratings convergence was in the area of Social Problems, where teacher ratings did not improve over previous years. Questions that could be addressed in future studies include whether adult rating scales adequately assess children's socioemotional competence and what impact peer evaluations may have on children with language impairment as children age. Results of this study add support to the belief that social aspects of language disorders vary according to observer and the developmental level of the child (Redmond & Rice, 2002).

The nature of relationships between children's language impairment and their social skills were explored in a study by Hart et al. (2004). Here, the authors inquired about whether children with specific language impairment show higher levels of withdrawal and poorer sociability than typically developing peers. They also questioned whether the severity of children's language problems predicted higher levels of withdrawal and poorer sociability. Children in this study (6–9 and 10–13 years old) had no problems with hearing, intellectual impairment, or formal diagnoses of emotional or behavioral disorders. At the time of the study, all of the children were enrolled in treatment for a language impairment that fell at least one standard deviation below the mean on a formal receptive-expressive language test. Teachers were asked to rate students' sociability (i.e., likeability and prosocial behavior) and withdrawal/solitary behaviors (i.e., reticence, solitary-passive withdrawal, solitary-active withdrawal) on an informal measure. These results were reviewed in light of the students' language abilities.

Hart et al. (2004) found that teachers rated children with language impairment as showing significantly higher levels of withdrawal than children without language impairment, particularly with regard to reticence, a behavior which they noted could lead to peer rejection. Ratings of solitary-passive withdrawal behaviors were also elevated in children with language impairment, but ratings of solitary-active withdrawal behaviors were rare. The high construct correlations that were derived statistically between reticence and solitary-passive withdrawal measures supported prior research suggesting that these characteristics "gradually merge into a single maladaptive type of behavior reflecting fear and anxiety across the school years" (Hart et al., 2004, p. 656). The children in this study did not demonstrate strong sociable behaviors that might have softened negative consequences of the withdrawal behaviors that were reported. That is, teachers' ratings of children with language impairment yielded significantly lower scores on both likeability and prosocial skills. Girls in the study used prosocial skills more than boys, but not at a significant level.

Not surprisingly, severity of the children's language disorders was related to their social problems, although use of a single standardized language test affected the authors' ability to capture some aspects of this association (e.g., likeability, withdrawal traits). With this caveat, children with severe receptive or severe expressive language disorders were rated as less prosocial than children whose scores fell in the moderate range of language disorder, as had been found in earlier work (Fujiki, Brinton, Morgan, & Hart, 1999). Other authors have described the presence of weak receptive language skills as predicting more serious SEBDs (Gallagher, 1999).

Studies reported above indicate that many children have both CDs and SEBDs. These high rates notwithstanding, either language disorders or SEBDs may remain undetected, particularly when one of the disorders draws attention to itself. Cohen et al. (1993) reviewed language and behavioral characteristics of children 4 to 12 years of age. The children, whose IQs fell at least in the low average range for age, had been referred for outpatient mental health services. The authors discovered that, in this population, the children whose language disorders had already been diagnosed had disorders that were more severe expressively and often coupled with internalizing mental health behaviors (e.g., anxiety, withdrawal). In children whose language disorders were present, but not diagnosed, the language disorders tended to be less severe expressively. The undiagnosed language disorders coexisted with mental health disorders that had strong externalizing symptoms (e.g., oppositionality or aggression).

This study illustrates some of the diagnostic challenges present when conditions co-occur. That is, the more obvious disorder may be the only one diagnosed. Less visible or less severe disorders may not be recognized, and their impairments may be attributed to the more striking disorder or to global developmental disorders, with the result that these needs may not be understood or addressed. Language impairments in the mild to moderate range that coexist with behavioral disorders might well suffer this fate (Camarata et al., 1988). Other factors that might result in language disorders remaining undetected in children with emotional and behavioral disorders include lack of continuity of care, children's lack of attention to incomprehensible language, and dependence on language in making psychiatric diagnoses (Cross, 1998). In an article about the status of research to work out nosological and epidemiological factors affecting children's social and emotional development, Egger and Angold (2006, p. 328) commented on "how late we are in recognizing the distress and impairment of . . . children and their families," an observation that holds true whether missed or misunderstood diagnoses fall into the communication or mental health domain.

Early work documented the coexistence of CD and SEBD clusters and described prevalence patterns within broadly defined patient groups. In an attempt to resolve some apparent conflicts from this body of work, a group of authors designed a study to incorporate variables that could affect trends over time: within-child variables, context (i.e., environment), and time (Lindsay, Dockrell, & Strand, 2007). In this study, the authors looked at how social, emotional, and behavioral difficulties in children 8–12 years of age related to their language disorders. Environmental factors were considered by looking at the degree to which the children's social-emotional-behavioral impairments and prosocial skills varied across home and school settings, based on parent- and teacher-rated questionnaires. The children ($n = 69$, 17 girls and 52 boys) entered the study at 8 years of age, then had follow-up assessments at 10 and 12 years of age. Language assessments included administration of standardized measures of vocabulary and grammar and completion of a pragmatic language checklist.

Study results supported previous research indicating the presence of higher levels of SEBDs in students diagnosed with language disorders and that the relation between these diagnoses held across the 8- to 12-year period. No gender differences were apparent.

Variability occurred in the type and intensity of students' reported SEBD difficulties and (positive) prosocial skills over time and across settings. One potential source of this variability is having a consistent parent rater across the study, whereas children's teachers changed annually. Some variability does likely stem from different demands placed on children in home versus school settings.

Teacher and parent ratings documented high levels of peer problems in children over the course of the study. Parent rating of peer problems remained high throughout the study, whereas teacher ratings decreased over time. Parent and teacher ratings documented the presence of hyperactivity; the parent ratings remained stable and the teacher ratings of hyperactivity decreased over time. Parents reported higher levels of social-emotional-behavioral difficulties across the study than the children's teachers did. Parent reports of children's conduct problems increased over time; teacher reports decreased. Teacher ratings of students' prosocial skills increased over time, but parent ratings did not.

In this study and others conducted by this group, the authors note that parents seemed to focus on their children's behavior independent of judgments about the children's language-literacy skills, whereas "teachers' judgments of behavioral difficulties may be mediated by the children's [deficient] language abilities," given the teachers' responsibility to teach (language-dependent) literacy skills over the course of the study (Lindsay et al., 2007, p. 824). In other words, teachers' perspectives may be more affected by academic factors, whereas parents may have a more holistic view of their children's abilities and needs. In sum, the authors cite the importance of looking at specific types of disorders within the SEBD cluster and of obtaining information across settings over time.

Longitudinal evidence yields valuable insights for treatment planning, as the following results illustrate. Language impairment diagnosed in young children predicted a significantly higher rate of psychiatric disorders in these children seen 14 years later (Beitchman et al., 2001). The significantly higher rate of anxiety disorders in young adults frequently took the form of social phobia, hypothesized to occur as a result of a history of difficulty communicating with others. Early diagnosis of speech impairment was not predictive of later psychiatric diagnosis. No gender differences were found, but having fewer young women in the follow-up study could have affected this result. In an earlier study with this same cohort, early language disorder diagnoses did not predict substance use disorders in adolescence, but adolescents with communication disorders who had problems with substance abuse "experienced a greater severity or a more global extent of disturbance" (Beitchman et al., 1999, p. 319).

Why Communication Disorder and Social-Emotional-Behavioral Disorder Diagnoses Coexist

A variety of hypotheses about why CDs and SEBDs co-occur have been offered explicitly or implicitly over the years. These hypotheses have implications for understanding communication and mental health disorders and for determining how services are provided to the children who have them (Angold, Costello, & Erkanli, 1999). Causal explanations for comorbid CDs and SEBDs have been proposed in both directions. At least on some level, several authors have posited the hypothesis that inadequate language skills result in increased emotional risk through compromised function and poor peer relationships, social rejection, and bullying (Cantwell & Baker, 1987; Conti-Ramsden & Botting, 2004; Fujiki, Brinton, Isaacson, & Summers, 2001). Social rejection can, in turn, result in reduced practice opportunities, which widen the gap in skills between children and their typically developing peers (Willinger et al., 2003).

Causality proposed in the opposite direction, with SEBDs causing CDs, has been ascribed to the negative impact of behavior disorders on parent–child interactions and, in turn, on language development (Willinger et al., 2003). Landry, Smith, Miller-Loncar, and Swank (1997) offer a more nuanced "causal" explanation, in the form of increased risk for children's language, cognitive, and social development stemming from low parental responsivity or negative parental affect when children display significant behavior problems. Given the complexity of CDs and SEBDs and the number of factors that affect them, describing their coexistence through "a direct causal chain will likely prove futile" (Brinton & Fujiki, 2005, p. 154).

Longitudinal studies have the potential to address questions about the developmental trajectory of these and other complex disorders. Knowledge about the life course of disorders should offer insights into changes over time that may currently be attributed to causality, such as an initial diagnosis of language impairment during adolescence (Im-Bolter & Cohen, 2007).

In contrast to a cascading model of deficit causing deficit, children's comorbid CD and SEBD diagnoses (and other comorbidities) have often been described as resulting from the broader common cause of a more pervasive, neurologically based developmental disorder (Accardo & Shapiro, 2005; Beitchman, 1985; Willinger et al., 2003). Neuroimaging studies of children who have developmental language disorders have revealed morphological differences in motor areas and language areas (Trauner, Wulfeck, Tallal, & Hesselink, 2000), supporting the picture of multiple diffuse effects that is observed clinically. The model of generalized neurological impairment does not preclude describing coexisting CDs and SEBDs as the result of combined genetic and environmental influences on development (Hayiou-Thomas, 2008; McGrath et al., 2007).

Assumptions one makes about why disorders coexist affect diagnostic practices and expectations regarding treatment outcomes. For example, an assumption of causal relationships between coexisting disorders might lead to the belief that treatment within the "causal domain" should resolve difficulties in both areas or prevent development of a "secondary" disorder. Some treatment findings contradict this perspective. For example, research relative to coexisting psychiatric and reading disabilities indicates that "effective treatments in one domain will not necessarily result in improvements in the other; as a result, each disorder needs separate treatment" (Maughan & Carroll, 2006, p. 353).

COMMUNICATION TREATMENT: COMMUNICATION DISORDERS AND SOCIAL-EMOTIONAL-BEHAVIORAL DISORDERS

In order to effect positive change, professionals who diagnose and treat children with coexisting CDs and SEBDs must consider the children's overall profile. Variability in children's profiles and changes in status due to development and overall health certainly complicate this process. Another source of variability in children's performance comes from demands that certain tasks place on them. Tasks vary in terms of attentional, cognitive, and linguistic demands and the options available for children to respond (e.g., length and complexity of response, verbal versus nonverbal options). Tasks also vary in terms of social-emotional and behavioral demands they place on children (e.g., known versus unfamiliar audience, emotional state, relative topic or activity). Of course, day-to-day variation may occur as a result of child-based characteristics that are not linked to their formal diagnoses. In sum, children's performance often varies across tasks, frequently but not exclusively, because of the way tasks are presented. Appropriate caution is needed in interpreting children's changes over time so that changes in response to treatment can be separated from change that occurs because of nontreatment factors (Yoder & Compton, 2004).

General Approaches to Speech-Language Treatment

Treatment techniques employed in the case studies presented in this chapter generally fall into the category called contemporary or naturalistic applied behavior analysis or milieu approaches. These treatment approaches share features with the didactic approaches from which they stem, in that adults develop the treatment goals and employ reinforcement to increase desired behaviors. There are some differences from traditional didactic treatment methods. One difference is an attempt to foster children's communicative initiation by granting access to a preferred object, contingent on some form of communicative output. Another difference is the tendency to reinforce behaviors intrinsically through "attainment of the child's desire or a social reinforcement" (Paul, 2008, p. 838). Example treatment strategies that fall under this milieu approach include mand-modeling, prompt-free training, and incidental teaching. Single-subject research studies document the success of such approaches in training children's communication skills.

Professionals who employ milieu treatment approaches conduct treatment in children's natural settings during regularly recurring activities using preferred objects and activities. They foster children's spontaneous verbal and nonverbal output by pausing at appropriate communicative junctures and allowing children to direct treatment through their indications of interest in objects or activities. In this approach, prompts and cues encourage expansions of children's output (instead of prompting to initiate communication). Reinforcement comes in the form of access to preferred objects or actions, in which the child has indicated interest. This approach respects that powerful learning occurs in meaningful communication contexts where ongoing interactions require children to integrate language content, form, and use. The inclusion of child-directed components makes this treatment approach dynamic and requires some "online" decision making, which depends on prior training, the sensitivity to recognize teaching opportunities, and the skill to execute child-directed training (Paul, 2008).

A variant of this approach, called responsivity education/prelinguistic milieu teaching, is useful with children whose developmental level falls approximately between 9 and 18 months of age, if they "have not yet become frequent, clear prelinguistic communicators" (Warren et al., 2006, p. 48). This approach is built on a transactional model of learning and social interaction, with the belief that adults (parents) and children influence each other's communication and both are influenced by features of the environment. Under positive circumstances, the child's initiations result in adult responses that serve as a scaffold onto which the child builds gradually more frequent and sophisticated language and social-emotional responses. Important elements of this treatment approach include provision of an environment that invites frequent communicative interaction, development of predictable social routines, and contingent responses that are adapted to the child's initiations. Training at this prelinguistic level increases the frequency, range, and clarity of children's nonverbal and verbal intentional communication and readies them for participation in an early language training program. Establishment of intentionality also fosters generalization across settings, conversational partners, and the early versions of "topics" children want to communicate. Benchmark skills indicating readiness for transition to a linguistic milieu treatment approach involve being able to "produce more than 1 or at most 2 spontaneous, intentional communication acts per minute in social play with an adult" (Warren et al., 2006, p. 49).

Language-Literacy Links

Spoken language serves as the basis for literacy development, so it is no surprise that tight links exist between the two forms of language. Impairments in spoken language are often mirrored in written language (Moats,

2000; Whitehurst & Lonigan, 2001). These language-literacy impairments are present in many children who have neurodevelopmental diagnoses (e.g., intellectual disability, autism spectrum disorders, and learning disabilities).

Children who have psychiatric diagnoses also have high rates of reading difficulties. Attentional problems were the most frequent type of psychiatric disorder to coexist with reading disability. Conduct problems, anxiety disorders, and poor self-esteem (but not clinical depression) also coexist with reading disabilities (Carroll, Maughan, Goodman, & Meltzer, 2005; Maughan & Carroll, 2006; Willcutt & Pennington, 2000). Further analysis suggested that reading difficulties present in children with conduct disorders may be mediated by attentional problems, but anxiety disorders coexisted with reading difficulties independent of attentional problems (Goldston et al., 2007).

Knowledge of the relationships between spoken language and literacy can be exploited to foster growth in spoken and print language modalities in fragile patient populations. Aspects of language that have received particular attention in relation to links between spoken language and literacy include phonological processing, vocabulary, comprehension, and the discourse skills needed for interpreting and generating expository text and narration. Inclusion of narratives in language intervention provides even wider language learning opportunities, like the chance to "explore processing limitations, create opportunities for using decontextualized language, facilitate social relationships, provide practice in constructive listening, improve reading comprehension, and identify language learning strengths and weaknesses" (Johnston, 2008, p. 98).

Lanter and Watson's (2008) suggestions for teaching literacy skills in children who have autism spectrum disorders are relevant to teaching children without autism whose developmental paths may not be typical. Coordination of spoken language and literacy training provides increased redundancy and practice opportunities. The authors note the importance of using literacy enhancement strategies that are adapted to the children's reading level (e.g., emergent reading → conventional reading → skilled reading). For early readers, strategies encourage cycling between spoken and print language (e.g., shared reading, story retelling, and talking about stories). Subsequent strategies emphasize techniques needed for increased reading efficiency (e.g., phonological awareness) and techniques designed to help students interpret meaning (e.g., using background knowledge and prior experiences to assist with comprehension). Comprehension of connected text, a multilevel linguistic and cognitive task, remains a challenge for many children after other language skills have been mastered (Boudreau, 2008; Kintsch, 2005; Nation & Norbury, 2005).

Augmentative and Alternative Communication: Literacy and Speech

In children who are nonverbal, the spoken-written language link may take the form of a link between augmentative and alternative communication (AAC) strategies and literacy. Children who do not communicate verbally cannot engage in verbal preliteracy activities (e.g., shared story reading), nor can they demonstrate other oral reading readiness skills (e.g., sound-letter matching, analysis and synthesis of speech sounds in novel words). Studies have shown that "the literacy learning experiences of students who use AAC are inconsistent and often of a lesser quality and quantity" in comparison with their non-AAC using peers (Fallon & Katz, 2008, p. 113). Children with "severe and multiple needs" may lack not only the speech skills through which literacy is often established, but they may have "less obvious and widely varying concomitant disorders in vision, hearing, health, cognition, or atten-

tion, [that] can lead to a sometimes exclusive, if understandable, focus by families on the child's medical, self-care, and therapeutic needs" (Koppenhaver, Hendrix, & Williams, 2007, p. 80). These children may have different early literacy experiences at home, and their teachers may feel ill equipped to adapt literacy training to the children's unique needs (Fallon & Katz, 2008). Because of literacy's importance in fostering independence and its power to allow communication that reaches beyond the immediate audience and context, there is considerable interest in developing the literacy skills of children who depend on AAC for expression. Training parents and professionals does appear to facilitate emergent literacy skills in children, as does the use of "student strengths and personal interests as a basis for developing wider language and literacy competence, often supported through integrated assistive technology" (Koppenhaver et al., 2007, p. 87).

Exposure to print language is often included as a component of speech and language treatment, then, for its potential benefits in development of speech, language, and literacy itself. Where needed, AAC systems can facilitate speech production and the acquisition of receptive and expressive vocabulary and increase positive behavior (Cafiero, 2001; Millar, Light, & Schlosser, 2006), as well as literacy. An example of an AAC system that is useful for fostering preliteracy skills and speech production skills is the Flip 'n Talk (FNT; Inman, 1999). The FNT consists of a manual core board with 16 high-frequency words and phrases, and an affixed spiral-bound flipchart of 30 semantic categories. Using "Natural Aided Language" (Cafiero, 1998) strategies to model language with this system, the FNT serves as a model for language generation. In this language stimulation approach the facilitator points out picture symbols on the child's communication display in conjunction with ongoing speech-language stimulation. Through the modeling process, the concept of using the pictorial symbols and speech interactively is demonstrated for the individual, along with specific examples of utterances the child might use to accomplish communicative goals.

Social Skills

Development of language for interaction is also a focus of the treatment goals and strategies in the case study presented here. Impaired social skills may stem from a number of factors, including insufficient knowledge, limited practice or feedback, limited cues or opportunities, insufficient reinforcement, and problem behaviors that interfere with social interactions (Elliot & Gresham, 1993). Impairments in social skills are a defining feature of autism spectrum disorders, but they also play a significant role in functional impairments that children with disorders like intellectual disability experience (Abbeduto & Hesketh, 1997). Children with behavior disorders, for example, have difficulty identifying intra- and interpersonal emotional cues and talking about emotional experiences, impairments that could not be attributed to their cognitive level (Cook, Greenberg, & Kusche, 1994).

CASE STUDY

A case study is presented next to illustrate the interplay of CDs and SEBDs in more detail. The child's family granted permission to tell her story. For privacy, the child's name was changed.

The child described has significant CDs and SEBDs. This child was selected to illustrate the treatment principles described in this chapter. The story illustrates the impact of comorbid CDs and SEBDs on the child's life and the factors that need to be considered in treatment programs to improve her functional skills.

The case study includes a description of the child's history and developmental levels, with special attention to her communication and social-emotional-behavioral profile. Speech-language treatment goals and strategies used to train the child are described. Because the settings in which children interact and their communication partners can have a powerful effect on outcome, these factors are also discussed as part of the child's treatment program. Primary communication partners typically include parents, sometimes siblings, and other professionals who work with the children. Training of critical adults in the child's world permits infusion of communication goals and strategies into daily routines. Attention to the environment may involve provision of external supports or accommodations, designed to provide structure or opportunity that children need in order to use newly learned communication skills. The child's progress in treatment and her continuing needs are presented.

Background

Ramona was born very prematurely, and her early life was dominated by her complex and fragile medical status. Her developmental diagnoses included mild-moderate intellectual deficiency, attention-deficit/hyperactivity disorder, pervasive developmental disorder-not otherwise specified with features of autism, motor coordination disorder, and speech-language disorder. Behaviorally, she exhibited noncompliance with medical and nonmedical demands, tantrums, hyperactivity, impulsivity, elopement, aggression, and pica. Ramona presented, at age 5, for an outpatient speech-language therapy because of concerns regarding limited communication skills and feeding concerns.

Born prematurely at 25 $\frac{1}{2}$ weeks, Ramona and her twin suffered multiple complications associated with their low birth weight and prematurity. Her neonatal course was remarkable for complications in respiration (e.g., bronchopulmonary dysplasia, tracheal malacia, subglottic stenosis, chronic lung disease, tracheotomy with ventilator dependency), vision (e.g., retinopathy of prematurity with subsequent laser surgery), and gastrointestinal function (e.g., Nissen fundoplication, gastrointestinal tube dependency). Additionally, she suffered bleeding in her brain (germinal matrix) and developed central sleep apnea.

During her first year, all spent in the hospital, Ramona had multiple pneumothoraces and surgeries for tracheal reconstruction with eventual tracheotomy and gastrointestinal tube (g-tube) placements. Her world involved tubes, pumps, and alarms. Her regimen included strict feeding and medication schedules. Ramona's life thus far was managed by routine and repetition. She was discharged from the neonatal nursery on a ventilator, requiring continuous nursing, and receiving all nutrition and hydration through her g-tube. She was cared for at home by her parents and a private nurse, but still required frequent scheduled and emergent doctor visits and hospital admissions. Ramona's care required a team of medical and developmental specialists. Her early life experiences involved having things done "to" her and "for" her more than "with" her. Nevertheless, she was eventually able to attend a full-day, special education preschool class for children with multiple impairments. She received speech-language services within her academic program. Ramona's private nurse rode the bus and attended school with her.

Ramona's tracheotomy tube was removed when she was 4 ½ years old. During this procedure, the surgeons removed scar tissue that had formed over her vocal folds and sealed them together. For the first time, Ramona could breathe through her mouth. Her stoma did not close spontaneously and required surgical closure. This procedure was completed 7 months after the decannulation surgery and 4 months into her speech-language treatment.

As her medical problems lessened, Ramona's behavioral and developmental concerns became more apparent and the next needs on the long list to be addressed. As seen in other children with neurodevelopment disorder profiles, Ramona demonstrated atypical registration and integration of sensory input. She had a high tolerance for pain, seldom crying. She resisted light touch and recoiled from objects (i.e., toothbrush, spoon) that were associated with touch to her head and mouth. She often hid in the corners of rooms after being physically guided through tasks. Ramona's parents learned to complete many daily tasks (e.g., hair and teeth brushing) immediately following bath time, while Ramona was still swaddled in a bath towel. Interactions with Ramona often involved avoiding events that might result in her complete withdrawal, incite a behavioral response (e.g., aggression, disruption), or aggravate her already vulnerable respiratory system.

Before she began speech-language services, Ramona received behavioral psychology treatment for noncompliance with medical procedures (e.g., bi-pap, electroencephalogram). As she tolerated more medical procedures, Ramona's behavioral psychology services addressed other concerns. These concerns included increased temper tantrums, aggressive behaviors (e.g., hitting, pushing, kicking), destructive behaviors (e.g., throwing objects, tearing), and elopement. Ramona's impulsive actions and the subsequent safety risk often led caregivers to restrict Ramona to "safer" areas within her house. Because many of her behaviors were dangerous, they resulted in quick responses from adults. Her cognitive and communication impairments in combination with her behavioral disorder limited the opportunities for teaching "online" or interrupting Ramona just prior to her engaging in a risk-taking behavior. Ramona could not be left unsupervised for any length of time. Ultimately, Ramona required intensive outpatient neurobehavioral services.

Speech-Language Evaluation Results

The results of Ramona's initial speech-language evaluation revealed solid receptive language skills at the 12- to 15-month range and solid expressive language skills at the 6- to 9-month range, with scattered skills in the 9- to 12-month range. She did not express herself through pictures or use a voice output system. Ramona did not use word approximations or vocalizations to communicate. Ramona most often attempted to reach an object (e.g., toy, video) she wanted rather than request it. Behaviorally, Ramona presented with restricted acceptance of tactile input (e.g., oral-motor exam, touching hand), a limited variety of interests, minimal play skills, and significant impulsivity.

At the time of initial meeting, Ramona could walk, run, jump, climb, kick, and pick up the smallest speck from the floor. With her emerging independence, Ramona explored her environment more consistently, seizing each of these moments head-on and with abundant energy. This independence offered incredible and frequent opportunities to learn, as well as inevitable safety risks. Ramona did not have the "toolbox" she needed to navigate this world, a world that required language as a principal means for interaction, learning, and independence. Ramona could not ask for what she wanted, request assistance, or share her knowledge. She could not invite other people to play with her or take her someplace. She did not understand or know how to share in the process. She could not seek information about novel things. Ramona could not express fear, anger, or joy verbally. She could not wait or stop herself from going after what she saw.

What Ramona did have was a small set of tools (e.g., behaviors, gestures) to use when interacting in her world. Her limited responses served both communicative and noncommunicative functions. The intent of her responses was often unclear, with interpretation left to familiar adults. When outward expressions occur in the form of aggressive, noncompliant, or disruptive behaviors, the communication intent of the action may be present but not be fully understood. With so few responses, each action had more than one meaning (e.g., aggression → pain, frustration, anxiety). The severity of her actions (e.g., pushing her twin down the stairs, running down the street) had clear and dire consequences. Ramona's interest and willingness to interact was blooming, yet she could not do it alone, consistently, or in socially acceptable ways.

Though decannulation occurred 4 months prior to her first visit, Ramona's stoma had not closed. She vocalized on inhalation and exhalation and had decreased breath support and poorly coordinated respiration for speech. She produced some vocalizations that resembled vowels but were word approximations. An oral-motor evaluation could not be completed because of Ramona's hypersensitivity.

Speech-language goals were developed, and weekly treatment began 1 month after the evaluation. Ramona needed to learn new ways of learning, behaving, and interacting. She needed to learn to "communicate" with the world around her. She needed to continue tolerating new tactile stimuli, so that her potential for speaking and eating could be further assessed. Her family needed to learn to interact with Ramona in different ways and to recognize subtle changes in her communication. Ramona's speech-language treatment program focused on increasing communication with the use of a multimodal approach (e.g., pictures, sign, gestures, speech) and evaluating potential for oral feeding.

Speech-Language Treatment: An Interdisciplinary Approach

For Ramona to make progress on any treatment goals, she needed to be challenged employing a systematic approach. Similar to her medical treatment desensitization, Ramona required desensitization to the techniques used to meet her communication and feeding goals. Many of the treatment techniques she needed required a hands-on approach. Therapy materials used for intervention included chewy tubes, toothbrushes, spoons, and an abdominal binder, as well as touch to her face and chest area.

Ramona's reactions to tactile stimuli were intense. Her heart rate and respiration significantly increased while anticipating, tolerating, and recovering from unwanted or new tactile sensory input. Ramona's oxygen levels decreased with prolonged reactions, to the point where her system shut down and she could no longer participate. She had to focus solely on breathing. At times, she engaged in rocking behaviors that may have further complicated her respiration status.

It was likely that a negative association between touch to the face and trunk and breathing difficulties was established early in her life. It appeared that Ramona associated unwanted touch with difficulty in breathing and difficulty in breathing with unwanted touch. Additionally, Ramona did not understand this bidirectional, tactile-respiratory connection, nor did she have the cognitive/language skills to indicate an aroused state. Ramona's respiratory status and her fear of compromised breathing were associated with anxiety, which often compounded her already poorly regulated sensory input system. A highly structured and predictable therapeutic routine was established, which included close monitoring of her respiratory status.

Ramona had the most difficulty tolerating oral input. Progress with feeding goals was slow. Ramona began accepting the presence of oral-motor tools or spoons on the table near her while she was engaged in a different activity. Her continued participation during activities that included intra-oral stimulation was reinforced by allowing access to her most preferred activity (e.g., toys, videos).

As with any intervention program, Ramona's active participation in the treatment process was paramount in minimizing the effect of anxiety and maintaining steady progress. Using picture communication, Ramona selected the "tool" or "food" for each oral-motor or feeding activity. Ramona's behavioral psychologist often helped with the selection, step-up, and implementation of reinforcement with increasing interdisciplinary treatment as the day for her swallow study approached. Eventually she selected the number of trials to be completed and which reinforcement she received. Pictures of the tools and food items were then added to Ramona's personalized communication system. Ramona learned to trust the process and challenge herself. She was experiencing the power of communication.

After 4 months of therapy, Ramona was ready for a modified barium swallow study to evaluate her swallowing skills radiographically. The study was completed with limited results, as only one swallow was recorded. Confirming clinical observations, Ramona had significant oral-phase dysphagia. She lacked the oral-motor skills to form a bolus of food and efficiently transport the bolus to her throat. The pharyngeal portion of the swallow was quick and efficient, with no penetration or aspiration. These results were promising, as the reflexive swallow and airway protection were intact. Two weeks later, Ramona had her stoma surgically closed. Ramona's system was now structurally capable of developing the pressure-dependent valving system for speech and swallowing that most children learn without intervention.

Ramona's speech production skills were now a primary focus. Very simply stated, speech production involves a series of precisely timed and pressure-sensitive movements of the vocal folds and articulators (e.g., lips, tongue) during a controlled exhalation. After Ramona's stoma was closed, further diagnostic evaluations of breathing and speech patterns were completed. Her harsh and hoarse vocal quality and low speaking volume suggested potential vocal fold damage. Direct examination of her vocal fold structure and function was not considered a priority at this time. Ramona demonstrated limited trunk rotation and flexibility. She used her "belly muscles" to breathe rather than her diaphragm. Ramona's chest wall evidenced multiple scars from surgeries, a possible factor in her limited trunk and chest wall mobility. Ramona had to learn a different breathing pattern and strengthen weak muscles, in order to improve respiratory control for speech.

To learn new breathing patterns, Ramona was required to tolerate therapy techniques that challenged her tactile defensiveness and had previously led to increased cardiovascular/respiratory stress. The same refusal behaviors emerged as were seen in the oral-motor treatment program. Using an approach similar to the one previously described, Ramona learned to tolerate passive stretches, wore an abdominal binder to facilitate diaphragmatic breathing, and learned vocabulary she needed (e.g., *deep breath, breathe out, hold breath, lean*) to complete therapy activities.

With each gained skill and subsequent change in speech production, Ramona's interactions with her family, familiar adults, and professionals improved. Ramona learned to say "Mom," "Dad," and her siblings' names. She learned to say "no" and to ask for help. She spontaneously told her mother, "I love you, Mom." Though her speech was not clear, she was definitely making progress in using speech to communicate and interact with her environment.

Ramona's receptive/expressive language and play skills had been addressed throughout all sessions. Language goals included improving vocabulary for items and concepts related to academic and social needs, introducing early literacy skills (e.g., story structure, letter-sound correspondence), and completing multiple step directions. These goals were embedded in daily activities (e.g., washing her face, setting the table for food trials), structured concept-focused activities, and simple play schemes.

Following success with general picture communication (e.g., activity specific boards, schedules, picture exchange), a personalized

Figure 11.1. Flip 'n Talk, Ramona's personalized manual core board. Flip 'n Talk is a manual communication board (Inman, 1999; reprinted by permission.) It consists of a manual core board with 16 high-frequency words and phrases and an affixed spiral-bound flipchart of 30 semantic categories. The flipchart was attached to the top of the communication board.

low-tech, picture-based system, the FNT, was developed for Ramona (see Figure 11.1). Ramona's system contained a static base board with an attached spiral-bound component, the "flipper," for additional vocabulary. Ramona quickly paired speech with her picture pointing. Though not designed for such use, Ramona's augmentative communication system also served as a pacing board for speech production. Through this process, she produced more consecutive words on exhalation, eventually up to the limits of her syntactic complexity. With input from family, school, and other therapists, specific semantic categories were developed and included therapy activities, therapists, and reinforcers.

Ramona used the FNT to make requests, answer simple *wh-* questions, and assist in adult-led communication repair (e.g., forced choice questions). From the beginning, Ramona's FNT was used within the context of preferred and nonpreferred therapy activities. This strategy reinforced the process of using language to mediate unfavorable situations by requesting "Stop" or "My turn" or "Can I have + object?" (i.e., sentence completion format) and more quickly replaced some maladaptive behaviors with socially appropriate communication skills. This process continues.

Ramona's parents, extended family, and private nurses learned how to use her FNT in structured and spontaneous interactions. Ramona reviewed picture names and locations as part of her daily routine at home. The FNT was also incorporated into her school day activities. With respect to feeding goals and continued oral-motor desensitization, Ramona's caregivers were trained in specific oral stimulation exercises and praise procedure used in therapy, educated on clinical signs of distress, and modeled appropriate language to request discontinuation. Ramona was eventually evaluated for and accepted into an interdisciplinary, inpatient feeding program.

Treatment Outcomes and Future Plans

Ramona made slow and steady progress across all goals, but had brief setbacks with each cold or illness. Various medication trials altered Ramona's sleep patterns; her sleep patterns also affected overall progress. As a direct result of interdisciplinary communications, Ramona's changing medical, behavioral, and developmental needs were identified, and the associated language concepts were incorporated into therapy. The interdisciplinary team interactions permitted updating of treatment progress and tracking of status (e.g., sleepiness, activity level).

After 15 months of therapy, Ramona participated in a follow-up speech-language evaluation. Results revealed an approximate 15- to 18-month gain in receptive language skills and an 18- to 21-month gain in expressive language skills. The gap between what she understood and what she could communicate was closing. Overall, Ramona's language skills were roughly commensurate with those of a typically developing 2½-year-old child. She showed emerging responsibility for her FNT, and she initiated more communication in therapy sessions, at home, and at school. Ramona combined three to six units of

meaning in combination or sequence (e.g., "Can I" + "play" + "big" + "Henry") to ask questions, stop an activity, or answer questions. She spoke using consonant-vowel and consonant-vowel-consonant words and word approximations. Ramona's consonant repertoire now included *m, n, p, b, t, d, w, l* and *j* (as in "yellow"). Ramona omitted final consonants in words (e.g., "bat" → "ba") and used "h" and "j" as substitutions for many speech sounds not yet in her repertoire (e.g., *f, v, s, z*). With respect to oral-motor skills, Ramona accepted multiple flavors of pureed food and a large variety of nonnutritive oral stimulation. She continued to struggle when she ate food with lumps or soft solids (e.g., crackers). During feeding sessions, the time she needed to swallow food significantly decreased, making oral feeding more practical and setting the stage for intensive feeding treatment.

With respect to her behavior profile, Ramona remained seated for 50–60 minutes, grabbed and threw objects less, and engaged in almost no aggressive behaviors. She accepted redirection more easily. Ramona's significant communication and feeding successes within and outside of speech-language sessions reflected the benefits of the interdisciplinary approach used in treatment, in addition to maturation, improved health, and medication changes. For children with complex CD and SEBD profiles like Ramona's, a coordinated, interdisciplinary treatment approach with regular communication among the team and family members provides the best opportunity for success.

SUMMARY

Because of their centrality to function, disorders in communication and social-emotional-behavioral skills markedly hamper children's ability to meet their everyday needs. Children who have CDs have poorly developed, inefficient, or atypical speech and language skills with which to effect change in their environment and accomplish their goals. When compounded with SEBDs, the complexity and severity of children's functional impairments increase significantly. Comprehensive treatment programs, designed to address the interplay between these functionally devastating disorders, are best equipped to foster positive outcomes for children, their families, and the professionals who work with them.

FUTURE DIRECTIONS TO EXPLORE

The survey of research and case study presented here illustrates the complex issues to be considered in the evaluation and treatment of children who have coexisting CDs and SEBDs. Following are suggested goals for training, research, and clinical practice.

Professionals across disciplines who serve children with these comorbid disorders should be given guidance about the importance of assessment across domains. For example, professionals who treat children with CDs "need to recognize that nearly one-half of their patients are likely to have a diagnosable psychiatric disorder, and that a substantial number of these patients are also likely to have developmental disorders" (Cantwell & Baker, 1987, p. 159). The same is true for primary care settings, where behavioral screening tools are needed, given the fact that "about 40% to 50% of office visits . . . involve behavioral, psychosocial, or educational problems" and "approximately 75% of children with psychiatric disturbances are first seen in primary care settings" (Miller, 2007, p. 178).

Training professionals about historically underreported diagnoses also seems valuable. Underreported diagnoses include those without obvious traits (e.g., receptive language disorders and internalizing emotional disorders) and subtle disorders of any type, particularly when they coexist with disorders that are severe or striking or include global developmental delay.

Clinical training should also include discussions about assumptions made on the basis of formal and informal assessment. In pondering why assessments were not automatically completed to rule out language problems in adolescents and young adults who had marked SEBDs, Brownlie et al. (2004) wondered if professionals saw fairly stable verbal intelligence quotient scores as predictive of low potential for improvement in language treatment, resulting in no referral being made.

Lines of research appear to be converging, in a way that should result in the development of improved screening and diagnostic tools and treatment programs. For example, the renewed interest in behavioral phenotyping, coupled with genetic research strategies, should yield insights about genetic profiles and genetic-environmental ties that would not have been possible a decade ago (Dykens & Hodapp, 2007; Feinstein & Singh, 2007; Mervis, 2004; Warren, 2004). Refined assessment tools and practices are needed where patient profiles are difficult to capture or professional controversy exists, as is the case for preschool children with mental health diagnoses (Egger & Angold, 2006) and adolescents, where "persisting language problems affect their personal relationships" and "little research has examined LI [language impairment] and psychopathology" (Im-Bolter & Cohen, 2007, p. 537). As noted earlier, tracking the developmental trajectory of these diagnoses, when they occur in isolation and in combination, will be helpful for all facets of clinical care.

Treatment efficacy studies continue to be important. In data compiled by the American Speech-Language-Hearing Association (2004), 75% of speech-language pathologists reported treating children who have pragmatic language and social skill disorders, but treatment studies in these areas were focused on younger children, were conducted with single subjects, or lacked sufficient rigor (Cirrin & Gillam, 2008). Research on intervention programs designed to address known and suspected weaknesses between CDs and SEBDs is also critical, given the functional consequences that children and their families experience. Treatment research could address effects of early intervention, including the possibility of preventing the onset of related disorders, as some pioneers in the study of these coexisting relationships hoped. Promising signs regarding young children's language, cognitive, and social-emotional-behavioral development come from treatment research designed to train parent responsiveness skills (Landry, Smith, & Swank, 2006; Warren & Brady, 2007) and determine the time in children's development at which this treatment might yield optimum results (Landry, Smith, Swank, & Guttentag, 2008).

REFERENCES

Abbeduto, L., & Hesketh, L.J. (1997). Pragmatic development in individuals with mental retardation: Learning to use language in social interactions. *Mental Retardation and Developmental Disabilities Research Reviews, 3,* 323–333.

Accardo, J., & Shapiro, B.K. (2005). Neurodevelopmental disabilities: Beyond the diagnosis. *Seminars in Pediatric Neurology, 12,* 242–249.

Achenbach, T.M. (1991a). *Manual for the Child Behavior Checklist/4–18.* Burlington: University of Vermont Press.

Achenbach, T.M. (1991b). *Manual for the Teacher Report Form.* Burlington: University of Vermont Press.

American Speech-Language-Hearing Association. (2004). *2004 ASHA Schools survey: Caseload characteristics.* Retrieved October 15, 2008, from www.asha.org.

Angold, A., Costello, E.J., & Erkanli, A. (1999). Comorbidity. *Journal of Child Psychology and Psychiatry, 40*(1), 57–87.

Beitchman, J.H. (1985). Speech and language impairment and psychiatric risk: Toward a model of neurodevelopmental immaturity. *Psychiatric Clinics of North America, 4,* 721–735.

Beitchman, J.H., Douglas, L., Wilson, B., Johnson, C.J., Young, A., Atkinson, L., et al. (1999). Adolescent substance use disorders: Findings from a 14-year follow-up of speech/language impaired and control children. *Journal of Clinical Child Psychology, 28,* 312–321.

Beitchman, J.H., Wilson, B., Johnson, C.J., Atkinson, L., Young, A., Adlaf, E., et al. (2001). Four-

teen-year follow-up of speech-language impaired and control children: Psychiatric outcome. *Journal of the American Academy of Child and Adolescent Psychiatry, 40,* 75–82.

Boudreau, D. (2008). Narrative abilities: Advances in research and implications for clinical practice. *Topics in Language Disorders, 28,* 99–114.

Brinton, B., & Fujiki, M. (2005). Social competence in children with language impairment: Making connections. *Seminars in Speech and Language, 26,* 151–159.

Brownlie, E.B., Beitchman, J.H., Escobar, M., Young, A., Atkinson, L., Johnson, C., et al. (2004). Early language impairment and young adult delinquent and aggressive behavior. *Journal of Abnormal Child Psychology, 32,* 453–467.

Cafiero, J. (1998). Communication power for individuals with autism. *Focus on Autism and Other Developmental Disabilities, 13,* 113–121.

Cafiero, J. (2001). The effect of an augmentative communication intervention on the communication, behavior, and academic program of an adolescent with autism. *Focus on Autism and Other Developmental Disabilities, 16,* 179–193.

Camarata, S.M., Hughes, C.A., & Ruhl, K.L. (1988). Mild/moderate behaviorally disordered students: A population at risk for language disorders. *Language, Speech and Hearing Services in Schools, 19,* 191–200.

Cantwell, D.P., & Baker, L. (1987). Prevalence and type of psychiatric and developmental disorders in three speech and language groups. *Journal of Communication Disorders, 20,* 151–160.

Cantwell, D.P., Baker, L., & Mattison, R. (1981). Prevalence, type, and correlates of psychiatric diagnoses in 200 children with communication disorder. *Developmental and Behavioral Pediatrics, 2,* 131–136.

Carroll, J., Maughan, B., Goodman, R., & Meltzer, H. (2005). Literacy difficulties and psychiatric disorders: Evidence for comorbidity. *Journal of Child Psychology and Psychiatry, 46,* 524–532.

Cirrin, F.M., & Gillam, R.B. (2008). Language intervention practices for school-age children with spoken language disorders: A systematic review. *Language, Speech, and Hearing Services in Schools, 39,* S110–S137.

Cohen, N., Davine, M., Horodessky, N., Lipsett, L., & Isaacson, L. (1993). Unsuspected language impairment in psychiatrically disturbed children: Prevalence and language and behavioral characteristics. *Journal of the American Academy of Child and Adolescent Psychiatry, 32,* 595–603.

Conti-Ramsden, G., & Botting, N. (2004). Social difficulties and victimization in children with

SLI at 11 years of age. *Journal of Speech, Language, and Hearing Research, 47,* 145–161.

Cook, E., Greenberg, M., & Kusche, C. (1994). The relations between emotional understanding, intellectual functioning, and disruptive behavior problems in elementary-school-aged children. *Journal of Abnormal Child Psychology, 22,* 205–219.

Cross, M. (1998). Undetected communication problems in children with behavioural problems. *International Journal of Language & Communication Disorders, 33,* 509–514.

Dykens, E.M. (1995). Measuring behavioral phenotypes: Provocations from the "new" genetics. *American Journal of Mental Retardation, 99,* 522–532.

Dykens, E.M., & Hodapp, R.M. (2007). Three steps toward improving the measurement of behavior in behavioral phenotype research. *Child and Adolescent Psychiatric Clinics of North America, 16,* 617–630.

Egger, H.L., & Angold, A. (2006). Common emotional and behavioral disorders in preschool children: Presentation, nosology, and epidemiology. *Journal of Child Psychology and Psychiatry, 47,* 313–337.

Elliot, S.N., & Gresham, F.M. (1993). Social skills interventions for children. *Behavior Modification, 17,* 287–313.

Fallon, K.A., & Katz, L.A. (2008). Augmentative and alternative communication and literacy teams: Facing the challenges, forging ahead. *Seminars in Speech and Language, 29,* 112–119.

Feinstein, C., & Singh, S. (2007). Social phenotypes in neurogenetic syndromes. *Child and Adolescent Psychiatric Clinics of North America, 16,* 631–647.

Fujiki, M., Brinton, B., Isaacson, T., & Summers, C. (2001). Social behaviors of children with language impairment on the playground: A pilot study. *Language, Speech, and Hearing Services in Schools, 32,* 101–113.

Fujiki, M., Brinton, B., Morgan, M., & Hart, C.H. (1999). Withdrawn and sociable behavior of children with language impairment. *Language, Speech, & Hearing Services in Schools, 30,* 183–195.

Gallagher, T.M. (1993). Clinical forum: Language and social skills in the school-age population, language skill and the development of social competence in school-age children. *Language, Speech, and Hearing Services in Schools, 24,* 199–205.

Gallagher, T.M. (1999). Interrelationships among children's language, behavior, and emotional problems. *Topics in Language Disorders, 19,* 1–15.

Gilmour, J., Hill, B., Place, M., & Skuse, D.H. (2004). Social communication deficits in conduct disorder: A clinical and community survey. *Journal of Child Psychology and Psychiatry, 45,* 967–978.

Glogowska, M., Roulstone, S., Peters, T.J., & Enderby, P. (2006). Early speech- and language-impaired children: Linguistic, literacy, and social outcomes. *Developmental Medicine and Child Neurology, 48,* 489–494.

Goldston, D.B., Walsh, A., Arnold, A.M., Reboussin, B., Daniel, S.S., Erkanli, A., et al. (2007). Reading problems, psychiatric disorders, and functional impairment from mid- to late adolescence. *Journal of the American Academy of Child and Adolescent Psychiatry, 46,* 25–32.

Gottesman, I.I., & Gould, T.D. (2003). The endophenotype concept in psychiatry: Etymology and strategic intentions. *American Journal of Psychiatry, 160,* 636–645.

Hart, H. (2004). Speech and language disorders and associated problems: Meeting children's needs. *Developmental Medicine and Child Neurology, 46,* 435.

Hart, K.I., Fujiki, M., Brinton, B., & Hart, C.H. (2004). The relationship between social behavior and severity of language impairment. *Journal of Speech, Language, and Hearing Research, 47,* 647–662.

Hayiou-Thomas, M.E. (2008). Genetic and environmental influences on early speech, language, and literacy development. *Journal of Communication Disorders, 41,* 397–408.

Hill, E.L. (2001). Non-specific nature of specific language impairment: A review of the literature with regard to concomitant motor impairments. *International Journal of Language and Communication Disorders, 36*(2), 149–171.

Im-Bolter, N., & Cohen, N. (2007). Language impairment and psychiatric comorbidities. *Pediatric Clinics of North America, 54,* 525–542.

Inman, N. (1999). *Flip 'n Talk manual augmentative communication system.* Available at http://www.inmaninnovations.com/flipntalk.aspx.

Johnston, J. (2008). Narratives: Twenty-five years later. *Topics in Language Disorders, 28,* 93–98.

Kintsch, E. (2005). Comprehension theory as a guide for the design of thoughtful questions. *Topics in Language Disorders, 25,* 51–64.

Koppenhaver, D.A., Hendrix, M.P., & Williams, A.R. (2007). Toward evidence-based literacy intervention for children with severe and multiple disabilities. *Seminars in Speech and Language, 28,* 79–90.

Landry, S.H., Smith, K.E., Miller-Loncar, C.L., & Swank, P.R. (1997). Predicting cognitive-language and social growth curves from early maternal behaviors in children at varying degrees of biological risk. *Developmental Psychology, 33,* 1040–1053.

Landry, S.H., Smith, K.E., & Swank, P.R. (2006). Responsive parenting: Establishing early foundations for social, communication, and independent problem-solving skills. *Developmental Psychology, 42,* 627–642.

Landry, S.H., Smith, K.E., Swank, P.R., & Guttentag, C. (2008). A responsive parenting intervention: The optimal timing across early childhood for impacting maternal behaviors and child outcomes. *Developmental Psychology, 44,* 1335–1353.

Lanter, E., & Watson, L.R. (2008). Promoting literacy in students with ASD: The basics for the SLP. *Language, Speech, and Hearing Services in Schools, 39,* 33–43.

Lindsay, G., Dockrell, J.E., & Strand, S. (2007). Longitudinal patterns of behavior problems in children with specific speech and language difficulties: Child and contextual factors. *British Journal of Educational Psychology, 77,* 811–828.

Maughan, B., & Carroll, J. (2006). Literacy and mental disorders. *Current Opinion in Psychiatry, 19,* 350–354.

McGrath, L.M., Pennington, B.F., Willcutt, E.G., Boada, R., Shriberg, L.D., & Smith S.D. (2007). Gene x environment interactions in speech sound disorder predict language and preliteracy outcomes. *Development and Psychopathology, 19,* 1047–1072.

Mervis, C. (2004). Cross-etiology comparison of cognitive and language development. In M.L. Rice & S.F. Warren (Eds.), *Developmental language disorders: From phenotypes to etiologies* (pp. 153–185). Mahwah, NJ: Lawrence Erlbaum Associates.

Millar, D.C., Light, J.C., & Schlosser, R.W. (2006). The impact of augmentative and alternative communication intervention on the speech production of individuals with developmental disabilities: A research review. *Journal of Speech, Language, & Hearing Research, 49,* 248–264.

Miller, J.W. (2007). Screening children for developmental behavioral problems: Principles for the practitioner. *Primary Care: Clinics in Office Practice, 34,* 177–201.

Moats, L. (2000). *Speech to print: Language essentials for teachers.* Baltimore: Paul H. Brookes Publishing Co.

Nation, K., & Norbury, C.F. (2005). Why reading comprehension fails: Insights from developmental disorders. *Topics in Language Disorders, 25,* 21–32.

Paul, R. (2008). Interventions to improve communication in autism. *Child and Adolescent Psychiatric Clinics of North America, 17,* 835–856.

Raitano, N.A., Pennington, B.F., Tunick, R.A., Boada, R., & Shriberg, L.D. (2004). Pre-literacy skills of subgroups of children with speech sound disorders. *Journal of Child Psychology and Psychiatry 45*(4), 821–835.

Redmond, S.M., & Rice, M.L. (1998). The socioemotional behaviors of children with SLI: Social adaptation or social deviance? *Journal of Speech, Language, and Hearing Research, 41,* 688–700.

Redmond, S.M., & Rice, M.L. (2002). The stability of behavioral ratings of children with SLI. *Journal of Speech, Language, and Hearing Research, 45,* 190–201.

Rescorla, L. (2002). Language and reading outcomes to age 9 in late-talking toddlers. *Journal of Speech, Language, and Hearing Research, 45,* 360–371.

Tager-Flusberg, H. (1999). An introduction to research on neurodevelopmental disorders from a cognitive neuroscience perspective. In H. Tager-Flusberg (Ed.), *Neurodevelopmental disorders* (pp. 3–24). Cambridge, MA: MIT Press.

Trauner, D., Wulfeck, B., Tallal, P., & Hesselink, J. (2000). Neurological and MRI profiles of children with developmental language impairment. *Developmental Medicine and Child Neurology, 42,* 470–475.

Warren, S.F. (2004). Intervention as experiment. In M.L. Rice & S.F. Warren (Eds.), *Developmental language disorders: From phenotypes to etiologies* (pp. 187–206). Mahwah, NJ: Lawrence Erlbaum Associates.

Warren, S.F., & Brady, N.C. (2007). The role of maternal responsivity in the development of children with intellectual disabilities. *Mental Retardation and Developmental Disabilities, 13,* 330–338.

Warren, S.F., Bredein-Oja, S.L., Fairchild Escalante, M., Finestack, L.H., Fey, M.E., & Brady, N.C. (2006). Responsivity education/prelinguistic milieu teaching. In R.J. McCauley & M.E. Fey (Eds.), *Treatment of language disorders in children* (pp. 47–75). Baltimore: Paul H. Brookes Publishing Co.

Webster, R.I., Majnemer, A., Platt, R.W., & Shevell, M.I. (2005). Motor function at school age in children with a preschool diagnosis of developmental language impairment. *Journal of Pediatrics, 146,* 80–85.

Whitehurst, G.J., & Lonigan, C.J. (2001). Emergent literacy: Development from prereaders to readers. In S.B. Neuman & D.K. Dickensen (Eds.), *Handbook of early literacy research* (pp. 11–29). New York: Guilford Press.

Willcutt, E.G., & Pennington, B.F. (2000). Psychiatric comorbidity in children and adolescents with reading disability. *Journal of Child Psychology and Psychiatry, 41,* 1039–1048.

Willinger, U., Brunner, E., Diendorfer-Radner, G., Sams, J., Sirsch, U., & Eisenwort, B. (2003). Behavior in children with language development disorders. *Canadian Journal of Psychiatry, 48,* 607–614.

Yoder, P., & Compton, D. (2004). Identifying predictors of treatment response. *Mental Retardation and Developmental Disabilities Research Reviews, 10,* 162–168.

12

A Clinical Approach to the Pharmacological Management of Behavioral Disturbance in Intellectual Disability

Sarah Risen, Pasquale J. Accardo, and Bruce K. Shapiro

Behavioral disturbances are frequent concomitants of intellectual disability, with estimates ranging from 40% to 70% (Syzmanski et al., 1990). Behavioral disturbances are a major reason for out-of-home placement and for failed treatment programs. Behavioral disturbances in people with intellectual disabilities have been a significant public health issue for quite some time. In 2001, the National Institutes of Health (NIH) (the National Institute of Neurological Disorders and Stroke [NINDS], the National Institute of Child Health and Human Development [NICHD], the National Institute of Mental Health [NIMH], and the NIH Office of Rare Diseases) and the Joseph P. Kennedy, Jr. Foundation sponsored a conference that focused on the issues of people with intellectual disability and behavioral disturbance (National Institutes of Health, 2001). Work groups addressed epidemiology, diagnosis and assessment, ethical considerations, interventions research, research design, and research training needs. One of the things highlighted in that conference was the inability of the mental health system to address behavioral disturbances in people with intellectual disability. This chapter addresses some of the clinical difficulties of diagnosis and then discusses pharmacological management of behavioral disturbances in children with intellectual disability.

WHAT IS THE NATURE OF THE BEHAVIORAL DISTURBANCE?

People with intellectual disability have deficient adaptive behavior. Indeed, this is a requirement for the diagnosis of intellectual disability. Adaptive behavior reflects the individual's ability to master his or her environment. Although associated with cognition, measured adaptive behavior may be discrepant from cognition. For example, people with receptive-expressive language disorders may have adaptive behavior skills that are substantially lower than their cognitive ability. People with autism, attention-deficit/hyperactivity disorder (ADHD), or cerebral palsy also may have adaptive behavior that is depressed relative to cognition.

Maladaptive behavior is not simply the deficiency of adaptive behavior. Maladaptive behavior is behavior that is never appropriate. It may be general, as in cases of aggression, self-injury, hyperactivity, anxiety, psychosis, or depression. On the other hand, maladaptive behavior in certain neurobehavioral disorders may be as specific as self-mutilation in

Lesch-Nyhan syndrome, food stealing in Prader-Willi syndrome (PWS), or the hand wringing seen in Rett syndrome.

THE DIFFICULTY OF DIAGNOSING AND TREATING BEHAVIORAL DISTURBANCE IN INTELLECTUAL DISABILITY

People with intellectual disability have higher rates of maladaptive behavior than the general population. Like the general population, children with intellectual disability experience disorders such as anxiety, depression, psychosis, and adjustment disorders. It is difficult to treat behavioral disturbance in intellectual disability for many reasons (Table 12.1).

Adjustment disorders result from a mismatch between the demands of an environmental situation and the abilities of an individual. Adjustment disorders are commonly seen when an individual is incompletely diagnosed: the intellectual disability is not appreciated, the person's function and measured intellect are discrepant, or associated dysfunctions are inadequately characterized. A case that was seen recently illustrates some of these issues concerning adjustment disorders in children with intellectual disability:

John is an 8½-year-old boy who was seen by specialists because of escalat-

ing behavioral disturbances. He seemed unhappy. He made self-deprecating remarks such as "I wish I were dead," and he seemed to lack the spark that he had previously. This year, John was more withdrawn in his special education class. John had been previously diagnosed as having intellectual disability. His IQ, measured by a standard test, was 60, and adaptive behavior measures were similar. John was in a second-grade class and receiving inclusion services. The class was working on reading and regrouping with addition.

John had been seen by a child psychiatrist, who initially felt that John might have ADHD, inattentive type. He was started on stimulants and the dose was titrated appropriately, but the response to the stimulants was modest at best. The possibility of depression was raised so John was started on a selective serotonin reuptake inhibitor (SSRI) agent; however, this also did not lift his mood.

An interdisciplinary evaluation revealed that John was able to perform academics at a kindergarten level rather than at the second-grade curriculum of his current classroom. On further questioning, it was noted that he spent 3 hours each night working on his homework. Adjustment of John's

Table 12.1.   Reasons for difficulties diagnosing behavioral disturbance in intellectual disability

There is a mismatch between demands and abilities (adjustment disorder).

The same behavior may be called different names.

The same behavior may have different mechanisms.

Comorbidities are frequently present.

The functional capacity of the child needs to be considered.

Medical or somatic causes may present similarly.

Behaviors may be innocent, unintentional, or imitative.

The patient may have poor communicative abilities.

The patient may be unable to alter responses according to social cues.

Impaired executive function leads to responses out of proportion to the stimulus.

academic expectations yielded a positive result, and medication was discontinued.

John's case highlights a number of the difficulties encountered when behavioral disturbances are diagnosed and treated in children with intellectual disabilities. In this case, the demands placed upon John were beyond his ability (adjustment disorder). As John was not able to express his frustration regarding his situation, he began to act out in manners seen in multiple behavioral disorders, including ADHD and depression.

Another reason for the difficulty of diagnosing behavior disorders in intellectual disability is that the same behavior may be interpreted differently or called by different names. For instance, a child may be constantly aware of the surrounding environment. Parents may see this behavior as creative or alert. As lack of focus may be impairing in a classroom setting, however, a teacher may relate this behavior to distractibility associated with ADHD. The child who shows excessive behavioral responses to loud noise, tags, rumpled socks, or food textures may be said to have sensory integration disorder. Others may point to his or her rolling on the floor in response to a joke as signs of dysmodulation, whereas others associate these behaviors with a diagnosis of anxiety. A child with a motor coordination disorder and intellectual disability may show hand-flapping behavior when he or she is excited or happy. Neurologists may call this overflow or synkinesis, while the psychiatrist may call this stereotypic behavior. As in the previous case, the psychiatrist interpreted John's behavior as a disorder of attention rather than a secondary phenomenon due to the difficulty of his schoolwork.

A third complicating factor that makes treating behavioral disturbance in intellectual disability difficult is that the same behavior may have different mechanisms. In our first example, John's self-deprecating statements and voiced desire "to be dead" were not the result of depression but his way of communicating that the current situation was intolerable. Another example is aggression, which may have a number of inciting causes. Aggression may be a result of medical illness, drug toxicity, preseizure irritability, postictal confusion, mania, depression, organic mood disorder, rage attacks, task-related anxiety, paranoid delusions, or an inability to express needs, as a means of obtaining positive reinforcement, or as avoidance behavior in the absence of an underlying dysfunction (Pary, Silka, & Blaha, 1995).

A fourth reason is that children with intellectual disability may have significant comorbid conditions that alter the expression of the behavioral disturbance.

Peter is a 10-year-old boy with cerebral palsy. He has typical cognitive abilities and attends a general education classroom with modest accommodations. When in new situations, Peter talks about his school in excruciating detail. He reviews the floor plan of his school completely and details all four wings of his school. He then goes into similar detail about his schedule. While people raise the question as to whether Peter had Asperger syndrome in addition to cerebral palsy, his perseverative behavior was substantially improved by a small dose of an SSRI agent that minimized his anxiety.

Children with both deafness and intellectual disability often exhibit poor eye contact, not as social avoidance or unawareness, but because these children are visually scanning their environment. Comorbid disorders are certainly increased in developmental disorders, and each behavioral disturbance may affect the appearance of other disorders.

A fifth reason for difficulty in treating behavioral disturbance in children with intellectual disabilities is the need to evaluate behavior in the context of the child's functional capacities.

William was a 4-year-old child with Down syndrome. He never was able to sit through an entire meal without getting up and walking around. He had similar behavior during story time in the school setting. His pediatrician raised the question as to whether William had ADHD and commenced treatment with stimulants. However, medication only made William's behavior worse. Developmental evaluation revealed that William's problem-solving/functional level was that of a 2-year-old child, with his language level being slightly lower.

In the context of his cognitive level, William's behavior is actually age appropriate.

Undiagnosed somatic illnesses may also affect behavior and are often not appreciated in a person with intellectual disability.

Kendra was a 14-year-old girl with Down syndrome. Her parents noticed that she had become more irritable and was less outgoing during her first year of high school. Many of the social outlets that Kendra enjoyed were no longer open to her because she was too old. They were concerned that she might be becoming depressed. On evaluation, she was noted to have a pulse of 65, a narrow pulse pressure, dull skin, depressed reflexes, and slow responses to questions. Laboratory studies confirmed the diagnosis of hypothyroidism. Treatment with thyroid hormone resulted in substantial improvement in Kendra's mood and behavior.

Kendra's case exemplifies a challenge particular to children with developmental syndromes. In addition to increased comorbid behavioral disorders, often multiple organ systems may be involved as well. Children with Down syndrome have higher rates of thyroid disease, are at increased risk for depression, and may develop early-onset dementia. All of these diagnoses should be considered in young adults with Down syndrome who show functional regression.

Another somatic illness that may mimic behavioral disorders is pain. Pain may cause aggression and screaming. Esophagitis as a result of gastric reflux is a not uncommon cause for irritable behavior in children who are quite young or at very low functional levels. Identifying the source of pain is very challenging in children who are nonverbal or at low levels of function.

Children with intellectual disability may also unintentionally or naively exhibit maladaptive behaviors. For example, Jonas asked his brother Ryan to "do him a favor" and hold a bicycle for him. When the police came to investigate a stolen bicycle, Ryan was the one arrested because the bicycle was in his possession. Lacking the higher cognitive skills needed to judge a troublesome situation, people with intellectual disability may imitate their peers or follow others without question. In their desire to be accepted, children with intellectual disabilities are at risk of choosing to imitate behavior that is not socially acceptable. Adolescents may be easily led into behaviors that are socially inappropriate, including early sexual experiences, drug use, and theft. Similarly, children with intellectual disability may simply imitate the maladaptive behavior of others with no particular intent.

Finally, many of the constructs used to establish psychiatric diagnoses are based on the ability of the individual to communicate. This poses obvious problems for people who are nonverbal, as the cause of their behavioral disturbance must be inferred by others. The difficulties faced by people with intellectual disability who are verbal are less obvious but may nevertheless be substantial. Often, children with intellectual disability have a limited ability to verbally communicate abstractions such as feelings, alter responses based on social cues, may overgeneralize, have difficulty in appreciating exceptions, and may not be able to make the subtle distinctions that govern social interaction. Furthermore, impaired executive function in children with intellectual disability frequently leads to responses that are out of proportion to the stimulus.

SCHEMA THAT ARE USED TO CLASSIFY BEHAVIORAL DISTURBANCE IN PEOPLE WITH INTELLECTUAL DISABILITY

Currently, there are a number of different approaches to classifying behavioral disturbance in people with intellectual disability. The applied behavioral techniques (functional assessment of behavior, outlined in Chapter 9) are independent of the diagnosis. Instead, the assessment evaluates the antecedents and consequences of a specific behavior. As tested in a functional analysis, avoidance, escape/attention, and internal/nonsocial stimuli are common precipitating factors for maladaptive behavior. Once the cause of the behavioral disturbance is determined, interventions are designed to alter the environment and ideally improve the maladaptive behavior.

The International Classification of Diseases (ICD) was designed by the World Health Organization to promote international comparability in the collection, processing, classification, and presentation of mortality statistics. It has become the international standard diagnostic classification for epidemiological purposes and clinical use. The ninth edition (ICD-9) was published in 1977; it assigns numeric codes to diagnoses and procedures. The U.S. National Center for Health Statistics modified the system to capture additional information; this modified version was called ICD-9 CM (Clinical Modification). The acceptance of ICD-9-CM was fostered by the requirement of ICD-9-CM codes for Medicare and Medicaid claims. ICD-10 was adopted in 1992, and the clinical modification will be implemented in 2013.

The most commonly used classification system is the *Diagnostic and Statistical Manual of the American Psychiatric Association, Fourth Edition, Text Revision* (*DSM-IV-TR;* American Psychiatric Association, 2000). The *DSM-IV-TR* is a multiaxial classification system that addresses a number of domains of behavior (p. 27)[1]:

Axis   I. Clinical Disorders
           Other Conditions That May Be a
           Focus of Clinical Attention
Axis II. Personality Disorders
           Mental Retardation
Axis III. General Medical Conditions
Axis IV. Psychosocial and Environmental
           Problems
Axis   V. Global Assessment of Functioning

The *DSM-IV-TR* provides specific criteria for diagnoses, and attempts have been made to make these descriptive definitions as objective as possible. Unfortunately, the criteria used for diagnosis of mental health disorders in typically developing populations may not be appropriate criteria for those with intellectual disabilities.

The National Association for the Dually Diagnosed and the American Psychiatric Association developed the *Diagnostic Manual–Intellectual Disability* (*DM-ID*) to address the issues related to diagnosis in people with intellectual disability, the limitations in applying *DSM-IV-TR* criteria to people with intellectual disability, and adaptations of the diagnostic criteria (Fletcher, Loschen, Stavrakaki, & First, 2007a). A companion volume (Fletcher, Loschen, Stavrakaki, & First, 2007b) offers a broad examination of the topic, including a description of each disorder, a summary of the *DSM-IV-TR* diagnostic criteria, a review of research including an evaluation of the strength of evidence supporting the literature conclusions, a discussion of the etiology and pathogenesis of the disorders, and adaptations of the diagnostic criteria for people with intellectual disability.

[1]*Reprinted with permission from the Diagnostic and Statistical Manual of Mental Disorders, Text Revision, Fourth Edition. (Copyright 2000). American Psychiatric Association.*

The *DM-ID* has been available for only a short time and is not widely used.

SPECIFIC DIAGNOSTIC CONSIDERATIONS

As previously discussed, psychiatric diagnosis is difficult in people with intellectual disability (Sovner, 1986). ADHD, depression, anxiety, and oppositional defiant disorder are some of the most frequent diagnoses in children with intellectual disability. However, each of these diagnoses requires particular consideration in children with intellectual disability.

ADHD is one of the most common diagnoses made in childhood. This is an extremely difficult diagnosis to make in an individual with intellectual disability. Currently there are no concrete measures to define inattention or hyperactivity specifically in intellectual disability. The clinician must consider whether the constructs of inattention or hyperactivity are related to the child's mental age or chronological age.

Depression and mood disorders are similarly difficult to discern. This, in part, is related to the social changes associated with aging. As children with intellectual disability grow, they are less able to cognitively relate to age-matched peers. However, their age-appropriate physical size often precludes these children from interacting with younger children. Thus, many individuals with intellectual disabilities seem more withdrawn as adolescents and mirror the phenotype of mild depression. Bipolar disorder is frequently considered and is difficult to clarify in people with intellectual disability. The difficulties with modulating behavior (hyperactivity), coupled with frequent disturbances of sleep, may be misconstrued as mania.

Anxiety is most often overlooked in people with developmental disorders and may be primary versus situational. Situational anxiety is much more common in children with intellectual disability. The anxiety often results from an inability to express concerns, difficulty in predicting the environment, and difficulties in modulating responses. Despite the cause, anxiety may be manifested as aggression, withdrawal, or high levels of activity.

Children and young adults with intellectual disabilities may be inappropriately diagnosed with oppositional defiant disorder. However, this diagnosis requires the assumption that the individual has the ability to perform the task requested yet deliberately chooses not to do so. In the case of children with intellectual and other developmental disabilities, often the child is unable to perform the task. Thus the issue is that the child cannot comply, rather than chooses not to.

PSYCHOPHARMACOLOGICAL AGENTS

Though the difficulties in accurately diagnosing behavioral disorders in children with developmental disabilities are clear, often the most effective treatment for these disorders includes a combination of behavioral and/or psychological therapy and medication management. Wachtel and Hagopian (2006) describe the "neurobehavioral model" combining applied behavioral analysis and psychopharmacology. The authors further provide a case series illustrating the effectiveness of this approach for severe problem behaviors in children and adolescents with intellectual disability. For further details on behavioral assessment and behavior treatment plans, see Chapter 13.

The major purpose of psychotropic medication is to decrease maladaptive behavior. This may be anxiety, hyperactivity/inattention, aggression, psychosis, or other disruptive behaviors. Psychotropic medications do not resolve impairments in cognition, generalization, abstraction, or judgment. Nevertheless, by decreasing unwanted behavior, psychopharmacological agents may have an indirect effect of increasing prosocial behavior. Historically, misinterpretation of the use of psychotropic medication has been described as a chemical straitjacket that turns children into zombies.

However, such a reduction of interaction or behavior occurs when medication is used without regard to target behaviors and without close monitoring of effects and side effects. When treated with the proper medication at the proper dose, the child should be largely unchanged, other than improvement of the targeted behavior and associated functions.

Specific Agents

We discuss here five main classes of medications as well as miscellaneous treatment options. The classes of psychotropic medications include stimulants, neuroleptics, antidepressants, anxiolytics, and mood stabilizers. Applications to adults and the general population are only mentioned in brief, as our primary focus is children with intellectual disability. There is an overall paucity of large, randomized controlled trials for many of these medications, particularly in children with intellectual disability. This is due in part to the difficulty of developing and completing large, masked, randomized, controlled drug trials in children. Nevertheless, available studies are presented to assess the overall data on specific psychotropics used in children with intellectual and developmental disabilities.

Stimulants

ADHD behaviors are more prevalent in children with intellectual disability as compared with the general population. Two studies evaluating the prevalence of behavior disorders in intellectual disability revealed a diagnosis of hyperkinesia for 8.7%–16% of children and adolescents (Emerson, 2003; Stromme & Diseth, 2000). In comparison, the general worldwide pediatric population prevalence of ADHD is estimated to be 5% (Polanczyk, de Lima, Horta, Biederman, & Rohde, 2007). As impulsivity, hyperactivity, and inattention can be safety risks and functionally impairing across many settings, stimulants are commonly prescribed in this population.

Mechanism of Action There are two categories of stimulants: methylphenidate and amphetamine, both of which are first-line treatments in ADHD. Both types of stimulants have short-, intermediate-, and long-acting forms. Liquid, chewable tablets, sprinkles, and patch preparations are available for and useful in the pediatric population. The methylphenidate category includes methylphenidate (Ritalin, Ritalin SR and LA, Concerta, Methylin, Metadate ER and CD, Daytrana) and dexmethylphenidate (Focalin and Focalin XR). Amphetamines include dextroamphetamine (Dexedrine, Dextrastat, and Vyvanse) and mixed amphetamine salts (Adderall and Adderall XR). Atomoxetine (Strattera) is a "nonstimulant" medication that is also used in the treatment of ADHD behaviors (Table 12.2).

Although the underlying pathophysiology of ADHD has yet to be fully elucidated (and likely involves complex interactions among many systems), abnormalities in the dopaminergic neurotransmitter pathways have been related to symptoms of ADHD (as discussed in a review by Solanto, 1998). Dopamine is a prominent neurotransmitter in the frontostriatal circuitry, which regulates attention, organization, planning, motivation, and reward. Neurophysiological data have been supported by neuroimaging data. For instance, a positron emission tomography (PET) study evaluating the effects of methylphenidate on striatal dopamine concentrations revealed a correlation between methylphenidate-induced striatal dopamine concentrations and tested measures for impulse control, attention, information processing, and variability (Rosa-Neto et al., 2005). As discussed in a review on brain imaging, genetic data, and the dopamine theory of ADHD, many of the genes implicated to date are dopamine receptors, including DAT1, DRD4, and DRD5 (Swanson et al., 2007). Thus research to date supports a role for dopamine in the behavioral abnormalities of ADHD.

Table 12.2. Psychopharmacological agents used to manage behavioral disturbance accompanying intellectual disability

Medication class	Type	Examples (brands)	Indications (per class)	Side effects	Notes
Stimulants	Methylphenidate	Concerta, Daytrana, Focalin, Methylin, Metadate, Ritalin	Inattention, impulsivity, hyperactivity	Appetite suppression, irritability, insomnia, tics, weight loss, anxiety, moodiness, nausea, dizziness	Black box: abuse potential Warning: cardiovascular effects and suicidal ideation/risk
	Amphetamine	Adderall, Dexedrine, Vyvanse	See above	See above	See above
Norepinephrine reuptake inhibitors	Atomoxetine	Strattera	Inattention, impulsivity, hyperactivity	Appetite suppression, weight loss, nausea, constipation	Black box: suicide ideation/risk Warning: liver risks
Neuroleptics	Typical (use rare)	Haldol, Prolixin, Thioridazine, Thorazine	Aggression, impulsivity, self-injury, disruptive behavior	Weight gain, sedation, extrapyramidal syndromes, increased prolactin	Laboratory monitoring per medication
	Atypical	Clozaril, Risperdal, Seroquel, Zyprexa	See above	Lower rate of extrapyramidal syndromes and prolactin effects	See above
Antidepressants	Selective serotonin reuptake inhibitors	Lexapro, Luvox, Paxil, Prozac, Zoloft	Depression, obsessive-compulsive disorder, anxiety disorders, ASDs, behavioral disorders	Possible "activation," mania, restlessness, irritability, agitation, rash, weight loss, nausea, anxiety	Black box: suicide ideation/risk
	Serotonin norepinephrine reuptake inhibitors	Effexor, Cymbalta, Remeron	See above	See above	See above
	Bupropion hydrochloride	Wellbutrin	See above	Hypertension, tachycardia, headache, weight change, rash, sleep changes, constipation	N/A
	Tricyclic antidepressants	Elavil, Norpramin, Pamelor, Tofranil	See above	Weight gain, bloating, constipation, headache, dizziness, sedation	Black box: suicide ideation/risk

Category	Class	Examples	Uses	Side effects	Monitoring
Anxiolytics	Benzodiazepines	Ativan, Valium, Xanax, Klonopin	Anxiety	Sedation, increased behavior side effects, dizziness, depression	N/A
	Buspirone	Buspar	See above	Headache, nausea, confusion, dizziness	N/A
Mood stabilizers	Lithium	Eskalith	Mania, fragile X syndrome	Weight gain, polyuria, polydipsia, GI effects, tremor, goiter, muscle irritability and weakness	Monitor for toxicity
	Antiepileptics	Depakote, Lamictal, Tegretol	Mania, aggression, impulsivity, repetitive behaviors, ASDs	Sedation, weight gain, GI effects, headache, tremor	Precautions and lab monitoring per medication
Miscellaneous	Alpha-adrenergic agents	Catapres, Tenex	ADHD, tics	Sedation, irritability, hypotension	N/A
	Acetylcholinesterase inhibitors	Aricept	ADHD, traumatic brain injury	GI effects, sedation, anorexia, muscle cramps	
	Glutamate modulators	Riluzole, Memantine	Obsessive compulsive disorder, anxiety, mood, ASDs	GI effects, dizziness, somnolence, headache	Lab monitor per medication
	Hormone	Melatonin	Sleep disorders	GI effects, irritability, headache	
	Opioid antagonist	Naltrexone	ASDs	Headache, GI effects, sleep changes, anxiety	

Key: ADHD, attention-deficit/hyperactivity disorder; ASDs, autism spectrum disorders; GI, gastrointestinal.

Stimulants are thought to modulate the dopamine and norepinephrine neurochemical imbalance through unclear, complex mechanisms. Solanto (1998) provides an extensive review of the neuropsychopharmacology of stimulant medication and ADHD. Both methylphenidate and dextroamphetamines are thought to increase levels of dopamine in the brain by blocking reuptake and stimulating the release of dopaminergic presynaptic vesicles.

Dysfunction of the noradrenergic mediated pathways connecting the prefrontal cortex and locus coeruleus are also thought to contribute to symptoms of ADHD, thereby supporting the possible noradrenergic effects of stimulants and more selective agents such as the alpha-adrenergic agonists (Arnsten, 1998). Levy and Swanson (2001) discussed the roles of both dopamine and norepinephrine in a comprehensive review article. In addition, Arnsten (1998) specifically addressed the role of the prefrontal cortex in working memory and the role of both norepinephrine and dopamine in ADHD pharmacological management. Both classes of stimulants are thought to also increase noradrenergic activity via alpha-2 receptors. Atomoxetine (Strattera) is thought to act as a norepinephrine reuptake inhibitor. Clonidine and guanfacine, other treatments for ADHD discussed in the "Miscellaneous" section of this chapter, are both alpha-adrenergic agonists.

Indications for the General Population

The average rate of response to stimulant medications in the general population is 85% (Wolraich et al., 1990). In a 14-month randomized controlled trial comparing medication management, behavioral therapy, combined treatment, and community care, the methylphenidate or combined (medication and behavior therapy) groups were significantly more effective in treating ADHD (MTA Cooperative Group, 1999). Thus stimulant medications are considered the first-line treatment for ADHD.

Indications Specific to Children with Intellectual Disability

Several studies have evaluated the efficacy of stimulants in children with intellectual disability. One double-blind crossover control study of children with mild–borderline intellectual disability and ADHD noted a 75% rate of response to methylphenidate with significant improvement in behavior, work output, restlessness, and interest (Handen, Breaux, Gosling, Ploof, & Feldman, 1990). An additional study of children with IQ 48–74 and ADHD revealed a 67% response to methylphenidate with significant reduction in activity level, irritability, and anxiety relative to placebo (Handen, Feldman, Gosling, Breaux, & McAuliffe, 1991).

Another placebo-controlled double-blind crossover trial of methylphenidate in children with ADHD and mild–moderate intellectual disability reported a 55% "substantial response" (defined as a 30% improvement) on behavior scales and 46% "substantial gain" in cognitive tasks at the highest dose of methylphenidate (0.60 mg/kg) (Pearson et al., 2004). In a prior double-blind crossover study, Pearson et al. (2003) revealed significant treatment effects on hyperactivity, inattention, aggression, and asocial behavior in children with intellectual disability and ADHD. In this study, teacher ratings emphasized treatment effects, whereas parent reports indicated side effects, primarily appetite suppression and sleeping problems, though notably no increase in anxiety. Both the positive treatment effects and side effects correlated with increasing methylphenidate dose.

In the largest study population of 90 children (4–17 years) with intellectual disability (mean IQ 58.5) and ADHD, Aman, Buican, and Arnold (2003) compiled three similar studies over an 11-year period. A 30% improvement in behaviors (attention, overactivity, and conduct problems) was noted in 44% of subjects on 0.4 mg/kg methylphenidate. Cognitive testing also revealed significant improvements, such as greater accuracy and shorter response times. Further subanalysis revealed a significant association between IQ and clinical response to methylphenidate. Only 19.4% of subjects with IQ < 50 had a significant clinical response compared with 53.6% of children with an IQ > 50.

There are no published trials on the efficacy of atomoxetine in children with intellectual disability. A large placebo-controlled and double-blind trial comparing methylphenidate, atomoxetine, and placebo in children with ADHD (IQ unknown) described a significant reduction (by 40%) in the ADHD Rating Scale (DuPaul, Power, Anastopoulos, & Reid, 1998) total score in both the methylphenidate and atomoxetine treatment groups compared with placebo. However, methylphenidate overall was more successful than atomoxetine. Of the patient group that did not respond to one medication or the other, treatment with the other ADHD medication proved to be effective in nearly 45% of subjects. The authors suggested this finding as support for a preferential response pattern to stimulant medication (Newcorn et al., 2008).

In summary, stimulants have been found to be effective overall treatment options for behaviors associated with ADHD in children with intellectual disability, though the response rate is slightly lower than that of typical peers. A higher dose of methylphenidate is often more effective, and children with developmental disabilities may be more sensitive to side effects. Most studies examine the effects of methylphenidate (not amphetamines), with doses ranging from 0.15 mg/kg to 0.6 mg/kg. For clinical trials, an "effective response" is most often based on 30% improvement in target behaviors. Though many studies attempt to isolate familial intellectual disability by excluding known diagnostic causes, the diagnoses of intellectual disability and ADHD and measures of outcome are nevertheless variable among the trials.

Specific Syndromes and Stimulants   Reviewing pharmacological effects on multiple neurodevelopmental disabilities, Hagerman (1999) reported that children with fragile X

syndrome and ADHD behaviors have a 65% response rate to stimulants. Furthermore, given the increased prevalence of ADHD symptoms in fetal alcohol syndrome, Hagerman (1999) proposed that dextroamphetamine may be the more effective class of stimulant, as amphetamines are proposed to be more specific to the D1 receptor in the mesolimbic system, which is particularly damaged by alcohol exposure.

Rates of ADHD in autism spectrum disorders (ASDs) have been noted to be as high as 78% in a chart review of 84 children with diagnosed ASDs (Lee & Ousley, 2006). Data analysis of secondary measures from the Research Units on Pediatric Psychopharmacology (RUPP) Autism Network revealed significant improvement in symptoms of ADHD, particularly impulsivity and hyperactivity, in children with an ASD treated with methylphenidate (Posey et al., 2007). In a placebo-controlled crossover trial, Arnold et al. (2006) report significant efficacy of atomoxetine in treating symptoms of ADHD in 16 children with autism.

Side Effects/Contraindications   Typical side effects of stimulants include appetite suppression, gastrointestinal complaints, headache, sleep changes, and irritability (which usually subside early in treatment initiation). Though stimulants were effective in treating target behaviors, Handen et al. (1990) reported "staring" in 6 of 12 subjects, 4 with associated drowsiness. In this study, one child also developed severe social withdrawal. The authors discuss these side effects as likely results of closer observation during the trial or a consequence of "less well developed cortical functioning" causing this population to be more susceptible to effects of stimulants. In a later study, Handen et al. (1991) reported no statistically significant increase in side effects, though, importantly, 22% of subjects (6 of 14) had to reduce the dose or discontinue stimulants altogether.

In a large randomized control trial of methylphenidate in 72 children with pervasive developmental disorder (PDD), despite the 49% effectiveness rate in the treatment group, 18% of subjects dropped out of the trial because of adverse events. The most common side effect causing trial withdrawal was irritability, though other reported adverse events included decreased appetite, sleep problems, gastrointestinal issues, increased emotional outbursts, and mood or affective changes (Research Units on Pediatric Psychopharmacology–Autism Network, 2005).

The cardiovascular risks of stimulant medication have been a prevalent concern. Conflicting statements regarding the necessary cardiovascular screening were released by prominent medical organizations. The American Heart Association recommends electrocardiograms (EKGs) for all children prior to starting medications, citing the increased incidence of ADHD in children with congenital heart disease and the possible likelihood of a baseline electrocardiogram detecting abnormalities (in the general pediatric population). However, the American Academy of Pediatrics and the American Academy of Child and Adolescent Psychiatry do not recommend an EKG prior to starting stimulants. A policy statement by the American Academy of Pediatrics (O'Keefe, 2008) challenged the lack of evidence that EKGs prevent cardiac death, as well as the lack of "compelling evidence" indicating that the risk of sudden cardiac death is higher in children receiving ADHD medications compared with the general population (Perrin, Friedman, Knilans, Black Box Working Group, & Section of Cardiology and Cardiac Surgery, 2008). Despite the conflict, a careful cardiac history should be taken, including, but not limited to, a history of cardiac disease including congenital heart disease, symptoms of cardiac disease, family history of sudden death in children, hypertrophic cardiomyopathy, or prolonged QT syndrome.

Neuroleptics

The physiological basis for prescribing neuroleptics, which are dopamine antagonists, presumes a significant role of dopamine dysregulation in disruptive and psychotic behaviors. In a review of atypical antipsychotics in developmental disabilities, Aman and Madrid (1999) discuss the history of the "dopamine hypothesis" of psychosis by first citing prior studies which indicated that dopaminergic agonists (e.g., amphetamines) could induce maladaptive behaviors. Further evidence, such as the induced movement disorders associated with neuroleptic use, bolstered the presumed role of dopamine dysfunction in disruptive or psychotic behavior (as dopamine receptors are prominent in movement control systems in addition to limbic/behavioral regions). Thus it is not surprising that neuroleptics are prescribed to treat disruptive, maladaptive behaviors. Given the potential side-effect profile, however, patients who have been prescribed these medications require careful monitoring by a physician.

Mechanism of Action   Neuroleptics (also known as antipsychotic medications) are divided into two general groups, typical and atypical. Typical neuroleptics are subdivided into high potency (increased extrapyramidal side effects [EPS] with reduced histaminergic, alpha-adrenergic, and anticholinergic effects) and low potency (reduced EPS with increased histaminergic, alpha-adrenergic, and anticholinergic effects). Typical high-potency neuroleptics include haloperidol (Haldol) and fluphenazine (Prolixin). Typical low-potency neuroleptics include chlorpromazine (Thorazine). Atypical neuroleptics are structurally and mechanically different from typical neuroleptics and include clozapine (Clozaril), olanzapine (Zyprexa), risperidone (Risperdal), quetiapine (Seroquel), ziprasidone (Geodon), aripiprazole (Abilify), and paliperidone (Invega) (Table 12.2). Both typical and atypical neuroleptics act to variable degrees as dopamine (specifically D_2) receptor antagonists, though atypical antipsychotics have more diverse and differing mechanism of actions as well as a reduced side-effect profile.

Reviewing both pharmacological laboratory and clinical data, including neuroimaging studies, Seeman (2002) discusses a theory on the mechanism of action of antipsychotics. The typical antipsychotics were noted to bind more tightly than dopamine to D_2 receptors. This lower dissociation constant explains the increased EPS side effects, increased serum prolactin, and tardive dyskinesia with chronic use of typical neuroleptics. Seeman then notes that the atypical antipsychotics bind more loosely to the D_2 receptor when compared with dopamine and have higher dissociation constants (release from the receptors more rapidly). His conclusion was that these factors explain the lower rate of EPS, prolactin levels, and reduced effects on cognition noted with the use of atypical neuroleptics. He further disputes the role of $5HT_{2A}$ receptor blockade thought to be associated with atypical neuroleptics by noting that the serotonin receptors are blocked at doses that exert no clinical effect on psychosis (Seeman, 2002).

Nevertheless, the theoretical role of serotonin receptors in antipsychotic effects has been discussed in the literature. Aman and Madrid (1999) explain the theories supporting the role of serotonin in the mechanism of action of atypical neuroleptics, from the $5HT_{2A}$:D_2 receptor affinity ratio to the possible role of serotonin in modulating dopamine release. Though possibly not related to the clinical antipsychotic effects, the variable histamine antagonism and anticholinergic nature of some neuroleptics discussed by the authors explain much of the side-effect profile of neuroleptics. Other neurotransmitters possibly modulated through neuroleptic use include GABA and glutamate.

Indications for the General Population

Antipsychotics are commonly used to treat schizophrenia, bipolar disorder, and mania, though each has specific Food and Drug Administration (FDA) approvals. There have been numerous lawsuits related to off-label use or false marketing of the neuroleptics. Neuroleptics are commonly prescribed in children as well, often off label. In a chart review of nearly 100 children being initiated on either neuroleptics or SSRIs, 75% of the newly prescribed medications were atypical antipsychotics, most often risperidone (Alacqua et al., 2008). Aggression, irritability, self-injury, and other such disruptive behaviors are common behavioral treatment targets in children prescribed neuroleptics.

Indications Specific to Children with Intellectual Disability

Risperidone was the first atypical neuroleptic the FDA approved for use in children, as treatment of irritability in autism, schizophrenia, manic episodes, and bipolar disorder. Aripiprazole recently obtained FDA approval for mania and Bipolar I in children 10–17 years old as well as teens with schizophrenia. Diagnoses of conduct disorder, hyperkinesis, and PDD are all increased in children with intellectual disability compared with matched peers without intellectual disability (Emerson, 2003). In an epidemiological study of 862 children with intellectual disability in the Netherlands, one in ten children used psychotropic medications, most commonly antipsychotics (3.9%) and stimulants (2.3%). Neuroleptic use was associated most often with a diagnosis of PDD (de Bildt, Mulder, Scheers, Minderaa, & Tobi, 2006).

One long-term, placebo-controlled crossover study examined the effects of risperidone on aggression, property destruction, and self-injury in children and adults with intellectual disability (Hellings et al., 2006). The 40 patients who completed the study were ages 8–56 years and had an even distribution of mild to profound intellectual disability. A majority of patients had also been diagnosed with ASDs. This trial revealed a 57.5% response rate to risperidone with 50% improvement on the Aberrant Behavior Checklist–Community Irritability subscale (Aman, Singh, Stewart, & Field, 1985a, 1985b). Increased appetite, weight gain (mean of 7.9 kg for children over 46 weeks), sedation, and gastrointestinal complaints were the most frequent side effects.

Risperidone was also found to be effective in a larger, 6-week randomized, double-blind, placebo-controlled trial. Snyder et al. (2002) evaluated the effects of risperidone on conduct and disruptive behavior disorder in 110 children with an IQ range of 36–84. In addition to a significant reduction on the conduct subscale (primary outcome), the risperidone treatment group also had a significant improvement in clinical global impression, reduced hyperactivity and social withdrawal, and improved adaptive skills and compliant/calm behavior (Snyder et al., 2002). In the open-label one-year follow-up study extension, Turgay, Binder, Snyder, and Fisman (2002) evaluated the long-term safety and efficacy of risperidone in 77 children with borderline to moderate intellectual disability and *DSM-IV*-diagnosed disruptive behavior disorder. Seventy-eight percent of children completed the full year of study. Baseline scores on Conduct Problem Subscales were significantly different between those patients already on risperidone during the preceding double-blind trial and those subjects who had been receiving placebo. Reassuringly, the subjects beginning the open-label trial risperidone-naïve had a significant, rapid improvement in target behaviors within 1 week of starting risperidone. Both test groups maintained the positive results throughout the one-year follow-up.

In a 4-week parallel group treatment trial, Filho et al. (2005) compared the effects of risperidone versus methylphenidate for ADHD behaviors in 45 children with moderate intellectual disability. Both groups had reduced symptoms on all measures of the ADHD scales (total score, hyperactive, inattentive, and oppositional defiant disorder). Though there was no significant difference between the two treatment groups, the effect of risperidone was more pronounced. Not surprisingly, children in the methylphenidate group had a significant weight reduction, whereas the risperidone group had significant weight gain.

Given the concerns of sedation, the effects of risperidone on cognitive function in children with intellectual disability have also been commonly reported. In an extensive review of the literature, Pandina et al. (2007) conclude that risperidone does not affect cognitive performance in children with intellectual disability in the short term and that cognitive gains are maintained, with improvement in attention and memory, while the children are on risperidone in long-term clinical trials.

Olanzapine has been studied in adolescents with intellectual disability and refractory disruptive behavior. Handen and Hardan (2006) reported an open-label, prospective trial of olanzapine in 16 adolescents with a mean IQ of 55 and diagnoses of disruptive behavior confirmed by the Aberrant Behavior Checklist. Most of the subjects were on multiple additional medications, including stimulants, SSRIs, alpha-adrenergic agonists, and/or mood-stabilizing agents. Four patients dropped out of the study, due to either side effects or worsening behavior. Of those completing the study, 10 of 15 had a significant reduction in behaviors ($\geq 50\%$), including aggression, agitation, impulsivity, overactivity, and inattention, with associated clinical improvement noted as well. Significant side effects included substantial weight gain (67% of subjects with > 10 pounds over the 8-week trial) and increased serum prolactin levels without associated clinical symptoms.

In contrast to prior literature, a recent randomized controlled trial of adults with intellectual disability indicated a lack of significant medication effect. Tyrer et al. (2008) compared the effects of risperidone, haloperidol, and placebo on recent aggressive behaviors. The results by 4 weeks revealed reduction of aggression in all three groups, with the most improvement on the modified overt aggression scale observed in the placebo group. Furthermore, there were no significant differences among the three treatment groups on any of the outcome measures. Strengths and weaknesses of the study, however, were further discussed in multiple editorials critiquing this study.

Specific Syndromes and Neuroleptics

Numerous studies support the efficacy of neuroleptics in ASDs. In an 8-week placebo-controlled trial of 101 children with autism and severe aggression, tantrums, or self-injurious behavior, 56.9% of participants treated with risperidone had a significant reduction in irritability based on the subscale of the Aberrant Behavior Checklist. Furthermore, 69% of those receiving risperidone had "much improved" to "very much improved" ratings on the Clinical Global Impressions Severity Scale (Guy, 1976), compared with only 12% of the placebo group (McCracken et al., 2002). In an open-label trial of 12 patients with autism or pervasive developmental disorder-not otherwise specified started on ziprasidone for unspecified maladaptive behaviors, 50% (6 of 12) responded to drug treatment based on improvements in the Clinical Global Impressions Severity scale (McDougle, Kem, & Posey, 2002). Though the trial was small, open label, and fairly nonspecific in target behaviors/outcomes, there was, notably, no weight gain associated with ziprasidone treatment over the 6 weeks.

Another open-label study evaluated the effects of risperidone in children with Down syndrome, ASDs, and disruptive behaviors and self-injury. Capone, Goyal, Grados, Smith, and Kammann (2008) report a significant reduction in scores on all five subscales of the Aberrant Behavior Checklist. Consistent with prior studies, the most common side effects in this patient population were hyperphagia (48%) and weight gain (70%).

Side Effects/Contraindications

Though somewhat controversial, typical antipsychotics are generally thought to produce more EPS (acute dystonic reactions, akathesias, subacute parkonsinism, tardive dyskinesia) than atypical antipsychotics. Withdrawal dyskinesias are common in both classes of neuroleptics, and thus all medications should be tapered slowly. Clozapine specifically carries a risk of agranulocytosis and requires specific laboratory monitoring. Furthermore, given the likelihood of hyperphagia and weight gain with many neuroleptics, these medications should be used with caution in patients with PWS (Hagerman, 1999) or other syndromes associated with obesity or metabolic syndrome. In a chart review of the initiation of psychotropic medications, 68% of those started on neuroleptics reported adverse events, with 31% discontinuing the therapy because of these adverse events. The most common side effects reported in this study included hyperprolactinemia, weight gain, sleepiness, and increased appetite (Alacqua et al., 2008).

Side effects of risperidone have been well reported, most commonly weight gain and somnolence. In a double-blind, placebo-controlled trial of 101 children with autism, the risperidone treatment group had no significant changes in laboratory testing but did have significant weight gain compared with the placebo group. Other side effects reported included somnolence, enuresis, increased appetite, rhinitis, and problems walking (Aman et al., 2005). Supporting previously discussed findings on risperidone and normal cognitive function, subjects from this trial who were able to complete cognitive testing revealed no decline in cognitive performance while on risperidone for 8 weeks (Aman et al., 2008).

In a conduct disorder treatment trial in children with intellectual disability previously discussed, Snyder et al. (2002) reported a mean 2.0-kg weight gain in the risperidone treatment group. Other common side effects included somnolence, gastrointestinal disturbance, upper respiratory infections, headache, increased appetite, and dyspepsia. The incidence of EPS was not significantly increased in the treatment group in this study.

The long-term study extension by Turgay et al. (2002) reported a notable percentage of adverse events in the risperidone treatment group. Though considered clinically mild–moderate by the investigators, 76 of 77 subjects experienced a new, reemergent, or worsening adverse event, most commonly somnolence (51.9%), headache (37.7%), and weight gain (36.4%). Prolactin levels were elevated in 19.5% of participants, but only two experienced associated clinical manifestations. Furthermore, 26% of subjects reported mild–moderate EPS, including hypertonia, hyperkinesias, involuntary muscle contractions, and tremor.

Overall, atypical neuroleptics have been found to be effective in the management of many disruptive behavior disorders; however, adverse effects such as significant weight gain, sedation, potential movement disorders, and elevated prolactin levels require careful consideration prior to prescription and throughout the treatment course with neuroleptics.

Antidepressants

Serotonin is a neurotransmitter thought to be involved in the regulation of mood, cognition, sleep, and other essential behaviors. In a review of serotonin knockout mice studies, Gingrich and Hen (2001) discuss the role of serotonin in anxiety, affect, aggression, and drug abuse across multiple mouse model trials. The link between serotonin and mood disorders is enhanced by the effectiveness of SSRIs in treating mood disorders. However, as discussed in a review on the pharmacology of SSRIs in children and adolescents, the medication effects are immediate, but the clinical effects often take 2–4 weeks (Leonard, March, Rickler, & Allen, 1997). Thus the role of SSRIs in modulating psychiatric behaviors likely involves more indirect pathways as well. In an early review of the mechanism of action of antidepressants, Charney, Menkes, and Heninger (1981) conclude that modulation of receptor sensitivity explains the long-term effects of antidepressants.

Antidepressants include the SSRIs fluoxetine (Prozac), sertraline (Zoloft), paroxetine (Paxil), citalopram (Celexa), escitalopram (Lexapro), and fluvoxamine (Luvox). Tricyclic antidepressants (TCAs) are an older class of antidepressants including amitiptyline (Elavil), nortriptyline (Pamelor), clomipramine (Anafranil), desipramine (Norpramin), and imipramine (Tofranil). Additional medications include multiple receptor antidepressants such as venlafaxine (Effexor), bupropion (Wellbutrin), trazodone (Desyrel), and nefazodone (Serzone) (Table 12.2).

Mechanism of Action   SSRIs increase the amount of available serotonin by blocking reuptake into presynaptic cells. TCAs block reuptake of norepinephrine and/or serotonin but also block histaminergic, cholinergic, and alpha-adrenergic receptors. Newer antidepressants may more selectively block serotonin and/or noradrenergic receptors without affecting other neurotransmitter systems. Largely, the mechanism of action of each antidepressant is unknown and is thought to be variable. Nevertheless, by modulating levels of neurotransmitters, notably serotonin, antidepressants have been successful medications in the treatment of mood disorders.

Indications for the General Population   The SSRIs have been approved by the FDA for treatment of depression in adults, but refer to the medication information for each specific antidepressant for precise FDA approval. Many antidepressants are also approved for the treatment of anxiety, phobias, obsessive-compulsive disorder (OCD), and other psychiatric diagnoses. Currently, the only FDA-approved SSRI for children (8 and older) with depression is fluoxetine. Fluoxetine, sertraline, fluvoxamine, and clomipramine have been approved by the FDA for the treatment of OCD in children. In a review of randomized, placebo-controlled trials on psychopharmacotherapy in children with anxiety disorders, Reinblatt and Riddle (2007) conclude there is good evidence to support the use and safety of SSRIs, primarily fluoxetine and fluvoxamine, in treating pediatric anxiety disorders. The authors also note one trial supporting the use of clomipramine in treating OCD behaviors, but the other trials evaluating tricyclic antidepressants in pediatric anxiety yielded mixed results and could not be supported without further research. Thus antidepressants are commonly prescribed to treat both mood and anxiety disorders.

Indications Specific to Children with Intellectual Disability
Racusin, Kovner-Kline, and King (1999) reviewed the link between serotonin and intellectual disability through primary serotonin dysregulation versus secondary effects. Based on this association, the authors described the clinical uses of SSRIs in the management of intellectual disability. Furthermore, the article noted frequent reports of effective management of maladaptive behaviors in children with intellectual disability treated with SSRIs.

An open trial of fluoxetine in 21 patients with severe to profound intellectual disability and aggression/self-injurious behavior reported marked improvement in self-injury, agitation, emotional lability, and aggression in 13 of 21 patients. Six additional patients had mild–moderate improvement on fluoxetine; there were only two nonresponders in this study (Markowitz, 1992). In a case series, four adults with intellectual disability and self-injurious behavior were treated with fluoxetine in conjunction with a behavioral therapy program. Self-injury levels decreased from 20% to 88% compared with baseline (Ricketts et al., 1993). In an initial evaluation of paroxetine treatment in seven adolescents with depression and mild intellectual disability, Masi, Marcheschi, and Pfanner (1997) observed a 41% improvement in the total score on the Montgomery-Asberg Depression Rating Scale (Montgomery & Asberg, 1979), and 4 of the 7 patients no longer met *DSM-IV* criteria for depression at the end of the trial.

Despite the paucity of large, controlled trials, many of the data support the theory that antidepressants are effective medications for particular behavioral disorders in children with intellectual disabilities. More research, particularly large, randomized, double-blind, placebo-controlled trials, is clearly needed in this area and particular patient population

Specific Syndromes and Antidepressants
While the precise pathophysiology of autism remains unclear, abnormalities in the serotonergic pathway have been implicated. In a review of the literature to evaluate the effects of SSRIs in autism, Kolevzon, Mathewson, and Hollander (2006) reviewed three randomized controlled trials and 10 open-label or chart review studies. Their review indicated that SSRIs often led to improvement in global functioning, anxiety, and repetitive behaviors. Agitation and activation were common side effects.

Fluoxetine has been well studied in autism. In one randomized, double-blind placebo crossover trial in 45 children and adolescents with autism, Hollander et al. (2005) reported significant reductions in repetitive behaviors in subjects treated with fluoxetine, though there was no statistical difference in the global autism rating. Side effects in this group were also not significantly different between the two groups. Authors of another evaluation, this time of 129 children with ASDs treated with fluoxetine, reported that 17% had an "excellent" response (briefly, loss of most of the "aloofness" characteristic of autism and development of more advanced interpersonal skills), 52% had a "good" response (substantial response causing parents to choose to continue the medicine), and 31% had "fair to poor" response (mild to no benefit or worsened behaviors). Significantly, response to fluoxetine in this study correlated with a family history of major affective disorder (85% of excellent–good responders), unusual intellectual achievement, and hyperlexia in childhood (DeLong, Ritch, & Burch, 2002). One study evaluated 23 children with autism and 16 children with intellectual disability before and during treatment with fluoxetine for perseverative behaviors. Fifteen of twenty-three children in the autism cohort and 10 of 16 children in the group with intellectual disabilities had a significant improvement on the Clinical Global Impressions Severity Scale. Side effects, however, were limiting for some patients (Cook, Rowlett, Jaselskis, & Leventhal, 1992).

In a case study of a child with autism, the addition of clomipramine to clonazepam, propranolol, and sertraline was associated with significant reductions in severe aggression and a need for crisis intervention (Luiselli, Blew, Keane, Thibadeau, & Holzman, 2000).

Anxiety is also commonly treated with antidepressants. A case series reported that 8 of 9 children with autism and transition-associated anxiety or agitation demonstrated a clinically significant response to low-dose sertraline (Steingard, Zimnitzky, DeMaso, Bauman, & Bucci, 1997).

Examining psychiatric illness in PWS, Soni et al. (2007) extensively evaluated 119 individuals with PWS. The results revealed that psychiatric illness in PWS mimics affective disorders, and SSRIs are the most common effective treatments. Of note, mood-stabilizing medications were not found to be effective in this group. Importantly, a few published case reports note potential significant adverse effects of SSRIs in PWS. One case report described an increase in aggressive hyperphagia and weight gain, but an improvement in obsessive-compulsive behaviors, in a 14-year-old boy after treatment with both fluvoxamine and fluoxetine was started (Kohn, Weizman, & Apter, 2001). Another editorial case report discussed a 13-year-old female with onset of psychosis 3 weeks after starting fluoxetine and rapid resolution of symptoms after discontinuing the medication (Herguner & Mukaddes, 2007). Larger trials are needed

to further evaluate such potential negative adverse events specific to PWS.

Down syndrome is another specific syndrome in which SSRIs are commonly used to treat comorbid conditions. In a case series, Sutor, Hansen, and Black (2006) report significant reduction in compulsive behaviors in three of four adults with Down syndrome when they are started on a SSRI (with or without risperidone).

Side Effects/Contraindications   Despite the reduced safety profile of TCAs compared with SSRIs, SSRIs do have noted side effects. In addition to side effects such as nausea, insomnia, dry mouth, headache, restlessness, agitation, and drowsiness, the FDA mandated a black box warning for risk of suicide (thoughts and tendency) in children and adolescents with major depression starting treatment with a SSRI.

Side effects significantly impaired function or outweighed benefits of treatment in 6 of 23 children with autism and 3 of 16 children with intellectual disability in a study evaluating fluoxetine for perseverative behavior. The impairing side effects reported in this study included restlessness, hyperactivity, agitation, decreased appetite, and insomnia (Cook et al., 1992). Though EPS symptoms are not common adverse events with antidepressants, Sokolski, Chicz-Demet, and Demet (2004) presented a case series of four children with autism who developed significant EPS after starting low-dose SSRIs. EPS symptoms included tongue protrusions, lip smacking, orofacial dyskinesia, jaw dystonia, and bruxism; all symptoms resolved within weeks after the SSRI was discontinued.

Given the increased incidence of behavioral disturbance, SSRIs are one of the most commonly prescribed classes of medications in children with intellectual disability. Evidence supports the efficacy of SSRIs in reducing maladaptive behaviors and anxiety. As the pathophysiology of specific syndromes such as autism is explored, SSRIs may have more specific applications as well.

Anxiolytics

Anxiety disorders include generalized anxiety disorder, phobias, panic attacks, OCDs, and situation-provoked anxiety. Anxiety is a frequent comorbid and often underdiagnosed behavior disorder in children with intellectual disability. In a review of the literature on anxiety and developmental disabilities, Davis, Saeed, and Antonacci (2008) summarize multiple studies, all of which confirm increased anxiety in intellectual disability alone as well as in specific syndromes such as ASDs.

The precise pathophysiology of anxiety is not understood. Based on both clinical features of anxiety and medication response, serotonin, dopamine, norepinephrine, and GABA neurotransmitters are likely to be involved in various aspects of anxiety symptoms. The most common treatment for anxiety is SSRIs and other atypical antidepressants (see prior section on antidepressants) combined with psychotherapy. In a review of pharmacological treatment of pediatric anxiety disorders, Reinblatt and Riddle (2007) conclude that there is "good evidence" to support the use of SSRIs.

Other treatment options include benzodiazepines, though this class of medications is recommended for only short-term or acute use (with an exception being as adjunctive medication in refractory epilepsy syndromes). Side-effect profile, development of tolerance, withdrawal, sensitivity to medication effects, and interactions with other medications can all limit the use of benzodiazepine use in children. Benzodiazepines include lorazepam (Ativan), alprazolam (Xanax), clonazepam (Klonopin), triazolam (Halcion), diazepam (Valium), and midazolam (Versed). Buspirone (Buspar) is an atypical anxiolytic that is also used and is not thought to have associated anticonvulsant, muscle relaxant, or sedative properties associated with benzodiazepines (Table 12.2).

Mechanism of Action To produce anxiolytic effects, benzodiazepines bind to the GABA-A α_2 receptor subunit and increase receptor GABA conductance. By increasing GABA activity, benzodiazepines inhibit action potentials. Anticonvulsant and sedative effects are mediated by the α_1 subunit of the GABA-A receptor (Johnston and Gross, 2008). Buspirone, however, is thought to act primarily as a serotonin agonist and modulates dopamine receptors (refer to package inserts for more information).

Indications for the General Population Though most commonly prescribed to treat anxiety, other therapeutic actions of benzodiazepines include hypnosis (sedation), muscle relaxants, anticonvulsants, and short-term amnestics. In a meta-analysis of benzodiazepines in the treatment of generalized anxiety disorders in the adult population, Martin et al. (2007) concluded there were no significant effects of benzodiazepines over placebo. The outcome measure in this meta-analysis was withdrawal from the study for any reason. Notably, however, the analysis also revealed that significantly fewer withdrawals were secondary to "lack of efficacy." These results are discussed in relation to heterogeneity of data, date of publication, and trial quality (Martin et al., 2007).

Data for children also cast doubt on the efficacy of benzodiazepines in the treatment of anxiety. In a small double-blind crossover study of 15 children with anxiety, primarily separation anxiety disorder, clonazepam was not significantly more effective than placebo (Graae, Milner, Rizzotto, & Klein, 1994). A placebo-controlled double-blind trial of 30 children with overanxious or avoidant disorders revealed no statistically significant difference between alprazolam and placebo despite a trend in improved efficacy of alprazolam (Simeon et al., 1992). However, SSRIs have less conflicting data (see above section on antidepressants as well). One open-label trial revealed significant reductions in anxiety, particularly separation anxiety and social phobia, in children treated with fluoxetine (Fairbanks et al., 1997). Buspar is another type of anxiolytic that has been approved for the treatment of generalized anxiety disorder and acute anxiety.

Indications Specific to Children with Intellectual Disability
Anxiety disorders in children with intellectual disability have a prevalence of 8.7% compared with 3.6% of peers without impairments (Emerson, 2003). One study evaluated behavioral equivalents of anxiety in children with fragile X syndrome (FXS). Based on parent and teacher questionnaires, this study concluded that children with FXS and anxiety demonstrate more observable behaviors, including arguing, staring, avoidance, compulsions, and specific phobias (Sullivan, Hooper, Hatton, 2007). Despite the increased prevalence, there have been no randomized controlled trials of benzodiazepine use in children with intellectual disability and anxiety.

Side Effects/Contraindications
Benzodiazepines are potent medications, producing neurochemical dependence and tolerance with even short-term use. Withdrawal symptoms are common if medication use is not carefully tapered. Furthermore, evidence indicates that adults with intellectual disability have a significant increase in behavioral side effects of benzodiazepines. In a study of 446 adults prescribed benzodiazepines, 13% of adults with intellectual disability had significant behavioral side effects (defined as agitation, aggression, anger, depression, euphoria, hostility, hyperactivity, irritability, schizophrenia, socially inappropriate behavior, or temper tantrums), with the highest incidence in those prescribed the medication for behavioral or psychiatric disorders (Kalachnik, Hanzel, Sevenich, & Harder, 2002).

In the small crossover trial of clonazepam in childhood anxiety, Graae and colleagues (1994) noted that although side effects of clonazepam did not differ from placebo statistically, three children did have to be removed from the trial because of significant side effects. The main side effects

noted were drowsiness, irritability, emotional lability, and oppositional defiant behaviors.

Overall, there are no data evaluating the effects of benzodiazepines in children with intellectual disability and behavioral disorders. Given the increased side effect profile in adults with intellectual disability, as well as side effects seen in the few trials of typical children, caution should be used when benzodiazepines are being prescribed. Medications listed in the antidepressant section are more efficacious and safer for the management of anxiety in children with disability. Buspar is another alternative, though more data are needed specifically for children with intellectual disability.

Mood Stabilizers

Mood stabilizers are used primarily to treat drastic cyclic alterations of behavior, such as mania, conduct disorder, or aggression. Lithium was the earliest mood-stabilizing agent used to treat mania. However, antiepileptic medications are now used frequently for behavior disorders beyond bipolar disorder. Typical mood-stabilizing agents include lithium, valproic acid (Depakote), lamotrigine (Lamictal), carbamazepine (Tegretol), oxcarbazepine (Trileptal), and possibly gabapentin (Neurontin), topiramate (Topamax), and levetiracetam (Keppra) (Table 12.2).

Mechanism of Action
As noted, lithium was the earliest mood-stabilizing agent. The pathophysiology underlying the mechanism of action of lithium remains unclear, though in general lithium augments abnormalities in brain signaling patterns primarily through second messengers. Valproic acid (VPA) is an antiepileptic that is now often used as a mood-stabilizing agent. VPA is presumed to exert its effects by altering levels of GABA, an inhibitory neurotransmitter, blocking glutamatergic NMDA receptors, and interfering with intracellular functioning. Carbamazepine is thought to inhibit cAMP formation and thus

impair neural excitation. These alterations, though mechanistically unclear at this time, have been shown clinically to regulate mood disorders and modulate disruptive behaviors.

Indications for the General Population

Lithium is FDA approved to treat bipolar disorder in children age 12 and older. Specific to behavior disorders, both carbamazepine and valproate are FDA approved in adults for the treatment of acute mania. None of the antiepileptics have been FDA approved for the treatment of behavior disorders in children, though many are commonly used by clinicians off label.

Indications Specific to Children with Intellectual Disability

In a thorough review of the literature on antiepileptics in the treatment of "nonepileptic disorders," Golden, Haut, and Moshe (2006) proposed recommendation ratings based on a study classification scheme by the American Academy of Neurology. Regarding behavior disorders specifically, the authors report that valproate is "probably effective" and carbamazepine is "probably ineffective" in the treatment of aggression, and lamotrigine is "possibly ineffective" in improving the core symptoms of PDD. Studies regarding the effectiveness of mood stabilizers in children have thus far evaluated specific syndromes such as ASDs rather than isolated intellectual disability.

Specific Syndromes and Mood Stabilizers

In an 8-week double-blind placebo-controlled trial of divalproex sodium (Depakote) in autism, divalproex correlated with a significant reduction in repetitive behaviors when compared with placebo (Hollander et al., 2006). Given its mild side-effect profile, levetiracetam would be an ideal mood stabilizer. However, in a sample of 20 children with autism in a placebo-controlled double-blind trial, levetiracetam showed no effect over placebo in any behaviors measured, including a global scale of

function, aggression, agitation, repetitive behaviors, impulsivity, and hyperactivity (Wasserman et al., 2006).

As the pathophysiology of FXS is elucidated, specific treatment targets are concomitantly being explored. For instance, lithium reduces mGLU receptor, which is thought to be dysregulated in FXS. In an open-label treatment trial of 15 children and adolescents with FXS, lithium was found to significantly improve total scores and certain subscales on multiple measures of behavioral function, thus warranting further placebo-controlled trials (Berry-Kravis et al., 2008).

Side Effects/Contraindications

Common side effects of lithium include nausea, diarrhea, acne, weight gain, sedation, tremor, goiter, polydipsia, and polyuria. Lithium levels require careful monitoring, as toxic levels are only twice therapeutic levels. Lithium toxicity causes tremor, ataxia, dysarthria, confusion, myoclonus, fasciculations, and potentially coma and death. Dehydration is a common, simple cause of lithium toxicity.

Typical side effects with VPA are nausea, increased appetite, weight gain, sedation, thrombocytopenia, transient hair loss, and tremor. A retrospective assessment of adults with intellectual disability on carbamazepine revealed a significantly increased risk of hyponatremia associated with high daily dose and high serum level. However, the sodium levels were reported not to correlate with clinical signs and thus are of uncertain significance (Kelly & Hillery, 2001). A historical review of 216 children (67 with developmental or intellectual disability) prescribed anticonvulsants revealed an increased rate of report of significant side effects in children with disability (40%) compared with those with typical cognition (20%) (Harbord, 2000).

Mood stabilizers are often used but inadequately studied medications in children with intellectual disability. Careful consideration of side effects is required before any of the mood stabilizers can be prescribed.

Miscellaneous

With the continuous advances in pharmacology and increased understanding of pathophysiological mechanisms of disease, new medications are being evaluated and old medications tested for additional purposes. This section is not a complete discussion of all of the miscellaneous medications being used in children with intellectual disability or specific developmental syndromes associated with behavioral disorders. Some of these drugs have been used for quite some time, whereas others are in the early, experimental phase. Discussed here include the alpha-adrenergic agents clonidine (Catapres) and guanfacine (Tenex); melatonin; naltrexone; and neurotropics such as donepezil (Aricept) (for the treatment of ADHD-like behaviors, autism, Down syndrome, and traumatic brain injury) and riluzole (Rilutek) (Table 12.2).

This section is organized by medication grouping rather than the prior outline of mechanism of action, indications, and side effects. The literature to date is summarized, but many of these medications do not have a plethora of data, and much remains unknown (such as side-effect profile or indications in children with intellectual disability).

Alpha-2 Agonists   Clonidine and guanfacine are second-line ADHD management medications, though they are often used first in children with tics (or Tourette syndrome) and ADHD. Both are alpha-2 adrenergic agonists used to treat hypertension and insomnia, as sedation is a prominent side effect. Arnsten, Scahill, and Findling (2007) provide a comprehensive review of the pathophysiology of ADHD and the corresponding mechanism of action of alpha-2 agonists in the treatment of ADHD. Briefly, this class of medications binds to postsynaptic alpha-2 receptors in the prefrontal cortex and thus alter prefrontal cortical functioning associated with symptoms of ADHD. Arnsten further discusses differences between the actions of clonidine and guanfacine, leading to the possibility of more specialized uses based on physiology and symptomatology (Arnsten

et al., 2007). As these agents are antihypertensives, hypotension, dizziness, and possibly fainting are potential adverse effects. Neither adrenergic agent can be stopped abruptly (including a missed dose), as rebound hypertension may result. Another common side effect is sedation, though guanfacine is thought to be less sedating than clonidine. Though disputed by some studies, alpha-adrenergic agents are thought to be associated with tachyphylaxis due to downregulation of the receptors. This may result in reduced medication effect over time and the need to increase the dose.

In a 12-week randomized placebo-controlled trial of clonidine in children with ADHD and intellectual disability, clonidine was found to be effective on both parents' and clinicians' rating scales with a dose-related response. The primary side effect reported was drowsiness (Agarwal, Sitholey, Kumar, & Prasad, 2001). A recent 16-week randomized controlled trial comparing methylphenidate, clonidine, and placebo in typical children with ADHD, however, revealed questionable efficacy of clonidine in the treatment of ADHD, with teacher ratings showing clonidine to be less efficacious than methylphenidate. However, both parent and clinical scales indicated significant positive effects in the clonidine-treated group (Palumbo et al., 2008).

Guanfacine has also recently been evaluated in children with autism and intellectual disability. In a double-blind, randomized, placebo-controlled trial, 11 children with or without autism and IQs ranging from severe to average were evaluated by Handen, Sahl, and Hardan (2008). Teacher and parent scales of hyperactivity both showed significant improvement in the guanfacine-treated group. The global improvement ratings also significantly improved on guanfacine. Five of eleven subjects experienced a 50% reduction in hyperactivity on the Aberrant Behavior Checklist. Finally, side effects limited maximum dose only. In this study, side effects did not differ significantly between the two treatment groups and included drowsiness, irritability, diarrhea, constipation, enuresis, and social withdrawal (Handen et al., 2008).

Melatonin   Sleep disorders are a common comorbidity in children with intellectual disability. Melatonin is a hormone synthesized by the pineal gland and linked to the circadian rhythm and natural sleep onset. Melatonin is an attractive treatment option to many families, given the relatively few side effects and lack of known interaction with other medications, especially when compared with alternatives like antihistamines or benzodiazepines.

In a randomized, double-blind, placebo crossover trial, 25 children and young adults with intellectual disability, seizures, and sleep disorder were evaluated on melatonin. Sleep latency was significantly reduced on melatonin, with no reported side effects (Coppola et al., 2004). In a review of the literature on melatonin used specifically in patients with intellectual disability, Sajith and Clarke (2007) concluded that melatonin thus far appears effective in reducing both sleep latency and total sleep time with minimal side effects. Importantly, melatonin can be purchased over the counter as a nutritional supplement and thus is not regulated for impurities or quality by the FDA.

Naltrexone   Based on the controversial and unproved theory that increased endogenous (or exogenous) opioid activity is associated with much of the symptomatology of autism, the opiate antagonist naltrexone has been studied as a potential treatment option. In a review of the literature, both case series and controlled trials, Elchaar, Maisch, Augusto, and Wehring (2006) conclude that naltrexone is likely to be effective in reducing self-injurious behavior in autism and possibly beneficial in the treatment of hyperactivity, agitation, aggression, and a few other symptoms. Side effects were also minimal and included hypoactivity and drowsiness. However, the controlled trials to date demonstrate many conflicting data; some studies show no effect, whereas other studies show clinical efficacy correlated with biochemical data.

Neurotropics   The frontal lobe and its intricate connections play a prominent role in executive function, attention, impulse control, cognitive organization, motor function, and sensorimotor integration (Malloy & Richardson, 1994). Acetylcholine, serotonin, and dopamine are major neurotransmitters in the frontal lobe (Mega & Cummings, 1994). We previously discussed therapeutic agents that alter both serotonin and dopamine in behavioral disorders. Logically, medications that effect acetylcholine levels are also being explored. For instance, donepezil is an acetylcholinesterase inhibitor (increases the amount of available acetylcholine) used to treat Alzheimer's disease. Given the impaired frontal lobe function in ADHD or after traumatic brain injury, donepezil has been studied as a potential therapeutic agent in these conditions as well.

One case series revealed that 7 of 8 children with PDD and ADHD-like symptoms had improved Clinical Global Impressions (CGI) ADHD subscale scores on donepezil. Furthermore, all 8 subjects on donepezil had improvement in PDD CGI global severity and global improvement scale scores (Doyle et al., 2006). However, an open trial of donepezil used as adjunctive treatment (to stimulants) for ADHD revealed no significant improvement on the outcomes measured in the 7 of 13 subjects who completed the trial (Wilens et al., 2005).

A 24-week double-blind, randomized, placebo-controlled crossover study evaluated the efficacy of donepezil in 18 patients within 1 year of sustaining traumatic brain injury. Outcome measures, sustained attention and short-term memory, both significantly improved on donepezil (Zhang, Plotkin, Wang, Sandel, & Lee, 2004). In a review of the literature on donepezil use in traumatic brain injury rehabilitation, Ballesteros, Guemes, Ibarra, and Quemada (2008) concluded that donepezil may be effective in cognitive impairments, but no strong conclusions can be drawn, given the paucity of large, controlled, well-designed trials.

Glutamate is the most common and ubiquitous neurotransmitter in the central nervous system. Tsapakis and Travis (2002) summarize the emerging theories regarding the role of glutamate in many psychiatric disorders, including schizophrenia, mood and anxiety disorders, and cognitive disorders. Riluzole is a glutamate antagonist and thus may modulate symptoms associated with various psychiatric disorders. In an open-label trial of riluzole in 6 children and adolescents with refractory OCD, 4 of 6 subjects had nearly 50% reduction of symptoms and improvement on the CGI improvement scale (Grant, Lougee, Hirschtritt, & Swedo, 2007). In an extensive review of the role of riluzole in psychiatric disorders, including discussion of the physiology of glutamate, Pittenger et al. (2008) concluded that the open-label trials indicate a positive effect of riluzole on mood, anxiety, and other psychiatric disorders. Naturally, large, controlled trials are needed to confirm these early reports. Pittenger et al. (2008), however, do note nausea and sedation as common side effects and the necessity of monitoring liver enzymes during treatment with riluzole. Other glutamate modulators, such as memantine and amantadine, are also being explored in various developmental disabilities, commonly ASDs.

GENERAL PRINCIPLES

The use of psychotropic medications to affect behavior is a relatively new phenomenon. Effective agents to treat psychiatric and behavioral symptoms are barely more than half a century old. Although the use of dextroamphetamine salts to treat ADHD behavior in children with "brain damage syndromes" dates back to the early 1930s, effective antidepressants and antipsychotics only became available in the 1960s and 1970s. Behavioral medications are one of the triumphs of 20th-century medicine. Nevertheless they are viewed by the public with a degree of concern that is out of proportion to their safety and efficacy.

There are four major rules for the use of drugs to affect children's behavior:

1. The behavior should be serious, that is, it should present risk of bodily harm to the child or others, cause the child to struggle or fail in schoolwork, result in frequent suspensions, or significantly affect the child's social interactions to the point of inducing other psychiatric symptoms.

2. Reasonable attempts at behavioral interventions (including class accommodations and counseling) have been tried and failed. A significant percentage of drug-treated challenging behaviors could probably be effectively managed with a positive behavioral strategies plan derived from a functional behavioral assessment properly performed. Pharmacologic management, however, should not always be the last approach used to treat challenging behaviors.

3. The ideal medication used should produce a dramatic improvement in the target behavior.

4. The ideal medication treatment program should have no side effects; even mild side effects should be relieved by adjustment of the dose, the delivery system, or the medication.

Psychotropic medications should not be instituted unless a comprehensive evaluation is undertaken. Frequently, all behavioral disturbances may be attributed to the child's intellectual disability, also known as diagnostic overshadowing. A complete assessment will recognize other contributing factors or diagnoses related to the behavior. Given that failure to identify clinically important comorbidities is the most common reason for failed treatment programs, it is not sufficient to address the intellectual disability alone.

Psychotropic medication is part of a comprehensive management program that addresses the child, family, and environmental needs. The problem behavior should be analyzed within the context of the child's intellectual, communicative, and adaptive function. Several other aspects of behavior need to be explored to ensure accurate diagnosis and to optimize the treatment plan. For instance, time of onset, chronicity, severity, moderating factors, and parental responses should be noted. Primary psychiatric disorders need to be distinguished from adjustment issues. The communicative intent of the behavior should be considered. Functional behavioral analysis should be undertaken. The possibility of physical discomfort or somatic illness as an etiology also should be considered. Comorbidities need to be identified and their contribution to the clinical picture clarified.

It is also important to ascertain the family's understanding of intellectual limitation. Additionally, the family views of the behavior and their methods for dealing with behavioral disturbances are important aspects of the history. Methods that have been tried successfully should be noted, as should those that have failed. Family function and the use of respite service also provide important information. School, work, or play functions may suggest information about possible environmental triggers for the problem behavior. Changes in the environment should be considered, as change may provoke anxiety, disrupt routine, or exacerbate other maladaptive behaviors.

Following the evaluation, hypotheses should be generated about the function of the problem behavior and serve as the basis from which to select potential treatments. Treatment goals need to be tailored to the individual, with plans for close monitoring of behavioral effects and side effects.

Psychotropic medication may be an important part of the management program. One of the limitations of psychotropic medication is that the criteria for use in children with intellectual disability are not well established. The reasons for this are outlined above. There are three potential approaches to selecting a psychotropic medication:

1. Analysis of the relevant neuropsychopharmacological pathways (however, this approach is mainly theoretical, with only marginal relevance to clinical use)

2. Identification of the most serious symptom and utilizing a drug that specifically targets that symptom

3. Starting with the mildest drug (in terms of potential side effects) and switching to drugs with potentially more serious side effects only if needed

The latter two approaches are empirical and require individualization, good communication with parents or caregivers, and clear criteria for success or failure.

The decision to use psychotropic medication is subjective. Psychotropic medication should be used for psychiatric disease and disruptive behavior. Psychotropic medication should not be used for doping to improve performance, to change behavior to suit the desires of the caregivers, or in the absence of clear goals, close monitoring, and ongoing communication with caregivers.

Some have advocated psychotropic medication as the first choice, whereas others have suggested that these agents should be used only as a last resort. We would support Eisenberg's (1971, p. 379) assertion that "pharmacologic methods provide neither the passport to a brave new world nor the gateway to the inferno. With thoughtful selection, careful regulation of dosage, and close scrutiny for toxicity, they can add a significant component to total patient care."

Dosing

Expected pediatric dosages for psychotropic medications are determined by mg/kg formulas. Calculating dose per weight is the most appropriate method for estimating the effect on typically developing brains. However, atypical brains are less predictable in their response to medications. The safest approach in using psychotropic drugs is to start with a low dose that is unlikely to be effective. The medication can then be increased in small, defined increments. As children have variable responses to medications, it is ideal to start with short-acting formulations. This approach will enable the physician and family to best estimate dose response and only then switch to longer-acting preparations for ease of use. An occasional child will respond positively for the entire day with a single short-acting morning dose.

Parental anxiety over side effects can be addressed by starting the medication and making all dose increments on Saturdays so that the parents can be the first to observe the drug's (side) effects. This approach also enables parents to stop a medication in the case of any suspected negative impact. Maintaining a trial (low starting) dose for at least a week allows the family and physician to better account for the frequently variable behavioral patterns that are the target of treatment.

Another advantage to titrating doses slowly is to minimize or avoid transient side effects. In response to new medications, any number of mild and transient behavioral, somatic, or psychiatric effects may occur. These effects include symptoms such as headache, stomachache, jitteriness, weepiness, anxiety, irritability, or sleep disturbance. The initial response often resolves in 1–2 weeks. Typically if the dose is titrated over several weeks to the ideal, effective level, these worrisome albeit transient symptoms can be avoided.

Effective psychopharmacologic intervention should have substantial positive effects; otherwise, the medication should be discontinued. If necessary, another agent should be considered and carefully instituted. Layered psychopharmacology (adding one agent to another agent despite suboptimal response) should be avoided.

Treatment Failures

As discussed previously, treatment failures may result from a number of different causes. Poor response to medication is noted when only one aspect of the problem is being addressed. Additionally, psychopharmacologic treatment without environmental intervention (behavioral modification, accommodations, and so forth) is usually unsuccessful. Another common cause of treatment failure is misdiagnosis, as is incom-

plete diagnosis. Diagnostic overshadowing, where all of the child's behaviors are attributed to intellectual disability, may cause an incomplete diagnosis and thus treatment failure. Insufficient medication titration or prematurely discontinuing the medication often results in persistent behavioral disturbance without medication effect. Drug toxicity may also contribute to treatment failure. Whereas a dose that is too small will not show the effect of a drug, too high a dose may actually exacerbate target behaviors. Pervasive side effects, such as excessive sedation, may diminish function and require discontinuance of the treatment. Finally, unrealistic expectations of psychopharmacologic agents are a common reason for treatment failures. Psychotropic medications are not curative and address specific behavioral symptoms in a nonspecific fashion.

Monitoring

Monitoring is a critical aspect of all programs that treat children with intellectual disability. Treatment programs need to be reviewed regularly to ensure that the objectives are being met. Modification of the plan may be necessary to meet the changing needs that accompany natural childhood maturation. A final aspect of monitoring is forward planning. Proactively preparing for the future enables the family and management team to set new objectives and transmit new information to parents.

In addition to routine treatment monitoring described above, beginning psychopharmacologic agents requires further evaluation. For instance, the treatment team must assess whether the medication addresses the target behaviors sufficiently. To adequately answer this question requires collection and interpretation of data from school, home, and other areas. Another concern is whether the child experiences toxicity from the medication. Typical and atypical side effects associated with the medication should be reviewed and appropriate monitoring studies obtained. Tics, sleep disturbances, irritability, and constipation are obvious effects when severe, but they may not be obvious initially. A child who is receiving multiple medications should be monitored, even when those medications are purported not to have psychopharmacologic effects. The treatment team should carefully and consistently consider the necessity of the prescribed medication or medications.

The management plan is typically interconnected; thus it is difficult to modify only one aspect of a behavioral treatment program. Most often, when psychopharmacologic agents are used, other programmatic changes are attempted concomitantly. While the medication often is credited for the positive effects of altering the behavioral disturbance, sometimes the child's need for psychopharmacologic agents diminishes with adjustments to the total program. Therefore, drug holidays or retitration should be tried regularly to ensure that medication is still required.

Monitoring the effects of psychopharmacology and an overall treatment plan requires frequent visits. This increased demand is associated with missed work and missed school and may also interfere with other aspects of the child's program (e.g., extracurricular activities). Fortunately, not all monitoring interactions require face-to-face interactions. Depending on the medication, the treatment objectives, the age of the child, compliance, and the family's observational and reporting capacities, some medication monitoring evaluations may be accomplished via telephone. We would not endorse using e-mail to monitor the progress of a behavioral treatment plan. In our experience, e-mail does not allow us to adequately measure responses and often is associated with incomplete data.

CONCLUSIONS

Pharmacologic agents may be useful adjuncts in a comprehensive management program that is used to address behavioral disturbance in people with intellectual disability. Psychopharmacologic medications are most useful in children with comorbid intellectual disability and psychiatric disorders. Even in the absence of psychiatric diagnosis, these agents may diminish troublesome behavioral symptoms. Typically, psychotropic medications address specific symptoms in a nonspecific manner.

There is little comprehensive evidence to support the use of most agents that are used for children with intellectual disability and behavioral disturbance. Consequently, the prescription of these agents is often empirical. Implementation of a comprehensive evaluation and treatment program is most likely to produce positive outcomes. The plan should address the capability of the individual, family, and the environment in which the child lives. The addition of psychopharmacology to the treatment plan requires individualization, clear treatment objectives, adequate titration, and close monitoring. Treatment failure necessitates reevaluation of the diagnosis, consideration of comorbidities, discussion of outcome expectations, ensurance of medication levels, and avoidance of polypharmacy. The primary goal of the treatment plan is to reduce maladaptive behaviors and enable the child to have an appropriate level of function.

REFERENCES

Agarwal, V., Sitholey, P., Kumar, S., & Prasad, M. (2001). Double-blind, placebo-controlled trial of clonidine in hyperactive children with mental retardation. *Mental Retardation, 39*(4), 259–267.

Alacqua, M., Trifiro, G., Arcoaci, V., Germano, E., Magazu, A., Calarese, T., et al.(2008). Use and tolerability of newer antipsychotics and antidepressants: A chart review in a paediatric setting. *Pharmacy World & Science, 30,* 44–50.

Aman, M.G., Arnold, L., McDougle, C.J., Vitiello, B., Scahill, L., Davies, M., et al. (2005). Acute and long-term safety and tolerability of risperidone in children with autism. *Journal of Child and Adolescent Psychopharmacology, 15,* 869–884.

Aman, M.G., Buican, B., & Arnold, L.G. (2003). Methylphenidate treatment in children with borderline IQ and mental retardation: Analysis of three aggregated studies. *Journal of*

Child and Adolescent Psychopharmacology, 13(1), 29–40.

Aman, M.G., Hollway, J.A., McDougle, C.J., Scahill, L., Tierney, E., McCracken, J.T., et al. (2008). Cognitive effects of risperidone in children with autism and irritable behavior. *Journal of Child and Adolescent Psychopharmacology, 18*(3), 227–236.

Aman, M.G., & Madrid, A. (1999). Atypical antipsychotics in persons with developmental disabilities. *Mental Retardation and Developmental Disabilities Research Reviews, 5,* 253–263.

Aman, M.G., Singh, N.N., Stewart, A.W., & Field, C.J. (1985a). The Aberrant Behavior Checklist: A behaviour rating scale for the assessment of treatment effects. *American Journal of Mental Deficiency, 89,* 485–491.

Aman, M.G., Singh, N.N., Stewart, A.W., & Field, C.J. (1985b). Psychometric characteristics of the Aberrant Behavior Checklist. *American Journal of Mental Deficiency, 89,* 492–502.

American Psychiatric Association. (1991). *Psychiatric services to adult mentally retarded and developmentally disabled persons: Task force report 30.* Washington, DC: Author.

American Psychiatric Association. (2000). *Diagnostic and statistical manual of mental disorders* (4th ed., text rev.). Washington, DC: Author.

Arnold, L.E., Aman, M.G., Cook, A.M., Witwer, A.N., Hall, K.L., Thompson, S., et al. (2006). Atomoxetine for hyperactivity in autism spectrum disorders: Placebo-controlled crossover pilot trial. *Journal of the American Academy of Child and Adolescent Psychiatry, 45*(10), 1196–1205.

Arnsten, A.F.T. (1998). Catecholamine modulation of prefrontal cortical cognitive function. *Trends in Cognitive Science, 2,* 436–447.

Arnsten, A.F., Scahill, L., & Findling, R.L. (2007). Alpha-2 adrenergic receptor agonists for the treatment of attention-deficit/hyperactivity disorder: Emerging concepts from new data. *Journal of Child and Adolescent Psychopharmacology, 17*(4), 393–406.

Ballesteros, J., Guemes, I., Ibarra, N., & Quemada, J.I. (2008). The effectiveness of donepezil for cognitive rehabilitation after traumatic brain injury: A systematic review. *Journal of Head Trauma Rehabilitation, 23*(3), 171–180.

Berry-Kravis, E., Sumis, A., Hervey, C., Nelson, M., Porges, S.W., Weng, N., et al. (2008). Open-label treatment trial of lithium to target the underlying defect in fragile X syndrome. *Journal of Developmental and Behavioral Pediatrics, 29,* 293–302.

Capone, G.T., Goyal, P., Grados, M., Smith, B., & Kammann, H. (2008). Risperidone use in children with Down syndrome, severe intellectual disability, and comorbid autistic spectrum

disorders: A naturalistic study. *Journal of Developmental and Behavioral Pediatrics, 29,* 106–116.

Charney, D.S., Menkes, D.B., & Heninger, G.R. (1981). Receptor sensitivity and the mechanism of action of antidepressant treatment. *Archives of General Psychiatry, 38*(10), 1160–1180.

Cook, E.H., Rowlett, R., Jaselskis, C., & Leventhal, B.L. (1992). Fluoxetine treatment of children and adults with autistic disorder and mental retardation. *Journal of the American Academy of Child and Adolescent Psychiatry, 31*(4), 739–745.

Coppola, G., Iervolino, G., Mastrosimone, M., La Torre, G., Ruiu, F., & Pascotto, A. (2004). Melatonin in wake-sleep disorders in children, adolescents, and young adults with mental retardation with or without epilepsy: A double-blind, cross-over, placebo-controlled trial. *Brain and Development, 26,* 373–376.

Davis, E., Saeed, S.A., & Antonacci, D.J. (2008). Anxiety disorders in persons with developmental disabilities: Empirically informed diagnosis and treatment. *Psychiatry Quarterly, 79,* 249–263.

de Bildt, A., Mulder, E.J., Scheers, T., Minderaa, R.B., & Tobi, H. (2006). Pervasive developmental disorder, behavior problems, and psychotropic drug use in children and adolescents with mental retardation. *Pediatrics, 118*(6), e1860–ee1866.

DeLong, G.R., Ritch, C.R., & Burch, S. (2002). Fluoxetine response in children with autistic spectrum disorders: Correlation with familial major affective disorder and intellectual achievement. *Developmental Medicine and Child Neurology, 44,* 652–659.

Doyle, R.L., Frazier, J., Spencer, T.J., Geller, D., Biederman, J., & Wilens, T. (2006). Donepezil in the treatment of ADHD-like symptoms in youths with pervasive developmental disorder: A case series. *Journal of Attention Disorders, 9,* 543–549.

DuPaul, G.J., Power, T.J., Anastopoulos, A.D., & Reid, R. (1998). *ADHD Rating Scale–IV: Checklists, norms, and clinical interpretation.* New York: The Guilford Press.

Eisenberg, L. (1971). Principles of drug therapy in child psychiatry with special reference to stimulant drugs. *American Journal of Orthopsychiatry, 41*(3), 371–379.

Elchaar, G.M., Maisch, N.M., Augusto, L.M.G., & Wehring, H.J. (2006). Efficacy and safety of naltrexone in pediatric patients with autistic disorder. *Annals of Pharmacotherapy, 40*(6), 1086–1095.

Emerson, E. (2003). Prevalence of psychiatric disorders in children and adolescents with and without intellectual disability. *Journal of Intellectual Disability Research, 47*(1), 51–58.

Fairbanks, J.M., Pine, D.S., Tancer, N.K., Dummit, E.S., Kentgen, L.M., Martin, J., et al. (1997). Open fluoxetine treatment of mixed anxiety disorders in children and adolescents. *Journal of Child and Adolescent Psychopharmacology, 7*(1), 17–29.

Filho, A.C.G., Bodanese, R., Silva, T.L., Alvares, J.P., Aman, M., & Rohde, L.A. (2005). Comparison of risperidone and methylphenidate for reducing ADHD symptoms in children and adolescents with moderate mental retardation. *Journal of the American Academy of Child and Adolescent Psychiatry, 44*(8), 748–755.

Fletcher, R., Loschen, E., Stavrakaki, C., & First, M. (Eds.). (2007a). *Diagnostic Manual–Intellectual Disability (DM-ID): A clinical guide for diagnosis of mental disorders in persons with intellectual disability.* Kingston, NY: NADD Press.

Fletcher, R., Loschen, E., Stavrakaki, C., & First, M. (Eds.). (2007b). *Diagnostic Manual–Intellectual Disability (DM-ID): A textbook of diagnosis of mental disorders in persons with intellectual disability.* Kingston, NY: NADD Press.

Gingrich, J.A., & Hen, R. (2001). Dissecting the role of the serotonin system in neuropsychiatric disorders using knockout mice. *Psychopharmacology, 155,* 1–10.

Golden, A.S., Haut, S.R., & Moshe, S.L. (2006). Nonepileptic uses of antiepileptic drugs in children and adolescents. *Pediatric Neurology, 34*(6), 421–432.

Graae, F., Milner, J., Rizzotto, L., & Klein, R.G. (1994). Clonazepam in childhood anxiety disorders. *Journal of the American Academy of Child and Adolescent Psychiatry, 33*(3), 372–376.

Grant, P., Lougee, L., Hirschtritt, M., & Swedo, S. (2007). An open-label trial of riluzole, a glutamate antagonist, in children with treatment-resistant obsessive-compulsive disorder. *Journal of Child and Adolescent Psychopharmacology, 17*(6), 761–767.

Guy, W. (1976). Clinical Global Impressions. In *ECDEU assessment manual for psychopharmacology* (rev., pp. 217–222). Rockville MD: National Institute of Mental Health.

Hagerman, R.J. (1999). Psychopharmacological interventions in fragile X syndrome, fetal alcohol syndrome, Prader-Willi syndrome, Angelman syndrome, Smith-Magenis syndrome, and velocardiofacial syndrome. *Mental Retardation and Developmental Disabilities Research Reviews, 5,* 305–313.

Handen, B.L., Breaux, A.M., Gosling, A., Ploof, D.L., & Feldman, H. (1990). Efficacy of methylphenidate among mentally retarded children with attention deficit hyperactivity disorder. *Pediatrics, 86,* 922–930.

Handen, B.L., Feldman, H., Gosling, A., Breaux, A.M., & McAuliffe, S. (1991). Adverse side effects of methylphenidate among mentally retarded children with ADHD. *Journal of the American Academy of Child and Adolescent Psychiatry, 30,* 241–245.

Handen, B.L., & Hardan, A.Y. (2006). Open-label prospective trial of olanzapine in adolescents with subaverage intelligence and disruptive behavior disorders. *Journal of the American Academy of Child and Adolescent Psychiatry, 45*(8), 928–935.

Handen, B.L., Sahl, R., & Hardan, A.Y. (2008). Guanfacine in children with autism and/or intellectual disabilities. *Journal of Developmental and Behavioral Pediatrics, 29*(4), 303–308.

Harbord, M.G. (2000). Significant anticonvulsant side-effects in children and adolescents. *Journal of Clinical Neurosciences, 7,* 213–216.

Hellings, J.A., Zarcone, J.R., Reese, R.M., Valdovinos, M.G., Marquis, J.G., Fleming, K.K., et al. (2006). A crossover study of risperidone in children, adolescents, and adults with mental retardation. *Journal of Autism and Developmental Disorders, 36*(3), 401–411.

Herguner, S., & Mukaddes, N.M. (2007). Psychosis associated with fluoxetine in Prader-Willi syndrome. *Journal of the American Academy of Child and Adolescent Psychiatry, 46*(8), 944.

Hollander, E., Phillips, A., Chaplin, W., Zagursky, K., Novotny, S., Wasserman, S., et al. (2005). A placebo controlled crossover trial of liquid fluoxetine on repetitive behaviors in childhood and adolescent autism. *Neuropsychopharmacology, 30,* 582–589.

Hollander, E., Soorya, L., Wasserman, S., Esposito, K., Chaplin, W., & Anagnostou, E. (2006). Divalproex sodium vs. placebo in the treatment of repetitive behaviours in autism spectrum disorder. *International Journal of Neuropsychopharmacology, 9*(2), 209–213.

Johnston, M.V., & Gross, R.A. (2008). *Principles of drug therapy in neurology* (2nd ed.). New York: Oxford University Press.

Kalachnik, J.E., Hanzel, T.E., Sevenich, R., & Harder, S.R. (2002). Benzodiazepine behavioral side effects: Review and implications for individuals with mental retardation. *American Journal on Mental Retardation, 107*(5), 376–410.

Kelly, B.D., & Hillery, J. (2001). Hyponatremia during carbamazepine therapy in patients with intellectual disability. *Journal of Intellectual Disabilities Research, 45*(2), 152–156.

Kohn, Y., Weizman, A., & Apter, A. (2001). Aggravation of food-related behavior in an adolescent with Prader Willi syndrome treatment with fluvoxamine and fluoxetine. *International Journal of Eating Disorders, 30,* 113–117.

Kolevzon, A., Mathewson, K.A., & Hollander, E. (2006). Selective serotonin reuptake inhibitors in autism: A review of efficacy and tolerability. *Journal of Clinical Psychiatry, 67*(3), 407–414.

Lee, D.O., & Ousley, O.Y. (2006). Attention-deficit hyperactivity disorder symptoms in a clinic sample of children and adolescents with pervasive developmental disorders. *Journal of Child and Adolescent Psychopharmacology, 16*(6), 737–746.

Leonard, H.L., March, J., Rickler, K., & Allen, A.J. (1997). Pharmacology of the selective serotonin reuptake inhibitors in children and adolescents. *Journal of the American Academy of Child and Adolescent Psychiatry, 36*(6), 725–736.

Levy, F., & Swanson, J.M. (2001). Timing, space and ADHD: The dopamine theory revisited. *Australian and New Zealand Journal of Psychiatry, 35*(4), 504–511.

Luiselli, J.K., Blew, P., Keane, J., Thibadeau, S., & Holzman, T. (2000). Pharmacotherapy for severe aggression in a child with autism: "Open label" evaluation of multiple medications on response frequency and intensity of behavioral intervention. *Journal of Behavior Therapy and Experimental Psychiatry, 31*(3–4), 219–230.

Malloy, P.F., & Richardson, E.D. (1994). Assessment of frontal lobe functions. *Journal of Neuropsychiatry and Clinical Neurosciences, 6*(4), 399–410.

Markowitz, P.I. (1992). Effect of fluoxetine on self-injurious behavior in the developmentally disabled: A preliminary study. *Journal of Clinical Psychopharmacology, 21*(1), 27–31.

Martin, J.L.R., Sianz-Pardo, M., Furukaw, T.A., Martin-Sanchez, E., Seoane, T., & Galan, C. (2007). Review: Benzodiazepines in generalized anxiety disorder: Heterogeneity of outcomes based on a systematic review and meta-analysis of clinical trials. *Journal of Psychopharmacology, 21*, 774–782.

Masi, G., Marcheschi, M., & Pfanner, P. (1997). Paroxetine in depressed adolescents with intellectual disability: An open label study. *Journal of Intellectual Disability Research, 41*(3), 268–272.

McCracken, J.T., McGough, J., Shah, B., Cronin, P., Hong, D., Aman, M.G., et al., for Research Units on Pediatric Psychopharmacology Autism Network. (2002). Risperidone in children with autism and serious behavioral problems. *New England Journal of Medicine, 347*, 314–321.

McDougle, C.J., Kem, D.L., & Posey, D.J. (2002). Case series: Use of ziprasidone for maladaptive symptoms in youths with autism. *Journal of the American Academy of Child and Adolescent Psychiatry, 41*(8), 921–927.

Mega, M.S., & Cummings, J.L. (1994). Frontal-subcortical circuits and neuropsychiatric disorders. *Journal of Neuropsychiatry and Clinical Neurosciences, 6*(4), 358–370.

Montgomery, S.A., & Asberg, M. (1979). A new depression scale designed to be sensitive to change. *British Journal of Psychiatry, 134*, 382–389.

MTA Cooperative Group. (1999). A 14-month randomized clinical trial of treatment strategies for ADHD. *Archives of General Psychiatry, 56*(12), 1073–1086.

National Institutes of Health. (2001, November 29–December 1). *Emotional and behavioral health in persons with mental retardation/developmental disabilities: Research challenges and opportunities.* Retrieved February 5, 2009, from http://www.ninds.nih.gov/news_and_events/proceedings/Emotional_Behavioral_Health_2001.htm

Newcorn, J.H., Kratochvil, C.J., Allen, A.J., Casat, C.D., Ruff, D.D., Moore, R.J., et al. (2008). Atomoxetine and osmotically released methylphenidate for the treatment of attention deficit hyperactivity disorder: Acute comparison and differential response. *American Journal of Psychiatry, 165*(6), 721–730.

O'Keefe, L. (2008). ECGs for all ADHD patients? AAP-AHA release joint 'clarification' on AHA recommendation. *AAP News, 29*(6), 1.

Palumbo, D., Sallee, F., Pelham, W., Bukstein, O., Daviss, W.B., McDermott, M., & the CTA Study Team. (2008). Clonidine for attention deficit/hyperactivity disorder: Efficacy and tolerability. *Journal of the American Academy of Child and Adolescent Psychiatry, 47*(2), 180–188.

Pandina, G.J., Bilder, R., Harvey, P.D., Keefe, R.S.E., Aman, M.G., & Gharabawi, G. (2007). Risperidone and cognitive function in children with disruptive behavior disorders. *Biological Psychiatry, 62*, 226–234.

Pary, R.J., Silka, V.R., & Blaha, S.J. (1995). Mental retardation. In O.J. Thienhaus (Ed.), *Manual of clinical hospital psychiatry* (pp. 287–309). Washington, DC: American Psychiatric Press.

Pearson, D.A., Lane, D.M., Santos, C.W., Casat, C.D., Jerger, S.W., Loveland, K.A., et al. (2004). Effects of methylphenidate in children with mental retardation and ADHD: Individual variation in mediation response. *Journal of the American Academy of Child and Adolescent Psychiatry, 43*(6), 686–698.

Pearson, D.A., Santos, C.W., Roache, J.D., Casat, C.D., Loveland, K.A., & Lachar, D. (2003). Treatment effects of methylphenidate on behavioral adjustment in children with mental retardation and ADHD. *Journal of the American*

Academy of Child and Adolescent Psychiatry, 42(2), 209–216.

Perrin, J.M., Friedman, R.A., Knilans, T.K., Black Box Working Group, & Section of Cardiology and Cardiac Surgery. (2008). Cardiovascular monitoring and stimulant drugs for attention-deficit/hyperactivity disorder. *Pediatrics, 122,* 451–453.

Pittenger, C., Coric, V., Bansar, M., Block, M., Krystal, J.H., & Sanacora, G. (2008). Riluzole in the treatment of mood and anxiety disorders. *CNS Drugs, 22*(9), 761–786.

Polancyzk, G., de Lima, M.S., Horta, B.L., Biederman, J., & Rohde, L.A. (2007). The Worldwide Prevalence of ADHD: A systematic review and metaregression analysis. *American Journal of Psychiatry, 164,* 942–948.

Posey, D.J., Aman, M.G., McCracken, J.T., Scahill, L., Tierney, E., Arnold, L.E., et al. (2007). Positive effects of methylphenidate on inattention and hyperactivity in pervasive developmental disorders: An analysis of secondary measures. *Biological Psychiatry, 61,* 538–544.

Racusin, R., Kovner-Kline, K., & King, B.H. (1999). Selective serotonin reuptake inhibitors in intellectual disability. *Mental Retardation and Developmental Disabilities Research Reviews, 5,* 264–269.

Reinblatt, S.P., & Riddle, M.A. (2007). The pharmacological management of childhood anxiety disorders: A review. *Psychopharmacology, 191,* 67–86.

Research Units on Pediatric Psychopharmacology–Autism Network. (2005). *Archives of General Psychiatry, 62*(11), 1266–1274.

Ricketts, R.W., Goza, A.B., Ellis, C.R., Singh, Y.N., Singh, N.N., & Cooke, J.C. (1993). Fluoxetine treatment of severe self-injury in young adults with mental retardation. *Journal of the American Academy of Child and Adolescent Psychiatry, 32*(4), 865–869.

Rosa-Neto, P., Lou, H.C., Cumming, P., Pryds, O., Karrebaek, H., Lunding, J., et al. (2005). Methylphenidate-evoked changes in striatal dopamine correlate with inattention and impulsivity in adolescents with attention deficit hyperactivity disorder. *Neuroimage, 25*(3), 868–876.

Sajith, S.G., & Clarke, D. (2007). Melatonin and sleep disorders associated with intellectual disability: A clinical review. *Journal of Intellectual Disability Research, 51*(1), 2–13.

Seeman, P. (2002). Atypical antipsychotics: Mechanism of action. *Canadian Journal of Psychiatry, 47,* 27–38.

Simeon, J.G., Ferguson, H.B., Knott, V., Roberts, N., Gauthier, B., Dubois, C., et al. (1992). Clinical, cognitive, and neurophysiological effects of alprazolam in children and adolescents with overanxious and avoidant disorders. *Journal of American Academy of Child and Adolescent Psychiatry, 31,* 29–33.

Snyder, R., Turgay, A., Aman, M., Binder, C., Fisman, S., & Carroll, A. (2002). Effects of risperidone on conduct and disruptive behavior disorders in children with subaverage IQs. *Journal of the American Academy of Child and Adolescent Psychiatry, 41*(9), 1026–1036.

Sokolski, K.N., Chicz-Demet, A., & Demet, E.M. (2004). SSRI-related extrapyramidal symptoms in autistic children: A case series. *Journal of Child and Adolescent Psychopharmacology, 14*(1), 143–147.

Solanto, M.V. (1998). Neuropsychopharmacological mechanisms of stimulant drug action in attention-deficit hyperactivity disorder: A review and integration. *Behavioural Brain Research, 94,* 127–152.

Soni, S., Whittington, J., Holland, A.J., Webb, T., Maina, E., Boer, H., et al. (2007). The course and outcome of psychiatric illness in people with Prader-Willi syndrome: Implications for management and treatment. *Journal of Intellectual Disabilities Research, 51*(1), 32–42.

Sovner, R. (1986). Limiting factors in using DSM-III criteria with mentally ill/mentally retarded persons. *Psychopharmacology Bulletin, 22,* 1057–1059.

Steingard, R.J., Zimnitzky, B., DeMaso, D.R., Bauman, M.L., & Bucci, J.P. (1997). Sertraline treatment of transition-associated anxiety and agitation in children with autistic disorder. *Journal of Child and Adolescent Psychopharmacology, 7,* 9–15.

Stromme, P., & Diseth, T.H. (2000). Prevalence of psychiatric diagnoses in children with mental retardation: Data from a population based study. *Developmental Medicine and Child Neurology, 42,* 266–270.

Sullivan, K., Hooper, S., & Hatton, D. (2007). Behavioural equivalents of anxiety in children with fragile X syndrome: Parent and teacher report. *Journal of Intellectual Disability Research, 51*(1), 54–65.

Sutor, B., Hansen, M.R., & Black, J.L. (2006). Obsessive compulsive disorder treatment in patients with Down syndrome: A case series. *Down Syndrome Research and Practice, 10*(1), 1–3.

Swanson, J.M., Kinsbourne, M., Migg, J., Lanphear, B., Stefanatos, G.A., Volkow, N., et al. (2007). Etiological subtypes of attention-deficit/hyperactivity disorder: Brain imaging, molecular genetics, and environmental factors and the dopamine hypothesis. *Neuropsychology Review, 17*(1), 39–59.

Szymanski, L., Madow, L., Mallory, G., Menolascino, F., Pace, L., & Eidelman, S. (1990). *Report of the task force on psychiatric services to adult mentally retarded and developmentally disabled persons.* Washington, DC: American Psychiatric Association.

Tsapakis, E.M., & Travis, M.J. (2002). Glutamate and psychiatric disorders. *Advances in Psychiatric Treatment, 8,* 189–197.

Turgay, A., Binder, C., Snyder, R., & Fisman, S. (2002). Long-term safety and efficacy of risperidone for the treatment of disruptive behavior disorders in children with subaverage IQs. *Pediatrics, 110*(3): e34.

Tyrer, P., Oliver-Africano, P.C., Ahmed, Z., Bouras, N., Cooray, S., Deb, S., et al. (2008). Risperidone, haloperidol, and placebo in the treatment of aggressive challenging behavior in patients with intellectual disability: A randomized controlled trial. *Lancet, 371,* 57–63.

Wachtel, L.E., & Hagopian, L.P. (2006). Psychopharmacology and applied behavioral analysis: Tandem treatment of severe problem behaviors in intellectual disability and a case series. *Israel Journal of Psychiatry and Related Sciences, 43*(4), 265–274.

Wasserman, S., Iyengar, R., Chaplin, W.F., Watner, D., Waldoks, S.E., Anagnostou, E., et al. (2006). Levetiracetam versus placebo in childhood and adolescent autism: A double-blind placebo-controlled study. *International Clinics of Psychopharmacology, 21*(6), 363–367.

Wilens, T.E., Waxmonsky, J., Scott, M., Swezey, A., Kwon, A., Spencer, T.J., et al. (2005). An open trial of adjunctive donepezil in attention-deficit/hyperactivity disorder. *Journal of Child and Adolescent Psychopharmacology, 15*(6), 947–955.

Wolraich, M.L., Lindgren, S., Stromquist, A., Milich, R., Davis, C., & Watson, D. (1990). Stimulant medication use by primary care physicians in the treatment of attention deficit hyperactivity disorder. *Pediatrics, 86*(1), 95–101.

World Health Organization. (1977). *International classification of diseases, ninth revision* (ICD-9). Geneva: World Health Organization.

World Health Organization. (1992). *International classification of diseases, tenth revision* (ICD-10). Geneva: World Health Organization.

Zhang, L., Plotkin, R.C., Wang, G., Sandel, E., & Lee, S. (2004). Cholinergic augmentation with donepezil enhances recovery in short-term memory and sustained attention after traumatic brain injury. *Archives of Physical Medicine and Rehabilitation, 85,* 1050–1055.

13

Integrating Behavioral and Pharmacological Interventions for Severe Problem Behavior Displayed by Children with Neurogenetic and Developmental Disorders

Louis P. Hagopian and Mary E. Caruso-Anderson

We discuss here the assessment and treatment of severe problem behavior such as self-injury and aggression displayed by individuals with neurogenetic disorders and intellectual disabilities. Individuals with intellectual disabilities form a heterogeneous and highly complex population because genetic and brain abnormalities can profoundly disrupt all aspects of functioning. Problem behavior in this population is equally complex and can have multiple determinants, which can be broadly categorized as biological and environmental. Biological variables include but are not limited to genetic abnormalities, psychiatric conditions, neurological dysfunction, and medical conditions that affect behavior. Environmental variables include the level and type of stimulation and reinforcement available, the social environment (i.e., the behavior of other people that occasions and reinforces the individual's problem behavior), and the history of behavior–environment interactions (i.e., the individual's behavioral history). Problem behavior may stem primarily from biological *or* environmental determinants; however, it is typically the product of the interaction of these factors to varying degrees.

Models for designing interventions based on the classification of problem behavior with consideration of the full range of biological and environmental controlling variables have been in existence for some time (Engel, 1977; Mace & Mauk, 1995; Pyles, Muniz, Cade, & Silva, 1997; Thompson, Egli, Symons, & Delaney, 1994; Wachtel & Hagopian, 2006). These models, termed "biopsychosocial," "biobehavioral," or "neurobehavioral," recognize the need for interdisciplinary collaboration to identify (and rule out) the range of possible determinants of problem behavior in each case and to design and apply the appropriate intervention. Generally speaking, this integrative approach advocates the use of behavioral interventions to address problems that stem from social and environmental variables, behavioral histories of reinforcement for problem behavior, and skill deficits; whereas pharmacological agents are appropriate for addressing problems stemming from biochemical dysfunction. Obviously there are exceptions to this general approach, as when self-injury is secondary to an unalterable biological condition such as Lesch-Nyhan syndrome (Anderson & Ernst, 1994; Nyhan, 1998), certain forms of traumatic brain injury

(Tonkonogy, 1991), or sensory neuropathy (Kirman & Bicknell, 1968; Roach, Abramson, & Lawless, 1985). In such cases, behavioral intervention and/or medication cannot directly address causal factors, but nevertheless may be applied (sometimes successfully) in an attempt to manage the problematic behavior (Kuhn, Hagopian, & Terlonge, 2008; Wurtele, King, & Drabman, 1984). There is general consensus among researchers and clinicians working with individuals with developmental and intellectual disabilities that the interdisciplinary approach described here represents current best practices (Plauche-Johnson, Myers, & American Academy of Pediatrics Council on Children with Disabilities, 2007; Reiss & Aman, 1998; Rush & Frances, 2000); however, actual research and clinical practices often fall short of this ideal. This chapter discusses how the determinants of problem behavior can be identified and how behavioral and pharmacological interventions can be combined in a complementary way. It should be noted at the outset, however, that before embarking on a costly and time-consuming process of assessment and treatment, it is important to first ensure that the individual is in a supportive environment where there are appropriate levels of structure, social interaction, stimulation, reinforcement, and opportunities for choice (Thompson, Moore, & Symons, 2007). In some cases, the initial focus of treatment should be on simply meeting these essential needs.

SEVERE PROBLEM BEHAVIOR: RISK FACTORS AND PREVALENCE

Individuals with intellectual and developmental disabilities display clusters of problem behavior that are not often seen in the typically developing population. These problem behaviors include self-injurious behavior (SIB; e.g., hitting, biting, scratching oneself), aggression directed toward others (e.g., hitting, pinching, kicking,

pulling hair), destructive behavior (e.g., breaking or throwing items), pica (eating inedible objects), elopement (running away from caregivers), noncompliance, and screaming (Holden & Gitlesen, 2003, 2006; Totsika, Toogood, Hastings, & Lewis, 2008). Typically, individuals display multiple types of problem behavior, and the levels of severity can range from relatively minor and short duration to highly severe, chronic, and potentially life-threatening. Injuries secondary to self-injury can include contusions and lacerations, retinal detachment and blindness, infections, and loss of tissue from self-biting, particularly of the tongue, lips, and hands. Aggressive behavior can result in tissue damage and broken bones to others and is associated with increased service costs, high rates of caregiver turnover, and placement in restrictive settings (Allen, 2000). Problem behavior is considered to be severe when it occurs frequently (e.g., on a daily basis), causes injury to self or others that requires medical treatment, requires more than one person to manage, and ultimately restricts participation in regular activities. For many individuals with developmental disabilities, problem behavior may represent the greatest barrier to integration and participation in community activities (Lowe et al., 2007). In addition, these behaviors are costly to society (National Institutes of Health, 1989); can result in the overuse of medication, mechanical restraint, or placement in restrictive settings; and can have a negative impact on care providers and families (Emerson et al., 2000).

Studies generally suggest that 5% to 10% of individuals with intellectual disabilities engage in highly severe and potentially life-threatening problem behavior. This prevalence rate rises to 50% when less severe problem behaviors are included (Dekker, Koot, van der Ende, & Verholst, 2002). Certain neurogenetic syndromes are associated with higher prevalence rates of problem behavior (e.g., fragile X syndrome, Prader-Willi syndrome), and some include self-injury as a characteristic of the behavioral phenotype (e.g., Smith-Magenis syndrome, Cornelia de Lange syndrome, Smith-Lemli-Opitz syndrome).

Epidemiological research has identified several individual variables that correlate with problem behavior in individuals with intellectual disabilities, which can be considered risk factors, although the direction of causality is not certain. These risk factors include the diagnosis of autism, the level of intellectual disability, the degree of receptive and expressive communication deficits, and presence of sensory impairments (Ando & Yoshimura, 1979; Bhaumik, Branford, McGrother, & Thorp, 1997; Chadwick, Walker, Bernard, & Taylor, 2000; Davidson et al., 1994; Holden & Gitlesen, 2006; Kiernan & Alborz, 1996; Lowe et al., 2007; McClintock, Hall, & Oliver, 2003). It is possible that these neurologically based deficits may predispose a developmental trajectory wherein problem behavior is more likely to occur, to be inadvertently reinforced by care providers, and to ultimately interfere with the development of adaptive behaviors.

CLASSIFICATION AND DIAGNOSIS

Classification of problem behavior and the selection of interventions requires identifica-tion of the determinants of problem behavior, including but not limited to organic/somatic factors, psychiatric conditions, psychosocial/environmental factors, and behavioral history (Kalachnik et al., 1998; Rush & Frances, 2000; Szymanski et al., 1998). We focus on two major taxonomies in this discussion: functional behavioral classification, which categorizes problem behavior in terms of its functional properties (i.e., its controlling antecedents and consequences), and psychiatric diagno-sis, which seeks to identify psychiatric conditions that may co-occur with or underlie problem behavior.

Functional Behavioral Classification

The dominant behavioral approach for the assessment and treatment of severe problem behavior displayed by individuals with developmental disabilities including autism is applied behavior analysis (ABA). ABA is concerned with the application of behavioral science in applied settings with the aim of addressing socially important issues such as behavior problems and learning (Baer, Wolf, & Risley, 1968). ABA is widely viewed as a scientifically supported approach for addressing learning and behavioral difficulties in individuals with a range of developmental disabilities, most notably autism (U.S. Department of Health and Human Services, 1999). The hallmarks of ABA include the application of operant learning principles to understand and change behavior, the precise measurement of behavior, and the use of experimental analysis methodologies (typically single-subject designs) to identify behavior–environment relations. Incidentally, these latter two features—precise measurement of behavior and analysis methodologies—can be readily applied to the objective and systematic evaluation of pharmacological interventions.

Functional behavioral assessment (FBA) represents a range of techniques designed to identify the environmental antecedents that occasion, as well as the consequences that reinforce and maintain, problem behavior. In some cases, careful interview of care providers and informal observations of the individual over time and across settings may lead to the identification of these controlling variables. However, in other cases, a more formal and intensive functional analysis may be required. Functional analysis is the most scientifically rigorous methodology used to identify the environmental determinants of behavior and involves examining how the behavior of interest changes with systematic manipulation of a variety of antecedent and consequent events. For example, to determine whether problem behavior is maintained by escape from academic demands (a common function of problem behavior; see below), an academic "work" environment is simulated wherein demands are presented and then briefly terminated contingent upon problem behavior. With repeated observation and analysis of behavioral data, this rigorous data-based method can produce objective and reliable conclusions about the environmental variables controlling problem behavior. There is general consensus that FBA represents best practice for assessment of problem behavior in this population and should be used to guide the development of behavioral interventions as well as inform psychiatric diagnosis (Kalachnik et al., 1998; Rush & Frances, 2000).

Functional behavioral classification seeks to characterize problem behavior in terms of its functional properties (i.e., controlling antecedents and consequences) rather than its structural or topographical properties (e.g., self-injury). Problem behavior can be functionally categorized within one of two broad categories, socially mediated and non-socially mediated. The socially mediated classification refers to control by antecedents and consequences that are mediated by other individuals, such as parents, teachers, other care providers, siblings, and peers. That is, other individuals may behave in a way that sets the occasion for problem behavior to occur and/or respond to problem behavior in a way that reinforces it (often inadvertently) and thereby increases its future probability. The research literature indicates that in approximately two-thirds of cases, self-injury can be shown to be directly maintained by socially mediated variables such as access to preferred items, access to attention, and escape from demands, to name a few (Hanley, Iwata, & McCord, 2003;

Iwata, Dorsey, Slifer, Bauman, & Richman, 1994).

The nonsocially mediated classification refers to control by factors that do not directly involve other people. Reinforcement by sensory consequences is inferred based on certain behavioral patterns and can be further subdivided into sensory-positive or sensory-negative reinforcement. The former refers to problem behavior that produces positive sensory stimulation (e.g., repeatedly banging an object to produce a sound), and the latter refers to problem behavior that is believed to attenuate a negative state (e.g., pain). Research to date suggests that approximately one-fourth of individuals display SIB maintained by some form of sensory reinforcement (Iwata et al., 1994).

Many individuals with severe problem behavior have a complex presentation, with multiple topographies of problem behavior with sometimes multiple functions. Within a given case where multiple forms of problem behavior are displayed (e.g., aggression, self-injury, and disruptive behavior), each form of behavior may have different functions (Derby, 2000); a single form can have multiple functions (Iwata et al., 1994; Kennedy, Meyer, Knowles, & Shukla, 2000); or multiple forms may have the same function (Magee & Ellis, 2000). Various forms of socially mediated problem behavior can also be linked in a hierarchical fashion, characterized by an escalating pattern from mild to more severe forms of problem behavior when mild problem behavior is ineffective (Lieving, Hagopian, Long, & O'Connor, 2004; Richman, Wacker, Asmus, Casey, & Andelman, 1999).

Functional behavioral classification and psychiatric diagnosis represent different but not necessarily incompatible classification schemes; the former focuses on identification of environmental controlling variables and the latter on identification of constellations of behaviors that represent signs and symptoms of biologically based conditions. A definitive functional behavioral classification does not necessarily mean that there is no co-occurring psychiatric condition, and vice versa. As noted at the outset, the two classification schemes may inform one another and often provide a more complete account of problem behavior than either does alone. However, it should also be noted that problem behavior may be primarily a function of environmental variables, and there may be no psychiatric disorder. Alternatively, problem behavior may stem largely from a psychiatric disorder or neurochemical dysfunction and may not be under the influence of environmental variables to any great extent. Finally, problem behavior may be maintained by a combination of environmental variables, as well as related to a psychiatric condition or other biological variable. Making these determinations is challenging and requires specialized expertise in behavior analysis, psychiatric assessment, and psychopharmacology.

Psychiatric Diagnosis

Psychiatric diagnosis of individuals with intellectual disabilities does not require the presence of problem behavior (Rojahn, Matson, Naglieri, & Mayville, 2004). That is, an individual with a neurogenetic disorder and intellectual disabilities may not display problem behavior, yet may have a diagnosable psychiatric condition that warrants treatment (such cases are not the focus of the current discussion). As noted earlier, the presence of problem behavior does not necessarily indicate the presence of psychiatric diagnosis (other than the diagnoses that simply describe the problem behavior itself, such as stereotypic movement disorder with self-injurious behavior). Diagnostic classification systems designed and validated for use with typically developing adults and children (*Diagnostic and Statistical Manual, Fourth Edition, Text Revision [DSM-IV-TR]*, American Psychiatric Association [APA], 2000; and the *International Statistical Classification of Diseases*

and Related Health Problems, Tenth Edition, Revised [ICD-10], World Health Organization, 1997) are also used for the classification of problem behavior in individuals with intellectual disabilities. The term *dual diagnosis* is used to describe individuals with intellectual disabilities who have comorbid psychiatric disorders—whether or not they engage in problem behavior of the sort discussed here. Challenges in using these diagnostic classification systems with this population have been discussed in detail elsewhere (Szymanski et al., 1998) and are only briefly summarized here.

There is general consensus that individuals with intellectual and developmental disabilities are at increased risk for psychiatric disorders (Borthwick-Duffy, 1994; Dykens, 2000; Einfeld & Aman, 1995; Emerson, 2003; Nøttestad & Linaker, 1999; Witwer & Lecavalier, 2008). In fact, some have suggested that an individual with intellectual disabilities is three to four times more likely to develop a psychiatric disorder than his or her typically developing counterpart (Borthwick-Duffy, 1994). Estimates of the prevalence of psychiatric disorders among individuals with intellectual disabilities vary widely across studies, ranging from 10% to 70% (Borthwick-Duffy, 1994; Göstason, 1985; Holden & Gitlesen, 2008). Variation in prevalence rates appears to be due to a number of factors, particularly the population being assessed (e.g., children, adults, inpatient, outpatient), subject characteristics (e.g., IQ, etiology), and the types of measures used to assess psychopathology (e.g., *DSM* or ICD diagnosis, rating scales of psychiatric symptomology; Dykens, 2000). However, it should be noted that estimates of *DSM-IV-TR* and ICD-10 psychiatric disorders are approximately 40% (Chadwick, Kusel, Cuddy, & Taylor, 2005; Dykens, 2000; Emerson, 2003).

The *DSM* and ICD systems were designed for typically developing individuals and therefore rely heavily on self-report and define impairment in functioning with reference to populations without disabilities.

Application to individuals with intellectual disabilities is difficult because communication deficits often preclude self-report, and restricted behavioral repertoires make it difficult to ascertain the extent to which impairments in functioning are due to developmental/intellectual disabilities versus a psychiatric condition (Borthwick-Duffy, 1994; Einfeld & Aman, 1995; Sturmey, 1995). Therefore, psychiatric diagnosis of this population utilizes behavioral referents (e.g., irritability, crying, hyperactivity, sleep disturbance, avoidance, facial expressions indicative of fear) believed to parallel the established criteria for a particular psychiatric disorder. For example, an individual with severe intellectual disabilities and communication deficits may not be able to verbalize feelings of worthlessness, guilt, or recurring thoughts of death as required for a diagnosis of depressive disorder. However, such a diagnosis may be appropriate if the individual cries easily or for no apparent reason, is restless or agitated much of the time, has difficulty sleeping at night, ignores attempts at social interaction, or displays unprovoked tantrums or rage, and if the role of environmental factors or medical conditions explaining these behaviors has been ruled out via behavioral assessment or medical examination, respectively (Matson et al., 1999). Although the individual does not display behavior (or verbalize feelings) that match standard diagnostic criteria, the observed behavior patterns parallel the criteria sufficiently to validate a working diagnosis of depressive disorder. Based on this working diagnosis, appropriate pharmacological treatment would be indicated. Szymanski et al. (1998) illustrated this nicely in their description of how the *DSM-IV* (APA, 1994) criteria for Major Depressive Episode and Major Manic Episode can be applied to individuals with intellectual disabilities. Hagopian and Jennett (2008) also discussed how anxiety disorders can be assessed and diagnosed in this population based on behavioral observations of avoidance and other

observable indicators of fear. Moreover, clinicians must attempt to determine whether impairments in functioning go beyond what one might expect given the individual's developmental and intellectual disabilities.

Although this approach is reasonable and represents current best practice, it is based on the assumption that individuals with intellectual disabilities exhibit the same psychiatric disorders as typically developing individuals. This assumption is somewhat tenuous, as there is evidence to suggest that symptoms of psychiatric disorders in this population usually manifest in altered or masked ways (Reiss, 1994). This is especially the case as the severity of intellectual disability increases, communication and adaptive skills decrease, and behavioral repertoires become more restricted (Emerson, 2003; Moss et al., 2000; Witwer & Lecavalier, 2008). Epidemiological studies indicate that individuals with milder intellectual disabilities are more likely to be given conventional psychiatric diagnoses such as anxiety or mood disorders and may show improvement in symptoms over time relative to those with greater intellectual impairments (Chadwick et al., 2005; Dykens, 2000).

The Relationship Between Psychopathology and Problem Behavior

Psychiatric conditions and problem behavior can co-occur; however, research on the relationship between psychiatric disorders and problem behavior has produced mixed findings (Holden & Gitlesen, 2003; Jenkins, Rose, & Jones, 1998; Paclawskyj, Matson, Bamburg, & Baglio, 1997; Rojahn, Borthwick-Duffy, & Jacobson, 1993). Studies with positive findings have revealed correlations between problem behavior and depression, anxiety, and psychosis (Holden & Gitlesen, 2003; Moss et al., 2000) and between problem behavior and psychopathology in general

(Moss et al., 2000; Rojahn et al., 2004). Furthermore, Hemmings and his colleagues found a specific relation between affective symptoms and SIB (Hemmings, Gravestock, Pickard, & Bouras, 2006). However, caution should be taken in the interpretation of the research on the relation between psychiatric conditions and problem behavior, given that psychiatric symptoms are often rated as present based on the existence of problem behavior (Holden & Gitlesen, 2009).

BEHAVIORAL AND PHARMACOLOGICAL TREATMENT

Function-Based Behavioral Interventions

Although the immediate purpose of the FBA is to identify the environmental variables controlling problem behavior, its ultimate purpose is to guide the design of function-based interventions. Though a wide range of behavioral interventions for severe behavior have been described in the literature (Didden, Duker, & Korzilius, 1997; Kahng, Iwata, & Lewin, 2002), function-based interventions share some core features. They generally involve the use of operant conditioning procedures to establish and maintain alternative adaptive behavior and to extinguish problem behavior. To replace problem behavior with appropriate behavior such as communication, cooperation, and social and leisure skills, these interventions require engineering the environment (including the behavior of care providers) to minimize the probability that problem behavior will occur and will contact reinforcement (for reviews, see Hanley et al., 2003; Pelios, Morren, Tesch, & Axelrod, 1999).

Over the past 40 years an extensive body of literature has demonstrated the effectiveness of ABA-based procedures for assessing and treating problem behavior and for increasing appropriate skills (i.e., early intensive behavioral intervention) for individuals with intellectual disabilities, autism, and related disorders. Several review articles and meta-analyses have been published that summarize the large body of literature on behavioral treatments for these problems. Six of these (DeMyer, Hingtgen, & Jackson, 1981; Herbert, Sharp, & Gaudiano, 2002; Hingtgen & Bryson, 1972; Kahng et al., 2002; Matson, Benavidez, Compton, Paclawskyj, & Baglio, 1996; Sturmey, 1995) collectively reviewed published studies spanning the years 1946 to 2001. Each of these reviews supported the efficacy of ABA-based procedures in the assessment and treatment of problem behavior associated with autism, intellectual disability, and related disorders. Similarly, three meta-analyses (Didden et al., 1997; Lundervold & Bourland, 1988; Weisz, Weiss, Han, Granger, & Morton, 1995) collectively analyzed hundreds of studies published between 1968 and 1994. The results of these meta-analyses indicate that treatments based on operant principles of learning were more effective than alternative treatments (e.g., psychotherapy, sensory integration therapy) in reducing problem behavior displayed by individuals with intellectual disabilities as well as individuals who do not have delays. Many scientific, governmental, and professional organizations, including the National Institutes of Health (1989), the American Association on Intellectual and Developmental Disabilities (formerly the American Association on Mental Retardation; Rush & Frances, 2000), the U.S. Surgeon General (U.S. Department of Health and Human Services, 1999), and the American Academy of Pediatrics (Plauche-Johnson et al., 2007), have characterized ABA-based procedures as empirically supported and as representing best practice for individuals with autism and developmental disabilities.

Functional communication training, for example, involves teaching the individual an appropriate communication response to access reinforcers that have historically maintained problem behavior (Carr & Durand, 1985, 1991). Once this new learning history is established, the focus of treatment becomes teaching the individual to tolerate waiting and not always gaining access to a requested item or activity, while maintaining reductions in problem behavior. This intervention has been widely studied over the past 20 years in over 60 investigations, including three large-*n* case studies of consecutive cases (Hagopian, Fisher, Sullivan, Acquisto, & LeBlanc, 1998; Kurtz et al., 2003; Wacker et al., 2005). Functional communication training has been demonstrated to be highly effective in reducing problem behavior by at least 80% in 77% of cases. Another type of function-based intervention, noncontingent reinforcement, involves providing free access to the reinforcer responsible for maintaining problem behavior (Hagopian, Crockett, van Stone, DeLeon, & Bowman, 2000; Lalli, Casey, & Kates, 1997; Vollmer, Iwata, Zarcone, Smith, & Mazaleski, 1993). Subsequently, access to reinforcement is gradually decreased while low levels of problem behavior are maintained. This intervention has also been widely studied and is effective for this population. Other classes of interventions involve providing reinforcement for incompatible or alternative responses such as compliance with instructions (Carr, Robinson, Taylor, & Carlson, 1990; Mulick, Schroeder, & Rojahn, 1980) or providing stimuli and activities that have the property of competing with reinforcement maintaining problem behavior (DeLeon, Toole, Gutshall, & Bowman, 2005; Hagopian, Contrucci-Kuhn, Long, & Rush, 2005; Piazza et al., 1998). For many individuals with intellectual disabilities, communication deficits necessitate the use of supplemental discriminative stimuli to signal transitions, activities, and the rules in effect at any given time. In combination with countless skill-building strategies (e.g., shaping, chaining, discrimination training, leisure skills training, compliance training), these ABA-derived interventions enable individuals with even the most severe intellectual

disabilities and behavioral problems to learn appropriate replacement behaviors and express their preferences and choices.

Pharmacological Interventions in Intellectual Disabilities

The prescription rate for psychotropic medication classes such as antipsychotics often exceeds the expected prevalence of associated psychiatric disorders because they are heavily used to treat problem behaviors and subdiagnostic tendencies (Clarke, Kelley, Thinn, & Corbett, 1990; Robertson et al., 2000). Similarly, the majority of research on the use of psychotropic medications in individuals with intellectual disabilities has examined their effect on the behaviors and subdiagnostic tendencies for which they are most commonly prescribed. The number of studies examining the effects of various psychotropic medications on definitively diagnosed psychiatric disorders in individuals with intellectual disabilities (i.e., individuals with dual diagnosis) varies greatly by drug class.

Given the challenges of obtaining an accurate psychiatric diagnosis in individuals with intellectual disabilities, it is not surprising that pharmacological interventions as treatments for problem behavior and psychiatric disorders are generally not as effective as for the typically developing population (Aman, 1996, 2004; Matson et al., 2000). This may be due to a number of factors, including the lack of standardized prescription practices, differences in symptom presentation in individuals with intellectual disabilities, and, perhaps most important to the current discussion, the failure to consider the behavioral function of problem behavior (this latter issue will be discussed shortly).

Our review of the literature on the use of psychotropic medications with this population identified over 500 studies. There is good support for the use of some psychotropic medications for treating psychiatric conditions in individuals with intellectual disabilities and psychiatric disorders. By far, the largest body of literature in this area is on the effects of stimulants. In addition to numerous uncontrolled studies, there are more than one dozen large-group controlled trials on the use of stimulants in individuals with intellectual disabilities and formally diagnosed attention-deficit/hyperactivity disorder (Arnold, Gadow, Pearson, & Varley, 1998; Sprague & Werry, 1971). There are also a significant number of studies examining the effects of antidepressants on depression (Cook, Rowlett, Jaseslos, & Leventhal, 1992; Davies, 1961; Field, Aman, White, & Vaithianathan, 1986; Gordon, Rapoport, Hamburger, State, & Mannheim, 1992; Hamdan-Allen, 1991; Howland, 1992; Masi, Marcheschi, & Pfanner, 1997; Pary, 1989; Ruedrich & Wilkinson, 1992; Sovner, Fox, Lowry, & Lowry, 1993) and obsessive-compulsive disorder (Bodfish & Madison, 1993; Cook, Terry, Heller, & Leventhal, 1990; McDougle, Price, & Goodman, 1990; Mehlinger, Scheftner, & Poznanski, 1990; Todd, 1991) in individuals with intellectual disabilities. However, only two of these studies demonstrated adequate methodological control (Aman, White, Vaithianathan, & Teeham, 1986; Gordon et al., 1992).

In contrast, there are very few studies on the effects of mood-stabilizing and neuroleptic medications on psychiatric disorders in individuals with intellectual disabilities, despite the widespread use of these medications (Valdovinos, Caruso, Roberts, Kim, & Kennedy, 2005). Several clinical trials evaluated the efficacy of lithium in individuals with a variety of *DSM* diagnoses, including bipolar disorder, depression, psychosis, and schizophrenia, producing equivocal results (Naylor, Donald, LePoidevin, & Reid, 1974; Worrall, Moody, & Naylor, 1975). Finally, we found no methodologically sound clinical trials validating the use of neuroleptics with people who have intellectual disabilities

and psychotic disorders. Moreover, there are a limited number of uncontrolled studies (e.g., case studies), and until recently, package inserts for neuroleptic medications contained disclaimers warning that there is an absence of research on the efficacy of these medications in people with intellectual disabilities (Aman & Madrid, 1999; Baumeister, Sevin, & King, 1998; Duggan & Brylewski, 1999; Thalayasingam, Alexander, & Singh, 2004).

There is also converging evidence suggesting that psychotropic medications may have positive effects on a number of subdiagnostic behavioral tendencies (that is, clinical targets that warrant treatment even though the individual does not meet diagnostic criteria for a *DSM* disorder) or specific behavioral targets. For example, in addition to standard uses of psychotropic medications, selective serotonin reuptake inhibitors have been found to decrease repetitive behavior and rituals (Aman, Arnold, & Armstrong, 2000; McDougle, Kresch, & Posey, 2000); psychostimulants and α-agonists have been found to decrease hyperactivity and impulsivity (Aman, 1996, 2004; Handen, Feldman, Lurier, & Murray, 1999); mood stabilizers such as lithium have been found to help with cyclicity in mood and irritability in people with intellectual disabilities (Carta, Hardoy, Dessi, Hardoy, & Carpiniello, 2001; Ruedrich, Swales, Fossaceca, Toliver, & Rutkowski, 2001), and atypical antipsychotics and α-agonists have been found to decrease aggression and SIB (Aman, 2004; Aman et al., 2002; McDougle et al., 1998; Zarcone et al., 2001).

The atypical antipsychotic risperidone deserves specific mention for several reasons. This is the first drug to be approved by the FDA for the symptomatic treatment of irritability, aggression, self-injury, and temper tantrums in children and adolescents with autism (Food and Drug Administration [FDA], 2006). In addition, 11 clinical trials on 580 children and adults with autism spectrum disorders support the use of risperidone in this population (Aman et al., 2002; Buitelaar, van der Gaag, Cohen-Kettenis, & Melman, 2001; Hellings et al., 2006; McDougle et al., 1998, 2005; Research Units on Pediatric Psychopharmacology Autism Network [RUPP], 2002; Shea et al., 2004; Snyder et al., 2002; Van Bellinghen & De Troch, 2001; Vanden Borre et al., 1993; Zarcone et al., 2001). While all of these studies produced positive results, nine specifically reported statistically significant reductions in ratings on the Aberrant Behavior Checklist, particularly for the irritability subscale. Decreases in reports of self-injury, aggression, property destruction, hyperactivity, and stereotypic behavior were also noted.

Of the over 500 studies we identified in our review, only 8 reported functional behavioral assessment findings of participants (Crosland et al., 2003; Dicesare, McAdam, Toner, & Varrell, 2005; Fisher, Piazza, & Page, 1989; Garcia & Smith, 1999; Swanson, 2000; Valdovinos, Ellinger, & Alexander, 2007; Valdovinos et al., 2002; Zarcone et al., 2004). Surprisingly, even the more sophisticated, large-scale randomized clinical trials did not attempt to determine whether problem behavior was maintained by environmental variables via functional behavioral assessment (Arnold et al., 2003; Research Units on Pediatric Psychopharmacology [RUPP] Autism Network, 2002). Failure to consider this potentially critical individual variable may result in an underestimate of the beneficial effects of psychotropic medications and is inconsistent with the professional consensus on what constitutes best clinical practice (Kalachnik et al., 1998; Plauche-Johnson et al., 2007; Rush & Frances, 2000). There is limited but compelling evidence to suggest that problem behavior under the influence of environmental variables may be less responsive to medication alone relative to problem behavior that appears to have a more biological basis (Mace & Mauk, 1995; Sandman,

Hetrick, Taylor, & Chicz-DeMet, 1997; Symons & Thompson, 1998).

INTEGRATING BEHAVIORAL AND PHARMACOLOGICAL APPROACHES

Integration of behavioral and pharmacological interventions for problem behavior in individuals with neurogenetic and developmental disorders might have the best outcomes when the determinants of problem behavior are successfully identified and interventions are systematically applied in a targeted fashion. Although the integration of behavioral and pharmacological approaches is regarded as best clinical practice, direct scientific evidence supporting the efficacy of this approach is quite limited. There are, however, several studies that provide indirect support.

Mace and Mauk (1995) developed a diagnostic system that classifies self-injury as "operant," "possibly biologic," or "mixed operant and possibly biologic," based on functional analysis results. Self-injury with "possible biologic" etiology is further subtyped according to a variety of factors (e.g., the frequency, intensity, and duration of bouts, the extent of tissue damage) to guide the selection of psychotropic medication. They found that matching treatment to the individual's biobehavioral diagnosis (e.g., behavioral treatment for operant SIB type, psychotropic medication for possibly biologic SIB type, and both treatments for the mixed SIB type) results in better clinical outcomes and reduces the incidence of nonresponders to both medication and behavioral treatment. Similarly, Symons and Thompson (1998) found that a combination of the opioid blocker naltrexone and functional communication training (FCT) most effectively reduce self-injury that appears to have both a biological and environmental basis compared with rates of SIB during placebo, naltrexone alone, and combined placebo/FCT phases. Finally, Sandman and colleagues found that some subjects with SIB show increased levels of the peptide beta-endorphin (which stimulates the release of endogenous opioids) immediately after an episode of SIB as compared with other times during the day when self-injury did not occur (Sandman, Hetrick, Taylor, & Chicz-DeMet, 1997). Furthermore, the opioid blocker naltrexone was found to be most effective in reducing self-injury in these subjects relative to subjects with SIB who did not show evidence of this biomarker (Sandman, Hetrick, Taylor, Marion, & Chicz-DeMet, 2000). To our knowledge, this is the first report of a biological marker for self-injury, and it supports the distinction made between operant and biological SIB.

Assessment as an Ongoing, Multimodal, Hypothesis-Testing Process

The field of applied behavior analysis has developed precise behavioral measurement techniques and systematic analysis methods that can be used to facilitate psychiatric diagnosis and the evaluation of pharmacological interventions. Determining the extent to which some or all of the individual's problem behavior is biologically or environmentally based is challenging and requires 1) converging evidence from multiple sources, including functional behavioral assessment findings; 2) repeated observation of behavior over time and across settings; and 3) observation and analysis of the impact of interventions in problem behavior. Pyles et al. (1997) provide a detailed assessment and treatment algorithm for integrating behavioral and pharmacological interventions that is consistent with the conceptual model described here.

Assessment and classification of problem behavior, whether through psychiatric diagnosis or the use of functional behavioral classification systems, must be viewed as an ongoing hypothesis testing process. Although hypotheses may be formulated after an initial

assessment, additional information should be obtained via supplementary assessments and through the observation of changes in behavior over time. Caution should be taken to avoid circular, nonfalsifiable hypotheses. For example, concluding that a child is displaying self-injury because she has a diagnosis of autism is of little value with regard to treatment selection (unfortunately this type of explanation is one that we encounter frequently). Likewise, an initial diagnosis of depression should be open to modification if behavioral assessment findings reveal that behavioral patterns thought to suggest depression are primarily due to environmental factors, or when additional data obtained over time reveal cyclical patterns in problem behavior and mood that are suggestive of bipolar disorder.

Psychiatric diagnosis should be based, in part, on a consideration of the results of FBA, given that FBA can determine the extent to which problem behavior is socially mediated, that is, influenced by environmental variables (Kalachnik et al., 1998). As noted previously, behavioral assessment must also consider indicators of the suspected psychiatric diagnoses that are not directly related to the specific behavioral targets, including sleep patterns, food intake, mood and range of affect, irritability, activity level, and so forth. However, definitive psychiatric or behavioral assessment findings should not be taken to mean that other controlling variables are not at play, as problem behavior can be multifaceted and under the joint influence of behavioral and biological variables. For example, it is possible that certain topographies of problem behavior (e.g., aggression) within an individual may be under environmental control (e.g., maintained by attention), whereas other topographies (e.g., self-injury associated with mood lability) may be secondary to a psychiatric disorder. It is also possible that an individual can display problem behavior that is primarily environmentally controlled but also have a psychiatric condition that interacts with the environmental variables (e.g., as in the case of an individual with escape-maintained behavior who is less tolerant of demands when depression is more severe). Again, because these interactions may be revealed subsequent to intervention, it is important to follow changes in behavior over time and in response to treatment.

Where the functional behavioral assessment findings are definitive, a behavioral intervention may be the best first course of action for several reasons. First, an intervention that directly addresses an identified causal factor of problem behavior will be more effective than an intervention that reduces problem behavior through another mechanism. For example, if a functional behavioral assessment reveals that the individual engages in aggression to access preferred activities, teaching him or her an appropriate way to express preferences directly addresses the controlling variable. In addition, this type of intervention would be beneficial beyond reducing aggression in that it broadens the individual's repertoire of appropriate behavior and adaptive skills. In this case, a behavioral treatment would be preferred over medication, which even if effective in reducing aggression, would neither directly address the cause nor teach an adaptive communication skill.

Second, the outcome of the behavioral intervention that addresses one causal factor may reveal the presence of other controlling factors, some of which may not be environmental. For example, a fully or partially successful treatment for problem behavior maintained by escape from demands may reduce high levels of aggression to the extent that problems with inattention or hyperactivity are made more apparent. Once they are identified, these problems can be effectively targeted with stimulant medication.

When functional behavioral assessment findings are not definitive, but there is evidence supporting a psychiatric diagnosis, then a pharmacological intervention is the best course of action. Simply put, mood fluctuations, hyperactivity, disorganized and impulsive behavior, irritability, and depression that occur independently of environmental variables are not likely to be addressed by behavioral interventions. Control by environmental variables can be subtle and difficult to detect; consequently, ruling out the role of environmental variables with some level of confidence makes it necessary to obtain large samples of behavior over time and across settings. That is, caution should be taken not to quickly "rule out" environmental causes without having performed a formal functional analysis.

As is the case with partially successful behavioral interventions, a partial response to medication may provide additional information that suggests the role of other variables, including the presence of a behaviorally based problem. For example, aggression that occurs indiscriminately across settings or conditions of a functional analysis suggests a lack of environmental control. Following treatment for mood instability, subsequent FBA findings may reveal that problem behavior occurs primarily during academic instruction and is maintained by escape from demands. Escape-maintained problem behavior can then be targeted with behavioral treatment.

In cases where there is insufficient or mixed evidence to support a psychiatric disorder, the presence of subdiagnostic tendencies that impair functioning may be appropriate targets for medication. Impulsivity, low tolerance for frustration, disinhibition, and ritualistic behavior can be isolated problems that warrant pharmacological treatment with medication even when they are not sufficient to meet *DSM* criteria for any diagnosis. Addressing such tendencies has the potential to make behavioral interventions more effective, even when the primary target behavior (e.g., aggression) is primarily under the influence of behavioral variables. For example, where problem behavior such as aggression is occasioned by the interruption of ritualistic behavior (Hagopian, Bruzek, Bowman, & Jennett, 2007; Murphey,

Macdonald, Hall, & Oliver, 2000), medication that reduces ritualistic behavior may make behavioral interventions targeting aggression more efficacious. Likewise, in cases where problem behavior is maintained by escape from academic instruction, medication reducing activity level may enhance the behavioral treatment by reducing out-of-seat behavior and fidgeting with instructional materials (Carlson, Pelham, Milich, & Dixon, 1992).

ADDITIONAL POTENTIAL BENEFITS OF INTEGRATING PHARMACOLOGICAL AND BEHAVIORAL INTERVENTIONS

Although the targeted and coordinated application of behavioral and pharmacological interventions represents best practice, we know little about the mechanisms by which psychotropic medications produce their clinical effects (Neef, Bicard, Endo, Coury, & Aman, 2005; Schaal & Hackenberg, 1994; Schroeder, Lewis, & Lipton, 1983; Thompson et al., 1994; Witkin & Katz, 1990) and less about how they affect behavioral interventions. Behavioral interventions require the reliable functioning of basic behavioral processes such as reinforcement, punishment, and stimulus control, and therefore it is important to know how psychotropic medications affect and alter the functioning of these processes. Although highly intensive behavioral interventions are needed to establish adaptive behaviors in this population, even individuals with the most severe disabilities can respond to such interventions, suggesting that these behavioral processes are somewhat intact. Ultimately, these basic behavioral processes can be reduced to neurobehavioral processes. For example, lesion and neuroimaging research over the past few decades has implicated midbrain areas in the involvement of operant processes such as reinforcement (ventral striatum; Robbins & Everitt, 1996; Schlund, Rosales-Ruiz, Vaidya, Glenn, & Staff, 2008), operant extinction

(pars compacta, ventral temental area; Pan, Schmidt, Wickens, & Hyland, 2008), and stimulus control (amygdala and ventral striatum; Schultz, 1999). If psychotropic medications affect the functioning of these neurobehavioral systems, they can alter the effects of behavior analytic interventions. Understanding how these medications affect more fundamental processes such as reinforcement, extinction, and stimulus control has the potential to extend our knowledge of how to best integrate behavioral and pharmacological interventions beyond what has been described thus far. Specifically, knowledge of how these agents affect basic behavioral processes could be used to both enhance the effects of behavioral interventions and avoid countertherapeutic interactions (Schaal & Hackenberg, 1994; Schroeder et al., 1983; Thompson et al., 1994).

Drug Effects on Basic Operant Processes in Humans

Although our understanding of drug–behavior interactions is in its infancy, some progress has been made in elucidating how drugs affect basic operant processes. The majority of research in this area has primarily involved two classes of medications: stimulants and neuroleptics. Basic research on human subjects has shown that stimulants increase preference for large delayed rewards over smaller, more immediate rewards, as well as the amount of effort a person exerts to obtain reinforcement (deWit, Engagasser, & Richards, 2002; Pietras, Cherek, Lane, Tcheremissine, & Steinberg, 2003; Wilkinson, Kircher, McMahon, & Sloane, 1995). In general, stimulants appear to increase the response to positive reinforcement by increasing reward value, although there is some evidence to suggest that this effect may be diminished with edible reinforcers (Cardinal, Robbins, & Everitt, 2000; Northup, Fusilier, Swanson, Roane, & Borrero, 1997) and social attention (Boelter & Hagopian, 2008; Dicesare et al., 2005). Research findings on the effects of stimulants on operant processes have interesting clinical implications. For example, this body of research suggests that psychostimulants could enhance the efficacy of a token system by facilitating tolerance to delays in token exchange and by increasing the reinforcing effects of tokens as well as the amount of schoolwork a child will do to earn them. This research also suggests that because stimulants enhance the reward value of certain reinforcers, the consistent use of extinction procedures will be important in avoiding instances where problem behavior may result in unintended access to reinforcement. Alternatively, if stimulants decrease the reward value of edible and social reinforcers, then their use may adversely affect behavioral interventions that rely on the use of those classes of reinforcers.

Neuroleptics have been found to enhance simple discrimination learning in individuals with intellectual disabilities at moderate doses (Campbell et al., 1982; Cutmore & Beniger, 1990), but perhaps the most interesting and consistent finding with regard to neuroleptic medications is their ability to selectively decrease avoidance responding (Fischman & Schuster, 1979). These findings have been widely replicated, and subsequent research has shown that a drug's effect on avoidance behavior is highly correlated with its antipsychotic properties (Cook & Weidley, 1957; Fischman & Schuster, 1979; Fischman, Smith, & Schuster, 1976). Basic findings on avoidance behavior are supported by clinical studies in which functional analyses of problem behavior were conducted on placebo and neuroleptic treatment. These clinical studies also demonstrated decreases in avoidance-related problem behavior with neuroleptic treatment (Crosland et al., 2003; Zarcone et al., 2004). Thus, it is possible that neuroleptics could selectively reduce problem behavior that is maintained by avoidance of academic tasks. Conversely, neuroleptics may also decrease the efficacy of response-cost programs that rely on an individual refraining from problem behaviors to avoid loss of points, chips, or tokens. Furthermore, the ability of neuroleptics to facilitate discrimination learning may enhance treatments that rely on stimulus control, such as the use of supplemental discriminative stimuli to signal rules or the availability of reinforcement.

Finally, three clinical studies that have examined the effects of the neuroleptic medication risperidone on socially mediated problem behavior (i.e., maintained by escape from demands, access to attention or tangible items) and nonsocially mediated problem behavior (i.e., problem behavior that has been shown not to serve a social function and is therefore hypothesized to be maintained by internal, biological events or by sensory reinforcement) are worth noting (Crosland et al., 2003; Valdovinos et al., 2002; Zarcone et al., 2004). When data from these three studies were aggregated across 18 participants, the data revealed that problem behavior maintained by social functions was reduced by 50% on the low dose of risperidone and by 67% on the high dose. Problem behavior maintained by nonsocial functions was reduced by 80% on the low dose of risperidone and by 100% on the high dose. These data are preliminary, and all three studies are limited in that many of the participants displayed low levels of problem behavior; however, these findings suggest that risperidone may be more effective for problem behavior maintained by nonsocial consequences.

Collectively, these limited but important findings suggest that the potential for integrating behavioral and pharmacological interventions extends beyond using them to more precisely target clinical problems under the joint control of environmental and behavioral variables. That is, knowledge of how medications alter fundamental behavioral processes has the potential to enhance behavior analytic interventions and avoid countertherapeutic interactions. This approach could help achieve the goal of selecting the correct therapy, or combination of therapies, and apply them in the proper amounts and sequences, to maximize the individual's overall functioning.

CONCLUSIONS AND RECOMMENDATIONS

Behavior analytic and pharmacological interventions for problem behavior in individuals with intellectual disabilities have historically represented different and sometimes competing approaches. Although quite different in terms of their empirical foundations and methods, there is increasing recognition that these approaches can be combined in a complementary fashion. Problem behavior can be under the joint influence of biological and environmental variables, and the development of interventions targeting those controlling determinants hinges on the integration of FBA and psychiatric assessment findings. Behavioral analytic interventions are highly effective for addressing problems that stem from social and environmental variables, behavioral histories of reinforcement for problem behavior, and skills deficits. Although there is strong evidence indicating that behavioral interventions are highly effective for certain anxiety disorders (Davis, Saeed, & Antonacci, 2008; Jennett & Hagopian, 2008), there is no basis for using behavior analytic interventions to treat mood disorders, psychosis, or personality disorders in individuals with intellectual disabilities. Likewise, pharmacologic interventions have proved to be helpful in addressing psychiatric conditions; however, medication will not change social and environmental factors that occasion or reinforce problem behavior, nor will it increase an individual's communication, social, and play skills. Nevertheless, medications have the potential to make behavioral interventions more effective, and behavioral interventions can help manage problems that have a biological basis. Behavioral and pharmacological interventions have their own specific applications and limitations; however, informed integration of the two has the potential to capitalize on the strengths and offset the limitations of each approach.

With the widespread use of FBA and the application of function-based interventions (which more directly target the controlling variables of problem behavior relative to previously used behavior modification techniques), the use of more restrictive behavioral interventions has decreased (Blakeslee, Sugai, & Gruba, 1994; Kahng et al., 2002). Likewise, the expanded use of multidisciplinary teams (which presumably resulted in more comprehensive, multipronged behavioral and pharmacological interventions) has been associated with reduced use of medication (see Davies, 1998). Arguably, the use of targeted interventions to directly address causal factors of problem behavior may be more effective at lower doses and therefore more sustainable over time relative to interventions that reduce problem behavior by other mechanisms. Certainly, a highly intensive behavioral intervention (via crisis management techniques or restraint) may be able to control problem behavior that is more biologically based and secondary to a psychiatric disorder, just as high doses of certain medications may control problem behavior maintained by social reinforcement (via sedative effects). However, control by such methods will likely require these interventions to be applied at higher doses, which will pose greater risks to the individual being treated and will ultimately make them less sustainable over time. For example, a more restrictive behavioral intervention may limit opportunities for community participation and increase isolation, which can in turn increase problem behavior. Similarly, higher doses of medication that produce sedative effects may increase the probability of side effects, limit options for additional medications that may be needed to address a psychiatric condition, and interfere with the acquisition of new skills. In short, interventions that mask problems via intensive behavioral control or sedation may result in other problems that make them unsustainable relative to interventions that directly target the causes of problem behavior.

In light of the complex medical and behavioral health care needs of this population, assessment and treatment of these problems often extends beyond the scope of any single discipline. The practitioner who first sees the individual must initiate the assessments that he or she is qualified to perform and then make a determination whether referral to another professional, including professionals of other disciplines, is indicated. Practitioners must recognize that problem behavior can be multifaceted and under the joint influence of biological and environmental variables, develop an awareness of current behavioral and pharmacological interventions that represent best practices, consider the boundaries of their own expertise and clinical discipline, and make referrals when appropriate. Although inpatient or hospital-based outpatient programs permit relatively easier interdisciplinary collaboration, independent practitioners serving this population should work toward identifying qualified professionals from other disciplines with whom they can collaborate. Before embarking on a costly and time-consuming process of assessment and treatment, however, it is important to first ensure that the individual is in a supportive environment where there are appropriate levels of structure, social interaction, stimulation, reinforcement, and opportunities for choice. In short, practitioners may be able to address some problems simply by meeting these basic needs.

REFERENCES

Allen, D. (2000). Recent research on physical aggression in persons with intellectual disability: An overview. *Journal of Intellectual & Developmental Disability, 25,* 41–57.

Aman, M.G. (1996). Stimulant drugs in the developmental disabilities revisited. *Journal of Developmental and Physical Disabilities, 8,* 347–365.

Aman, M.G. (2004). Management of hyperactivity and other acting-out problems in patients with autism spectrum disorder. *Seminars in Pediatric Neurology, 11,* 225–228.

Aman, M.G., Arnold, L.E., & Armstrong, S.C. (2000). Review of serotonergic agents and preservative behavior in patients with developmental disabilities. *Mental Retardation and Developmental Disabilities Research Reviews, 5,* 279–289.

Aman, M.G., De Smedt, G., Derivan, A., Lyons, B., Findling, R.L., & Risperidone Disruptive Behavior Study Group. (2002). Double-blind, placebo-controlled study of risperidone for the treatment of disruptive behaviors in children with subaverage intelligence. *American Journal of Psychiatry, 159,* 1337–1346.

Aman, M.G., & Madrid, A. (1999). Atypical antipsychotics in persons with developmental disabilities. *Mental Retardation and Developmental Disabilities, 5,* 253–263.

Aman, M.G., White, A.J., Vaithianathan, C., & Teehan, D.J. (1986). Preliminary study of imipramine in profoundly retarded residents. *Journal of Autism and Developmental Disorders, 16,* 263–273.

American Psychiatric Association. (1994). *Diagnostic and statistical manual of mental disorders* (4th ed.). Washington, DC: Author.

American Psychiatric Association. (2000). *Diagnostic and statistical manual of mental disorders* (4th ed., text rev.). Washington, DC: Author.

Anderson, L.T., & Ernst, M. (1994). Self-injury in Lesch-Nyhan disease. *Journal of Autism and Developmental Disorders, 24,* 67–81.

Ando, H., & Yoshimura, I. (1979). Comprehension skill levels and prevalence of maladaptive behaviors in autistic and mentally retarded children: A statistical study. *Child Psychiatry and Human Development, 9,* 131–136.

Arnold, L.E., Bitielo, B., McDougle, C., Scahill, L., Shah, B., Gonzalez, N.M., et al. (2003). Parent-defined target symptoms respond to risperidone in RUPP autism study: Customer approach to clinical trials. *Journal of the American Academy of Child and Adolescent Psychiatry, 42* (12), 1443–1450.

Arnold, L.E., Gadow, K., Pearson, D., & Varley, C.K. (1998). Stimulants. In S. Reiss & M.G. Aman (Eds.), *Psychotropic medication and developmental disabilities: The International Consensus Handbook* (pp. 229–258). Columbus: The Ohio State University Nisonger Center.

Baer, D.M., Wolf, M.M., & Risley, T.R. (1968). Some current dimensions of applied behavior analysis. *Journal of Applied Behavior Analysis, 1,* 91–97.

Baumeister, A.A., Sevin, J.A., & King, B.H. (1998). Neuroleptic medications. In S. Reiss

& M.G. Aman (Eds.), *Psychotropic medication and developmental disabilities: The international consensus handbook.* Columbus: The Ohio State University Nisonger Center.

Bhaumik, S., Branford, D., McGrother, C., & Thorp, C. (1997). Autistic traits in adults with learning disabilities. *British Journal of Psychiatry, 170,* 502–506.

Blakeslee, T., Sugai, G., & Gruba, J. (1994). A review of functional assessment use in data-based intervention studies. *Journal of Behavioral Education, 4,* 397–413.

Bodfish, J.W., & Madison, J.T. (1993). Diagnosis and fluoxetine treatment of compulsive behavior disorder of adults with mental retardation. *American Journal on Mental Retardation, 98,* 360–367.

Boelter, E.W., & Hagopian, L.P. (2008). *Combining medication and behavioral procedures to decrease problem behavior maintained by negative reinforcement.* Poster presented at the Annual Convention of the Association for Behavior Analysis, Chicago, IL.

Borthwick-Duffy, S.A. (1994). Epidemiology and prevalence of psychopathology in people with mental retardation. *Journal of Consulting and Clinical Psychology, 62,* 17–27.

Buitelaar, J.K., van der Gaag, R.J., Cohen-Kettenis, P., & Melman, C.T.M. (2001). A randomized controlled trial of risperidone in the treatment of aggression in hospitalized adolescents with subaverage cognitive abilities. *Journal of Clinical Psychiatry, 62,* 239–248.

Campbell, M., Anderson, L.T., Small, A.M., Perry, R., Green, W.H., & Caplan, R. (1982). The effects of haloperidol on learning and behavior in autistic children. *Journal of Autism and Developmental Disorders, 12,* 167–175.

Cardinal, R.N., Robbins, T.W., & Everitt, B.J. (2000). The effects of d-amphetamine, chlordiazepoxide, α-flupenthixol and behavioural manipulations on choice of signaled and unsignaled delayed reinforcement in rats. *Psychopharmacology, 152,* 362–375.

Carlson, C.L., Pelham, W.E., Milich, R., & Dixon, J. (1992). Single and combined effects of methylphenidate and behavior therapy on the classroom performance of children with attention-deficit hyperactivity disorder. *Journal of Abnormal Child Psychology, 20,* 213–220.

Carr, E.G., & Durand, V. (1985). Reducing behavior problems through functional communication training. *Journal of Applied Behavior Analysis, 18,* 111–126.

Carr, E.G., & Durand, V. (1991). Functional communication training to reduce challenging behavior: Maintenance and application in new settings. *Journal of Applied Behavior Analysis, 24,* 251–264.

Carr, E.G., Robinson, S., Taylor, J.C., & Carlson, J.I. (1990). *Positive approaches to the treatment of severe behavior problems in persons with developmental disabilities: A review and analysis of reinforcement and stimulus-based procedures.* Seattle: The Association for Persons with Severe Handicaps.

Carta, M.G., Hardoy, M.C., Dessi, I., Hardoy, M.J., & Carpiniello, B. (2001). Adjunctive gabapentin in patients with intellectual disability and bipolar spectrum disorders. *Journal of Intellectual Disability Research, 45,* 139–145.

Chadwick, O., Kusel, Y., Cuddy, M., & Taylor, E. (2005). Psychiatric diagnoses and behavior problems from childhood to early adolescence in young people with severe intellectual disabilities. *Psychological Medicine, 35,* 751–760.

Chadwick, O., Walker, N., Bernard, S., & Taylor, E. (2000). Factors affecting the risk of behaviour problems in children with severe intellectual disability. *Journal of Intellectual Disability Research, 44,* 108–123.

Clarke, D.J., Kelley, S., Thinn, K., & Corbett, J.A. (1990). Psychotropic drugs and mental retardation: 1. Disabilities and the prescription of drugs for behaviour and for epilepsy in three residential settings. *Journal of Mental Deficiency Research, 34,* 385–95.

Cook, E.H., Rowlett, R., Jaseslos, C., & Leventhal, B.L. (1992). Fluoxetine treatment of children and adults with autistic disorder and mental retardation. *Journal of the American Academy of Child and Adolescent Psychiatry, 31,* 739–745.

Cook, E.H., Terry, E.J., Heller, W., & Leventhal, B.L. (1990). Fluoxetine treatment of borderline mentally retarded adults with obsessive-compulsive disorder. *Journal of Clinical Psychopharmacology, 10,* 228–229

Cook, L., & Weidley, E., (1957). Behavioral effects of some psychopharmacological agents. *Annals of the New York Academy of Sciences, 66,* 740–752.

Crosland, K.A., Zarcone, J.R., Lindauer, S.E., Valdovinos, M.G., Zarcone, T.J., Hellings, J.A., et al. (2003). Use of functional analysis methodology in the evaluation of medication effects. *Journal of Autism and Developmental Disorders, 33,* 271–279.

Cutmore, T.R.H., & Beniger, R.J. (1990). Do neuroleptics impair learning in schizophrenic patients? *Schizophrenia Research, 3,* 173–186.

Davidson, P.W., Cain, N.N., Sloane-Reeves, J.E., Van Speybroech, A., Segel, J., Gutkin, J., et al. (1994). Characteristics of community-based individuals with mental retardation and aggressive behavioral disorder. *American Journal on Mental Retardation, 98,* 704–716.

Davies, M. (1998). Treating severe self-injury in a community setting: Constraints on assessment and intervention. *Child Psychology & Psychiatry Review, 3,* 26.

Davies, T.S. (1961). A monoamine oxidase inhibitor (Niamid) in the treatment of the mentally subnormal. *Journal of Mental Science, 107,* 115–118.

Davis, E., Saeed, S.A., Antonacci, D.J. (2008). Anxiety disorders in persons with developmental disabilities: Empirically informed diagnosis and treatment. *Psychiatry Quarterly, 79,* 249–263.

Dekker, M.C., Koot, H.M., van der Ende, J., & Verhulst, F. (2002). Emotional and behavioral problems in children and adolescents with and without intellectual disability. *Journal of Child Psychology and Psychiatry and Allied Disciplines, 43,* 1087–1098.

DeLeon, I.G., Toole, L.M., Gutshall, K.A., & Bowman, L. (2005). Individualized sampling parameters for behavioral observations: Enhancing the predictive validity of competing stimulus assessments. *Research in Developmental Disabilities, 26,* 440–455.

DeMyer, M.K., Hingtgen, J.N., & Jackson, R.K. (1981). Infantile autism: A decade of research. *Schizophrenia Bulletin, 7,* 388–451.

Derby, K.M. (2000). Functional analysis of aberrant behavior through measurement of separate response topographies. *Journal of Applied Behavior Analysis, 33,* 113–117.

deWit, H., Engagasser, J.L., & Richards, J.B. (2002). Acute administration of d-amphetamine decreases impulsivity in healthy volunteers. *Neuropsychopharmacology, 27,* 813–825.

Dicesare, A., McAdam, D.B., Toner, A., & Varrell, J. (2005). The effects of methylphenidate on a functional analysis of disruptive behavior: A replication and extension. *Journal of Applied Behavior Analysis, 38,* 125–128.

Didden, R., Duker, P., & Korzilius, H. (1997). Meta-analytic study on treatment effectiveness of problem behavior with individuals who have mental retardation. *American Journal on Mental Retardation, 101,* 387–399.

Duggan, L., & Brylewski, J. (1999). Effectiveness of antipsychotic medication in people with intellectual disability and schizophrenia: A systematic review. *Journal of Intellectual Disability Research, 43,* 94–104.

Dykens, E.M. (2000). Annotation: Psychopathology in children with intellectual disability. *Journal of Child Psychology and Psychiatry, 41,* 407–417.

Einfeld, S.L., & Aman, M. (1995). Issues in the taxonomy of psychopathology in mental retardation. *Journal of Autism and Developmental Disorders, 25,* 143–167.

Emerson, E. (2003). Prevalence of psychiatric disorders in children and adolescents with and without intellectual disability. *Journal of Intellectual Disability Research, 47,* 51–58.

Emerson, E., Kiernan, C., Alborz, A., Reeves, D., Mason, H., Swarbrick, R., et al. (2001). The prevalence of challenging behaviors: A total population study. *Research in Developmental Disabilities, 22,* 77–93.

Emerson, E., Robertson, J., Gregory, N., Hatton, C., Kessissoglou, S., Hallam, A., et al. (2000). Treatment and management of challenging behaviours in residential settings. *Journal of Applied Research in Intellectual Disabilities, 13,* 197–215.

Engel, G.L. (1977). The need for a new medical model: A challenge for biomedicine. *Science, 196,* 129–136.

Field, C.J., Aman, M.G., White, A.J., & Vaithianathan, C. (1986). A single-subject study of imipramine in a mentally retarded woman with depressive symptoms. *Journal of Mental Deficiency Research, 30,* 191–198.

Fischman, M.W., & Schuster, C.R. (1979). The effects of chlorpromazine and pentobarbital on behavior maintained by electric shock or point loss avoidance in humans. *Psychopharmacology, 66,* 3–11.

Fischman, M.W., Smith, R.C., & Schuster, C.R. (1976). Effects of chlorpromazine on avoidance and escape responding in humans. *Pharmacology Biochemistry and Behavior, 4,* 111–114.

Fisher, W., Piazza, C.C., & Page, T.J. (1989). Assessing independent and interactive effects of behavioral and pharmacologic interventions for a client with dual diagnoses. *Journal of Behavior Therapy and Experimental Psychiatry, 20,* 241–250.

Food and Drug Administration. (2006, October). *FDA approves the first drug to treat irritability associated with autism, Risperdal.* Retrieved December 18, 2008, from http://www.fda.gov/bbs/topics/news/2006/new01485.html

Garcia, D., & Smith, R.G. (1999). Using analog baselines to assess the effects of naltrexone on self-injurious behavior. *Research in Developmental Disabilities, 20,* 1–20.

Gordon, C.T., Rapoport, J.L., Hamburger, S.D., State, R.C., & Mannheim, G.B. (1992). Differential response of seven subjects with autistic disorder to clomipramine and dispramine. *American Journal of Psychiatry, 149,* 363–366.

Göstason, R. (1985). Psychiatric illness among the mentally retarded: A Swedish population study. *Acta Psychiatrica Scandinavica, 318*(Suppl. 71), 1–117.

Hagopian, L.P., Bruzek, J.L., Bowman, L.G., & Jennett, H.K. (2007). Assessment and treatment of problem behavior occasioned by interruption of free-operant behavior. *Journal of Applied Behavior Analysis, 40,* 89–103.

Hagopian, L.P., Contrucci-Kuhn, S.A., Long, E.S., & Rush, K.S. (2005). Schedule thinning following communication training: Using competing stimuli to enhance tolerance to decrements in reinforcer density. *Journal of Applied Behavior Analysis, 38,* 177–193.

Hagopian, L.P., Crockett, J.L., van Stone, M., DeLeon, I.G., & Bowman, L. G. (2000). Effects of noncontingent reinforcement on problem behavior and stimulus engagement: The role of satiation, extinction, and alternative reinforcement. *Journal of Applied Behavior Analysis, 33,* 433–449.

Hagopian, L.P., Fisher, W.W., Sullivan, M.T., Acquisto, J., & LeBlanc, L.A. (1998). Effectiveness of functional communication training with and without extinction and punishment: A summary of 21 inpatient cases. *Journal of Applied Behavior Analysis, 31,* 211–235.

Hagopian, L.P., & Jennett, H.K. (2008). Behavioral assessment and treatment of anxiety on individuals with intellectual disabilities and autism. *Journal of Developmental and Physical Disabilities, 20,* 467–483.

Hamdan-Allen, G. (1991). Brief report: Trichotillomania in an autistic male. *Journal of Autism and Developmental Disorders, 21,* 79–82.

Handen, B.L., Feldman, H.M., Lurier, A., & Murray, P.J.H. (1999). Efficacy of methylphenidate among preschool children with developmental disabilities and ADHD. *Journal of the American Academy of Child and Adolescent Psychiatry, 38,* 805–812.

Hanley, G.P., Iwata, B.A., & McCord, B.E. (2003). Functional analysis of problem behavior: A review. *Journal of Applied Behavior Analysis, 36,* 147–185.

Hellings, J.A., Zarcone, J.R., Reese, R.M., Valdovinos, M.G., Marquis, J.G., Fleming, K.K., et al. (2006). A crossover study of risperidone in children, adolescents and adults with mental retardation. *Journal of Autism and Developmental Disorders, 36,* 401–411.

Hemmings, C.P., Gravestock, S., Pickard, M., & Bouras, N. (2006). Psychiatric symptoms and problem behaviours in people with intellectual disabilities. *Journal of Intellectual Disability Research, 50,* 269–276.

Herbert, J., Sharp, B., & Gaudiano, B. (2002). Separating fact from fiction in the etiology and treatment of autism. *The Scientific Review of Mental Health Practice, 1,* 23–43.

Hingtgen, J.N., & Bryson, C.Q. (1972). Recent developments in the study of early childhood psychoses: Infantile autism, childhood schizophrenia and related disorders. *Schizophrenia Bulletin, 8,* 8–54.

Holden, B., & Gitlesen, J.P. (2003). Prevalence of psychiatric symptoms in adults with mental retardation and challenging behaviour. *Research in Developmental Disabilities, 24,* 323–332.

Holden, B., & Gitlesen, J.P. (2006). A total population study of challenging behaviour in the county of Hedmark, Norway: Prevalence, and risk markers. *Research in Developmental Disabilities, 27,* 456–465.

Holden, B., & Gitlesen, J.P. (2008). The relationship between psychiatric symptomatology and motivation of challenging behaviour: A preliminary study. *Research in Developmental Disabilities, 29,* 408–413.

Holden, B., & Gitlesen, J.P. (2009). The overlap between psychiatric symptoms and challenging behaviour: A preliminary study. *Research in Developmental Disabilities, 30,* 210–218.

Howland, R.H. (1992). Fluoxetine treatment of depression in mentally retarded adults. *Journal of Nervous and Mental Disease, 180,* 202–205.

Iwata, B.A., Dorsey, M.F., Slifer, K.J., Bauman, K.E., & Richman, G.S. (1994). Toward a functional analysis of self-injury. *Journal of Applied Behavior Analysis, 27,* 197–209.

Jenkins, R., Rose, J., & Jones, T. (1998). The checklist of challenging behavior and its relationship with the psychopathology inventory for mentally retarded adults. *Journal of Intellectual Disability Research, 42,* 273–278.

Jennett, H.K., & Hagopian, L.P. (2008). Identifying empirically supported treatments for phobic avoidance in individuals with intellectual disabilities. *Behavior Therapy, 32,* 151–161.

Kahng, S.W., Iwata, B.A., & Lewin, A.B. (2002). Behavioral treatment of self-injury, 1964 to 2000. *American Journal on Mental Retardation, 107,* 212–221.

Kalachnik, J.E., Levethal, B.L., James, D.H., Sovner, R., Kastner, T.A., Walsh, K., et al. (1998). Guidelines for the use of psychotropic medication. In S. Reiss & M.G. Aman (Eds.), *Psychotropic medication and developmental disabilities: The International Consensus Handbook.* Columbus: The Ohio State University Nisonger Center.

Kennedy, C.H., Meyer, K.A., Knowles, T., & Shukla, S. (2000). Analyzing the multiple functions of stereotypical behavior for students with autism: Implications for assessment and treatment. *Journal of Applied Behavior Analysis, 33,* 559–571.

Kiernan, C., & Alborz, A. (1996). Persistence and change in challenging and problem behaviours of young adults with intellectual disability living in the family home. *Journal of Applied Research in Intellectual Disabilities, 9,* 181–193.

Kirman, B.H., & Bicknell, J. (1968). Congenital insensitivity to pain in an imbecile boy. *Developmental Medicine and Child Neurology, 10,* 57–63.

Kuhn, D., Hagopian, L., & Terlonge, C. (2008). Treatment of life-threatening self-injurious behavior secondary to hereditary sensory and autonomic neuropathy type II: A controlled case study. *Journal of Child Neurology, 23,* 381–388.

Kurtz, P.F., Chin, M.D., Huete, J.M., Tarbox, R.S.F., O'Connor, J.T., & Paclawskyj, T.R. (2003). Functional analysis and treatment of self-injurious behavior in young children: A summary of 30 cases. *Journal of Behavior Analysis, 36,* 205–219.

Lalli, J.S., Casey, S.D., & Kates, K. (1997). Noncontingent reinforcement as treatment for severe problem behavior: Some procedural variations. *Journal of Applied Behavior Analysis, 30,* 127–137.

Lieving, G.A., Hagopian, L.P., Long, E.S., & O'Connor, J.T. (2004). Response-class hierarchies and resurgence of severe problem behavior. *Psychological Record, 54,* 621–634.

Lowe, K., Allen, D., Jones, E., Brophy, S., Moore, K., & James, W. (2007). Challenging behaviours: Prevalence and topographies. *Journal of Intellectual Disability Research, 51,* 625–636.

Lundervold, D., & Bourland, G., (1988). Quantitative analysis of treatment of aggression, self-injury, and property destruction. *Behavior Modification, 12,* 590–617.

Mace, F.C., & Mauk, J.E. (1995). Bio-behavioral diagnosis and treatment of self-injury. *Mental Retardation and Developmental Disabilities Research Reviews, 1,* 104–110.

Magee, S.K., & Ellis, J. (2000). Extinction effects during the assessment of multiple problem behaviors. *Journal of Applied Behavior Analysis, 33,* 313–316.

Masi, G., Marcheschi, M., & Pfanner, P. (1997). Paroxetine in depressed adolescents with intellectual disability: An open label study. *Journal of Intellectual Disability Research, 41,* 268–272.

Matson, J.L., Bamburg, J.W., Mayville, E.A., Pinkston, J., Bielecki, J., Kuhn, D., et al. (2000). Psychopharmacology and mental retardation: A 10 year review (1990–1999). *Research in Developmental Disabilities, 21,* 263–296.

Matson, J.L., Benavidez, D.A., Compton, L.S., Paclawskyj, T., & Baglio, C. (1996). Behavioral treatment of autistic persons: A review of research from 1980 to the present. *Research in Developmental Disabilities, 17,* 433–465.

Matson, J.L., Rush, K.S., Hamilton, M., Anderson, S.J., Bamburg, J.W., & Baglio, C.S. (1999). Characteristics of depression as assessed by the diagnostic assessment for the severely handicapped-II (DASH-II). *Research in Developmental Disabilities, 20,* 305–313.

McClintock, K., Hall, S., & Oliver, C. (2003). Risk markers associated with challenging behaviours in people with intellectual disabilities: A meta-analytic study. *Journal of Intellectual Disability Research, 47,* 405–416.

McDougle, C.J., Holmes, J.P., Carlson, D.C., Pelton, G.H., Cohen, D.J., & Price, L.H. (1998). A double-blind, placebo-controlled study of risperidone in adults with autistic disorder and other pervasive developmental disorders. *Archives of General Psychiatry, 55,* 633–641.

McDougle, C.J., Kresch, L.E., & Posey, D.J. (2000). Repetitive thoughts and behavior in pervasive developmental disorders: Treatment with serotonin reuptake inhibitors. *Journal of Autism and Developmental Disorders, 30,* 427–435.

McDougle, C.J., Price, L.H., & Goodman, W.K. (1990). Fluvoxamine treatment of coincident autistic disorder and obsessive compulsive disorder: A case report. *Journal of Autism and Developmental Disorders, 20,* 537–543.

McDougle, C.J., Scahill, L., Aman, M.G., McCracken, J.T., Tierney, E., & Davies, M. (2005). Risperidone for the core symptom domain of autism: Results from the study by the autism network of the research units on pediatric psychopharmacology. *American Journal on Psychiatry, 162,* 1142–1148.

Mehlinger, R., Scheftner, W., & Poznanski, E. (1990). Fluoxetine and autism. *Journal of the American Academy of Child and Adolescent Psychiatry, 29,* 985.

Moss, S., Emerson, E., Kiernan, C., Turner, S., Hatton, C., & Alborz, A. (2000). Psychiatric symptoms in adults with learning disability and challenging behaviour. *British Journal of Psychiatry, 177,* 452–456.

Mulick, J.A., Schroeder, S.R., & Rojahn, J. (1980). Chronic ruminative vomiting: A comparison of four treatment procedures. *Journal of Autism and Developmental Disorders, 10,* 203–213.

Murphey, G., Macdonald, S., Hall, S., & Oliver, C. (2000). Aggression and the termination of "rituals": A new variant of the escape function for challenging behavior? *Research in Developmental Disabilities, 21,* 43–59.

National Institutes of Health. (1989, September). *Treatment of destructive behaviors in persons with developmental disabilities.* NIH Consensus Statement online. Retrieved December 15,

2008, from ftp://nlmpubs.nlm.nih.gov/hstat/nihcdcs/75destr.txt

Naylor, G.J., Donald, J.M., LePoidevin, D., & Reid, A.H. (1974). A double-blind trial of long-term lithium therapy in mental defectives. *British Journal of Psychiatry, 124,* 52–57.

Neef, N.A., Bicard, D.F., Endo, S., Coury, D.L., & Aman, M.G. (2005). Evaluation of pharmacological treatment of impulsivity in children with attention deficit hyperactivity disorder. *Journal of Applied Behavior Analysis, 38,* 135–146.

Northup, J., Fusilier, I., Swanson, V., Roane, H., & Borrero, J. (1997). An evaluation of methylphenidate as a potential establishing operation for some common classroom reinforcers. *Journal of Applied Behavior Analysis, 30,* 615–625.

Nøttestad, J.A., & Linaker, O.M. (1999). Psychiatric health needs and services before and after complete deinstitutionalization of people with intellectual disability. *Journal of Intellectual Disability Research, 43,* 523–530.

Nyhan, W.L. (1998). Lessons from Lesch-Nyhan syndrome. In S.R. Schroeder, M.L. Oster-Granite, & T.T. Thomson (Eds.), *Self-injurious behavior: Gene-brain-behavior relationships* (pp. 251–267). Washington, DC: American Psychological Association.

Paclawskyj, T.R., Matson, J.L., Bamburg, J.W., & Baglio, C.S. (1997). A comparison of the diagnostic assessment for the severely handicapped-II (DASH-II) and the Aberrant Behavior Checklist (ABC). *Research in Developmental Disabilities, 18,* 289–298.

Pan, W., Schmidt, R., Wickens, J.R., & Hyland, B.I. (2008). Tripartite mechanism of extinction suggested by dopamine neuron activity and temporal difference model. *The Journal of Neuroscience, 28,* 9619–9631.

Pary, R.J. (1989). Pretreatment systolic orthostatic blood pressure depression in Down's syndrome. *Journal of Clinical Psychopharmacology, 9,* 146–147.

Pelios, L., Morren, J., Tesch, D., & Axelrod, S. (1999). The impact of functional analysis methodology on treatment choice for self-injurious and aggressive behavior. *Journal of Applied Behavior Analysis, 32,* 185–195.

Piazza, C.C., Fisher, W.W., Hanley, G.P., LeBlanc, L.A., Worsdell, A.S., Lindauer, S.E., et al. (1998). Treatment of pica through multiple analyses of its reinforcing functions. *Journal of Applied Behavior Analysis, 31,* 165–189.

Pietras, C.J., Cherek, D.R., Lane, S.D., Tcheremissine, O.V., & Steinberg, J.L. (2003). Effects of methylphenidate on impulsive choice in adult humans. *Psychopharmacology, 170,* 390–398.

Plauche-Johnson, C., Myers, S.M., & American Academy of Pediatrics Council on Children with Disabilities. (2007). Identification and evaluation of children with autism spectrum disorders. *Pediatrics, 120,* 1183–1215.

Pyles, D.A.M., Muniz, K., Cade, A., & Silva, R. (1997). A behavioral diagnostic paradigm for integrating behavior-analytic and psychopharmacological interventions for people with a dual diagnosis. *Research in Developmental Disabilities, 18,* 185–214.

Qureshi, H., & Alborz, A. (1992). Epidemiology of challenging behaviour. *Mental Handicap Research, 5,* 130–145.

Reiss, S. (1994). *Handbook of challenging behavior: Mental health aspects of mental retardation.* Worthington, OH: IDS Publishing.

Reiss, S., & Aman, M.G. (1998). Preface. In S. Reiss & M.G. Aman (Eds.), *Psychotropic medication and developmental disabilities: The International Consensus Handbook.* Columbus: The Ohio State University Nisonger Center.

Research Units on Pediatric Psychopharmacology Autism Network. (2002). Risperidone in children with autism and serious behavioral problems. *The New England Journal of Medicine, 347,* 314–320.

Richman, D.M., Wacker, D.P., Asmus, J.M., Casey, S.D., & Andelman, M. (1999). Further analysis of problem behavior in response class hierarchies. *Journal of Applied Behavior Analysis, 32,* 269–283.

Roach, E.S., Abramson, J.S., & Lawless, M.R. (1985). Self-injurious behavior in acquired sensory neuropathy. *Neuropediatrics, 16,* 159–161.

Robbins, T.W., & Everitt, B.J. (1996). Neurobehavioural mechanisms of reward and motivation. *Current Opinion in Neurobiology, 6,* 228–236.

Robertson, J., Emerson, E., Gregory, N., Hatton, C., Kessissoglou, S., & Hallam, A. (2000). Receipt of psychotropic medication by people with intellectual disability in residential settings. *Journal of Intellectual Disability Research, 44,* 666–676.

Rojahn, J., Borthwick-Duffy, S.A., & Jacobson, J.W. (1993). The association between psychiatric diagnoses and severe behavior problems in mental retardation. *Annals of Clinical Psychiatry, 5,* 163–170.

Rojahn, J., Matson, J.L., Naglieri, J.A., & Mayville, E. (2004). Relationships between psychiatric conditions and behavior problems among adults with mental retardation. *American Journal on Mental Retardation, 109,* 21–33.

Ruedrich, S., Swales, T.P., Fossaceca, C., Toliver, J., & Rutkowski, A. (2001). Effect of divalproex sodium on aggression and self-injurious behaviour in adults with intellectual disability: A retrospective review. *Journal of Intellectual Disability Research, 43,* 105–111.

Ruedrich, S., & Wilkinson, L. (1992). Atypical unipolar depression in mentally retarded patients: Amoxapine treatment. *Journal of Nervous and Mental Disease, 180,* 206–207.

Rush, A.J., & Frances, A. (2000). Expert consensus guideline series: Treatment of psychiatric and behavioral problems in mental retardation. *American Journal on Mental Retardation, 105*(3), 161–228.

Sandman, C.A., Hetrick, W., Taylor, D.V., & Chicz-DeMet (1997). Dissociation of POMC peptides after self-injury predicts responses to centrally acting opiate blockers. *American Journal on Mental Retardation, 102,* 182–199.

Sandman, C.A., Hetrick, W., Taylor, D., Marion, S., & Chicz-DeMet, A. (2000). Uncoupling of proopiomelancortin (POMC) fragments is related to self-injury. *Peptides, 21,* 785–791.

Schaal, D.W., & Hackenberg, T. (1994). Toward a functional analysis of drug treatment for behavior problems of people with developmental disabilities. *American Journal on Mental Retardation, 99,* 123–140.

Schlund, M.W., Rosales-Ruiz, J., Vaidya, M., Glenn, S.S., & Staff, D. (2008). Experience-dependent plasticity: Differential changes in activation associated with repeated reinforcement. *Neuroscience, 155,* 17–23.

Schroeder, S.R., Lewis, M.H., & Lipton, M.A. (1983). Interactions of pharmacotherapy and behavior therapy among children with learning and behavioral disorders. *Advances in Learning and Behavioral Disabilities, 2,* 179–229.

Schultz, W. (1999). The reward signal of midbrain dopamine neurons. *News in Physiological Science, 14,* 249–255.

Shea, S., Turgay A., Carroll, A., Schulz, M., Orlik, H., Smith, I., et al. (2004). Risperidone in the treatment of disruptive behavioral symptoms in children with autistic and pervasive developmental disorders. *Pedatrics, 114,* 634–641.

Snyder, R., Turgay, A., Aman, M.G., Binder, C., Fisman, S., & Carroll, A. (2002). Effects of risperidone on conduct and disruptive behavior disorders in children with subaverage IQs. *Journal of the American Child and Adolescent Psychiatry, 41,* 1026–1036.

Sovner, R., Fox, C.J., Lowry, M.J., & Lowry, M.A. (1993). Fluoxetine treatment of depression and associated self-injury in two adults with mental retardation. *Journal of Intellectual Disability Research, 37,* 301–311.

Sprague, R.L., & Werry, J.S. (1971). Methodology of psychopharmacological studies with the retarded. In N.R. Ellis (Ed.), *International Review of Research in Mental Retardation,* (Vol. 5, pp. 147–219). New York: Academic Press.

Sturmey, P. (1995). Diagnostic-based pharmacological treatment of behavior disorders in persons with developmental disabilities: A review and a decision-making typology. *Research in Developmental Disabilities, 16,* 235–252.

Swanson, V. (2000). A comprehensive functional assessment of the effects of methylphenidate on the disruptive behavior of children with severe mental retardation (Doctoral Dissertation, The Louisiana State University, 2000). *Dissertation Abstracts International, 60*(9-B), 4871.

Symons, F.J., & Thompson, T.T. (1998). Functional communication training and naltrexone treatment of self-injurious behaviour: An experimental case report. *Journal of Applied Research in Intellectual Disabilities, 11,* 273–292.

Szymanski, L.S., King, B., Goldberg, B., Reid, A., Tonge, B., & Cain, N. (1998). Diagnosis of mental disorders in people with mental retardation. In S. Reiss & M.G. Aman (Eds.), *Psychotropic medication and developmental disabilities: The International Consensus Handbook.* Columbus: The Ohio State University Nisonger Center.

Thalayasingam, S., Alexander, R.T., & Singh, I. (2004). The use of clozapine in adults with intellectual disability. *Journal of Intellectual Disability Research, 48,* 572–579.

Thompson, T., Egli, M., Symons, F., & Delaney, D. (1994). Neurobehavioral mechanisms of drug action in developmental disabilities. In T. Thompson & D.B. Gray (Eds.), *Destructive behavior in developmental disabilities: Diagnosis and treatment* (pp. 133–188). Thousand Oaks, CA: Sage.

Thompson, T., Moore, T., & Symons, F. (2007). Psychotherapeutic medications for positive behavior support. In S. Odom, R. Horner, M. Snell, & J. Blacker (Eds.), *Handbook on developmental disabilities.* New York: Guilford Publications.

Todd, R.D. (1991). Fluoxetine in autism. *American Journal of Psychiatry, 148,* 1089.

Tonkonogy, J.M. (1991). Violence and temporal lesion, head CT and MRI data. *Journal of Neuropsychiatry and Clinical Neurosciences, 3,* 189–196.

Totsika, V., Toogood, S., Hastings, R.P., & Lewis, S. (2008). Persistence of challenging behaviours in adults with intellectual disability over a period of 11 years. *Journal of Intellectual Disability Research, 52,* 446–457.

U.S. Department of Health and Human Services. (1999). Children and mental health. In *Mental health: A report of the Surgeon General* (pp. 124–194). Rockville, MD: U.S. Department of Health and Human Services, Substance Abuse and Mental Health Services Administration, Center for Mental Health Services, National Institutes of Health, National Institutes of Mental Health.

Valdovinos, M., Caruso, M., Roberts, C., Kim, G., & Kennedy, C.H. (2005). Medical and behavioral symptoms as potential medication side effects in adults with developmental disabilities. *American Journal on Mental Retardation, 110,* 164–170.

Valdovinos, M.G., Ellinger, N.P., & Alexander, M.L. (2007). Changes in the rate of problem behavior associated with the discontinuation of the antipsychotic medication quetiapine. *Mental Health Aspects of Developmental Disabilities, 10,* 64–67.

Valdovinos, M.G., Napolitano, D.A., Zarcone, J.R., Hellings, J.A., Williams, D.C., & Schroeder, S.R. (2002). Multimodal evaluation of risperidone for destructive behavior: Functional analysis, direct observations, rating scales, and psychiatric impressions. *Experimental and Clinical Psychopharmacology, 10,* 268–275.

Van Bellinghen, M., & De Troch, C. (2001). Risperidone in the treatment of behavioral disturbances in children and adolescents with borderline intellectual functioning: A double-blind, placebo-controlled pilot trial. *Journal of Child and Adolescent Psychopharmacology, 11,* 5–13.

Vanden Borre, R., Vermote, R., Buttiens, M., Thiry, P., Dierick, G., Geutjens, J., et al. (1993). Risperidone as add-on therapy in behavioural disturbances in mental retardation: A double-blind placebo-controlled cross-over study. *Acta Psychiatrica Scandinavica, 87,* 167–171.

Vollmer, T.R., Iwata, B.A., Zarcone, J.R., Smith, R.G., & Mazaleski, J.L. (1993). The role of attention in the treatment of attention-maintained self-injurious behavior: Noncontingent reinforcement (NCR) and differential reinforcement of other behavior (DRO). *Journal of Applied Behavior Analysis, 26,* 9–26.

Wachtel, L.E., & Hagopian, L.P. (2006). Psychopharmacology and applied behavior analysis: Tandem treatment of severe problem behavior in intellectual disability and a case series. *The Israel Journal of Psychiatry and Related Science, 43,* 265–274.

Wacker, D.P., Berg, W.K., Harding, J.W., Barretto, A., Rankin, B., & Gazner, J. (2005). Treatment effectiveness, stimulus generalization, and acceptability to parents of functional communication training. *Educational Psychology, 25,* 233–256.

Weisz, J.R., Weiss, B., Han, S.S., Granger, D.A., & Morton, T. (1995). Effects of psychotherapy with children and adolescents revisited: A meta-analysis of treatment outcome studies. *Psychological Bulletin, 117,* 450–468.

Wilkinson, P.C., Kircher, J.C., McMahon, W.M., & Sloane, N.H. (1995). Effects of methylphenidate on reward strength in boys with attention-deficit hyperactivity disorder. *Journal of the American Academy of Child Adolescent Psychiatry, 34,* 897–901.

Witkin, J.M., & Katz, J.L. (1990). Analysis of behavioral effects of drugs. *Drug Development Research, 20,* 389–409.

Witwer, A.N., & Lecavalier, L. (2008). Psychopathology in children with intellectual disability: Risk markers and correlates. *Journal of Mental Health Research in Intellectual Disabilities, 1,* 75–96.

World Health Organization. (1997). *International statistical classification of diseases and related health problems* (10th ed., revised) (ICD-10). Geneva: World Health Organization.

Worrall, E.P., Moody, J.P., & Naylor, G.J. (1975). Lithium in non-manic depressives: Antiaggressive effect and red blood cell lithium values. *British Journal of Psychiatry, 126,* 464–468.

Wurtele, S.K., King, A.C., & Drabman, R.S. (1984). Treatment package to reduce SIB in a Lesch-Nyhan patient. *Journal of Mental Deficiency, 28,* 227–234.

Zarcone, J.R., Hellings, J.A., Crandall, K., Reese, R.M., Marquis, J., Fleming, K., et al. (2001). Effects of risperidone on aberrant behavior of persons with developmental disabilities: I. A double-blind crossover study using multiple measures. *American Journal on Mental Retardation, 106,* 525–538.

Zarcone, J.R., Lindauer, S.E., Morse, P.S., Crosland, K.A., Valdovinos, M.G., McKerchar, T.L., et al. (2004). Effects of risperidone on destructive behavior of persons with developmental disabilities: III. Functional analysis. *American Journal on Mental Retardation, 4,* 310–321.

Future Implications

14

New Genetic Techniques

Implications for Neurobehavioral Syndromes

Lisa T. Emrick

It is a known fact that both physical and mental characteristics are passed down through families. The National Human Genome Research Institute describes a timeline for the Human Genome Project that starts in 1859. In 1859 Charles Darwin published his pinnacle work, *The Origin of Species.* In 1865, Gregor Mendel described how physical qualities of a species are inherited within units of material. In 1909, a Danish botanist named Wilhelm Johannsen named those units of material genes after the Greek word *genos,* meaning "birth." In 1953 Francis Crick and James Watson described the backbone of genetic information and what we now know as DNA, deoxyribonucleic acid. There was a flurry of discoveries in the 1990s with the advent of new technologies and computer-based programs called bioinformatics to better understand the human genome. In 2001, the International Human Genome Sequencing Consortium (International Human Genome Sequencing Consortium, 2001, 2004) and a separate group headed by Craig Venter (Venter et al., 2001) collaborated to publish the human genome. Since the publication of the human genome in 2001 researchers have continued to make discoveries to further our understanding of the human genome and its role in human behavior and disease.

Neurobehavioral syndromes are complex disorders. Most types of behavior are generated from the interaction between a complex group of genes and the environment. As researchers learn more about genetics, they recognize that these interactions are complex (Peltonen & McKusick, 2001). This chapter presents basic concepts in genetics and molecular biology, as well as an overview of genetic techniques that are involved in characterizing the complex etiologies of neurobehavioral syndromes. The goal is for the reader to achieve a better understanding of the various genetic technologies available and of the strengths and weaknesses of each technique. Moreover, by understanding the technologies the reader will have a better understanding of the clinical information provided by these techniques thus far.

GENETICS PRIMER

To understand what goes wrong with the human genome to cause genetic disorders one needs to learn the basics of genetics and how the genome works typically. In 1865, when Mendel described the concept of inheritable traits, now referred to as genes, he described the concepts of dominant and recessive alleles. The word *allele* is also derived from a Greek word, *allelos,* which means each "other." An allele is one set of a DNA sequence that usually will produce a gene. Offspring normally receive one allele for a DNA sequence from each parent for most of their genome. Mendel's work with pea pods demonstrated dominance for certain alleles. In other words, if a dominant allele is present it will be fully expressed in the phenotype (observable trait) of the offspring no matter what the other allele is. Recessive alleles are expressed only if both of the alleles are recessive. If a person has two of the same alleles, he or she is said to be homozygous for that allele. If they have two different alleles, then they are said to be heterozygous. Mendel also described the concept of incomplete dominance when he saw pink pea-pod flowers when he mixed red and white flowers. The pink flowers, he theorized, were generated by the heterozygous offspring. Incomplete dominance does not occur for all genes.

What are genes made of? Genes are made up of DNA sequences that code for a protein product. The DNA sequence makes up the genotype for the protein product. DNA is a chain of smaller pieces called nucleotide bases. There are four nucleotide bases, adenosine (A), cytosine (C), guanine (G), and thymine (T), that make up our genetic code (i.e., genome). James Watson and Francis Crick described the DNA chain's preferred configuration with two separate strands wound together to form a double helix. The nucleotide bases A, C, T, and G bind to each other in a specific manner (A-T and G-C) to form a stable structure. Researchers are able to exploit this preferred structure by creating single-stranded pieces of DNA that will bind to other pieces of DNA in certain conditions for stability. This is a simplified explanation behind the genetic technique called hybridization. Hybridization is used in various molecular genetic techniques such as fluorescent in situ hybridization and comparative genomic hybridization explained later in the chapter. DNA is involved in two major processes: making functional protein products and replication. As stated previously, gene products that make up functional proteins have their origins in a DNA sequence.

Figure 14.1 illustrates how a piece of DNA is turned into RNA (ribonucleic acid), an intermediate code, through a process called transcription. RNA is then translated into amino acids and then posttranslational processing to yield a final protein product. Errors can arise during transcription, translation, or the final step of posttranslational processing to alter the function of the protein product. A wrong DNA or RNA nucleotide can be incorporated into a sequence, causing a mutation during transcription or replication that then may alter the overall function of a protein product.

Not all DNA becomes a gene. The genome is made up of 2.85 billion nucleotide bases and approximately 20,000–25,000 protein products. The coding region is only approximately 1.2% of the total genome. More is known about the coding regions (exons) than the noncoding regions (introns and intragenic DNA). The exons are the primary functional pieces of DNA, and the introns and intragenic DNA sequences appear to play a more structural role; however, they may also play an important yet currently poorly defined functional role. Figure 14.1 demonstrates how promoter regions can

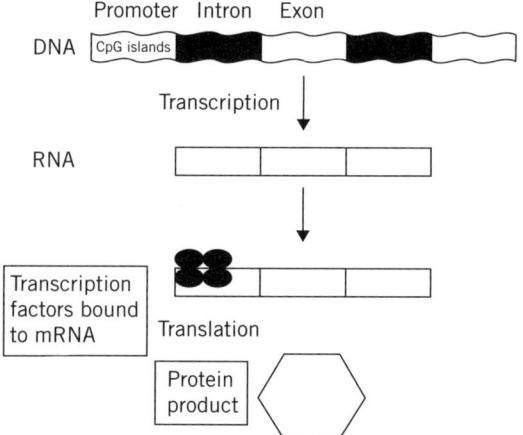

Figure 14.1. A simplified version of the central dogma theory for production of a protein product from a DNA molecule. A DNA sequence is made up of two DNA strands that form coding regions (exons): regulatory regions such as the promoter region and noncoding regions (introns). A DNA strand undergoes transcription, whereby its code becomes a single-stranded RNA code (mRNA). The mRNA is then translated into a protein product. Translation is partially regulated by transcriptional factors that bind to the RNA sequence and turn on or off the further processing to a protein product.

occur upstream prior to a sequence for a gene. Promoters act like an on/off switch for the expression of the gene. Modifications of the promoter region, such as the addition of a chemical group (e.g., a methyl group), can change the promoter switch, usually to "off." Methylation of a specific promoter region called CpG islands (because of the high level of the nucleotides cytosine and guanine) can decrease the level of expression for its protein product. Therefore, phenotypic changes can result from a change in the genotype, such as in a change in the DNA sequence, or the genotype can remain the same but a change occurs because of a difference in postprocessing of the DNA sequence, such as the methylation of the promoter region for a gene. This basic concept is discussed in further detail later in the chapter.

Most of a cell's DNA is contained within the cell's nucleus. The DNA organizes itself into single pieces of DNA wrapped around protein molecules called histones to form a more stable structure called chromatin. The chromatin structures are wound even further into organized three-dimensional structures called chromosomes during one phase of replication. There are two types of DNA replication, meiosis and mitosis. The goal of mitosis is to generate two exact replicas of the original cell. Therefore, the cellular DNA must first double, then line up in an orderly fashion, forming chromosomes. The chromosomes then divide into two equal parts to become part of their respective cells. Meiosis, as stated previously, is the replication of chromosomes to form the gametes (eggs in females and sperm in males). It undergoes steps similar to those of mitosis, but has additional steps as well. The chromosomes line up as in mitosis; however, in meiosis the two sets of chromosomes undergo crossover events to form genetically new cells. The crossover events are referred to as recombination events. Recombination is a key concept in the world of genetic variability. It is why full siblings may look and act alike but they are not exactly alike, even though they have the same mother and father. The process of meiosis also undergoes an additional separation of cells to form haploid cells (23 chromosomes = N). An egg with one set of chromosomes and a sperm cell with one set of chromosomes come together to form an embryo with two sets of chromosomes (two sets of 23, therefore $46 = 2N$).

Numerical errors occur when a set of chromosomes fails to separate during either meiosis or mitosis, which is referred to as chromosomal nondisjunction. Therefore, instead of producing two cells with either one or two chromosomes depending on whether the error occurs in meiosis (N) or mitosis ($2N$), the offspring of the replication will be one cell with an extra chromosome and another cell with a missing chromosome.

An error that occurs in the DNA sequence during meiosis will occur in all of the cell lines of an embryo that forms from a mutant egg or a mutant sperm cell. An error that occurs after fertilization during mitosis will only be a part of that one cell and future generations of that cell line. Genetic errors that occur during mitosis explain the clinical phenomenon of mosaicism.

Errors in the genome can come from having too little DNA, as in deletions, or too much DNA in the case of duplications or rearrangement of DNA, as in translocations or inversions of the DNA sequence. The changes to the genetic sequence can be as small as one nucleotide or as large as a whole chromosome. Any of these changes in the genome may or may not result in different levels of expression of genes and result in varied phenotypes.

As stated previously, the human genome is set up for the chromosomes to have two alleles, one inherited from each parent. This allows for balanced rearrangements; for example, in most cases a deletion in one allele can be balanced by a normal allele in unaffected individuals. However, in the case of the sex chromosomes, males have only one X chromosome and are therefore more susceptible to unbalanced arrangements on the X chromosome.

Along with the discovery of the human genome has come the discovery of the proteome, the proteins that are the expression of the genome; proteomics is the study of their identification and characterization. A discussion of the proteome is beyond the scope of this chapter; however, it is important to recognize the importance of understanding the various types of protein products that can affect phenotypic variability. Just as DNA has promoter regions that act as on/off switches, transcription factors, one type of protein product, can also act as on/off switches in the cell. Proteins are also involved in complex pathways within the cell, and therefore different mutations affecting various proteins will have different impacts, depending on what that protein product is and what role it plays within a cell.

This concludes the primer on basic genetic concepts; the next sections describe in more detail the theories behind errors in the genome, how they may lead to phenotypic changes, and the genetic techniques used to detect them.

CHROMOSOMES

The existence of chromosomes has been known since at least the mid-1800s. They have a short arm, the p arm, and a long arm, the q arm, connected by the centromere. The centromere acts as a docking station for the cell to attach to and separates chromosomes during meiosis or mitosis. The ends of the chromosomes are called telomeres. The telomeres and the area adjacent to them, referred to as subtelomeric regions, are rich in genes compared with the rest of the chromosome (Shaw-Smith et al., 2004).

There are 22 pairs of chromosomes (referred to as autosomal chromosomes) and two sex-linked chromosomes (X, Y). A karyotype is the end product of conventional cytogenetic analysis. It is a display of the chromosomes arranged in numerical order based roughly on size and is a part of the initial workup for developmental delay. For example, in 1959 researchers detected three copies of chromosome 21 (trisomy 21; Down syndrome) by cytogenetic analysis (Bejjani & Shaffer, 2008). This example shows cytogenetics to be a good technique for detecting the error in chromosomal number called aneuploidy. In the majority of cases these errors occur because of nondisjunction of chromosome 21 during meiosis.

In the 1970s researchers developed a staining technique that causes light and dark banding patterns that further characterize chromosomes. Initially, cytogenetic analysis provided a resolution between 400 and 500 bands. A region (locus) in a chromosome is designated by its number in the karyotype, and then arm, region, and band location. High-resolution techniques that increase the resolution to 650 to 1000 bands were created shortly after the development of traditional banding techniques. Therefore with the addition of these banding techniques, a karyotype is not only able to detect chromosomal numerical abnormalities such as trisomy, but can also detect chromosomal structural abnormalities. Structural abnormalities include deletions, duplications, inversions, ring formations, translocations, insertions, mosaicism, uniparental disomy, isochromosomes, and complex rearrangements involving multiple structural abnormalities. See Table 14.1 for definitions of these genetic rearrangements detected by routine cytogenetic analysis.

The high-resolution banding technique can detect changes between 2 million to 5 million bases (Mb) of DNA. The number of genes on any given chromosome is variable. There can be as few as 10 genes within a 3 Mb sequence of DNA or as many as 100 genes. One group of clinical disorders caused by a variable deleted region of chromosome 22 is now designated as 22q11 deletion syndromes. DiGeorge syndrome is a clinical syndrome first described in the 1950s

Table 14.1. Overview of chromosomal abnormalities

Structural abnormality	Definition
Deletion	A loss of DNA sequence with variable sizes within the chromosomes. Deletions may occur in the setting of ring formation (see below); inversions or unbalanced translocations with an overall loss of genetic material.
Duplication	A gain of DNA sequence with variable sizes within the chromosomes. Duplication may occur in the setting of an unbalanced translocation, insertion, or an isochromosomal formation of an additional DNA sequence.
Insertion	A break of DNA sequences in two separate chromosomes with linkage of at least one set of the DNA sequences into the other chromosome. This can be balanced in one generation but lead to an unbalanced duplication in the next generation.
Inversion	A flip in the DNA sequence order in a chromosome of variable sizes.
Isochromosomes	A chromosome produced by horizontal splitting at the centromere; therefore, both arms are from the same side of the centromere, are of equal length, and possess identical genes.
Mosaicism	A chromosomal abnormality that occurred after meiosis and therefore affects only certain cell lines.
Ring formation	Chromosomal breaks can occur then link together to form a ring versus a linear formation. Clinical examples are chromosomal ring 20, 13, 14, 15 and the X chromosome.
Translocation	A break of a DNA sequence usually from the telomere of one chromosome and then linked to the end of another end of a different chromosome.
Uniparental disomy	A chromosomal abnormality where the chromosomal number is correct but the material is inherited from only one parent.

Source: Korf (2001).

and 1960s. The clinical phenotypes include cardiac, endocrine, immunologic, cognitive, behavioral, and facial abnormalities. In 1978 Robert Shprintzen described a constellation of symptoms, including abnormal palates and unique facial features, cardiac abnormalities, hypernasal speech, and developmental delay, including cognitive and specific learning disabilities. This syndrome was initially called Shprintzen syndrome but was also known as velocardiofacial syndrome. In the 1990s researchers discovered that the majority of cases of DiGeorge syndrome, velocardiofacial syndrome, Shprintzen syndrome, and other, phenotypically similar syndromes share an approximately 2- to 3-Mb deletion on chromosome 22q and are now grouped under 22q11 deletion syndrome. There are approximately 30 to 40 genes within the deleted region. Researchers have not yet characterized all of the genes within this area. However, it is clear from this example that more than just the size of deletion affects the

clinical phenotype, as there is considerable phenotypic variation among children with the same-size deletion (Motzkin, Marion, Goldberg, Shprintzen, & Saenger, 1993).

A translocation occurs when a portion of DNA from one chromosome may be located on a different chromosome. Translocations may be balanced or unbalanced, depending on whether the DNA is also on the original chromosome, implying an increase in the total amount of DNA. It is unclear what causes translocations to occur. Approximately 3%–5% of Down syndrome cases are caused by translocations of a part of chromosome 21, either to itself or to another chromosome. It is important to determine whether the syndrome is caused by a nondisjunction of chromosome 21 in one egg or whether there is a balanced translocation on either a maternal or paternal chromosome that is present in 50% of their eggs or sperm and therefore is more likely to affect future children.

Fluorescent in situ hybridization (FISH) is a form of molecular cytogenetics used to detect or confirm chromosomal deletions, duplications, or translocations. FISH involves the hybridization of specific fluorescently labeled DNA probes to patients' DNA with a suspected congenital disorder. The concept of hybridization, as described earlier, is based on the fundamental knowledge of DNA structure and base pairing discovered by Watson and Crick. The probe is a piece of DNA with a known DNA sequence that will pair with another sequence on the chromosome if present. The detection of a chromosomal abnormality is revealed by a color change on the chromosome, either in the known location, as in a deletion of a piece of DNA, or in a different location, as in a translocation, compared with a control set of chromosomes. FISH can detect submicroscopic changes of less than 3 million to 5 million base pairs (limit of high-resolution banding karyotype). It is a powerful method for the detection and diagnosis of a specific disorder when used in the context of certain clinical phenotypes, such as Williams-Beuren syndrome. Williams-Beuren syndrome is characterized by intellectual disability and expressive language ability that appears to be better than intellectual ability. One needs to know the DNA sequences or chromosomal regions of interest to design specific probes designed to hybridize to those regions. FISH is not a good general screening tool because the conventional probes are for specific disorders.

There are FISH probes specific for subtelomeric regions of chromosomes that are used as a general screening tool for people with intellectual disabilities. As stated earlier in the chapter, subtelomeric regions of the chromosomes are gene-rich areas located at the ends of chromosomes. Studies have shown that subtelomeric screening with FISH may be positive in approximately 5% of cases of patients with intellectual disabilities of unknown etiology (Inoue & Lupski, 2002). Subtelomeric screening may not be as useful

now, in the era of complete genomic hybridization arrays (discussed later in this chapter). FISH analysis used to be limited to detecting 2 to 3 Mb. New methods can detect changes ranging from 35 to 100 kb.

Aneuploidy is defined as an abnormal chromosomal number. Trisomy 21 is a common chromosomal numerical abnormality, but there are many more types. Aneuploidy can occur at multiple stages of chromosomal replication, meiosis, or mitosis. Nondisjunction of chromosomes was discussed in the previous section. Many of these abnormalities are lethal and result in spontaneous abortion. The cells may try to salvage themselves by losing one of the extra chromosomes early in development. This is called trisomy rescue and is one of the mechanisms of uniparental disomy. Uniparental disomy refers to the situation where one has the correct number of chromosomes ($2N$), but they are inherited from the same parent. Uniparental disomy may be clinically benign and therefore more prevalent than is currently understood. However, it is not always benign.

Prader-Willi and Angelman syndromes illustrate the concept of genetic imprinting. DNA can be different if it is inherited from the mother or the father. The genome has specific parental genomic modifications (genomic imprinting; Li, Beard, & Jaenisch, 1993). A deletion in the region of chromosome 15q11-15q13 results in two very different phenotypes, depending on whether the missing DNA is from the mother or the father. Missing DNA from the father results in Prader-Willi syndrome, which is characterized by hypotonia, poor feeding as an infant that progresses to hyperphagia in childhood, intellectual disability, short stature, and hypogonadism. If the missing material is from the mother it results in Angelman syndrome, which is characterized by significant language impairment, intellectual disability, movement disorders, infantile hypotonia, and later spasticity and seizures. There are multiple causes of missing DNA in these two syndromes. The majority of the cases (approximately 70%)

arise from a deletion in one part of the chromosome from one of the parents. Other mechanisms of inheritance include uniparental disomy (approximately 28% Prader-Willi syndrome and between 3% and 5% Angelman syndrome) and changes in methylation (approximately 2% Prader-Willi and 25% Angelman syndrome) (American Society of Human Genetics/American College of Medical Genetics Technology Transfer Committee, 1996) of a DNA sequence that effectively shuts off the expression of the 15q11-15q.13 region on that allele. Prader-Willi and Angelman syndromes can be caused by different methods of inheritance, and therefore there are multiple ways that they can be diagnosed. In 1996 the American Society of Human Genetics and the American College of Medical Genetics put out a joint report proposing two separate algorithms for the diagnosis of either Prader-Willi or Angelman syndrome (American Society of Human Genetics/ American College of Medical Genetics Technology Transfer Committee, 1996). Each algorithm incorporates methylation testing, using two different probes that have been linked to either paternal or maternal inheritance. If both are present, then it is very unlikely to make the diagnosis of either syndrome. If, however, the test is positive for either the presence of only either maternally or paternally inherited DNA, then one could do further testing with FISH or polymerase chain reaction (PCR) specific for either Prader-Willi or Angelman syndrome. A high-resolution cytogenetic karyotype should also be done to rule out a translocation.

Genomic variability may occur between different cells in different tissues for one person, a phenomenon called mosaicism. Genetic mosaics may arise either in an embryo with trisomy neurobehavioral syndromes from nondisjunction of chromosomes sometime shortly after fertilization or from a mutation in a cell line after replication. Therefore, a child may have cells in one type of tissue with three copies of a chromosome and then in other tissues have only two copies but may still be abnormal if the chromosomes are from the same parent. The presence of mosaics is important to recognize, as people may have a less affected phenotype and are more likely to have a false diagnostic test because the mutation is not expressed in all of the tissues. Therefore, if a diagnostic test is negative in blood, then one would consider testing skin cells because they come from a different developmental cellular lineage.

Prader-Willi and Angelman syndromes provide examples of disorders that are caused by the inheritance of only one parental allele. However, there are certain genes in humans, which are normally meant to have only one expressed allele, from either the mother or the father. One allele is essentially turned off, usually through methylation of the promoter region. This is seen in females, where many of the genes on one of the X chromosomes are turned off (X inactivation) through genetic imprinting (Robinson, 2000).

There are other mechanisms theorized to produce uniparental disomy, but they are beyond the scope of this chapter. The important concept is to be aware of the multiple modes of inheritance of genetic disorders and the concept of mosaics that may arise within disorders with chromosomal numerical abnormalities.

DNA

DNA, as described previously, is made up of four nucleotides, adenosine, thymine, cytosine, and guanine. Three nucleotides come together in the coding regions during transcription to form a RNA molecule. The RNA code then becomes an amino acid, and the amino acids are translated into protein products. There are various mechanisms, such as DNA mutations, formation of unstable DNA triple repeats, and DNA methylation, that can lead to either a change or a modification in a

Table 14.2. Neurobehavioral syndromes, associated known genotypic changes, and genetic techniques used to detect the disorders

Neurobehavioral syndrome	Chromosomal abnormality	Affected genes	Mechanism of action	Genetic technique used for detection
Down syndrome	Trisomy 21 Translocation Mosaicism		Polyploidy by nondisjunction, translocation, mosaicism	Routine cytogenetic analysis May need multiple tissues tested for mosaicism
Fragile X	CGG triple repeat on X chromosome	*FMR1*	Unstable triple repeat methylation	Methylation detection by Southern blotting or PCR
Rett syndrome		*MECP2* or *CDKL5* mutation	DNA sequence mutation (eight within *MECP2* defined thus far)	PCR
Williams syndrome	7q 11.23	Multiple genes *ELN*	Deletion duplication	FISH CGH array PCR assay
Prader-Willi syndrome	15q deletion		Methylation Uniparental maternal disomy Unbalanced translocations	Methylation test with specific maternal and paternal probes Confirm with PCR or FISH High-resolution cytogenetics for translocation
Angelman syndrome	15q deletion	*UBE3A*	De novo maternal deletion Uniparental paternal disomy Imprinting defect *UBE3A*	Methylation test with specific maternal and paternal probes Confirm with PCR or FISH High resolution cytogenetics for translocation
Velocardiofacial syndrome	22q deletion	*TBX1*	deletion	High resolution cytogenetic FISH CGH array

Sources: American Society of Human Genetics/American College of Medical Genetics Technology Transfer Committee, 1996; Amir et al., 1999; Kaufmann and Reiss, 1999; Motzkin, 1993; Scala et al., 2005; and Somerville et al., 2005.
Key: CGH, comparative genomic hybridization; FISH, fluorescent in situ hybridization; PCR, polymerase chain reaction.

protein product. See Table 14.2 for an overview of some of the neurobehavioral syndromes with known genetic etiologies and the genetic techniques used to diagnose them.

A single mutation that changes just one nucleotide can have significant clinical implications, depending on its location within the genome. Researchers know more about mutations that occur within the coding regions, exons, of the genome because they have sought them out based on an abnormal clinical phenotype. For example, in 1999 researchers discovered eight different mutations in the MECP2 gene that causes Rett

syndrome (Amir et al., 1999). Rett syndrome is a progressive neurodegenerative disease characterized by loss of language, slowed head growth, purposeful hand movements, intellectual disability, seizures, and intermittent hyperventilation. It is seen primarily only in girls and therefore was thought to occur as an X-linked dominant disorder that was lethal in boys. Researchers therefore began looking on the X chromosomes of affected patients to identify the gene and the subsequent mutations. The multiple mutations discovered illustrate the concept of allelic heterogeneity. Allelic heterogeneity

occurs when there is more than one mutation that can lead to a disorder. Rett
syndrome also illustrates the concept of
genomic heterogeneity. Genomic heterogeneity occurs when there is more than one
gene that is responsible for causing the
behavioral phenotype. After the discovery of
the MECP2 mutations, there were still
females known to have the clinical phenotype of Rett syndrome who were negative for
those mutations. In 2005 researchers discovered new mutations on a different gene,
CDKL5 (Scala et al., 2005), in some of these patients. These patients had a certain seizure disorder called infantile spasms that CDKL5
had already been associated with, and therefore researchers targeted this gene as
another possible cause of Rett syndrome.
There is both phenotypic variability and
genotypic variability within Rett syndrome;
however, they are not all directly linked.
There are some associations such as the
infantile spasms in the CDKL5 variation, and
there has been one mutation associated with
a milder clinical course (Young et al., 2008);
however, overall there is no strong link
between genotype and phenotype, implying
that there are other factors involved.

Molecular biology techniques such as
DNA cloning and PCR were instrumental in
sequencing the human genome. PCR is the
technique used to identify genetic disorders
with known mutations. There are specific
centers that provide PCR detection for certain disorders; they can be located on the
web site http://www.ncbi.nlm.nih.gov/sites/
GeneTests/?db=GeneTests.

In addition to chromosomal rearrangements and mutations, DNA expression can
be modified with the addition of chemical
side groups to a DNA nucleotide, cytosine.
The basic premise of methylation, as stated
earlier in the chapter, is the addition of a
methyl group, CH_3, to cytosine, which can
affect the level of expression of a gene. The
methylation usually occurs in the promoter
region for a gene. Methylation is thought to
be the primary method of genomic imprinting as described in the previous section on
chromosomes.

The premise behind detection studies
for methylation involves enzymes that specifically cleave DNA at sites with methylation.
Therefore, detection is based on the size of
the DNA sequence that can be detected by
Southern blotting, PCR, or most recently by
matrix-assisted laser desorption/ionization
mass spectroscopy. Southern blotting is a
molecular biology technique that utilizes the
concept of hybridization for detection of a
specific DNA sequence and a gel-based technology that segregates DNA pieces by size. It
is able to detect if there is an abnormal number of repeats by size but is not sensitive
enough to detect the exact number of
repeats. PCR is a more sensitive test that is
able to provide the exact sequence and therefore the exact number of repeats present.
Mass spectroscopy is a highly sensitive test for
detection of a size of a DNA sequence, and
one could use mathematics to calculate the
number of repeats based on the known
weight of each nucleotide.

There are certain patterns of DNA
sequences that were characterized with the
sequencing of the human genome. One pattern seen within the genome is the trinucleotide repeat. Trinucleotide repeats are
repetitive sequences of three nucleotide
bases, for example CGG, that occur normally. However, if an abnormal number of
trinucleotide repeats occurs, then they are
also associated with known genetic disorders. The exact mechanisms of the trinucleotide repeats are still not completely
understood. The repeats are associated with
unstable regions within the chromosome. It
is unclear whether the sequence itself is
unstable, or the complex of the DNA with
the chromatin is unstable, or a combination
of both. Genetic disorders associated with
trinucleotide repeats are associated with
allelic expansion and anticipation. In other
words, the number of trinucleotide repeats
may increase in each subsequent generation.
Anticipation is defined as the reduction in

age of onset of a disorder or the increase in phenotypic severity in subsequent generations. Fragile X is one of these disorders. Fragile X syndrome is characterized by mild to moderate intellectual disability, characteristic facial features, and connective tissue abnormalities. It is associated with a fragile site on the X chromosome (Xp27.3) (Kaufmann & Reiss, 1999). The gene FMR1 was identified as a cause of fragile X syndrome. The gene contains a trinucleotide repeat of nucleotides CGG. The CGG repeat number ranges between 5 and 50 repeats in the normal population, 50 to 200 repeats (referred to as a premutation) in unaffected transmitting male and female carriers, and >200 repeats in affected individuals. In general, the longer the repeat is, the more unstable the sequence and the more likely it is for expansion of the repeat to occur during either meiosis or mitosis. In fragile X, it is more likely for an allelic expansion to occur when inherited by a son from a mother than when it is inherited by a daughter from a father. The trinucleotide repeat for fragile X is unstable in the generation of eggs but is stable in sperm for an unknown reason. The identification of variable repeat numbers in different tissues that were likely secondary to unstable replication during mitosis led to the discovery of mosaicism within fragile X syndrome. The identification of a large population of mosaicism within boys with fragile X (20%–40%) explains some of the wide phenotypic variability in this syndrome (Kaufmann & Reiss, 1999). It was initially postulated that the number of trinucleotide repeats was directly related to phenotypic variability, but once again there is not a direct phenotype-to-genotype correlation. In 1994 (Kaufmann & Reiss, 1999) scientists discovered a significant role of methylation of the trinucleotide repeats and cognitive function. The researchers found that subjects with the full mutation without any methylation had higher IQs than those with

complete methylation, and those with intermediate methylation scored somewhere between the two groups.

SINGLE-NUCLEOTIDE POLYMORPHISMS AND COPY NUMBER VARIANTS

The latest discovery in understanding the human genome comes from revealing the extent to which structural variability plays a role in determining phenotypic variability. In 1998 single-nucleotide polymorphisms (SNPs) were described and thought to play the major role in phenotypic variation (Wang et al., 1998). An SNP is a nucleotide position where two alternative bases can occur in at least >1% of the human population. The difference between a single-nucleotide mutation and an SNP is the higher prevalence of the polymorphism in SNPs. The clinical significance is also still uncertain in SNPs. In 2001, the International SNP Map Working group identified 1.42 million SNPs based on the sequencing of genomes from 24 ethnically diverse individuals (Waterson & McPhereson, 2001). In 2007, there were an estimated 10 million to 15 million SNPs characterized in the human genome. The increase in the number of SNPs identified is secondary to the development of more sensitive detection methods such as comparative genomic hybridization array technology (see the following discussion) and the sequencing of more individual genomes. The genotyping, location, and clinical significance of SNPs continue to be revealed. Of the coding SNPs known to date, 0.5% result in premature stop codons and therefore directly affect a protein product (Ng et al., 2008). SNPs are used as markers in linkage and linkage disequilibrium studies (see the following discussion) to help identify locations of genes involved in diseases. SNPs also appear to have a low rate of mutation and therefore are also being used in human ancestry studies. Consortiums, such as the International HapMap Project, are continuing to map out SNPs and better clarify their clinical roles.

In 2004, researchers described large areas of either insertions or deletions of DNA sequence that occur in at least 1% of the population that are collectively termed copy number variants (CNVs) (Sebat et al., 2004). Since the initial discovery of CNVs, estimates for the total number of CNVs range from approximately 4 Mb to a less conservative estimate of 24 Mb. They range in size from 1,000 bases (1 kilobase [kb]) to 1,000,000 bases (1 Mb) in length (Cook & Scherer, 2008). Therefore, on a nucleotide basis they represent more variability within the genome than SNPs (Sebat, 2007). CNVs have multiple known and suspected clinical impacts. Initially, researchers thought that CNVs played a major role in phenotypic variability. CNVs can directly disrupt a genetic sequence by either deletion or duplication of genetic material, or they may play a role in making a gene more susceptible to disruption from an environmental stimulus, or they may be totally benign and have no phenotypic impact. Prader-Willi syndrome, Angelman syndrome, and 22q11 deletion are all examples of neurobehavioral syndromes caused by well-characterized CNVs. Increased copy number of the gene CCL3L1 is associated with reduced genetic susceptibility to HIV infection and the progression to AIDS (Gonzalez et al., 2005).

Researchers are continuing to enhance their understanding of the human genome and what may go wrong to form a chromosomal rearrangement. There are many syndromes that have been described thus far that are caused by chromosomal deletions. There are regions of the genome that appear to be more susceptible to deletions because on either side of them are similar sequences called low copy repeats (LCRs). The LCRs may misalign during meiosis because of their similar sequences and cause a deletion in one set of the chromosomes. Based on this theory of mechanism of action, the other set of

chromosomes that underwent meiosis should then have the extra chromosomal material that was deleted from the other chromosome. It is thought that deletions lend themselves to more severe phenotypes, and therefore more have been discovered. However, with the advancement in screening DNA technologies, more DNA duplications are being discovered.

Is there a quantitative phenotypic range between not having enough of a DNA sequence and having too much? Is there a dosage effect? An example of this is seen within the chromosomal loci for Williams-Beuren syndrome. Williams-Beuren syndrome, as stated privously, is characterized by intellectual disability, expressive language that appears to be better than intellectual abilities, and supravalvular aortic stenosis. It is caused by a deletion on chromosome 7q11.23. The sequencing of the genome around this locus demonstrated a set of low copy repeats. In 2005 researchers described a patient with severe expressive language delay, attention-deficit/hyperactivity disorder, and growth retardation. The person had a chromosomal duplication in 7q11.23, the same locus seen to be deleted in people with Williams-Beuren syndrome (Somerville et al., 2005). This had led the researchers to describe the concept of a dosage effect.

Structural variation is clinically important; however, the extent to which it occurs and its overall significance are not yet well understood (Conrad, Andrews, Carter, Hurles, & Pritchard, 2006). Researchers are generating genomic databases such as the International HapMap database to identify new CNVs as well as confirm and characterize previously described CNVs (Human Structural Variation Working Group, 2007; Jakobsson et al., 2008; Redon et al., 2006). In 2005, the National Human Genome Research Institute Large-Scale Genome Sequencing Program was initiated to characterize variability within the various human genomes to come up with a reference genome to be used to compare all other genomes (Human Structural Variation Working Group, 2007).

COMPARATIVE GENOMIC HYBRIDIZATION

Researchers were able to discover CNVs with the use of a new technology, comparative genomic hybridization (CGH) arrays (Carter, 2007). CGH, like FISH, utilizes fluorescent labeling of sample DNA and reference DNA. Unlike FISH, the two samples are then hybridized together with a probe and then scanned for the amount of fluorescence in all of the chromosomes. A computer detects a certain amount of fluorescence that correlates with an amount of DNA (i.e., a certain copy number) for that sequence of DNA. The theory is that if there is a deletion or if there is duplication of genetic material, then there will be a change, either a decrease or increase in the amount of fluorescence in that region.

Like conventional cytogenetics, CGH technology has developed a higher-resolution method, CGH arrays. Array technology utilizes probes of DNA fixed to a solid support array plate that not only provides higher resolution but also a faster method for detection. The detection level for CGH array depends on the type of probe used. The probes vary from their size and where they bind to within the genome. There are probes from bacterial artificial chromosomes (BACs), oligonucleotidess, and SNPs that are all based on the known human genome sequence. BAC clones vary in size and their resolution capabilities. The average size of most BAC clones is approximately 150 kb, with a resolution of less than 46 kb. Oligonucleotides are artificially created pieces of DNA (nucleotides) put together as specific probes to known sequences of the human genome to link to specific pieces of DNA. One oligonucleotide does not cover an area as large as a BAC clone. However, they can be as effective if one uses multiple overlapping oligonucleotides. In 2006 scientists began to link SNPs with CNVs, thereby showing the role of SNPs as markers for detection of larger genomic variations (Hinds, Kloek, Jen, Chen, & Frazer, 2006). The resolution of the probes depends on the size, the distance, and the location of the array probes on a sample genome. Oligonucleotide arrays should have at least 500,000 probes to prevent a higher signal-to-noise ratio than BAC array probes (Miller et al, 2008). It is important to ensure that the array probes from any source provide good coverage of the gene-rich subtelomeric regions (Stankiewicz & Beaudet, 2007). As stated earlier in the chapter, researchers have declared that subtelomeric FISH can provide a detection rate of up to 5% of unbalanced arrangements in patients with intellectual disabilities. Data from CGH arrays yield a detection rate between 10% and 15% in the same population (Shaw-Smith et al., 2004). CGH arrays can be used for either whole genome or specific regions of interest, such as the mitochondrial genome.

CGH arrays allow for detection of many types of variation, such as deletions and duplications in the range of 0.5 Mb to 15 kb, depending on the probe used. Table 14.3 lists the detection rates for various genetic diagnostic tests. It is useful to identify material that is not visible with routine or high-resolution cytogenetic techniques. CGH arrays may lead to an increase in detection of the number of novel duplications, such as in the Williams-Beuren duplication described earlier in the chapter versus novel deletions.

The technology is still limited by its detection rate, and therefore there are still rearrangements that are not detected. CGH does not detect balanced rearrangements such as inversions and balanced translocations because its detection method is based on overall change in fluorescence. CGH is a useful screening tool, and some labs still use a confirmatory test such as FISH to confirm the copy number variant. For instance, CGH can detect two areas of chromosomal imbalance due to a translocation, but then FISH is needed to show the location of the rearrangement.

Table 14.3. Overviews of various genetic techniques to detect chromosomal abnormalities and their associated resolution levels

Genetic technique	DNA resolution
Traditional cytogenetic banding	> 5 Mb
Pulse-field gel electrophoresis	> a few hundred kb to a few Mb
Comparative genomic hybridization arrays using bacterial artificial chromosomal probes (BacArrays)	500–1000 kb
FISH	> 35 kb
Comparative genomic hybridization arrays using SNPs as probes	15–40 kb
Southern blotting	10–40 kb

Source: Bejjani and Shaffer (2008).

Key: FISH, fluorescent in situ hybridization; kb, kilobases; Mb, million base pairs; SNPs, single-nucleotide polymorphisms.

One of the main weaknesses a clinician should be aware of is the variability in the various types of CGH arrays that are commercially available. The probes vary in the parts of the genome that they will bind to and therefore will detect rearrangements in different parts of the genome that may lead to false negatives.

Another weakness is a positive result of a rearrangement of unknown clinical significance. Scientist-generated public databases such as the Database of Genomic Variants (http://projects.tcag.ca/variation), the DECI-PHER database (http://www.sanger.ac.uk/PostGenomics/decipher), and the University of California at San Diego database (http://genome.ucsc.edu) are used to characterize genetic variants and assess the clinical significance of the results. If a CNV is identified, performing the same CGH array on the parents if known may be helpful. It is helpful to know if the parents have the same CNV and if they are clinically similar to or different from the patient. If the CNV is not detected in either parent, then it is considered to be de novo. De novo CNVs that occur in genotypically sensitive areas (e.g., coding regions or promoter sequences) may be clinically significant, as they are highly penetrant. Penetrance relates to the likelihood that a genetic sequence will be expressed phenotypically. A referral to a genetic specialist is helpful in interpreting the results. This is a case where the technology is more advanced than the current clinical knowledge.

LINKAGE AND ASSOCIATION STUDIES

Mendel looked at the end products of his experiments, the phenotype, and worked backward to describe what he thought was the science behind the pea-pod flowers' colors. Scientists continue to use this approach today when performing linkage studies.

The genes for many of the disorders described in this chapter were partially discovered through linkage analysis. Researchers use complex mathematical models to hypothesize relationships (i.e., linkages) between two parts of the genome to help create genomic maps. One part of the genome can be a marker with a known location, such as a mile marker on a highway, and the other part being analyzed is the area of interest. A fundamental purpose of linkage analysis is to see if a gene segregates independently according to Mendel's second law of independent segregation or if there is linkage with another part of the genome. Linkage is based on the likelihood of a recombinant crossover event between two chromosomes during meiosis.

Researchers then use the linkage analysis to determine locations of genes of interest by generating a map in relation to the known marker. The likelihood of a crossover event is directly related to the distance between the marker and the gene of interest. The unit of measure in this analysis is a centimorgan. One centimorgan (cm) is equal to the distance between genes for which one product of meiosis in 100 is a recombination event.

There are a few key concepts that should be reviewed to gain a better understanding of the variables used in calculating the likelihood of linkage. Does the genotype of the altered gene led to an altered phenotype, that is, a clinically significant result? To perform the calculations, one must be able to define the phenotype and characterize the genotype as accurately as possible. One would like to know how often the genotype occurs and, when it occurs, how often it results in an altered phenotype (Farrer, 2004).

The first part of a linkage analysis is defining the phenotype of interest. This includes defining characteristic features of the disease. The next step is to make up a pedigree for families who express the disease trait of interest. Second, if possible one needs to confirm the presence or absence of the diagnosis in the family members in the pedigree. The pedigree analysis will provide the investigator with the likely mode of inheritance and assist in generating a genetic model for that trait. When deciding what statistical model(s) to use and the number of families to study to calculate a linkage score between a gene and a marker, one needs to know the mode of inheritance (autosomal dominant versus autosomal recessive versus X-linked), gene frequency, rate of penetrance and whether the disease trait is qualitative versus quantitative. Qualitative traits are defined by either the presence or the absence of the trait. Quantitative traits have a range of possible phenotypes. Regarding the

marker, one needs to know the mode of inheritance, the number of alleles, and the allele frequency (Pericak-Vance, 1996b). Genetic heterozygosity calculates the relative frequencies of heterozygous (two different alleles) versus homozygous (two of the same allele) individuals in the population. SNPs and CNVs are becoming common markers for linkage analysis of complex genetic disorders such as autism spectrum disorders.

An altered genotype may not necessarily reflect an altered phenotype. As described before, penetrance is the probability that a mutant genotype will result in an altered phenotype. A mutant gene with incomplete penetrance may not be phenotypically expressed in one generation, but then that allele can be passed on and expressed in the next generation. Mutant genes can also have variable expression in different tissues either secondary to effects from the local environments or because of mosaicism. The former category is called pleiotropy. Penetrance and pleiotropic genes can explain varied phenotypes by a mutant genotype. It is not well understood why the mutant genes are not expressed.

Phenotypes arise because of different mutations of a genotype. As in the example of Rett syndrome, mutations can be varied within the same allele (called allelic heterogeneity) or a mutation can be on a different gene on different loci on either the same chromosome or different chromosomes. The latter is called locus heterogeneity. Allelic and locus heterogeneity are types of genomic heterogeneity. The concept of heterogeneity is key to linkage studies because it implies multiple genetic loci of interest for similar phenotypes. The goal of linkage studies is to locate an area in the genome based on a certain defined phenotype. If one phenotype can be caused by a mutation at more than one location it will confound the data. Therefore one needs to assume a certain level of genomic heterogeneity when performing the linkage analysis.

There are different forms of linkage analysis based on the mode of inheritance and what other variables are known. These types include parametric, nonparametric, and homozygosity linkage analysis. The preferred type of linkage analysis is parametric analysis. The genetic model including the mode of inheritance, the marker allele frequency, the mutation, and the penetrance are known in a parametric analysis. Parametric analysis is also known as lod score analysis. The lod score is the likelihood ratio of no linkage between a gene and a marker (either another gene or variable sequence) compared with a relationship assuming varying degrees of linkage. Statistical software runs the models and calculates a lod score. A lod score of >3 is considered significant; between >1.5 and <3 is promising; > -2 to <1.5 is inconclusive and indicates additional information is needed; < -2 indicates no linkage (Pericak-Vance, 1996a).

If a lod score is greater than 3, then follow-up studies must be performed to determine the gene sequence for additional markers to confirm the locus. If the score is between 1.5 and 3, then additional analysis can be performed to find additional markers that may be more precise for localization. The markers then can be used as known sequences to develop primers to perform PCR in that region and then sequence that region to obtain a possible genotype. Complex disorders may be caused by multiple genes (polygenic) acting together to form a clinical phenotype. Defining a genetic model to perform parametric analysis is usually not possible; therefore one can make a nonparametric analysis. Nonparametric linkage analysis is also known as model-free analysis because one does not need to know the genetic model including the mode of inheritance in order to perform the analysis. One does need to know the disease phenotype and the marker allele frequencies. Nonparametric analysis is not as powerful a tool as parametric analysis

because less is known about the genetic model and assumptions that can be made with known models. However, nonparametric analysis is a good tool for generating possible targets for complex disorders (Pericak-Vance, 1996a).

Linkage disequilibrium is the nonrandom association of alleles at two or more loci. Detecting linkage disequilibrium, contrary to what its name may imply, does not ensure that there is linkage or a lack of equilibrium. Like regular linkage studies, linkage disequilibrium is detected with the use of a statistical mathematical model to calculate the differences between the frequencies of alleles. Linkage studies try to better define a location for disease traits by studying specific families in sib pair or affected relative analysis. Linkage disequilibrium studies are also used in genome wide screens such as association and case control studies to look for associations between gene markers. It was originally used in the study of population genetics because it reflects the effects of natural selection, genetic mutation, and conversion rates that overall affect the frequency of gene surviving through evolution.

Linkage disequilibrium has been defined with the use of different equations to demonstrate the association between genes and/or genes and markers within the genome. These studies have shown patterns called haplotype blocks, which are hypothesized to be hotspots for genetic recombination. Researchers have found that the SNPs within these hotspots are in strong linkage disequilibrium with genetic alleles that may cause complex inherited disorders.

The neurobehavioral syndromes often have complex behavioral phenotypes. The further classification of these disorders to more specific behavioral phenotypes (defined as endophenotypes) could assist in decreasing the amount of genetic heterogeneity in a linkage analysis. Researchers are working on the ability to define specific endophenotypes in complex neurobehavioral syndromes, to perform linkage disequilibrium analysis to locate possible quantitative trait loci (QTL). QTL are DNA sequences, identified by linkage disequilibrium studies, that are thought to be associated with phenotypically quantitative traits. An illustration of this can be found in the study of identifying susceptibility genes in autism spectrum disorders. Autism spectrum disorders are a phenotypically and genotypically complex set of disorders. Various researchers have identified multiple genes associated with autism spectrum disorders, but each individual gene implicated to have a potential role is disordered in only 1% of the total population of people with autism spectrum disorders (Abrahams & Geschwind, 2008). Scientific groups are trying to define a specific endophenotype to make the association more robust. For example, Alarcon et al. (2002) performed a linkage disequilibrium analysis of variable endophenotypes relating to language to identify a QTL on chromosome 7. Then the same group went on to use SNP markers to perform linkage disequilibrium association studies within this QTL and identified a susceptible gene CNTNAP on chromosome 7 for language delay in autism spectrum disorders (Alarcon et al., 2008). This gene is a part of a family of proteins thought to play a role in neuronal synapses that are known to interact with other genes that are associated with autism spectrum disorders.

The last type of linkage analysis to discuss is homozygosity analysis. Homozygosity linkage studies look for possible gene targets between individuals with recent shared ancestry (consanguineous relationships). The goal is to enhance the search for inheritable factors in affected populations with less genotypic diversity and increase the yield for recessive inheritable traits. Researchers are utilizing this method to find possible targets involved in autism spectrum disorders (Morrow et al., 2008).

ENVIRONMENTAL FACTORS AND EPIGENETICS

As more genotypes for various disorders are being discovered and with the knowledge of additional variability of structural differences such as SNPs and CNVs within the genotypes, there is still a disconnect between genotypes to phenotypes. Questions remain, such as, what role does the environment play in determining our phenotype and genotype? How do genes know how to turn themselves on or off in certain cells?

What role does the environment play in genetic variation? The year 2009 is the 150th anniversary of the publication of Charles Darwin's pivotal book, *The Origin of Species*. In this book, Darwin discusses the process of natural selection of species. The theory is that the best adapted will survive in their respective environments and pass on their genotypes to subsequent generations. Researchers are still trying to better define this complex relationship. In 1997, the National Institute of Environmental Health Science began the Environmental Genome Project. Most of this chapter thus far has discussed genetic concepts from a structural perspective. When discussing environmental influences on genes, one first needs to know more about the functional roles of certain genes.

Scientists have generated a list of eight gene categories based on function that they feel are responsive to environmental stimuli such as temperature change, toxin/drug exposure, radiation, or other stress inducers. The gene categories are involved in the cell cycle, DNA repair, cell division, cell signaling, cell structure, gene expression, apoptosis, and metabolism (including drug metabolism). These categories are self-explanatory, with the exception of apoptosis. Apoptosis is the process by which a cell, if no longer viable or able to function appropriately, will shut itself down through a program of cell death. The categories can be defined by the singular notion of cell survival or grouped into subcategories based on cellular replication (cell cycle, cell division), cellular function (cell signaling and gene expression), cellular repair (DNA repair and apoptosis), and cellular being (cell structure and metabolism). A cell needs to be able to act and adapt to its environment in order to survive.

A consortium of scientists at the National Institute of Environmental Health Sciences are utilizing linkage disequilibrium association studies with the addition of other statistical methods such as regression analysis to investigate the environmental influence on genes in one of the eight categories listed previously and identify SNPs and their roles with these genes (Wang et al., 2004).

How does a cell know what genes to turn on and which ones to turn off? As stated earlier, the human genome is made up of approximately 20,000–25,000 genes, but not all of the genes are expressed in every cell. Only a few genes are turned on based on the cell's function. Scientists are gaining understanding of other kinds of varied genomic expression that do not involve changes within the actual DNA sequence. This area of science is called epigenetics (the Greek prefix *epi-* loosely means "in addition to"). Methylation, X chromosomal inactivation, and imprinting are a few of the methods that have already been discussed in this chapter. One major component of epigenetic changes is the interaction of DNA and the histone proteins, which affects the formation of chromatin. Scientists are discovering variation in chromatin states, which affects the ability of the DNA structures within the chromatin to undergo transcription and be expressed (Johannes, Colot, & Jansen, 2008). In 2006, Andrew Fire and Craig Mello won the Nobel Prize in Physiology and Medicine for their discovery in 1998 of doublestranded RNA molecules that silence gene expression. RNA interference, methylation, and varied chromatin structures are just a few ways a cell selects genes to be turned on from among the 20,000–25,000 possibilities in its genome. The complex interactions of epigenetic mechanisms with genotypic variation, leading to the wide array of phenotypes on this planet, has only just begun to be understood.

SUMMARY

Researchers discovered some of the genes causing or associated with neurobehavioral syndromes by using behavioral phenotypes to identify genotypes. However, there are no solid correlations between behavioral phenotypes and genotypes. Now with the development of new genetic techniques such as CGH arrays, researchers are describing and defining structural variants such as SNPs and CNVs within the genome. However, they find that these changes still do not explain the full clinically diverse picture. The study of epigenetic mechanisms in conjunction with further study of genotypic variation with the use of linkage analysis to look at relationships between genes and the environment will further advance our knowledge of this complex interrelationship between genotypes and behavioral phenotypes.

REFERENCES

Abrahams, B., & Geschwind, D. (2008). Advances in autism genetics: On the threshold of a new neurobiology. *Nature Reviews Genetics, 9,* 341–355.

Alarcon, M., Abrahams, B., Stone, J., Duvall, J., Perederiy, J., Bomar, J., et al. (2008). Linkage, association and gene expression analyses identify CNTNAP2 as an autism-susceptibility gene. *American Journal of Human Genetics, 82,* 150–159.

Alarcon, M., Cantor, R., Liu, J., Gillam, T., Autism Genetic Resource Exchange Consortium, & Geschwind, D. (2002). Evidence for a language quantitative trait locus on chromosome 7q in multiplex autism families. *American Journal of Human Genetics, 70,* 60–71.

American Society of Human Genetics/American College of Medical Genetics Technology Transfer Committee. (1996). Diagnostic testing for

Prader-Willi and Angelman syndromes: Report of the AHSG/ACMG Test and Technology Transfer Committee. *American Journal of Human Genetics, 58,* 1085–1088.

Amir, R., Van den Veyver, I., Wan, M., Tran, C., Francke, U., & Zoghbi, H. (1999). Rett syndrome is caused by mutations in X-linked MECP2, encoding methyl-CpG-binding protein 2. *Nature Genetics, 23,* 185–188.

Bejjani, B., & Shaffer, L. (2008). Clinical utility of contemporary molecular cytogenetics. *Annual Review of Genomics and Human Genetics, 9,* 71–86.

Carter, N. (2007). Methods and strategies for analyzing copy number variation using DNA microarrays. *Nature Genetics, 39,* S16–S21.

Conrad, D., Andrews, T.D., Carter, N., Hurles, M., & Pritchard, J. (2006). A high resolution survey of deletion polymorphism in the human genome. *Nature Genetics, 38,* 75–81.

Cook, E., & Scherer, S. (2008). Copy-number variations associated with neuropsychiatric conditions. *Nature, 455,* 919–923.

Farrer, L. (2004). Collection of clinical and epidemiological data for linkage studies. *Current Protocols in Human Genetics,40,* 1.1.1–1.1.17.

Gonzalez, E., Kulkarni, H., Bolivar, H., Mangano, A., Sanchez, R., Catano, G., et al. (2005). The influence of CCL3L1 gene-containing segmental duplications on HIV-1/AIDS susceptibility. *Science, 307,* 1434–1440.

Hinds, D., Kloek, A., Jen, M., Chen, X., & Frazer, K. (2006). Common deletions and SNPs are in linkage disequilibrium in the human genome. *Nature Genetics, 38,* 82–85.

Human Structural Variation Working Group. (2007). Completing the map of human genetic variation. *Nature, 447,* 161–165.

Inoue, K., & Lupski, J. (2002). Molecular mechanisms of genetic disorders. *Annual Review of Genomics and Human Genetics, 3,* 199–242.

International Human Genome Sequencing Consortium. (2001). Initial sequencing and analysis of the human genome. *Nature, 409,* 860–921.

International Human Genome Sequencing Consortium. (2004). Finishing the euchromatic sequence of the human genome. *Nature, 431,* 931–945.

Jakobsson, M., Scholz, S., Scheet, P., Gibbs, J.R., VanLiere, J., Fung, H., et al. (2008). Genotype, haplotype and copy number variation in worldwide human populations. *Nature, 451,* 998–1003.

Johannes, F., Colot, V., & Jansen, R.C. (2008). Epigenome dynamics: A quantitative genetics perspective. *Nature Reviews Genetics, 9,* 883–890.

Kaufmann, W., & Reiss, A. (1999). Molecular and cellular genetics of fragile X syndrome. *American Journal of Medical Genetics (Neuropsychiatric Genetics), 88,* 11–24.

Korf, B. (2001). Overview of clinical cytogenetics. *Current Protocols in Human Genetics* 8.1.1–8.1.10.

Li, E., Beard, C., & Jaenisch, R. (1993). Role for DNA methylation in genomic imprinting. *Nature, 366,* 362–365.

Miller, D., Shen, Y., & Bai-Lin, W. (2008). Oligonucleotide microarrays for clinical diagnosis of copy number variation. *Current Protocols in Human Genetics, 58,* 8.12.1–8.12.17.

Morrow, E., Yoo, S., Flavell, S., Kim T., Lin, Y., Hill, R., et al. (2008). Identifying autism loci and genes by tracing recent shared ancestry. *Science, 321,* 218–223.

Motzkin, B., Marion, R., Goldberg, R., Shprintzen, R., & Saenger, P. (1993). Variable phenotypes in velocardiofacial syndrome with chromosomal deletion. *Journal of Pediatrics, 123,* 406–410.

Ng, P., Levy, S., Huang, J., Stockwell, T.B., Walenz, B.P., Li, K., et al. (2008). Genetic variation in an individual human exome. *PLoS Genetics, 4,* 1–13.

Peltonen, L., & McKusick, V. (2001). Dissecting human disease in the postgenomic era. *Science, 291,* 1224–1229.

Pericak-Vance, M. (1996a). Analysis of genetic linkage data for Mendelian traits. *Current Protocols in Human Genetics,* 1.4.1–1.4.31.

Pericak-Vance, M. (1996b). Overview of linkage analysis in complex traits. *Current Protocols in Human Genetics,* 1.9.1–1.9.19.

Redon, R., Ishikawa, S., Fitch, K.R., Feuk, L., Perry, G.H., Andrews, T.D., et al. (2006). Global variation in copy number in the human genome. *Nature, 444,* 444–454.

Robinson, W. (2000). Mechanisms leading to uniparental disomy and their clinical consequences. *BioEssays, 22,* 452–459.

Scala, E., Ariani, F., Mari, F., Caselli, R., Pescucci, C., Longo, I., et al. (2005). CDKL5/STK9 is mutated in Rett syndrome variant with infantile spasms. *Journal of Medical Genetics, 42,* 103–105.

Sebat, J. (2007). Major changes in our DNA lead to major changes in our thinking. *Nature Genetics* (Suppl. 39), S3–S5.

Sebat, J., Lakshmi, B., Troge, J., Alexander, J., Young, J., Lundin, P., et al. (2004). Large-scale copy number polymorphism in the human genome. *Science, 305,* 525–528.

Shaw-Smith, C., Redon, R., Rickman, L., Rio, M., Willat, L., Fiegler, H., et al. (2004). Microarray based comparative genomic hybridisation (array-CGH) detects submicroscopic chromosomal deletions and duplications in patients with learning disability/mental retardation and dysmorphic features. *Journal of Medical Genetics, 41,* 241–248.

Somerville, M.J., Mervis, C.B., Young E.J., Seo, E.J., del Campo, M., Bamforth, S., et al. (2005). Severe expressive-language delay related to duplication of the Williams-Beuren locus. *New England Journal of Medicine, 353,* 1694–1791.

Stankiewicz, P., & Beaudet, A. (2007). Use of array CGH in the evaluation of dysmorphology, malformations, developmental delay, and idiopathic mental retardation. *Current Opinion in Genetics and Development, 17,* 182–192.

Venter, J.C., Adams, M.D., Myers, E.W., Li, P.W., Mural, R.J., Sutton, G.G., et al. (2001). The sequence of the human genome. *Science, 291,* 1304–1351.

Wang, D., Fan, J., Siao, C., Berno, A., Young, P., Sapolsky, R., et al. (1998). Large-scale identification, mapping, and genotyping of single nucleotide polymorphisms in the human genome. *Science, 280,* 1077–1082.

Wang, X., Tomso, D.J., Liu, X., & Bell, D.A. (2004). Single nucleotide polymorphism in transcriptional regulatory regions and expression of environmentally responsive genes. *Toxicology and Applied Pharmacology, 207,* S84–S90.

Waterson, R., & McPhereson, J. (2001). A map of human genome sequence variation containing 1.42 million single nucleotide polymorphisms. *Nature, 409,* 928–933.

Young, D.J., Bebbington, A., Anderson, A., Ravine, D., Ellaway, C., Kulkami, A., et al. (2008). The diagnosis of autism in a female: Could it be Rett syndrome? *European Journal of Pediatrics, 167,* 661–669.

15

Genetically Informative Phenotypes

Opportunities for Progress and Potential Pitfalls

Peter Szatmari

It is remarkable to realize how much we have learned in the past few years about the diagnosis and classification of the disorders known as the "developmental disabilities." In the *Diagnostic and Statistical Manual of Mental Disorders, Second Edition* (*DSM-II;* American Psychiatric Association [APA], 1968), the only developmental disability diagnosis listed was Mental Retardation, and autism was considered a type of "childhood psychosis." In the fourth edition (*DSM-IV;* APA, 1994), in addition to the diagnosis of Mental Retardation, different types of specific learning disabilities, Developmental Coordination Disorder, and the pervasive developmental disorders (more commonly referred to as autism spectrum disorders [ASDs]) are listed, along with specific diagnostic criteria for each. Although this progress has been impressive, the more we learn, the more we appreciate how much more needs to be accomplished to truly reduce the suffering of children with developmental disabilities.

In children's mental health research, *disorder* refers to a set of behaviors or skills and abilities that more or less hang together and form a *syndrome.* There is no stipulation that the set of behaviors or skills is associated with a specific etiology (Dykens & Hodapp, 2001). Those who formulated these criteria anticipated that specific etiologic risk factors would one day be uncovered. This disconnection between disorder and etiology was first proposed in *DSM-III* (1980), and it is a radical departure from previous classification systems.

A neurobehavioral disorder is one in which the concept of etiology makes a reappearance. The assumption here is that the behavioral signs and symptoms presumably have a neurological, or brain-based, etiology. The term *neurobehavioral* includes autism, the other developmental disabilities, and disorders such as attention-deficit/hyperactivity disorder (ADHD), Tourette syndrome, and genetic and chromosomal disorders that affect brain development. Anxiety, mood disorders, and conduct disorders are generally not considered to be neurobehavioral disorders, even though they too have a neurological origin, at least at some level.

The search for the etiologic factors responsible for these neurobehavioral disorders has been more or less successful, depending on the disorder in question. It is clear that all of the developmental disorders, including autism, specific learning disabilities, ADHD, Tourette syndrome, and childhood schizophrenia, have a strong genetic component (Gould & Manji, 2004; Inoue & Lupski, 2003; Rutter, 2002; Skuse, 1997; Stoltenberg & Burmeister, 2000). In fact, if one were to rely on twin studies alone, it would be possible to conclude that all of the developmental and psychiatric disorders of childhood and adolescence have a strong genetic component (Gillespie et al., 2004; Iacono, McGue, & Krueger, 2006; Kaprio & Koshenvuo, 2002; McGuffin, 2005). The difficulty is that twin studies can inform us only about the extent to which genetic factors are involved in the etiology of a disorder. Estimates of the amount of variance that can be accounted for by genetic and/or environmental factors are useful but rather imprecise. One would anticipate that the disorders with the greatest amount of variance attributable to genetic factors should be the disorders in which it has been easiest to detect susceptibility genes. However, this has not been the case. Heritability does not guarantee linkage or association (Rice & Borecki, 2001).

In fact, the field has been largely unsuccessful in identifying common genetic variants that account for neurobehavioral disorders. This is in stark contrast to the rapid progress that we have seen in identifying relatively rare genetic mutations that account for specific syndromes (Bearden, Reus, & Freimer, 2004; Raymond & Tarpey, 2006). Phenotypes associated with single gene mutations and chromosome abnormalities associated with Down syndrome, fragile X syndrome, tuberous sclerosis, Prader-Willi syndrome, Angelman syndrome, and Smith-Magenis syndrome have been well described (Dykens & Hodapp, 2001; Gabbett, Peters, Carmichael, Darmanian, & Collins, 2008; Gropman, Duncan, & Smith, 2006; Rachidi & Lopes, 2008). In spite of the fact that each of these disorders has a single cause, the variation in clinical presentation associated with these disorders is remarkable, but perhaps it is less than what is usually associated with disorders caused by more complex genetic mechanisms, which presumably involve multiple genetic variants.

This discrepancy between single and multiple genetic variants emphasizes the distinction between simple Mendelian and complex genetic disorders (Baron, 1995; Bearden, Reus, & Freimer, 2004; Nurnberger, 2002; Risch & Merikangas, 1996). As a group, Mendelian disorders are single-gene disorders associated with autosomal dominant, recessive, or X-linked transmission. They can also refer to disorders arising from deletions, duplications, or translocations in DNA segments that occur spontaneously or de novo and disrupt genes within those regions. This includes Williams-Beuren syndrome (Osborne & Mervis, 2007), Smith-Magenis syndrome (Gropman et al., 2006), velocardiofacial syndrome (Cohen, Chow, Weksberg, & Bassett, 1999) and several others. These disorders are caused by disruptions in chromosomal regions or genes either transmitted in an easily recognizable fashion from parent to child or arising spontaneously. In these situations, the genetically informative phenotype matches more or less exactly the signs and symptoms of the disorder.

A complex genetic disorder is caused by multiple genetic variants (either common or rare or both). Moreover, the clinical presentation can be complicated by incomplete penetrance, pleiotropy, or genetic heterogeneity (Zlotogora, 2003). *Incomplete penetrance* refers to situations in which a genetic variant associated with a disorder exists, but there is no observable clinical manifestation. *Pleiotropy* (or variable expressivity) refers to multiple clinical presentations arising from a single genetic variant, and *genetic heterogeneity* refers to the observation that the same disorder can be caused independently by multiple genetic variants, either at the same genetic locus or at different loci (Ott & Bhat, 1999). In a complex genetic disorder, these independent genetic variants are neither necessary nor sufficient to cause the disorder. Each genetic risk factor has a small to moderate effect size and accounts for a small amount of the variance in the presentation of the disorder, even though that risk factor may be quite common. It is true that rare genetic variants can be associated with complex disorders (e.g., copy number variants in intellectual disability [Slavotinek, 2008] and in autism [see Marshall et al., 2008]). In these circumstances, although the rare variant has a large effect, the low frequency of the copy number variant again means that it accounts for only a small amount of the variance in the population.

With complex genetic disorders, it is important to distinguish between a disorder and a phenotype. *Disorder* refers to the behavioral signs and symptoms that more or less hang together more frequently than they would by chance alone. *Phenotype* refers to an expression of the genotype (Rice, Saccone, & Rasmussen, 2001). Genotypes and phenotypes define each other reciprocally and cannot be understood in isolation. A genotype refers to a single DNA variant or a constellation of genetic variants (that arise through single-nucleotide polymorphisms, a chromosomal abnormality, or a copy number variant) that has a "product." A phenotype is the product of that genotype; it is part of the causal chain of that genotype. For example, a phenotype could refer to RNA, proteins, cells, systems of cells, neural networks, cognitive skills, or constellations of symptoms and behaviors and eventually to the components of the "true" manifestation of the disorder. The *DSM* categorization tries to capture the final phenotype associated with a specific genotype. What we classify as a disorder is only a proxy measure of the true phenotype. Each of the intermediate steps between genotype and disorder is part of the causal chain and is itself a phenotype. The disorder is made up of the final phenotype in that causal chain, as well as traits and behaviors caused by environmental risk factors.

Complex genetic models of causation then refer to multiple genes interacting in an additive or multiplicative fashion, probably in the context of environmental risk factors as well. One can think of a river with multiple tributaries coming together in a final common pathway (Cannon & Keller, 2006). The origin of each of those tributaries is a small stream that becomes a creek that becomes a river, and all of these individual rivers become a larger final common pathway that eventually leads to the disorder. The origin of each of those tributaries can be a specific genotype, and the more the products (i.e., the proteins or lack of proteins) of those individual genotypes come together and form a common causal chain, the greater the risk of developing the disorder.

Each of those tributaries is also a phenotype in its own right. It is useful to think of three different types of genetically informative phenotypes: intermediate (or endophenotypes), component, and covariate (Szatmari et al., 2007). Genetically informative phenotypes have a stronger association with the genotype than the disorder itself. These phenotypes tend to be simpler, less complex, and closer in the causal chain to the original genotype (Almasy, 2003; Grigorenko & Pauls, 2003; Tsuang, 2001; Tsuang & Faraone, 2000). They are more "up river," to follow our analogy. We now turn to a fuller description of these informative phenotypes, how they can be used in genetic studies, and the pitfalls that readers and researchers need to be aware of in utilizing these phenotypes (see Table 15.1 for a summary).

COMPONENT PHENOTYPES

Component phenotypes are the individual phenotypes that make up a disorder. This is like dissecting the disorder into its "component" parts. There are three requirements of component phenotypes: they should represent a dimension or trait that is part of the disorder but is independent of other dimensions or traits, they should be more common among affected individuals than among normal controls, and they should be familial (and under genetic control). As a result of these three requirements, a component phenotype should have greater heritability than the disorder itself, if heritability could be precisely measured and compared among phenotypes. Incorporating component phenotypes into a linkage or association study (rather than employing the disorder itself) should have a precise outcome; the linkage or association signal should be greater than if the disorder were used as the phenotype of analysis.

Some examples from recent genetic studies in ASDs might be helpful in illustrating these points. *ASDs* refers to a group of disorders characterized by impairments in

Table 15.1. Genetically informative phenotypes (GIPs)

Type of GIP	Criteria	Methodological challenges
Component phenotype	Component of *DSM-IV-TR* disorder Cases > controls Familial	Not an outcome or comorbidity Phenotypes often correlated Multiple testing
Intermediate phenotypes	Not part of *DSM-IV-TR* criteria Cases > controls Unaffected relatives > controls Familial Mediates genotype–phenotype causal chain	Measurement in affected and unaffected Regression to the mean
Covariates	Confounding Moderating Increases familiality, relative risk to family members, linkage or association signal	Multiple testing Need for replication

Key: *DSM-IV-TR, Diagnostic and Statistical Manual of Mental Disorders, Fourth Edition, Text Revision* (American Psychiatric Association, 2000).

social reciprocity and verbal and nonverbal communication, and by a preference for repetitive stereotyped behaviors (Skuse, 2007). ASDs almost invariably involve cognitive impairments ranging from severe intellectual disability to mild deficits in abstract problem solving (Szatmari, 2003). Onset is always prior to 36 months of age. The *DSM-IV* (American Psychiatric Association, 1994) has provided operational descriptions of this triad of autism; however, evidence of cognitive impairment is not part of the diagnostic criteria.

One of the important issues that has arisen in the last few years is whether it is possible to decompose this complex phenotype into simpler component phenotypes. Clinicians have recognized for years that there is enormous variability in the clinical expression of ASDs. This variability is captured by classifying individuals with an ASD as having the *DSM* diagnosis of Autistic Disorder, Asperger's Disorder, or Pervasive Developmental Disorder-Not Otherwise Specified (PDD-NOS). (Rett's Disorder and disintegrative disorder are not usually considered examples of ASDs.) However, the use of the term *spectrum* in reference to ASDs implies that these disorders represent an underlying unitary phenomenon that differs only in severity, with autism at one end and either Asperger syndrome or PDD-NOS at the other (it has never been clear which). However, this attempt to decompose the ASD phenotype has not been useful in terms of identifying etiologic differences that underlie these distinctions (Szatmari, 2003).

A more common distinction is to refer to "higher functioning" and "lower functioning" individuals with autism. This distinction is largely determined by the degree of intellectual disability comorbid with the ASD, but this distinction still represents a relatively arbitrary way of classifying individuals on the spectrum. Level of functioning is frequently used in research studies of cognitive and imaging profiles and is a useful marker of long-term outcome (Beglinger & Smith, 2001). People with lower functioning autism not only have more intellectual disability but also tend to exhibit more severe symptoms in terms of impairments in social reciprocity, more difficulties in nonverbal communication, and more examples of repetitive stereotyped behaviors, particularly those at the level of sensory-motor functioning (Szatmari, 2003; Szatmari et al., 2006).

However, closer examination of the ASD phenotype reveals greater complexity than this simple bifurcation. For example, it is not always true that greater cognitive impairment is accompanied by more symptoms of autism. For example, individuals with autism who have severe intellectual disability are most often nonverbal and hence cannot show some of the communication symptoms, such as echolalia, pronoun reversal, neologisms, and deficits in conversation, that one sees in higher functioning individuals. Similarly, those who are lower functioning may not be as aware of their environment and so may not be sensitive to changes in the environment and, hence, cannot show resistance to change. They may also not have the cognitive ability to demonstrate circumscribed interests or unusual preoccupations.

The possibility of decomposing the ASD phenotype has been examined more rigorously in a recent series of studies using statistical techniques such as factor analysis. Factor analysis attempts to decompose a heterogeneous phenotype into simpler independent dimensions. It essentially looks at the variability in symptoms between individuals, groups together items that have the highest correlation with each other, and separates them from items that have lower correlations.

For example, we conducted a factor analysis (Szatmari et al., 2002) of the three domains of symptoms of autism from the Autism Diagnostic Interview (ADI; Lord et al., 1997) and measures of level of functioning from the Vineland Adaptive Behavior Scales (Sparrow, Balla, & Cicchetti, 1984) (Communication, Socialization, and Activities of Daily Living). We found that measures of symptoms of autism in social reciprocity, communication, and repetitive stereotyped behaviors were all correlated with each other and loaded on one factor. In contrast, measures of adaptive functioning in socialization, communication, and activities of daily living were also correlated with each other but loaded on a separate factor. This suggests that symptoms and level of functioning are independent and orthogonal.

More recently, there has been great interest in looking in more detail at the triad of autism symptoms, ignoring for the moment level of functioning. Twin studies have looked at the distribution of behaviors characteristic of autism in a general population and have come up with conflicting results. John Constantino and colleagues have reported that all three aspects of the triad of autism load on a single factor and so should be seen as a unitary construct (Constantino & Todd, 2003). In contrast, in a sample of twins, again taken from the general population (but this time using a different instrument; the Childhood Asperger Syndrome Test), Ronald and Happé and colleagues reported that the triad should be thought of as comprising two dimensions: social communication and repetitive stereotyped behaviors (Ronald et al., 2006).

These twin studies were conducted on individuals without ASDs, and the assumption was that the ASD phenotype is normally distributed in the general population. People with ASDs are simply at the extreme end of that distribution. A number of factor analytic studies have instead used the ADI (Lord et al., 1994) with clinical samples of individuals with ASDs. Although somewhat different results have been reported, depending on the sampling frame, the age of subjects, and the type of factor analysis used, it appears as if, to a large extent, items measuring social reciprocity and verbal and nonverbal communication are very highly correlated and load on one factor, whereas items measuring repetitive stereotyped behaviors load on another factor (Georgiades et al., 2007). There has also been interest in exploring the factor structure of the repetitive stereotyped behaviors domain by itself. Here too there is some agreement; two separate dimensions appear in the factor analyses: a "higher order" dimension involving circumscribed interests, rituals, and resistance to change, and a "lower order" dimension consisting of repetitive motor and sensory behaviors (Gabriels, Cuccaro, Hill, Ivers, & Goldson, 2005; Silverman et al., 2002; Szatmari et al., 2006). These two dimensions also appear to have different associations with clinical and etiologic markers. For example, the higher order dimension appears to be more familial than the lower order dimension (Silverman et al., 2002). Variation in the lower order dimension is more closely associated with IQ than the higher order dimension, which shows little correlation with IQ.

Some work has also been done in decomposing the cognitive phenotypes associated with ASDs. It has been reported that children with ASDs from the same family have similar language and IQ scores (Szatmari et al., 1996). Part of this "familiality" on language could be due to the fact that their similarity on IQ accounts for their similarity on language, since those two dimensions are so highly correlated. Using a general estimating equation, we found that the familial aggregation of IQ was independent of the familial aggregation of language scores and vice versa (Szatmari et al., 2008). Both language and IQ are familial dimensions in ASDs, and the familiality of language is independent of that of IQ. These studies, taken together, suggest that the ASD phenotype can be decomposed into several component phenotypes, including social-communication, insistence on sameness, sensory motor behaviors, IQ, and language.

An illustration of the usefulness of this decomposing approach in uncovering genetic variants for autism was illustrated by a recent paper from the Autism Genome Project (AGP). The AGP is a collaborative effort of many different research groups around the world. Pooling of individual samples resulted in a sample size of approximately 1,500 families (Autism Genome Project, 2007). Common genotyping and clinical measures were taken for all affected relatives. Several component phenotypes were used in this study: language delay, absence of intellectual disability, quantitative measures of social reciprocity, and repetitive behaviors from the ADI-R. The two highest linkage signals came from chromosomes 11 and 15 (Liu, Paterson, & Szatmari, 2008). On chromosome 15, individuals who had IQs above 70 had a lod score of 4.0. On chromosome 11, those with delayed language had a lod score of 3.4. These scores were larger than the highest lod score when ASD was used as the phenotype of classification. Although no genes have yet been identified in these regions, it has been demonstrated that using component phenotypes may be more productive and helpful in linkage studies than using a composite phenotype like ASD. There are several other examples where component phenotypes have been used in linkage analysis of autism (Alarcon, Cantor, Lieu, Gilliam, & Geschwind, 2002; Bradford et al., 2001; Duvall et al., 2007) and other disorders (Allan, Cardno, & McGuffin, 2008; Almasy, 2003; Barr, 2001; Faraone et al., 1995; Fiedorowicz, Epping, & Flaum, 2008; Francks, MacPhie, & Monaco, 2002; Rommelse, Arias-Vásquez, et al., 2008).

There are, however, a number of potential pitfalls to using component phenotypes in genetic studies, and these should be kept in mind. It is important to remember that such phenotypes should represent independent facets of a disorder. They should not be thought of as outcomes or as manifestations of comorbidity. For example, aggressive behavior is a common occurrence among individuals with ASDs. However, it is a comorbidity, often the result of intellectual disability, poor language skills, or an unaccommodating environment. It is not a part of the phenotypes that comprise the disorder. Individuals with ASDs who are also aggressive may represent a special subgroup that have a different genotype associated with them. In that case, aggression may represent a covariate or a proxy marker of a more homogeneous subgroup, but it is not by itself a component phenotype. It is also important to pay particular attention to the extent to which these component phenotypes correlate with each other. The results of factor analysis are very sensitive to the instrument used, the sample size available, and the sampling frame employed. Comparing results from different factor analytic studies thus becomes difficult, and as a result, nonreplication is all too common.

Finally, testing the usefulness of component phenotypes is difficult. Such a phenotype is defined by the fact that it is "closer" in the causal chain to the genotype than the disorder itself. The validity of the component phenotype is proved by the fact that the linkage signals should be "stronger" or the association signal should be "more significant." However, comparing linkage signals or significance values between studies is open to question, and there are no statistical tests of such comparisons. The ultimate value of a component phenotype is the extent to which it leads to quicker identification of actual genetic variants "causing" the disorder. To a certain extent, the size of the linkage signal proves nothing; it is the gene within that region that must be identified and that can still take much effort.

INTERMEDIATE PHENOTYPES

Endophenotypes is a term now commonly used in psychiatric genetics. Gottesman and Shields (1972) originally defined the term as a genetically informative phenotype that is not only more common in affected individuals than controls, but also more common in the unaffected relatives than controls. However, the term itself has been used somewhat uncritically in the literature and now often includes component phenotypes as discussed above. An endophenotype is not a characteristic of the disorder or a symptom of the disease. It usually occurs in an intermediate position in the causal chain, "up river" as it were, from a component phenotype. The key feature is that there is familial aggregation of the phenotype among affected *and* unaffected individuals within a pedigree

(Gottesman & Gould, 2003). It is not necessary that unaffected individuals score as high as affected individuals on a measure of such phenotypes. To avoid confusion with component phenotypes, we prefer to use the term *intermediate* phenotype (as defined by Carlson, Eberle, Kruglyak, & Nickerson, 2004). Examples of intermediate phenotypes include physiological, cognitive, or behavioral measures as long as the defining characteristics are present. In classic epidemiologic terms, an intermediate phenotype may be thought of as a mediating variable in the relationship between a risk factor (the genotype) and the outcome (the *DSM-IV* diagnosis). Again, the strength of relationship between the genotype and the intermediate phenotype should be greater than that between the genotype and the *DSM-IV* diagnosis.

The documentation and use of intermediate phenotypes have become more and more popular in recent years (Glahn, Bearden, Niendam, & Escamilla, 2004; Hasler, Drevets, Manji, & Charney, 2004; Keri & Janka, 2004; Lenox, Gould, & Manji, 2002). The hope is that through the detection of such intermediate phenotypes, it will be possible to gain a clearer understanding of underlying pathophysiology. In linkage and association studies, it would be possible to classify an unaffected relative (who has the intermediate phenotype) as "affected," thereby increasing the power to detect a significant linkage or association signal. This increase in power may become crucial in the context of susceptibility genes that may account for only a small amount of the variance in the etiology of psychiatric illness (Risch & Merikangas, 1996). Intermediate phenotypes can also identify specific subgroups that show a more familial version of the disorder. For example, investigators may identify specific pedigrees in which there are many "unaffected" individuals who have the intermediate phenotype. By looking at the affected individuals in these pedigrees it may be possible to identify a more familial form of the disorder. This may also increase the power to detect linkage or association.

There are two recent examples of the usefulness of employing intermediate phenotypes in detecting genetic susceptibility factors through linkage and association. One has been the broader autism phenotype among families with ASDs. It has been known for some time that parents and siblings of children with ASDs are not only at higher risk for the disorder itself, but also may be at higher risk for impairments in social reciprocity and language (Szatmari et al., 2000). These impairments, though, fall below the threshold for Asperger syndrome or PDD-NOS. For example, if one includes measures for the broader autism phenotype in twin studies, the concordance rate among monozygous twins goes up to 90%, versus 10% in dyzygous twins (Bailey et al., 1995; Le Couteur et al., 1996). The prevalence of this broader autism phenotype among unaffected siblings and parents is reported to be around 20% compared with roughly 10% in controls (Pickles et al., 1995; Szatmari et al., 2000). Duvall and colleagues applied these data to a linkage study (Duvall et al., 2007). Teachers were asked to fill out a quantitative measure of social reciprocity (the Social Responsiveness Scale, SRS; see Constantino et al., 2003) on unaffected siblings of individuals with ASDs. A quantitative trait linkage analysis was then carried out with these intermediate phenotypes, and a significant linkage signal was detected on chromosome 11 (a region similar to the one reported by the AGP in a different sample but with a much larger sample size; Szatmari et al., 2007). When the analysis was restricted to affected sib pairs only, no linkage signal was detectable, demonstrating the increased power available when a measure of the broader autism phenotype on unaffected relatives is included. The authors conclude that by incorporating intermediate phenotypes in the linkage analysis, they were able to detect a significant linkage signal that was not evident when they looked at sib pairs alone.

Another useful example of the use of intermediate phenotypes comes in the study of ADHD (Rommelse, Altink, et al., 2008). Genome-wide linkage analyses were performed in a Dutch sample with 238 ADHD probands and their 112 affected and 195 nonaffected siblings. Eight neuropsychological intermediate phenotypes and an overall score were used as quantitative traits. Significant genome-wide linkage signals were found for motor timing (on chromosome 2q21.1, lod score: 3.94) and for digit span (on 13q12.11, lod score: 3.95). Other suggestive linkage signals were found on other chromosomes as well. These results also suggest that incorporating intermediate phenotypes into linkage studies may increase the power to detect susceptibility loci in complex disorders.

One important issue in studies that use intermediate phenotypes is the degree of measurement error associated with the phenotypes. Every instrument has a certain amount of measurement error (i.e., sensitivity and specificity) associated with it, and the degree of measurement error may vary according to the population being studied. For example, the measurement error associated with assessing social reciprocity in people with autism may be quite different when their unaffected relatives are assessed. In linkage and association studies, specificity (the portion of individuals without the disorder who are identified as not having the disorder; see Sackett & Straus, 1998) is more important than sensitivity (Rice et al., 2001). Misclassifying truly unaffected individuals as affected will have an important effect on obscuring association and linkage signals, even though one gains increased power from an increased sample size (Faraone et al., 1995). Another example of this is the phenomenon of "regression to the mean"; this occurs when individuals with extreme scores on some measurement tool are selected for study and their scores "regress" to the population mean over time (Bland and Altman, 1994). A person who truly has an average score will on some occasions score at the extreme end of the distribution, simply as a reflection of random measurement error. When intermediate phenotypes are used to classify relatives, individuals are selected based on extreme scores. Some of those individuals will have true scores in the normal range, if they are to be assessed a second time. Therefore, those classified as affected with the intermediate phenotype would be misclassified in the linkage or association study and so obscure whatever relationships between genotype and phenotype might be observed. Studies that use intermediate phenotypes therefore must use measures with high specificity or sample individuals multiple times to ensure that only those who are truly affected with the intermediate phenotype are included in the analysis.

COVARIATES

The final type of informative phenotype we consider here is "covariates." Covariates are phenotypes that either obscure the relationship between genotype and disorder (unless their effects are accounted for) or identify specific subgroups in which the relationship between genotype and phenotype or disorder is particularly strong. In the classic epidemiologic literature, covariates can be either confounding or moderating variables. In a sense, a confounding variable cannot by itself be genetically informative. It does the opposite; it obscures the relationship between genotype and phenotype. But taking into account the confounding variable allows the significant relationship to become apparent in a linkage or association analysis. Age is often a covariate and might be a good example. For a particular disorder, variation in age might be associated with a variation in the phenotype that is not genetically relevant. For example, there are several linkage studies of the genetics of language ability with an affected sib-pair design (Bartlett et al., 2002; Caylak, 2007). A quantitative trait such as scores on a language

test might be used to provide more power than categorical classification as affected or unaffected. However, in this context, the similarity between sibs in their language ability will be influenced by age: older children will have better language ability than younger children. The sharing of genetic variants responsible for language variation will only become apparent once the effect of age is taken into account. Gender, treatment effects, and ethnicity are other important confounding covariates that may obscure genotype–phenotype relationships.

The more common use of a covariate as a genetically informative phenotype refers to its moderating influence. The classic example would be age of onset, such as in early-onset Alzheimer's disease or breast cancer, which becomes a phenotype for dealing with genetic heterogeneity. In this context, age of onset provides a marker for a more genetically homogeneous subgroup within a larger population of individuals with a disorder that might have many different genetic causes (Pastor & Goate, 2004). In fact, the genetic complexity of psychiatric disorders may be largely due to genetic heterogeneity. This can be either inter- or intralocus heterogeneity. *Intralocus heterogeneity* refers to the effect of different alleles at a single locus that have varying associations with the disease. The genetic locus associated with cystic fibrosis is a good example; there are many different alleles at this single locus that are associated with variation in the phenotypic expression of the disorder (Castellani et al., 2008). *Interlocus heterogeneity* refers to the possibility that two genes at distinct loci produce phenotypes that are very similar; the two different genes at loci on chromosomes 9 and 16 that cause tuberous sclerosis would be a classic example of this (Curatolo, Bombardieri, & Jozwiak, 2008). Both forms of heterogeneity are a real problem for association studies; linkage studies are not affected by intralocus heterogeneity, inasmuch as all affected relatives within a pedigree will have the same transmitted allele at the marker locus.

In ASDs, certain phenotypic features, such as age of onset, presence of epilepsy, intellectual disability, and repetitive stereotyped behaviors, may all identify specific subgroups within the wider population of children with ASDs that have a more direct relationship with the genotype than the wider, more heterogeneous phenotype. Covariates that act as moderators account for some of the variation *within* a disorder, not *between* those who are affected and unaffected. The criteria for determining whether a phenotype is a moderator covariate are somewhat different from those for the other examples of genetically informative phenotypes. With moderator covariates, there is no requirement that the phenotype be more common in cases than controls. However, a moderator still needs to be familial, that is, affected individuals from the same family or pedigree need to be more similar than affected individuals between pedigrees. To put it another way, the relative risk for the phenotype of interest among family members in this particular subgroup will be higher than in the general population. For example, the relative risk of Alzheimer's disease or breast cancer among early-onset individuals with that disease is greater than in other individuals with those disorders, or in those with later age of onset. Once again, the ultimate test of a covariate is in the pudding; either the linkage or the association signal would be stronger in that particular subset than in the general population of individuals with that disorder (Devlin et al., 2002; Hauser et al., 2004).

An example of the potentially informative covariate of this sort would be gender in children with autism. It is well known that autism is more common by about 4 to 1 among boys than among girls. There has never been an adequate explanation of this gender bias; it cannot be explained by autosomal or X-linked transmission (Szatmari, Jones, Zwaigenbaum, & MacLean, 1998). One hypothesis that has been put forward (Banach et al., 2008; Tsai, Stewart, & August, 1981) is that the liability to develop autism is normally distributed in the population, suggesting a multiple gene etiology with sex-specific thresholds. There might be epigenetic or environmental effects that raise the threshold needed to develop autism in girls, relative to boys. Thus girls require a greater genetic load to "cross" the threshold and become affected. As a result, affected sib pairs that contain a girl ought to be more genetically loaded than sib pairs that contain only boys. Similarly, affected sib pairs with two girls should be even more loaded and have greater genetic homogeneity. It appears that this might well be the case (Szatmari et al., 2007). The affected sib pair linkage genome scan reported by the Autism Genome Project did in fact report that there were more significant linkage signals apparent among sib pairs that contain a girl than among male-only sib pairs. Chromosomal regions on 5p, 9p, and 11p had higher linkage signals in sib pairs with an affected girl than in sib pairs with only males. Whether this will translate to the finding of specific susceptibility genes remains to be determined.

The most important methodological issue in the use of covariates is multiple testing. Subdividing a population of affected individuals into various subgroups and performing linkage or association analysis in each of those subgroups represents a classic example of multiple testing. There are no methodologically acceptable ways of correcting for multiple testing. The classic Bonferroni correction may be overly conservative, because none of the subgroups are independent of others. Nevertheless, replication becomes extremely important for ensuring that the identification of significant linkage signals within these subgroups is not due to chance alone.

CONCLUSION

There is real hope that with the remarkable advances in genotyping, bioinformatics, and statistical analysis, the next decade will see real progress in the identification of the genes and molecular mechanisms that are responsible for many neurobehavioral disorders. It must be admitted, however, that the first generation of genome scans has been somewhat disappointing. It is hoped that the identification of genes in the next generation of studies will come about through the accumulation of large data sets, the investigation of gene–gene and gene–environment interaction, and the use of novel samples such as dense pedigrees with many affected individuals. Regardless of these advances, however, the entire framework depends on having more reliable and valid measurement of disorders and phenotypes. Whether one starts from linkage studies or deep sequencing, accuracy of measurement and the genetic informativeness of phenotypes will be the key to success. There is currently a debate in the field over whether genetic studies of complex disorders should put a priority on collecting very large sample sizes and include only minimal phenotype information as a way of maximizing power and keeping costs down. There are others (this author included) who believe this is putting the cart before the horse. To continue with the metaphor, genetically informative phenotypes are the "horses" that will do the hard work in allowing scientists to make those exciting discoveries. Otherwise, we will not know which phenotype the genetic mechanism might explain. The careful and precise measurements of disorder, of component phenotypes, of intermediate phenotypes and covariates will continue to be essential stepping-stones as the field moves slowly (and, all too often, painfully) to success.

REFERENCES

Alarcon, M., Cantor, R., Lieu, J., Gilliam, T., & Geschwind, D. (2002). Evidence for a language quantitative trait locus on chromosome 7q in multiple autism families. *American Journal of Human Genetics, 70*(1), 60–71.

Allan, C.L., Cardno, A.G., & McGuffin, P. (2008). Schizophrenia: From genes to phenes to disease. *Current Psychiatry Reports, 10*(4), 339–343.

Almasy, L. (2003). Quantitative risk factors as indices of alcoholism susceptibility. *Annals of Medicine 35,* 337–343.

American Psychiatric Association. (1968). *Diagnostic and statistical manual of mental disorders* (2nd ed.). Washington, DC: Author.

American Psychiatric Association. (1980). *Diagnostic and statistical manual of mental disorders* (3rd ed.). Washington, DC: Author.

American Psychiatric Association. (1994). *Diagnostic and statistical manual of mental disorders* (4th ed.). Washington, DC: Author.

American Psychiatric Association. (2000). *Diagnostic and statistical manual of mental disorders* (4th ed., text rev.). Washington, DC: Author.

Autism Genome Project Consortium. (2007). *Nature Genetics, 39*(3), 319–328.

Bailey, A., Le Couteur, A., Gottesman, I., Bolton, P., Simonoff, E., Yuzda, E., et al. (1995). Autism as a strongly genetic disorder: Evidence from a British twin study. *Psychological Medicine, 25*(1), 63–77.

Banach, R., Thompson, A., Szatmari, P., Goldberg, J., Tuff, L., Zwaigenbaum, L., et al. (2008). Brief report: Relationship between non-verbal IQ and gender in autism. *Journal of Autism and Developmental Disorders, 39*(1), 188–193.

Baron, M. (1995). Searching for complex disease genes: Can it be made easier? *Psychiatric Genetics, 5*(2), 89–91.

Barr, W. (2001). Schizophrenia and attention deficit disorder. Two complex disorders of attention. *Annals of the New York Academy of Science, 931,* 239–250.

Bartlett, C.W., Flax, J.F., Logue, M.W., Vieland, V.J., Bassett, A.S., Tallal, P., et al. (2002). A major susceptibility locus for specific language impairment is located on 13q21. *American Journal of Human Genetics, 71*(1), 45–55.

Bearden, C., Reus, V., & Freimer, N. (2004). Why genetic investigation of psychiatric disorders is so difficult. *Current Opinion in Genetics and Development, 14,* 280–286.

Beglinger, L.J., & Smith, T.H. (2001). A review of subtyping in autism and proposed dimensional classification model. *Journal of Autism and Developmental Disorders, 31*(4), 411–422.

Bland, J., & Altman, D. (1994). Some examples of regression towards the mean. *British Medical Journal, 309,* 780.

Bradford, Y., Haines, J., Hutcheson, H., Gardiner, M., Braun, T., Sheffield, V., et al. (2001). Incorporating language phenotypes strengthens evidence of linkage to autism. *American Journal of Medical Genetics, 105*(6), 539–547.

Cannon, T.D., & Keller, M.C. (2006). Endophenotypes in the genetic analyses of mental disorders. *Annual Review of Clinical Psychology,* *2,* 267–290.

Carlson, C., Eberle, M., Kruglyak, L., & Nickerson, D. (2004). Mapping complex disease loci in whole-genome association studies. *Nature, 429,* 446–452.

Castellani, C., Cuppens, H., Macek, M., Jr., Cassiman, J.J., Kerem, E., Durie, P., et al. (2008). Consensus on the use and interpretation of cystic fibrosis mutation analysis in clinical practice. *Journal of Cystic Fibrosis, 7*(3), 179–196.

Caylak, E. (2007). A review of association and linkage studies for genetical analyses of learning disorders. *American Journal of Medical Genetics B Neuropsychiatric Genetics, 144B*(7), 923–943.

Cohen, E., Chow, E.W., Weksberg, R., & Bassett, A.S. (1999). Phenotype of adults with the 22q11 deletion syndrome: A review. *American Journal of Medical Genetics, 86*(4), 359–365.

Constantino, J.N., Davis, S.A., Todd, R.D., Schindler, M.K., Gross, M.M., Brophy, S.L., et al. (2003). Validation of a brief quantitative measure of autistic traits: Comparison of the social responsiveness scale with the Autism Diagnostic Interview-Revised. *Journal of Autism and Developmental Disorders, 33*(4), 427–33.

Constantino, J.N., & Todd, R.D. (2003). Autistic traits in the general population: A twin study. *Archives of General Psychiatry, 60*(5), 524–530.

Curatolo, P., Bombardieri, R., & Jozwiak, S. (2008). Tuberous sclerosis. *Lancet, 23;372*(9639), 657–668.

Devlin, B., Bacanu, S.-A., Klump, K., Bulik, C., Fichter, M., Halmi, K., et al. (2002). Linkage analysis of anorexia nervosa incorporating behavioral covariates. *Human Molecular Genetics, 11*(6), 689–696.

Duvall, J.A., Lu, A., Cantor, R.M., Todd, R.D., Constantino, J.N., & Geschwind, D.H. (2007). A quantitative trait locus analysis of social responsiveness in multiplex autism families. *American Journal of Psychiatry, 164*(4), 656–662.

Dykens, E.M., & Hodapp, R.M. (2001). Research in mental retardation: Toward an etiologic approach. *Journal of Child Psychology and Psychiatry, 42*(1), 49–71.

Faraone, S.V., Kremen, W.S., Lyons, M.J., Pepple, J.R., Seidman, L.J., & Tsuang, M.T. (1995). Diagnostic accuracy and linkage analysis: How useful are schizophrenia spectrum phenotypes? *American Journal of Psychiatry, 152*(9), 1286–1290.

Fiedorowicz, J.G., Epping, E.A., & Flaum, M. (2008). Toward defining schizophrenia as a more useful clinical concept. *Current Psychiatry Reports, 10*(4), 344–351.

Francks, C., MacPhie, I.L., & Monaco, A.P. (2002). The genetic basis of dyslexia. *Lancet Neurology, 1*(8), 483–490.

Gabbett, M.T., Peters, G.B., Carmichael, J.M., Darmanian, A.P., & Collins, F.A. (2008). Prader-Willi syndrome phenocopy due to duplication of Xq21.1-q21.31, with array CGH of the critical region. *Clinical Genetics, 73*(4), 353–359.

Gabriels, R.L., Cuccaro, M.L., Hill, D.E., Ivers, B.J., & Goldson, E. (2005). Repetitive behaviors in autism: Relationships with associated clinical features. *Research in Developmental Disabilities, 26*(2), 169–181.

Georgiades, S., Szatmari, P., Zwaigenbaum, L., Duku, E., Bryson, S., & Roberts, W. (2007). Structure of the autism symptom phenotype: A proposed multidimensional model. *Journal of the American Academy of Child and Adolescent Psychiatry, 46*(2), 188–196.

Gillespie, N., Kirk, K., Evans, D., Heath, A., Hickie, I., & Martin, N. (2004). Do the genetic or environmental determinants of anxiety and depression change with age? A longitudinal study of Australian twins. *Twin Research, 7,* 39–53.

Glahn, D., Bearden, C., Niendam, T., & Escamilla, M. (2004). The feasibility of neuropsychologial endophenotypes in the search for genes associated with bipolar affective disorder. *Bipolar Disorders, 61,* 171–182.

Gottesman, I., & Gould, T. (2003). The endophenotype concept in psychiatry: Etiology and strategic intention. *American Journal of Psychiatry, 160,* 636–645.

Gottesman, I., & Shields, J. (1972). *Schizophrenia and genetics: A twin study vantage point.* New York: Academic Press

Gould, T., & Manji, H. (2004). The molecular medicine revolution and psychiatry: Bridging the gap between basic neuroscience research and clinical psychiatry. *Journal of Clinical Psychiatry, 65,* 598–604.

Grigorenko, E., & Pauls, D. (2003). Analytical methods applied to psychiatric genetics. *Methods in Molecular Medicine, 77,* 23–61.

Gropman, A.L., Duncan, W.C., & Smith, A.C. (2006). Neurologic and developmental features of the Smith-Magenis syndrome (del 17p11.2). *Pediatric Neurology, 34*(5), 337–350.

Hasler, G., Drevets, W., Manji, H., & Charney, D. (2004). Discovering endophenotypes for major depression. *Neuropsychopharmacology, 29,* 1765–1781.

Hauser, E., Watanabe, R., Duren, W., Bass, M., Langefeld, C., & Boehnke, M. (2004). Ordered subset analysis in genetic linkage mapping of complex traits. *Genetic Epidemiology, 27*(1), 53–63.

Iacono, W.G., McGue, M., & Krueger, R.F. (2006). Minnesota Center for Twin and Family

Research. *Twin Research and Human Genetics,* *9*(6), 978–984.

Inoue, K., & Lupski, J. (2003). Genetics and genomics of behavioural and psychiatric disorders. *Current Opinion in Genetics and Development,* *13,* 303–309.

Kaprio, J., & Koshenvuo, M. (2002). Genetic and environmental factors in complex disease: The older Finnish Twin Cohort. *Twin Research, 5,* 358–365.

Keri, S., & Janka, Z. (2004). Critical evaluation of cognitive dysfunctions as endophenotypes of schizophrenia. *Acta Psychiatrica Scandinavica, 110,* 83–91.

Le Couteur, A., Bailey, A., Goode, S., Pickles, A., Robertson, S., Gottesman, I., et al. (1996). A broader phenotype of autism: The clinical spectrum in twins. *Journal of Child Psychology and Psychiatry, 37*(7), 785–801.

Lenox, R., Gould, T., & Manji, H. (2002). Endophenotypes in bipolar disorder. *American Journal of Medical Genetics, 114,* 391–406.

Liu, X.Q., Paterson, A.D., & Szatmari, P. (2008). Autism Genome Project Consortium. Genome-wide linkage analyses of quantitative and categorical autism subphenotypes. *Biological Psychiatry, 64*(7), 561–570.

Lord, C., Pickles, A., McLennan, J., Rutter, M., Bregman, J., Folstein, S., et al. (1997). Diagnosing autism: Analyses of data from the Autism Diagnostic Interview. *Journal of Autism and Developmental Disorders, 27*(5), 501–517.

Lord, C., Rutter, M., & Couteur, A.L. (1994). Autism Diagnostic Interview-Revised: A revised version of a diagnostic interview for caregivers of individuals with possible pervasive developmental disorders. *Journal of Autism and Developmental Disorders, 24,* 659–685.

Marshall, C.R., Noor, A., Vincent, J.B., Lionel, A.C., Feuk, L., Skaug, J., et al. (2008). Structural variation of chromosomes in autism spectrum disorder. *American Journal of Human Genetics, 82*(2), 477–488.

McGuffin, P. (2005). The impact of genetics on child psychiatry: A 20-year perspective. *Current Psychiatry Reports, 7*(2), 115–116.

Nurnberger, J., Jr. (2002). Implications of multifactorial inheritance for identification of genetic mechanisms in major psychiatric disorders. *Psychiatric Genetics, 12,* 121–126.

Osborne, L.R., & Mervis, C.B. (2007). Rearrangements of the Williams-Beuren syndrome locus: Molecular basis and implications for speech and language development. *Expert Reviews in Molecular Medicine, 9*(15), 1–16.

Ott, J., & Bhat, A. (1999). Linkage analysis in heterogeneous and complex traits. *European Child and Adolescent Psychiatry, 8*(Suppl. 3), 43–46.

Pastor, P., & Goate, A. (2004). Molecular genetics of Alzheimer's disease. *Current Psychiatry Reports, 6,* 125–133.

Pickles, A., Bolton, P., Macdonald, H., Bailey, A., Le Couteur, A., Sim, C.H., et al. (1995). Latent-class analysis of recurrence risks for complex phenotypes with selection and measurement error: A twin and family history study of autism. *American Journal of Human Genetics, 57*(3), 717–726.

Rachidi, M., & Lopes, C. (2008). Mental retardation and associated neurological dysfunctions in Down syndrome: A consequence of dysregulation in critical chromosome 21 genes and associated molecular pathways. *European Journal of Paediatric Neurology, 2*(3), 168–182.

Raymond, F.L., & Tarpey, P. (2006). The genetics of mental retardation. *Human Molecular Genetics, 15,* 110–116.

Rice, T., & Borecki, I. (2001). Familial resemblance and heritability. *Advances in Genetics, 42,* 35–44.

Rice, J., Saccone, N., & Rasmussen, E. (2001). Definition of the phenotype. *Advances in Genetics, 42,* 69–76.

Risch, N., & Merikangas, K. (1996). The future of genetic studies of complex human diseases. *Science, 273*(5281), 1516–1517.

Rommelse, N.N., Altink, M.E., Oosterlaan, J., Buschgens, C.J., Buitelaar, J., & Sergeant, J.A. (2008). Support for an independent familial segregation of executive and intelligence endophenotypes in ADHD families. *Psychological Medicine, 38*(11), 1595–1606.

Rommelse, N.N., Arias-Vásquez, A., Altink, M.E., Buschgens, C.J., Fliers, E., Asherson, P., et al. (2008). Neuropsychological endophenotype approach to genome-wide linkage analysis identifies susceptibility loci for ADHD on 2q21.1 and 13q12.11. *American Journal of Human Genetics, 83*(1), 99–105.

Ronald, A., Happé, F., Bolton, P., Butcher, L.M., Price, T.S., Wheelwright, S., et al. (2006). Genetic heterogeneity between the three components of the autism spectrum: A twin study. *Journal of the American Academy of Child and Adolescent Psychiatry, 45*(6), 691–699.

Rutter, M. (2002). The interplay of nature, nurture, and developmental influences: The challenge ahead for mental health. *Archives of General Psychiatry, 59*(11), 996–1000.

Sackett, D.L., & Straus, S. (1998). On some clinically useful measures of the accuracy of diagnostic tests. *ACP Journal Club, 129,* A17–A19.

Silverman, J.M., Smith, C.J., Schmeidler, J., Hollander, E., Lawlor, B.A., Fitzgerald, M., et al. (2002). Autism Genetic Research Exchange Consortium. Symptom domains in autism and related conditions: Evidence for familiality. *American Journal of Medical Genetics, 114*(1), 64–73.

Skuse, D.H. (1997). Genetic factors in the etiology of child psychiatric disorders. *Current Opinion in Pediatrics, 9*(4), 354–360.

Skuse, D.H. (2007). Rethinking the nature of genetic vulnerability to autistic spectrum disorders. *Trends in Genetics, 23*(8), 387–395.

Slavotinek, A.M. (2008). Novel microdeletion syndromes detected by chromosome microarrays. *Human Genetics, 124*(1), 1–17.

Sparrow, S., Balla, D., & Cicchetti, D. (1984). *Vineland Adaptive Behavior Scales (VABS)*. Circle Pines, MN: American Guidance Service.

Stoltenberg, S., & Burmeister, M. (2000). Recent progress in psychiatric genetics—some hope but no hype. *Human Molecular Genetics, 9*, 927–935.

Szatmari, P. (2003). The causes of autism spectrum disorders. *British Medical Journal, 326*(7382), 173–174.

Szatmari, P., Georgiades, S., Bryson, S., Zwaigenbaum, L., Roberts, W., Mahoney, W., et al. (2006). Investigating the structure of the restricted, repetitive behaviours and interests domain of autism. *Journal of Child Psychology and Psychiatry, 47*(6), 582–590.

Szatmari, P., Jones, M.B., Holden, J., Bryson, S., Mahoney, W., Tuff, L., et al. (1996). High phenotypic correlations among siblings with autism and pervasive developmental disorders. *American Journal of Medical Genetics, 67*(4), 354–360.

Szatmari, P., Jones, M., Zwaigenbaum, L., & MacLean, J. (1998). Genetics of autism: Overview and new directions. *Special Issue of the Journal of Autism and Developmental Disorders, 28*(5), 363–380.

Szatmari, P., Maclean, J., Jones, M., Bryson, S., Zwaigenbaum, L., Bartolucci, G., et al. (2000). The familial aggregation of the lesser variant in biological and non-biological relatives of PDD probands: A family history study. *Journal of Child Psychology and Psychiatry, 41*(5), 579–586.

Szatmari, P., Maziade, M., Zwaigenbaum, L., Mérette, C., Roy, M.A., Joober, R., et al. (2007). Informative phenotypes for genetic studies of psychiatric disorders. *American Journal of Medical Genetics B Neuropsychiatric Genetics, 144B*(5), 581–588.

Szatmari, P., Merette, C., Bryson, S., Thivierge, J., Roy, M.-A., Cayer, M., et al. (2002). Quantifying dimensions in autism: A factor analytic study. *Journal of the American Academy of Child and Adolescent Psychiatry, 41*, 467–474.

Szatmari, P., Mérette, C., Emond, C., Zwaigenbaum, L., Jones, M.B., Maziade, M., et al. (2008). Decomposing the autism phenotype into familial dimensions. *American Journal of Medical Genetics B Neuropsychiatric Genetics, 147B*(1), 3–9.

Tsai, L., Stewart, M.A., & August, G. (1981). Implication of sex differences in the familial transmission of infantile autism. *Journal of Autism and Developmental Disorders, 11*(2), 165–731.

Tsuang, M. (2001). Defining alternative phenotypes of genetic studies: What can we learn from studies of schizophrenia? *American Journal of Medical Genetics, 105*(1), 8–10.

Tsuang, M., & Faraone, S. (2000). The future of psychiatric genetics. *Current Psychiatry Reports, 2*(2), 133–136.

Zlotogora, J. (2003). Penetrance and expressivity in the molecular age. *Genets in Medicine, 5*(5), 347–352.

16

Social Phenotypes in Genetically Based Neurodevelopmental Disorders

Carl Feinstein and Shivani Verma

A rapidly growing body of clinical neuroscience research describes how specific gene defects may have a profound effect on human cognition and behavior. In particular, the study of genetically based neurodevelopmental disorders (GNDDs) has been instrumental in elucidating gene–brain–behavior relationships by examining the behavioral phenotypes found in people with known genetic variations. Several of these GNDDs have, indeed, been shown to present with distinctive patterns of cognitive and behavioral features, in addition to the medical sequelae originally used to make the diagnosis (Feinstein & Chahal, 2009). However, it has only recently become clear that unique social traits and disabilities can also be distinguished from more general patterns of behavior and mood in GNDDs and that these may also derive from specific gene or chromosomal deletions and mutations (Feinstein & Singh, 2007).

In this chapter we focus on the distinctive social traits of several GNDDs. Specifically, we review the genetic etiology and distinctive behavioral and social traits of seven genetically based syndromes: fragile X, Down, Prader-Willi, Smith-Magenis, Turner, Williams, and velocardiofacial syndromes, emphasizing the unique social features observed in people with these conditions. To set the stage for this type of review, it is necessary first to provide a short history of the development and controversies surrounding this research. In particular, we review how scientific advances in the genetic identification of GNDDs have resulted in a shift in how their behavioral phenotypes were studied and the resulting benefits and drawbacks of this shift.

PHENOTYPE AND BEHAVIORAL PHENOTYPE: DEFINITIONS

In this chapter we rely on the following general definition of *phenotype:* the "observable characteristics of an organism, which are the joint product of both genotypic and environmental influences" (Gottesman & Gould, 2003). The term *behavioral phenotype* as we use it is based on the definition provided by Dykens and Cassidy (1995): "The heightened probability or likelihood that people with a given syndrome will exhibit certain behavioral and developmental sequelae relative to those without the syndrome." This definition of behavioral phenotype is further elaborated by the Society for the Study of Behavioural Phenotype as follows: "A characteristic pattern of motor, cognitive, linguistic and/or social abnormalities which is consistently associated with a biological disorder" (1998).

We distinguish between the broader concept of behavioral phenotypes and a narrower version that we refer to as the *psychiatric phenotype*. The psychiatric phenotype is a description of a subset of behavioral/ developmental characteristics in groups of individuals that results when clinical data describing the behavior of people with a particular genotype are confined to descriptors found on standardized structured and semi-structured research diagnostic interviews that generate *Diagnostic and Statistical Manual of Mental Disorders, Fourth Edition, Text Revision* (*DSM-IV-TR;* American Psychiatric Association, 2000) and World Health Organization psychiatric diagnoses (Feinstein & Singh, 2007). As such, the psychiatric phenotype is a far more limited construct, restricted only to those descriptors and classifications of overall personality, behavior, and cognition codified in the standard diagnostic manual prevailing at the time a particular clinical research study was conducted.

DISTINCTIONS BETWEEN "PSYCHIATRIC PHENOTYPES" AND "BEHAVIORAL PHENOTYPES"

How did the psychiatric phenotype come to replace more comprehensive narrative descriptions of behavior? Before the discovery of the specific gene mutations, chromosomal deletions, or copy number variations that underlie the various GNDDs, clinical researchers studying developmental disabilities could rely only on careful observation and history-taking, documented by descriptive reports, to detect distinctive and recurring patterns of cognitive, medical, and behavioral traits among more homogeneous subgroups of these children. These more homogeneous clinical syndromes could then be studied with regard to their chromosomal and genetic makeup to determine the genetic etiology of the disorder (Feinstein & Singh, 2007). Finding the genetic etiology of these syndromes, in turn, made possible the biological validation of the disorders, so that the diagnosis could be made by a standardized, reliable laboratory test. However, when laboratory genetic testing became the basis of diagnosis, careful, observer-based behav-

ioral observations and descriptions was no longer required. The clinical neuroscience of behavioral neurogenetics became possible, in which the cognitive and behavioral sequelae of known genetic abnormalities could be studied and even associated with abnormalities in brain morphology and function (Reiss & Dant, 2003). A second-generation research approach became prevalent. Attention shifted to molecular genetics and neuroimaging, and it became more convenient to use reliable "off the shelf" rating scales or highly structured interviews to elicit behavior symptoms codified as psychiatric diagnostic criteria in the prevailing *DSM* or World Health Organization manuals. These symptom lists became the proxy for behavior in the study of people with GNDDs.

However, parents, caregivers, teachers, and specialized clinicians cannot escape the reality that children with genetically based developmental disorders commonly express perplexing and unique behavioral patterns that often include symptoms from several different standard psychiatric diagnoses, as well as distinctive behavior traits and symptoms that are not described by the psychiatric diagnostic nomenclature (Feinstein & Chahal, 2009; Feinstein & Singh, 2007; Skuse, 2000). In addition, Gould and Gottesman (2006) have reminded us that the disorders cataloged by our current psychiatric diagnostic system lack a biologically validated etiology, but instead still rely on clusters of behaviors. It is fundamentally illogical to use non–biologically validated, behaviorally defined psychiatric diagnoses to classify children who have a distinctive, biologically validated genetic condition. Too much data concerning the cognitive and behavioral profiles of children with biologically validated disorders are lost if only standard symptoms from the psychiatric nomenclature are used to describe these conditions.

Thus, reliance on a "psychiatric phenotype" to describe behavior and social traits in GNDDs has two distinct drawbacks: 1) It risks missing important behavioral and social traits found in GNDDs that are *not* represented in psychiatric symptom checklists (if you are not looking for it, you will probably fail to observe it), and 2) It attempts illogically to map the unique phenotype of a biologically validated genetic disorder to heterogeneous behaviorally defined syndromes that lack biological validation. Furthermore, as will become clear in this chapter, children with particular genetically based syndromes have complex and distinctive traits of social behavior. Use of the standard psychiatric terminology to describe the social phenotype in children with GNDDs enormously constrains the discussion to the single topic of whether the children have an autism spectrum disorder (ASD), even when it is evident to all concerned clinicians and caregivers that there are other prominent or distinct social behaviors present (Feinstein & Reiss, 1998).

DIAGNOSTIC OVERSHADOWING: ATTRIBUTING SOCIAL IMPAIRMENTS SOLELY TO COGNITIVE IMPAIRMENTS

Another type of bias, diagnostic overshadowing, must be addressed in the study of social behaviors in GNDDs (Jopp & Keys, 2001). Diagnostic overshadowing assumes that intellectual disability alone explains all of the variations from normal behavioral and social functioning found in people with GNDDs. Although limitations in cognitive capacity, along with a myriad of family and social forces, certainly are influential in the development of distinctive social and behavioral patterns in individuals with GNDDs, it is still important to determine the extent to which their unique genetic differences directly influence their social phenotype.

Fragile X Syndrome

Fragile X syndrome (FXS) is the most common known genetic cause of intellectual disability. It has been diagnosed in up to 3% of males with special needs (McConkie-Rosell et al., 2005), occurring in 1 in every 4,000 boys and 1 in every 6,000 to 8,000 girls (Hagerman, 1999). For decades, the cause of the syndrome was unknown, but in 1991 the genetic cause was found to be an expansion mutation in the form of a CGG triplicate repeat of the FMR1 gene of the X chromosome. Whereas the number of repeats is about 6–44 in unaffected individuals, at the upper limit of the normal range instability begins to appear, with an increased risk of expansion of the number of repeats from generation to generation into what is termed the "premutation range" (55–200) and then the full mutation range (more than 200 repeats). A full mutation often results in hypermethylation silencing of the gene and consequently no production of the FMR protein. This full mutation state results in the distinctive traits and disabilities associated with FXS (Reiss & Dant, 2003). More recently, the premutation has also been shown to be associated with similar syndromic features, as well as distinct medical and cognitive findings (Hagerman & Hagerman, 2008).

Children with fragile X often appear typical at birth, but soon grow to show a characteristic physical, behavioral, and social phenotype. Related to gender, imprinting, mosaicism, degree of penetrance of the partial mutation, and other factors, there are several phenotypic variants expressed in affected individuals. These variants present with different types and degrees of cognitive, behavioral, and social impairments (Reiss & Dant, 2003). Thus, some individuals may have typical intelligence and only mild impairments in behavior and socialization, whereas others have severe intellectual disability and autism.

In this chapter we focus first on the classic behavioral and social features of FXS, as expressed when the full mutation is present and penetrant, and then describe briefly some of the more common variants. The classic fragile X behavioral phenotype includes impairments in attention, impulsivity, organizational problems, mood instability, hand flapping, and hand biting (Eliez & Feinstein, 2001; Hagerman, 2002; Mazzocco, 2000).

People with the full fragile X phenotype also display a unique *social* phenotype, characterized by social anxiety, increased time to initiate social interaction, difficulty forming meaningful peer relationships, social withdrawal, and high emotionality. Eye contact, a pivotal element of human social interaction, is aberrant in FXS. Many people with FXS have poor eye contact, gaze aversion, and decreased accuracy in judging direction of gaze (Holsen, Dalton, Johnstone, & Davidson, 2008; Kates, Abrams, Kaufmann, Breiter, & Reiss, 1997). These highly distinctive gaze features may be related to underlying neurophysiologic and neuroendocrine abnormalities in the stress-related cortisol response for eye gaze and eye contact (Hessl, Glaser, Dyer-Friedman, & Reiss, 2006). A recent study using functional magnetic resonance imaging techniques found that boys with FXS process gaze differently than controls, with greater sensitization in the left amygdala and left insula in reaction to direct gaze when compared with controls (Watson, Hoeft, et al., 2008). Preliminary data have also suggested that social anxiety in FXS may be related to the inability to successfully recruit higher-level social cognition regions in the prefrontal cortex during face encoding (Holsen et al., 2008).

In addition, people with FXS have problems with "theory of mind," manifested by difficulties in processing and integrating the social context in which interactions occur and in reading the emotional and social responses of others (Cornish, Burack, Rahman, Russo, & Grant, 2005). These impairments are present even in relatively high-functioning males and mildly affected females (Kaufmann, Corell, Kau, et al., 2004; Mazzocco, Pennington, & Hagerman, 1994; Merenstein, Sobesky, Taylor, RIddle, Tran, & Hagerman, 1996). Individuals with FXS do appear to be interested in socialization, but their inability to maintain eye contact and their atypical social behaviors make this difficult outside of the family (Kau et al., 2004).

Females, in general, present with fewer and milder symptoms than males because they have two X chromosomes, one with the mutation and one without. Social withdrawal and anxiety are still relatively common among girls with fragile X, even in the presence of typical intelligence (suggesting that this social dysfunction is not due to intelligence, but rather to independent deficits on social cognition; Freund, Reiss, & Abrams, 1993; Lachiewicz, 1992; Mazzocco et al., 1994). There is increasing evidence that social impairment exists even among people without the full fragile X mutation ("premutation carriers"). Hessl et al. found that male (and to a lesser extent female) premutation carriers reported higher rates of anxiety, hostility, and obsessive-compulsive symptoms compared with age-matched controls (Hessl et al., 2005). This research suggests that social impairments exist in a spectrum, even among those without the full mutation (premutation carriers), and among those with one normal X chromosome (females), and that they are not solely a result of impairments in intelligence and cognitive skills. Additional studies are needed to confirm these findings (Van Esch, 2006).

A considerable literature has been devoted to the relationship between autism and FXS. However, the fragile X phenotype is a biologically validated syndrome caused by a well-described gene mutation, and it is associated with a highly unique behavioral phenotype, whereas the etiology of autism spectrum disorders (although many genetic and some environmental factors have been implicated) is still unclear and the social and behavioral phenotypes encompass a more diverse and heterogeneous set of symptoms (Feinstein & Reiss, 1998; Szatmari et al., 2007). Nevertheless, given that individuals with FXS frequently have social and/or communicative impairments that qualify them for a diagnosis of an ASD, it is clear that the FMR1 mutation is one of many genetic conditions that can lead to an autism spectrum disorder. In this sense, it is a biologically validated model of how genetic effects can lead to some of the cardinal features of autism.

Last, in addition to neuroimaging techniques and clinical studies, researchers have also attempted to use animal models to learn about the social phenotype of FXS (McNaughton et al., 2008; Mineur, Huynh, & Crusio, 2006; Spencer et al., 2005). In a review of these studies, Brodkin found that the evidence was limited, but that it did confirm increased anxiety among the FMR1 knockout mouse compared with the wild-type mouse. The approach–avoidance behaviors, however, showed conflicting results. These differences may have been due to differences in study design, as well as the types of mice used, and further research is needed (Brodkin, 2008).

Down Syndrome

Down syndrome is the most common chromosomal cause of intellectual disability; it occurs in 1 in 1,000 live births, and the incidence increases with advancing maternal age. The genetic origin lies in trisomy of all or a critical portion of chromosome 21 (Korenberg et al., 1994). Individuals with Down syndrome have distinctive facial features, duodenal stenosis, cardiac malformations, and intellectual disability (Epstein et al., 1991; Korenberg et al., 1994; Roizen, 2002).

Children with Down syndrome are generally described as engaging and affectionate, with social communication and relationships that are comparable to those of typically developing controls (Laws and Bishop, 2004). Compared with children with intellectual disability of mixed etiology, children with Down syndrome have less risk of significant psychopathology, although this risk is still higher than that of typically developing peers. The nature of the psychopathology in Down syndrome changes along with age— younger children showed more externalizing behaviors (e.g., opposing/refusing, impul-

siveness, and inattention), whereas adolescents and young adults showed more internalizing behaviors (e.g., shy/insecure, low self-confidence) (Nicham et al., 2003).

Even in the context of generally positive traits of social engagement, a growing body of research indicates that children with Down syndrome have notable underlying impairments in social cognition. These impairments appear to center on poor recognition of facial emotional expression in others, particularly fear and, to a lesser extent, surprise (Williams, Wishart, Pitcairn, & Willis, 2005). In a recent study, children with Down syndrome were also found to have difficulty recognizing neutral facial expressions, tending to interpret them as overly positive (Barisnikov, Hippolyte, & Van der Linden, 2008). Some researchers have suggested that this impaired ability to recognize neutral expressions may lead to socially inappropriate behavior and relationship difficulties (Soresi & Nota, 2000).

Another indication that Down syndrome is associated with underlying impairments in social cognition is the finding that a subgroup of up to 7%–10% of people with Down syndrome meets the diagnostic criteria for *DSM* autism (Bregman & Volkmar, 1988; Capone et al., 2005; Ghaziuddin, 1997; Howlin, Wing, & Gould, 1995; Kent, Evans, Moli, & Sharp, 1999; Rasmussen, Borjesson, Wentz, & Gillberg, 2001; Starr et al., 2005). This repeatedly confirmed finding indicates that some neurobiological feature of Down syndrome confers an approximately tenfold increase in risk for autism. Ghaziuddin found that first-degree relatives of individuals with Down syndrome and autism have a significantly higher rate of traits of the broader autism phenotype than relatives of Down syndrome without autism. This finding suggests that Down syndrome is a potent risk factor for autism in the coincidental presence of other genetic risk factors for autism in a given child with Down syndrome, or that something about the chromosomal abnormality expresses a predisposition toward traits characteristic of autism. It is important to be vigilant about this comorbidity, because a child with Down syndrome and autism requires a more specialized clinical approach to behavior management (Ghosh, Shah, Dhir, & Merchant, 2008).

Prader-Willi Syndrome

Prader-Willi syndrome is a chromosomal disorder that occurs in 1 in 10,000 to 15,000 births. It is the most common syndromic form of obesity, affecting about 350,000 individuals throughout the world, and results from a missing paternally imprinted portion of chromosome 15, usually (70% of the time) occurring in the region 15q11-13 (Goldstone, Holland, Hauffa, Hokken-Koelega, & Tauber, 2008). It can also result from maternal disomy, in which two intact copies of chromosome 15 of maternal origin are present, instead of the usual combination of one maternal and one paternal chromosome 15 (Dykens & Roof, 2008). Prader-Willi syndrome has a physical phenotype characterized by hyperphagia, hypotonia, hypogonadism, diminished fetal activity, muscular hypotonia, short stature, hypogonadotropic hypogonadism, small hands and feet, developmental delays, and distinct facial features (Wattendorf & Muenke, 2005).

The behavioral phenotype of Prader-Willi syndrome is most known for the insatiable appetite and excessive food intake that develop in childhood. Initially, however, infants with Prader-Willi syndrome may first present with failure to thrive and even require supplemental tube feedings. Eventually, they develop the characteristic insatiable polyphagia, likely arising from hypothalamic dysfunction in the satiety center. The food-seeking behavior becomes increasingly difficult for parents to control and is characterized by temper tantrums, begging, lying, stealing, taking food from the garbage, and attempts to eat frozen, raw, or even pet food. Other prominent behaviors include stubbornness, inflexibility, and an insistence on sameness. Children with Prader-Willi syndrome secondary to maternal uniparental disomy are at higher risk for more severe behavior problems such as bizarre rituals like playing with feces, skin picking, and anal and vaginal digging (Benarroch, Hirsch, Genstil, Landau, & Gross-Tasur, 2007). Individuals with the long type I deletion version were found to have more compulsions regarding cleanliness (i.e., excessive bathing or grooming). Those with the short type II deletion were more likely to have academic compulsions (i.e., rereading, erasing answers, and counting objects or numbers; Zarcone et al., 2007). In addition to the problematic behavioral features and obsessive/ritualistic traits outlined above, a subgroup of adults with Prader-Willi syndrome was found to have higher rates of psychotic disorders. Boer et al. (2002) found that prevalence rates of affective psychotic disorder are significantly greater in people with PWD due to uniparental disomy compared with those with PWD due to deletion. The psychotic symptoms consisted of both auditory hallucinations and persecutory delusions.

Certain aspects of the behavioral phenotype of Prader-Willi syndrome have such detrimental consequences for social adaptation that they verge on what could be more properly termed elements of the social phenotype. The imperative to seek food or engage in other unpleasant behavior, regardless of the social effect this has on others, leads to egocentric, oppositional, and externalizing demeanor, which includes antisocial acts, such as stealing and lying. Higher frequencies of stubbornness, tantrums, disobedience, and excessive talking are widely reported in people with Prader-Willi syndrome(Dykens & Cassidy, 1995). The antisocial behavior is often centered on attempts to obtain food (i.e., lying, stealing, and hiding; Akefeldt & Gillberg, 1999; Einfeld, Smith, Durvasula, Florio, & Tonge, 1999) and can result in physical attacks on others. This increased propensity to display problematic behaviors that are not responsive to social influences, consequences, or relationships cannot be viewed simply as due to intellectual disability (Clarke et al., 2002; Dykens & Kasari, 1997).

However, in addition to the poor social consequences of the egocentric drive for food or other subjective needs, people with Prader-Willi syndrome have underlying primary impairments in social cognition. A survey by Greenswag (1987) described solitary behavior, social withdrawal, and poor peer relations in Prader-Willi syndrome. A later study estimated that most patients preferred being alone, observing that many patients displayed argumentative, verbally abusive, and aggressive behavior toward others (Dykens & Kasari, 1997). In social situations, children with Prader-Willi have difficulty recognizing social cues and interpreting social situations (Koenig, Klin, & Schultz, 2004). These impairments in primary socialization overlap with symptoms of the autism spectrum, such that some individuals with Prader-Willi syndrome meet diagnostic criteria for an autism spectrum disorder. Interestingly, a higher prevalence of traits characteristic of autism has been found among Prader-Willi syndrome secondary to uniparental disomy compared with the deletion type (Dimitropoulos & Schultz, 2007). Thus, individuals with Prader-Willi syndrome have a sort of compound social deficit phenotype, combining a more primary deficit in social cognition with a secondary type that is the consequence of social consequences of driven unsocial and antisocial behaviors. These latter behaviors impede opportunities for social learning by necessitating environmental restrictions and removal from mainstream settings, leaving people with Prader-Willi syndrome fewer opportunities and more limited social context in which to develop and interact.

Smith-Magenis Syndrome

Smith-Magenis syndrome (SMS) occurs in between 1 in 15,000 to 25,000 births and is caused in 90% of cases by a de novo interstitial deletion in chromosome 17pll.2. About 10% of cases are caused by a mutation of the *RAI1* gene, which is located within the deletion site. Haploinsufficiency of *RAI1* is probably the cause of the behavioral and neurological aspects of SMS, as well as some of the associated anatomical and medical anomalies (Elsea & Girirajan, 2008; Girirajan et al., 2006). Children with SMS have brachycephaly, midface hypoplasia, prognathism, hoarse voice, speech delay (either with or without hearing loss), psychomotor and growth retardation, and behavior problems (Smith, McGavran, Robinson, et al., 1986).

The behavioral phenotype of SMS evolves from infancy, when many children with SMS are described as "perfect or beautiful babies" because of their social smile, evident responsiveness to one-to-one engagement with the caregiver, and infrequent crying. As these children reach the preschool and school years, obvious enjoyment of and the seeking out of one-to-one caregiver adult attention remains prominent. However, frequent, highly disruptive tantrums become a problem that greatly complicates adaptation, both at home and at school. These tantrums are noteworthy because of their intensity, their apparent unresponsiveness to intervention, and their attention-getting quality. As discussed later in this chapter, the tantrums lie at the interface between a behavior phenotype that is characterized on the one hand by a combination of mood dysregulation and aggressive/self-injurious outbursts and on the other hand by a very intense and dysphoric response to social frustrations.

The most salient antecedent to the tantrums of children with SMS is frustration in gaining adult one-to-one attention and dysregulated emotional outbursts in response to the withdrawal of that adult attention (Feinstein & Singh, 2007; Haas-Givler, 1994). The tantrum behaviors quickly escalate into aggression, destructiveness, and self-injury. However, these tantrums also include attention-demanding features such as public disrobing and a more deliberate type of self-injury, including nail-pulling (Dykens, Finucane, & Gayley, 1997; Dykens & Smith, 1998; Greenberg et al., 1996; Smith, Dykens, & Greenberg, 1998). Self-injury can be extensive and has been reported to result in erroneous investigations of the parents for child abuse (Smith et al., 1998). Furthermore, these tantrums often pose difficult classroom management problems for educators, in that they are extraordinarily disruptive and demanding of teacher attention and often include behaviors that may require the removal of other students from the classroom.

The admixture of social and behavioral phenotypic features for SMS is unique, and it highlights some of the more nuanced distinctions between immature social engagement and full, developmentally appropriate, reciprocal interactions. SMS children are very adult oriented, eagerly seeking one-to-one adult attention. This one-to-one attention generally involves activities of the child's choice, such as show and tell; simple, relatively noninteractive watching of the child engage in preferred activity; or the provision of kindly, enthusiastic, verbal support. Although this form of engagement is certainly social in its attention-seeking, it is minimally reciprocal and very one sided. It lacks age- or developmental-level social-cognition processes, such as taking into account the contextual exigencies or calibrating behavior to the needs of the participating adult. Furthermore, withdrawal of kindly attention by the adult, regardless of the reason, is the most common cause of the tantrums described above. To the behavior-oriented clinician, it is quite obvious that the tantrums themselves, as out of control as they seem, are powerfully attention seeking. However, they are so intense that it is difficult and sometimes impossible to ignore them. In this sense, the difficult-to-avoid social reinforcement of the tantrums appears to amplify further the delayed or negative aspects of the genetically based SMS social phenotype (Taylor & Oliver, 2008).

Turner Syndrome

Turner syndrome (TS) is a genetic disorder associated with partial or complete absence of one of the two X chromosomes in a phenotypic girl (Turner, 1938). Occurring in approximately 1 in 2,000 live female births, it results in the karyotype XO or a mosaic of 45 XO and 46 XX. The physical phenotype includes short stature, a webbed neck, abnormal pubertal development, gonadal dysgenesis, ovarian failure, renal dysgenesis, thyroid dysfunction, and cardiac malformations (Ranke & Saenger, 2001). Women with Turner syndrome are also at high risk of premature ovarian failure (Jones & Smith, 2006; Ross et al., 2004).

The behavioral phenotype of girls with TS has been shown to differ by age. Younger girls are often described as hyperactive and immature, whereas older girls are more anxious and depressed (McCauley, Ito, & Kay, 1986).

The social phenotype of girls with Turner syndrome consists of lower self-esteem, fewer friends, and fewer social activities compared with age-matched controls (McCauley et al., 1986). Whether these impairments are due to other aspects of the syndrome phenotype (i.e., short stature) remains unclear. Downey and colleagues (Downey, Ehrhardt, Gruen, Bell, & Morishima, 1989) found that women with TS have more social impairment than women with constitutional short stature alone. Schmidt and colleagues (Schmidt et al., 2006) found that women with premature ovarian failure and women who have Turner syndrome had significantly lower scores on scales of shyness, social anxiety, and self-esteem. However, they did not find a significant difference between the Turner syndrome group and the premature ovarian failure group. Reports have also found that girls with Turner syndrome who undergo hormone treatment typically experience an increase in self-concept through the course of adolescence (Christopoulos, Deligeroglou, Laggari, Christogioros, & Creatsas,

2008). Compared with their own sisters, women with Turner syndrome were found to have more social, thought, and attention problems and poor adaptive socialization skills (Mazzocco, Baumgardner, Freund, & Reiss, 1998). Recently there has been some evidence that girls with Turner syndrome have a higher risk for developing schizophrenia (Prior, Chue, & Tibbo, 2000; Roser & Kawohl, 2008). This may suggest the possibility that in some cases their social impairments are related to psychiatric disease. An investigation by Lawrence and colleagues (Lawrence et al., 2003) into gaze processing in Turner syndrome uncovered other impairments associated with social functioning. Although women who have Turner syndrome performed normally on facial recognition tasks, they showed significant impairment in the classification of expression from the upper face, particularly for expressions of "fear" from the eyes. Researchers hypothesized that this impairment is the result of overresponsiveness of the amygdala (because of its enlarged size) in detecting gaze or fear in faces. More research is needed to identify the etiology of the social phenotype of Turner syndrome. Thus far, research suggests that the social phenotype is a result of environmental factors (i.e., heightism, family environment, school environment) combined with genetic and medical factors that then affect intellectual ability and social/behavioral skills (Christopoulos et al., 2008; Rovet, 1993, 2004).

Williams Syndrome

Williams syndrome (WS) is a rare disorder, occurring in 1 in 20,000 (Morris, Demsey, Leonard, Dilts, & Blackburn, 1988) to 1 in 7,500 (Stromme, Bjornstad, & Ramstad, 2002) live births. The syndrome is caused by a hemizygous deletion of approximately 25 genes in chromosome band 7q11.23 on either paternal or maternal chromosome 7 (Jarvinen-Pasley et al., 2008). The physical phenotype consists of hypercalcemia, hyperacusis, distinctive facial features, and abnormalities of the heart, muscles, and kidneys. It is usually accompanied by mild to moderate intellectual disability (Greenberg, 1989).

The social phenotype of people with Williams syndrome is unique in that they have significant hypersociability, showing a strong interest in social interaction throughout their lives (Doyle, Bellugi, Korenberg, & Graham, 2004; Gosch & Pankau, 1994; Jarvinen-Pasley et al., 2008). This social predisposition exists despite initial delays in vocabulary acquisition, grammar use, and gesturing and is independent of cognitive impairment. For example, Doyle and colleagues found that people with Williams syndrome consistently scored higher than individuals who have Down syndrome and also have cognitive impairments but generally have well-developed sociability, implying that it is not merely a result of "lack of understanding of the social conventions governing contact with others."

Infants with WS are unusually interested in faces, sometimes staring and smiling at the experimenters rather than completing the assigned task (Laing, Butterworth, & Ansari, et al., 2002; Mervis & Robinson, 2000). They are commonly described as "overly friendly," often approaching strangers to engage them in conversation (Gosch & Pankau, 1994; Jones, Bellugi, Lai, et al., 2000). Studies have shown that individuals with Williams syndrome consistently rate unfamiliar faces as more approachable compared with typically developing controls and controls with intellectual disabilities, regardless of facial expression (Jarvinen-Pasley et al., 2008).

Paradoxically, the heightened sociability of people with WS does not lead to strong social relationships or stable friendships. Children with Williams syndrome are generally more socially anxious, and, by adulthood, most experience failure to develop and maintain friendships, suffering from social isolation and maladaptive and unsatisfying peer interactions (Gosch & Pankau, 1994;

Jarvinen-Pasley et al., 2008). This social failure stems from impairments in social cognition, which more than counterbalance the increased social-engagement-seeking drive of people with WS. The social cognitive impairments in WS manifest as impairments in theory of mind and limitations in capacity to interpret complex facial emotional expressions (Gagliardi et al., 2003; Laing & Jarrold, 2007; Plesa-Skwerer, Faja, Verbalis, Schofield, & Tager-Flusberg, 2006; Sullivan & Tager-Flusberg, 1999).

The pattern of strengths and impairments that make up the social phenotype in Williams syndrome illustrates that social functioning has several dimensions that can be dissociated from each other, including affiliativeness (which is a strength in Williams syndrome), empathy and intersubjective awareness (also a relative strength), and higher order theory of mind involving perspective taking and the incorporation of social contextual cues (a pronounced deficit in Williams syndrome; Plesa-Skwerer et al., 2006). Thus WS is an example of how high sociability does not automatically result in strong social relationships. Studies of Williams syndrome show that normal social human social adaptation is a function of multiple variables: sociability, awareness of other people's social motivation and cues, and an ability to moderate one's actions accordingly. Jarvinen-Pasley summarized the social paradox in Williams syndrome by saying,

> In sum, individuals with WS appear to show difficulties in interpreting others' behavior in terms of their mental states. Taken together with their overfriendly behavioral predisposition, it is perhaps not surprising that they have substantial problems in social adjustment, such as difficulty making and keeping friends. (Jarvinen-Pasley et al., 2008)

Velocardiofacial Syndrome

Velocardiofacial syndrome (VCFS) results from a microdeletion in the chromosome region 22q11.2 (Fine et al., 2005; Shprintzen,

2000) and occurs in approximately 1 in 4,000 people; most cases are de novo mutations (Papolos et al., 1996; Ryan et al., 1997). Between 6% and 28% are inherited from a parent with VCFS as an autosomal dominant trait (Gothelf et al., 2007). The physical phenotype includes cleft palate, velopharyngeal insufficiency, cardiac malformations, and distinctive facial features (Cohen, Chow, Weksberg, & Bassett, 1999; McDonald-McGinn et al., 1999; Shprintzen et al., 2005). Many individuals with VCFS have reduced intelligence (a mean IQ of 70); impaired language, with a dramatic delay in early language development as well as receptive and higher order language; difficulties with abstract reasoning; and visuospatial problems (Gerdes et al., 1999; Moss et al., 1999). What has drawn the most attention from psychiatry and clinical neuroscience, however, is that VCFS is the single most common known risk factor for schizophrenia (Gothelf, 2007; Gothelf, Schaer, & Eliez, 2008).

The behavioral phenotype for younger children with VCFS is characterized by mood lability, social withdrawal, poor social skills, shyness, attentional problems, oppositional defiant disorder, overactivity, disinhibition, anxiety disorders, and mood dysregulation (Feinstein & Eliez, 2000; Feinstein, Eliez, Blasey, & Reiss, 2002; Gothelf et al., 2008; Papolos et al., 1996; Swillen et al., 1999). Most toddler and preschool-age children with VCFS have mild gross motor delays and severe language delays. In school-age children, the IQ ranges in the borderline to mild intellectual disability range, with boys slightly more cognitively impaired than girls (Antshel, AbdulSabur, Roizen, & Kates, 2005; Moss et al., 1999). Children with VCFS show improvements in language ability as they grow older, although they retain impairments in higher order language skills (Moss et al., 1999).

Approximately 30% of individuals with VCFS develop psychosis—generally schizophrenia or schizoaffective disorder—by adolescence or young adulthood, accounting for approximately 2% of all cases of schizophrenia (Bassett & Chow, 1999; Gothelf et al., 2005; Gothelf et al., 2007; Murphy, Jones, & Owen, 1999). VCFS has been referred to as a neurodevelopmental model for schizophrenia because many characteristics of VCFS (attentional problems, language problems, social impairments, and learning disabilities) have also consistently been observed in studies of children and adolescents at high risk for schizophrenia (Eliez, 2004; Kravanti, Fearon, et al., 2004).

The social phenotype found in individuals with VCFS includes communication difficulties, withdrawn and shy behavior, difficulties initiating interactions, and decreased repertoire of facial expressions (Fine et al., 2005; Niklasson, Rasmussen, Oskarsdóttir, & Gillberg, 2002; Woodin et al., 2001). The degree of impairment in social-interactive function found in individuals with VCFS results in 14%–50% meeting criteria for an autism spectrum disorder (Fine et al., 2005; Vorstman et al., 2006). This clearly has important advantages for families who appropriately seek targeted services and interventions for their children with VCFS under the diagnosis of an autism spectrum disorder rather than fighting for services under the less familiar diagnosis of VCFS, which is still widely unrecognized by both health providers and educators. However, as in cases of FXS and other neurogenetic disorders, it remains an unsolved nosological inconsistency that a biologically validated, neurodevelopmental condition of known genetic etiology such as VCFS can be said to somehow "belong" to a behaviorally defined, more phenotypically heterogeneous group of disorders (autism spectrum) that likely has multiple complex and interacting etiologies. Furthermore, there are no reports of a significantly higher incidence of 22q11 deletion in individuals with autism from the general population (Gothelf et al., 2008). Even further complicating the issue of whether social impairments in VCFS should be viewed as characteristic of autism is the fact that many

of the early prodromal or premorbid clinical findings of schizophreniform psychosis are commonly described as schizotypal. There is little scientific research that distinguishes between these two overlapping clusters of symptoms, the autism spectrum and schizotypy. Indeed, some patients who are preschizophrenic or schizophrenic possibly would exceed the threshold for the diagnosis of autism if given an assessment instrument for autism.

SUMMARY

There are many subtle variations and complex features of social phenotypes that distinguish the various GNDDs from each other. The study of these genetically based variants in human sociability increasingly informs us that social functioning is not a monolithic trait, but a complex set of interacting operations that results from developmentally unfolding gene–environment interactions. The study of various GNDDs and their associated social phenotypes may help illuminate the neurodevelopmental underpinning of that most complex of all traits, human sociability, for all individuals, both with and without developmental disabilities.

REFERENCES

Akefeldt, A., & Gillberg, C. (1999). Behavior and personality characteristics of children and young adults with Prader-Willi syndrome: A controlled study. *Journal of the American Academy of Child and Adolescent Psychiatry, 38*(6), 761–769.

American Psychiatric Association. (2000). *Diagnostic and statistical manual of mental disorders* (4th ed., text rev.). Washington, DC: Author.

Antshel, K.M., AbdulSabur, N., Roizen, N., & Kates, W.R. (2005). Sex differences in cognitive functioning in velocardiofacial syndrome (VCFS). *Developmental Neuropsychology, 28*(3), 849–869.

Barisnikov, K., Hippolyte, L., & Van der Linden, M. (2008). Face processing and facial emotion recognition in adults with Down syndrome. *American Journal of Mental Retardation, 113*(4), 292–306.

Bassett, A.S., & Chow, E.W. (1999). 22q11 deletion syndrome: A genetic subtype of schizophrenia. *Biological Psychiatry, 46*(7), 882–891.

Benarroch, F., Hirsch, H.J., Genstil, L., Landau, Y.E., & Gross-Tsur, V. (2007). Prader-Willi syndrome: Medical prevention and behavioral challenges. *Child and Adolescent Psychiatric Clinics of North America, 16*(3), 695–708.

Boer, H., Holland, A., Whittington, J., Butler, J., Webb, T., & Clarke, D. (2002). Psychotic illness in people with Prader-Willi syndrome due to chromosome 15 maternal uniparental disomy. *Lancet, 359*(9301), 135–136.

Bregman, J.D., & Volkmar, F.R. (1988). Autistic social dysfunction and Down syndrome. *Journal of the American Academy of Child and Adolescent Psychiatry, 27*(4), 440–441.

Brodkin, E.S. (2008). Social behavior phenotypes in fragile X syndrome, autism, and the Fmr1 knockout mouse: Theoretical comment on McNaughton et al. (2008). *Behavioral Neuroscience, 122*(2), 483–489.

Capone, G.T., Grados, M.A., Kaufmann, W.E., Bernad-Ripoll, S., & Jewell, A. (2005). Down syndrome and comorbid autism-spectrum disorder: Characterization using the aberrant behavior checklist. *American Journal of Medical Genetics A, 134*(4), 373–380.

Christopoulos, P., Deligeoroglou, E., Laggari, V., Christogioros, S., & Creatsas, G. (2008). Psychological and behavioural aspects of patients with Turner syndrome from childhood to adulthood: A review of the clinical literature. *Journal of Psychosomatic Obstetrics and Gynaecology, 29*(1), 45–51.

Clarke, D.J., Boer, H., Whittington, J., Holland, A., Butler, J., & Webb, T. (2002). Prader-Willi syndrome, compulsive and ritualistic behaviours: The first population-based survey. *British Journal of Psychiatry, 180*, 358–362.

Cohen, E., Chow, E.W., Weksberg, R., & Bassett, A.S. (1999). Phenotype of adults with the 22q11 deletion syndrome: A review. *American Journal of Medical Genetics, 86*(4), 359–365.

Cornish, K., Burack, J.A., Rahman, A., Russo, N., & Grant, C. (2005). Theory of mind deficits in children with fragile X syndrome. *Journal of Intellectual Disability Research, 49*(Part 5), 372–378.

Dimitropoulos, A., & Schultz, R.T. (2007). Autistic-like symptomatology in Prader-Willi syndrome: A review of recent findings. *Current Psychiatry Reports, 9*(2), 159–164.

Downey, J., Ehrhardt, A.A., Gruen, R., Bell, J.J., & Morishima, A. (1989). Psychopathology and social functioning in women with Turner syndrome. *Journal of Nervous and Mental Disease, 177*(4), 191–201.

Doyle, T.F., Bellugi, U., Korenberg, J.R., & Graham, J. (2004). "Everybody in the world is my friend" hypersociability in young children with Williams syndrome. *American Journal of Medical Genetics A, 124A*(3), 263–273.

Dykens, E.M., & Cassidy, S.B. (1995). Correlates of maladaptive behavior in children and adults with Prader-Willi syndrome. *American Journal of Medical Genetics, 60*(6), 546–549.

Dykens, E.M., Finucane, B.M., Gayley, C. (1997). Brief report: Cognitive and behavioral profiles in persons with Smith-Magenis syndrome. *Journal of Autism and Developmental Disorders, 27*(2), 203–211.

Dykens, E.M., & Kasari, C. (1997). Maladaptive behavior in children with Prader-Willi syndrome, Down syndrome, and nonspecific mental retardation. *American Journal of Mental Retardation, 102*(3), 228–237.

Dykens, E.M., & Roof, E. (2008). Behavior in Prader-Willi syndrome: Relationship to genetic subtypes and age. *Journal of Child Psychology and Psychiatry, 49*(9), 1001–1008.

Dykens, E.M., & Smith, A.C. (1998). Distinctiveness and correlates of maladaptive behaviour in children and adolescents with Smith-Magenis syndrome. *Journal of Intellectual Disability Research, 42*(Part 6), 481–489.

Einfeld, S.L., Smith, A., Durvasula, S., Florio, T., & Tonge, B.J. (1999). Behavior and emotional disturbance in Prader-Willi syndrome. *American Journal of Medical Genetics, 82*(2), 123–127.

Eliez, S.F.C. (2004). Velo-cardio-facial syndrome (deletion 22q11.2): A homogeneous neurodevelopmental model for schizophrenia. In M. Keshavan, J. Kennedy, & R. Murray (Eds.), *Neurodevelopment and schizophrenia* (pp. 121–137). New York: Cambridge University Press.

Eliez, S., & Feinstein, C. (2001). The fragile X syndrome: Bridging the gap from gene to behavior. *Current Opinion in Psychiatry, 14*, 443–449.

Elsea, S., & Girirajan, S. (2008). Smith-Magenis syndrome. *European Journal of Human Genetics, 16*(4), 412–421.

Epstein, C.J., Korenberg, J.R., Annerèn, G., Antonarakis, S.E., Aymè, S., Courchesne, E., Epstein, L.B., et al. (1991). Protocols to establish genotype-phenotype correlations in Down syndrome. *American Journal Human Genetics, 49*(1), 207–235.

Feinstein, C., & Chahal, L. (2009). Psychiatric phenotypes associated with neurogenetic disorders. *Psychiatric Clinics of North America, 32*(1), 15–37.

Feinstein, C., & Eliez, S. (2000). The velocardiofacial syndrome in psychiatry. *Current Opinion in Psychiatry, 13*, 485–490.

Feinstein, C., Eliez, S., Blasey, C., & Reiss, A.L. (2002). Psychiatric disorders and behavioral problems in children with velocardiofacial syndrome: Usefulness as phenotypic indicators of schizophrenia risk. *Biological Psychiatry, 51*(4), 312–318.

Feinstein, C., & Reiss, A.L. (1998). Autism: The point of view from fragile X studies. *Journal of Autism and Developmental Disorders, 28*(5), 393–405.

Feinstein, C., & Singh, S. (2007). Social phenotypes in neurogenetic syndromes. *Child and Adolescent Psychiatric Clinics of North America, 16*(3), 631–647.

Fine, S.E., Weissman, A., Gerdes, M., Pinto-Martin, J., Zackai, E., McDonald-McGinn, D., et al. (2005). Autism spectrum disorders and symptoms in children with molecularly confirmed 22q11.2 deletion syndrome. *Journal of Autism and Developmental Disorders, 35*(4), 461–470.

Freund, L.S., Reiss, A.L., Abrams, M.T. (1993). Psychiatric disorders associated with fragile X in the young female. *Pediatrics, 91*(2), 321–329.

Gagliardi, C., Frigerio, E., Bur, D.M., Cazzaniga, I., Perrett, D.I., & Borgatti, R. (2003). Facial expression recognition in Williams syndrome. *Neuropsychologia, 41*(6), 733–738.

Gerdes, M., Solot, C., Wang, P.P., Moss, E., LaRossa, D., Randall, P., et al. (1999). Cognitive and behavior profile of preschool children with chromosome 22q11.2 deletion. *American Journal of Medical Genetics, 85*(2), 127–133.

Ghaziuddin, M. (1997). Autism in Down's syndrome: Family history correlates. *Journal of Intellectual Disability Research, 41*(Part 1), 87–91.

Ghosh, M., Shah, A.H., Dhir, K., & Merchant, K.F. (2008). Behavior in children with Down syndrome. *Indian Journal of Pediatrics, 75*(7), 685–689.

Girirajan, S., Vlangos, C.N., Szomju, B.B., Edelman, E., Trevors, C.D., Dupuis, L., et al. (2006). Genotype-phenotype correlations in Smith-Magenis syndrome: Evidence that multiple genes in 17p11.2 contribute to the clinical spectrum. *Genetics in Medicine, 8*, 417–427.

Goldstone, A.P., Holland, A.J., Hauffa, B.P., Hokken-Koelega, A.C., & Tauber, M. (2008). Recommendations for the diagnosis and management of Prader-Willi syndrome. *Journal of Clinical Endocrinology and Metabolism, 93*, 4183–4197.

Gosch, A., & Pankau, R. (1994). Social-emotional and behavioral adjustment in children with Williams-Beuren syndrome. *American Journal of Medical Genetics, 53*(4), 335–339.

Gothelf, D. (2007). Velocardiofacial syndrome. *Child and Adolescent Psychiatric Clinics of North America, 16*(3), 677–693.

Gothelf, D., Eliez, S., Thompson, T., Hinard, C., Penniman, L., Feinstein, C., et al. (2005). COMT genotype predicts longitudinal cognitive decline and psychosis in 22q11.2 deletion syndrome. *Nature Neuroscience, 8*(11), 1500–1502.

Gothelf, D., Feinstein, C., Thompson, E., Gu, L., Penniman, E., Van Stone, H., et al. (2007). Risk factors for the emergence of psychotic disorders in adolescents with 22q11.2 deletion syndrome. *American Journal of Psychiatry, 164*(4), 663–669.

Gothelf, D., Schaer, M., & Eliez, S. (2008). Genes, brain development and psychiatric phenotypes in velo-cardio-facial syndrome. *Developmental Disabilities Research Reviews, 14*(1), 59–68.

Gottesman, I.I., & Gould, T.D. (2003). The endophenotype concept in psychiatry: Etymology and strategic intentions. *American Journal of Psychiatry, 160*(4), 636–645.

Gould, T.D., & Gottesman, I.I. (2006). Psychiatric endophenotypes and the development of valid animal models. *Genes, Brain, and Behavior, 5*(2), 113–119.

Greenberg, F. (1989). Williams syndrome. *Pediatrics, 84*(5), 922–923.

Greenberg, F., Lewis, R.A., Potocki, L., Glaze, D., Parke, J., Killian, J., et al. (1996). Multi-disciplinary clinical study of Smith-Magenis syndrome (deletion 17p11.2). *American Journal of Medical Genetics, 62*(3), 247–254.

Greenswag, L.R. (1987). Adults with Prader-Willi syndrome: A survey of 232 cases. *Developmental Medicine and Child Neurology, 29*(2), 145–152.

Haas-Givler, B. (1994). Educational implications and behavior concerns of SMS: From the teacher's perspective. *Spectrum (Newsletter of PRISMS), 1*(2), 3–4.

Hagerman, R. (1999). Clinical and molecular aspects of fragile X syndrome. In H. Tager-Flusberg (Ed.), *Neurodevelopmental Disorders,* (pp. 27–41). Cambridge, MA: The MIT Press.

Hagerman, R. (2002). Physical and behavioral phenotype. In R. Hagerman and P. Hagerman (Eds.), *Fragile X syndrome: Diagnosis, treatment, and research* (pp. 3–109). Baltimore: Johns Hopkins University Press.

Hagerman, R.J., & Hagerman, P.J. (2008). Testing for fragile X gene mutations throughout the life span. *Journal of the American Medical Association, 300*(20), 2419–2421.

Hessl, D., Glaser, B., Dyer-Friedman, J., & Reiss, A.L. (2006). Social behavior and cortisol reactivity in children with fragile X syndrome. *Journal of Child Psychology and Psychiatry, 47*(6), 602–610.

Hessl, D., Tassone, F., Loesch, D.Z., Berry-Kravis, E., Leehey, M.A., Gane, L.W., et al. (2005). Abnormal elevation of FMR1 mRNA is associated with psychological symptoms in individuals with the fragile X premutation. *American Journal of Medical Genetics B Neuropsychiatric Genetics, 139B*(1), 115–121.

Holsen, L.M., Dalton, K.M., Johnstone, T., & Davidson, R.J. (2008). Prefrontal social cognition network dysfunction underlying face encoding and social anxiety in fragile X syndrome. *Neuroimage, 43*(3), 592–604.

Howlin, P., Wing, L., & Gould, J. (1995). The recognition of autism in children with Down syndrome—implications for intervention and some speculations about pathology. *Developmental Medicine and Child Neurology, 37*(5), 406–414.

Jarvinen-Pasley, A., Bellugi, U., Reilly, J., Mills, D.L., Galaburda, A., Reiss, A.L., et al. (2008). Defining the social phenotype in Williams syndrome: A model for linking gene, the brain, and behavior. *Development and Psychopathology, 20*(1), 1–35.

Jones, W., Bellugi, U., Lai, Z., Chiles, M., Reilly, J., Lincoln, A., et al. (2000). Hypersociability in Williams syndrome. *Journal of Cognitive Neuroscience, 12*(Suppl. 1), 30–46.

Jones, K., & Smith, D.W. (2006). *Smith's recognizable patterns of human malformation* (6th ed.). Philadelphia: Elsevier Saunders.

Jopp, D.A., & Keys, C.B. (2001). Diagnostic overshadowing reviewed and reconsidered. *American Journal of Mental Retardation, 106*(5), 416–433.

Kates, W.R., Abrams, M.T., Kaufmann, W.E., Breiter, S.N., & Reiss, A.L. (1997). Reliability and validity of MRI measurement of the amygdala and hippocampus in children with fragile X syndrome. *Psychiatry Research, 75*(1), 31–48.

Kau, A.S., Tierney, E., Bukelis, I., Stump, M.H., Kates, W.R., Trescher, W.H., et al. (2004). Social behavior profile in young males with fragile X syndrome: Characteristics and specificity. *American Journal of Medical Genetics A, 126A*(1), 9–17.

Kaufmann, W., Corell, R., Kau, A.S., et al. (2004). Autism spectrum disorder in fragile X syndrome: Communication, social interaction, and specific behaviors. *American Journal of Medical Genetics, 129A*, 225–234.

Kent, L., Evans, J., Moli, P., & Sharp, M. (1999). Comorbidity of autistic spectrum disorders in children with Down syndrome. *Developmental Medicine and Child Neurology, 41*(3), 153–158.

Koenig, K., Klin, A., & Schultz, R. (2004). Deficits in social attribution ability in Prader-Willi syndrome. *Journal of Autism and Developmental Disorders, 34*(5), 573–582.

Korenberg, J.R., Chen, X.N., Schipper, R., Sun, Z., Gonsky, S., Gerwehr, S., et al. (1994). Down syndrome phenotypes: The consequences of chromosomal imbalance. *Proceedings of the National Academy of Sciences USA, 91*(11), 4997–5001.

Kravanti E., D.P., Fearon P., et al. (2004). Can one identify preschizophrenia in children? In M. Keshavan, J. Kennedy, & R. Murray (Eds.), *Neurodevelopment and schizophrenia* (pp. 415–431). New York: Cambridge University Press.

Lachiewicz, A.M. (1992). Abnormal behaviors of young girls with fragile X syndrome. *American Journal of Medical Genetics, 43*(1–2), 72–77.

Laing, E., Butterworth, G., Ansari, D., et al. (2002). Atypical development of language and social communication in toddlers with Williams syndrome. *Developmental Science, 5*, 233–246.

Laing, E., & Jarrold, C. (2007). Comprehension of spatial language in Williams syndrome: Evidence for impaired spatial representation of verbal descriptions. *Clinical Linguistics and Phonetics, 21*(9), 689–704.

Lawrence, K., Campbell, R., Swettenham, J., Terstegge, J., Akers, R., Coleman, M., et al. (2003). Interpreting gaze in Turner syndrome: Impaired sensitivity to intention and emotion, but preservation of social cueing. *Neuropsychologia, 41*(8), 894–905.

Laws, G., & Bishop, D. (2004). Pragmatic language impairment and social deficits in Williams syndrome: A comparison with Down's syndrome and specific language impairment. *International Journal of Language and Communication Disorders, 39*(1), 45–64.

Mazzocco, M. (2000). Advances in research on the fragile X syndrome. *Mental Retardation and Developmental Disabilities Research Reviews, 6*, 96–106.

Mazzocco, M.M., Baumgardner, T., Freund, L.S., & Reiss, A.L. (1998). Social functioning among girls with fragile X or Turner syndrome and their sisters. *Journal of Autism and Developmental Disorders, 28*(6), 509–517.

Mazzocco, M.M., Pennington, B.F., & Hagerman, R.J. (1994). Social cognition skills among females with fragile X. *Journal of Autism and Developmental Disorders, 24*(4), 473–485.

McCauley, E., Ito, J., & Kay, T. (1986). Psychosocial functioning in girls with Turner's syndrome and short stature: Social skills, behavior problems, and self-concept. *Journal of the American Academy of Child Psychiatry, 25*(1), 105–112.

McConkie-Rosell, A., Finucane, B., Cronister, A., Abrams, L., Bennett, R.L., & Pettersen, B.J. (2005). Genetic counseling for fragile X syndrome: Updated recommendations of the

National Society of Genetic Counselors. *Journal of Genetic Counseling, 14*(4), 249–270.

McDonald-McGinn, D.M., Kirschner, R., Goldmuntz, E., Sullivan, K., Eicher, P. Gerdes, M., et al. (1999). The Philadelphia story: The 22q11.2 deletion: Report on 250 patients. *Genetic Counseling, 10*(1), 11–24.

McNaughton, C.H., Moon, J., Strawderman, M.S., Maclean, K.N., Evans, J., & Strupp, B.J. (2008). Evidence for social anxiety and impaired social cognition in a mouse model of fragile X syndrome. *Behavioral Neuroscience, 122*(2), 293–300.

Merenstein, S.A., Sobesky, W.E., Taylor, A.K., Riddle, J.E., Tran, H.X., & Hagerman, R.J. (1996). Molecular-clinical correlations in males with an expanded FMR1 mutation. *American Journal of Medical Genetics, 64*(2), 388–394.

Mervis, C.B., & Robinson, B.F. (2000). Expressive vocabulary ability of toddlers with Williams syndrome or Down syndrome: A comparison. *Developmental Neuropsychology, 17*(1), 111–126.

Mineur, Y.S., Huynh, L.X., & Crusio, W.E. (2006). Social behavior deficits in the Fmr1 mutant mouse. *Behavioural Brain Research, 168*(1), 172–175.

Morris, C.A., Demsey, S.A., Leonard, D.O., Dilts, C., & Blackburn, B.L. (1988). Natural history of Williams syndrome: Physical characteristics. *Journal of Pediatrics, 113*(2), 318–326.

Moss, E.M., Batshaw, M.L., Solot, C.B., Gerdes, M., McDonald-McGinn, D.M., Driscoll, D.A., et al. (1999). Psychoeducational profile of the 22q11.2 microdeletion: A complex pattern. *Journal of Pediatrics, 134*(2), 193–198.

Murphy, K.C., Jones, L.A., & Owen, M.J. (1999). High rates of schizophrenia in adults with velo-cardio-facial syndrome. *Archives of General Psychiatry, 56*(10), 940–945.

Nicham, R., Weitzdorfer, R., Hauser, M., Freidl, M., Schubert, E., Wurst, G., et al. (2003). Spectrum of cognitive, behavioural and emotional problems in children and young adults with Down syndrome. *Journal of Neural Transmisson* (67, Suppl.), 173–191.

Niklasson, L., Rasmussen, P., Oskarsdóttir, S., & Gillberg, C. (2002). Chromosome 22q11 deletion syndrome (CATCH 22): Neuropsychiatric and neuropsychological aspects. *Developmental Medicine and Child Neurology, 44*(1), 44–50.

Papolos, D.F., Faedda, G.L., Veit, S., Goldberg, R., Morrow, B., Kucherlapati, R., et al. (1996). Bipolar spectrum disorders in patients diagnosed with velo-cardio-facial syndrome: Does a hemizygous deletion of chromosome 22q11 result in bipolar affective disorder? *American Journal of Psychiatry, 153*(12), 1541–1547.

Plesa-Skwerer, D., Faja, S., Verbalis, A., Schofield, C., & Tager-Flusberg, H. (2006). Perceiving facial and vocal expressions of emotion in individuals with Williams syndrome. *American Journal of Mental Retardation, 111*(1), 15–26.

Prior, T.I., Chue, P.S., & Tibbo, P. (2000). Investigation of Turner syndrome in schizophrenia. *American Journal of Medical Genetics, 96*(3), 373–378.

Ranke, M.B., & Saenger, P. (2001). Turner's syndrome. *Lancet, 358*(9278), 309–314.

Rasmussen, P., Borjesson, O., Wentz, E., & Gillberg, C. (2001). Autistic disorders in Down syndrome: Background factors and clinical correlates. *Developmental Medicine and Child Neurology, 43*(11), 750–754.

Reiss, A., & Dant, C. (2003). The behavioral neurogenetics of fragile X syndrome: Analyzing gene-brain-behavior relationships in child developmental psychopathologies. *Developmental Psychopathology, 15*(4), 927–978.

Roizen, N.J. (2002). Medical care and monitoring for the adolescent with Down syndrome. *Adolescent Medicine, 13*(2), 345–358, vii.

Roser, P., & Kawohl, W. (2008). Turner syndrome and schizophrenia: A further hint for the role of the X-chromosome in the pathogenesis of schizophrenic disorders. *World Journal of Biological Psychiatry,* 1–4.

Ross, J.L., Stefanatos, G.A., Kushner, H., Bondy, C., Nelson, L., & Zinn, A. (2004). The effect of genetic differences and ovarian failure: Intact cognitive function in adult women with premature ovarian failure versus Turner syndrome. *Journal of Clinical Endocrinology and Metabolism, 89*(4), 1817–1822.

Rovet, J.F. (1993). The psychoeducational characteristics of children with Turner syndrome. *Journal of Learning Disabilities, 26*(5), 333–341.

Rovet, J. (2004). Turner syndrome: A review of genetic and hormonal influences on neuropsychological functioning. *Child Neuropsychology, 10*(4), 262–279.

Ryan, A.K., Goodship, J.A., Wilson, D.I., Philip, N., Levy, A., & Seidel, H. (1997). Spectrum of clinical features associated with interstitial chromosome 22q11 deletions: A European collaborative study. *Journal of Medical Genetics, 34*(10), 798–804.

Schmidt, P.J., Cardoso, G.M., Ross, J.L., Haq, N., Rubinow, D.R., & Bondy, C.A. (2006). Shyness, social anxiety, and impaired self-esteem in Turner syndrome and premature ovarian failure. *Journal of the American Medical Association, 295*(12), 1374–1376.

Shprintzen, R.J. (2000). Velo-cardio-facial syndrome: A distinctive behavioral phenotype.

Mental Retardation and Developmental Disabilities Research Review, 6(2), 142–147.

Shprintzen, R.J., Higgins, A.M., Antshel, K., Fremont, W., Roizen, N., & Kates, W. (2005). Velo-cardio-facial syndrome. *Current Opinion in Pediatrics, 17*(6), 725–730.

Skuse, D.H. (2000). Behavioural phenotypes: What do they teach us? *Archives of Disease in Childhood, 82*(3), 222–225.

Smith, A.C., Dykens, E., & Greenberg, F. (1998). Behavioral phenotype of Smith-Magenis syndrome (del 17p11.2). *American Journal of Medical Genetics, 81*(2), 179–185.

Smith, A.C., McGavran, L., Robinson, J., Waldstein, G., Macfarlane, J., Zonona, J., et al. (1986). Interstitial deletion of (17)(p11.2p11.2) in nine patients. *American Journal of Medical Genetics, 24*(3), 393–414.

Society for the Study of Behavioural Phenotype. (1998). Society for the Study of Behavioural Phenotypes 7th Annual Meeting. Cambridge, United Kingdom. November 13–14, 1997. Abstracts. *Genetic Counseling, 9*(2), 161–176.

Soresi, S., & Nota, L. (2000). A social skill training for persons with Down's syndrome. *European Psychologist, 5*, 34–43.

Spencer, C.M., Alekseyenko, O., Serysheva, E., Yuva-Paylor, L.A., & Paylor, R. (2005). Altered anxiety-related and social behaviors in the Fmr1 knockout mouse model of fragile X syndrome. *Genes, Brain, and Behavior, 4*(7), 420–430.

Starr, E.M., Berument, S.K., Tomlins, M., Papanikolaou, K., Pickels, A., Lord, C., et al. (2005). Brief report: Autism in individuals with Down syndrome. *Journal of Autism and Developmental Disorders, 35*(5), 665–673.

Stromme, P., Bjornstad, P.G., & Ramstad, K. (2002). Prevalence estimation of Williams syndrome. *Journal of Child Neurology, 17*(4), 269–271.

Sullivan, K., & Tager-Flusberg, H. (1999). Second-order belief attribution in Williams syndrome: Intact or impaired? *American Journal of Mental Retardation, 104*(6), 523–532.

Swillen, A., Devriendt, K., Legius, E., Prinzie, P., Vogels, A., Ghesquière, P., & Fryns, J.P. (1999). The behavioural phenotype in velo-cardio-facial syndrome (VCFS): From infancy to adolescence. *Genetic Counseling, 10*(1), 79–88.

Szatmari, P., Paterson, A.D., Zwaigenbaum, L., Roberts, W., Brian, J., Liu, X.Q., et al. (2007). Mapping autism risk loci using genetic linkage and chromosomal rearrangements. *Nature Genetics, 39*(3), 319–328.

Taylor, L., & Oliver, C. (2008). The behavioral phenotype of Smith-Magenis syndrome: Evidence for a gene-environment interaction. *Journal of Intellectual Disability Research, 52*(10), 830–841.

Turner, H. (1938). A syndrome of infantilism, congenital webbed neck and cubitus valgus. *Endocrinology, 28*, 566–574.

Van Esch, H. (2006). The fragile X premutation: New insights and clinical consequences. *European Journal of Medical Genetics, 49*(1), 1–8.

Vorstman, J., Morcus, M., Duijff, S.N., Klaassen, P.W., Heineman-de Boer, J.A., Beemer, F.A., et al. (2006). The 22q11.2 deletion in children: High ratio of autistic disorders and early onset of psychotic symptoms. *Journal of the American Academy of Child and Adolescent Psychiatry, 45*(9), 1104–1113.

Watson, C., Hoeft, F., Garrett, A.S., Hall, S.S., & Reiss, A.L. (2008). Aberrant brain activation during gaze processing in boys with fragile X syndrome. *Archives of General Psychiatry, 65*(11), 1315–1323.

Wattendorf, D.J., & Muenke, M. (2005). Prader-Willi syndrome. *American Family Physician, 72*(5), 827–830.

Williams, K.R., Wishart, J.G., Pitcairn, T.K., & Willis, D.S. (2005). Emotion recognition by children with Down syndrome: Investigation of specific impairments and error patterns. *American Journal of Mental Retardation, 110*(5), 378–392.

Woodin, M., Wang, P.P., Aleman, D., McDonald-McGinn, D., Zackai, E., & Moss, E. (2001). Neuropsychological profile of children and adolescents with the 22q11.2 microdeletion. *Genetics in Medicine, 3*(1), 34–39.

Zarcone, J., Napolitano, D., Peterson, C., Breidbord, J., Ferraioli, S., Caruso-Anderson, M., et al. (2007). The relationship between compulsive behaviour and academic achievement across the three genetic subtypes of Prader-Willi syndrome. *Journal of Intellectual Disability Research, 51*(6), 478–487.

17

Behavioral Phenotypes

Nature versus Nurture Revisited

Pasquale J. Accardo and Margie L. Jaworski

From homogeneity to heterogeneity.
　　　　　—*Kanner (1964)*

The understanding of intellectual disability and behavioral pathology has been a long journey. Each successive attempt, in isolation, has run into limitations in explaining observed behavior. Behavioral phenotypes are now interpreted as complex phenomena with interacting causes that are not easily delineated as exclusively genetic or environmental, and they have been found to be more responsive to a combined therapeutic approach of genetic testing and environmental management.

THE 19TH-CENTURY BACKGROUND

The earliest scientific description of severe intellectual impairment and its behavioral treatment was reported by Itard (1801/ 1806/1962). Lane (1977) reviewed the documents of this case and argued that a more correct diagnosis would have been autism. Although his reclassification of the case would probably not meet current diagnostic criteria for autism, his reading does suggest that the behavioral phenomenology of "mental retardation" was rather complex from the very beginning.

Down (1866) reported the first genetic syndrome associated with cognitive impairment. In his more detailed description of children with "mental affections," Down (1887/1990) included many behavioral characteristics of autism spectrum disorders: loss of language, echolalia, referring to oneself in the third person, living in a world of one's own, fascination with music, rocking, poking fingers in ears, high threshold to pain, tendency to automatisms and rhythmical actions, hyperlexia, astonishing mental arithmetic ability, pica, and licking. He did not, however, link these behaviors together or associate them with Down syndrome (Kent, Evans, Paul, & Sharp, 1999).

Down's reference to an "ethnic classification of idiots" and his use of the term *Mongolian* has generated anachronistic charges of racism. This is especially peculiar, inasmuch as he used the equal occurrence of Down syndrome within all racial groups as an argument for the unity of the human species— ethnic families were distinct only as varieties. He also referred to other "racial" groupings—Caucasian, Malay, Aztec, Ethiopian, and South Sea Islands—as syndrome descriptors. Later clinicians would use a wide variety of terms, such as *bird-head, cat-cry, cherubism, Cockayne, cyclops, elfin, Kabuki, leprechaun, marionette (puppet), Munchausen, Ondine, panic, Pickwickian, priapism, prune-belly, Pygmalion, St. Anthony, St. Vitus, satyriasis, straw Peter,* and *syphilis,* to characterize the physical and behavioral features of different syndromes. These mnemonics have always been subject to misinterpretation, as will their more neutral successors; in a teetotaling society the useful language descriptor *cocktail party chatter* will probably need to be retired.

Francis Galton's *Hereditary Genius* (1892/ 1979) founded the field of behavioral genetics.

The "genius" that he plotted through familial generations was a complex combination of and interaction among intelligence, talent, effort, and high achievement that also included many personality factors. The "intelligence" tests that he first invented were more measures of psychophysiological processing and motor skills than what would today be referred to as intelligence (Forrest, 1974). Galton's ideas were not widely accepted at first. However, the extension of a physicalist determinism from a mechanistic physics to human behavior (despite any evidence to support this transference to biological systems) produced a sense of despair in writers such as William James and Henry Adams; the one sought solace in the study of sex, the other in religion. In the next century the lost papers of Gregor Mendel would be rediscovered (Bateson, 1909), and with an impressive mathematical substructure, the power of Galton's "eugenics" would unfold in an escalating series of nightmare scenarios (Black, 2003; Chase, 1980).

20TH-CENTURY PROGRESS

In the first half of the 20th century the theories of Freud emphasized the importance of early experience for later personality development. Behaviorism and social learning theory were fellow travelers in this overemphasis on nurture. The popular models of intellectual impairment in the educational and psychological literature included Helen Keller and various reports of wolf children. With care and concern—and specialized teaching techniques—people who were thought to be "mentally retarded" could actually learn. (There remains a persistent straw man that *mental retardation* [global or generalized cognitive impairment] implies that an affected child will not progress but stay fixed at the cognitive level present at the time of diagnosis.) Unfortunately these paradigms were myths that a tenacious belief system found increasingly difficult to maintain.

Helen Keller did not have an intellectual disability; she was intellectually superior—but blind and deaf. There are no genuine feral children raised by wolves, apes, or any other fauna except in fairy tales, Disney movies, and Tarzan stories. The fascination of professionals with such obvious fabrications represents a delusion worthy of study (Newton, 2002).

As the century wore on, promising strategies and interventions such as glutamic acid therapy and megavitamins consistently failed, and these failures needed to be explained away. Instead of learning from the almost irremediable predictiveness of IQs (Spitz, 1986), professionals in the field obscured the nature of the debacle by attacking IQs as invalid, inaccurate, misleading, unhelpful, and otherwise irrelevant. The scores were a rigid measuring stick that simply refused to bend. They were then relegated to the status of a lesser (and not the most important) component of the definition of mental retardation.

Oddly, the only professionals to actually succeed in "increasing intelligence" were the physicians who "prevented" intellectual impairment with hormones (e.g., thyroid), special diets (as done for phenylketonuria), vitamin supplementation (e.g., pyridoxine), and various vaccines (e.g., pneumococcal). They remain the only hope for the development of true *noetics*—drugs used to increase intelligence—at present still a null class.

Somatotyping explored the relationship between body habitus and personality types (Kretschmer, 1936/1970; Sheldon, 1954): the endomorphic or pyknic body shape (on the obese side) had a "viscerotonic" temperament, the mesomorphic or athletic body shape had a "somatotonic" temperament, and the ectomorphic or asthenic/leptosomatic body type (on the anorectic side) had a "cerebrotonic" temperament. Rather than the body habitus causing the personality type, they were probably both outcomes of a certain life history. Somatotyping has joined phrenology and physiognomy as a quaint piece of poor science (Tytler, 1982; Wechsler, 1982).

Behavioral patterns were associated with a variety of medical conditions: specific personality types were described for juvenile rheumatoid arthritis, asthma, diabetes, dwarfism, and epilepsy (Alexander & French, 1948; Wright, Schaefer, & Solomons, 1979). Evidence to support such clinical phenotypes was mostly anecdotal, with a paucity of objective measures; most of these behavioral patterns declined with improved medical management of the underlying conditions.

Almost a century after the first descriptions of Down syndrome, it remained uncertain whether this "congenital" condition was due to a genetic, hormonal, or other etiology (Lejeune, 2008; Yanet, 1957). Moderate intellectual disability was so generally accepted as the rule in this condition, that into the 1970s and 1980s school-age children with Down syndrome would be referred for testing to explain their lack of spoken language. When their IQ tested in the 20s, it would be found that the school had neglected to test the children, because the IQ range of 50–55 associated with Down syndrome was taken as a given that did not need to be formally confirmed.

One of the first syndrome compendiums (Smith, 1967) focused more on poor growth (short stature) rather than intellectual disability; the second such collection (Gellis & Feingold, 1968), on the other hand, focused specifically on "mental retardation" syndromes. It became common in genetics textbooks in the latter half of the 20th century to associate IQ ranges with different genetic syndromes (e.g., Down syndrome = IQ 55!). "Being syndromic" or "having a syndrome" became almost synonymous with "being mentally retarded," so that syndromes not associated with cognitive impairment needed special notations. Indeed, the older research on mental retardation strictly divided it into IQ groups: borderline, mild (educable), moderate (trainable), severe (subtrainable), and profound. Now studies are reported according to etiologies: Down syndrome, Prader-Willi syndrome, fetal alcohol syndrome, and fragile X syndrome—all of which

are recognized to reflect a broad spectrum of IQ ranges.

Attempts to correlate intelligence with genetics became the subject of public debate, mostly conducted by people with no qualifications in the complex statistical basis of such arguments (Herrnstein, 1973; Herrnstein & Murray, 1994; Jencks, 1972; Jensen, 1973, 1980). Indeed, the complexity of the mathematics allowed many experts to pontificate without any possibility of rebuttal or even rational discussion (but see Scarr, 1981; Tobach & Proshansky, 1976). Similarly, research attempted to clarify the impact of sex differences, whether genetic, hormonal, or social (Maccoby, 1966; Maccoby & Jacklin, 1974). Deviations of sex chromosomes were also investigated. Attempts to associate specific genotypes with criminal behavior were at the least premature and suffered serious design flaws (Witkin et al., 1977). The debacle of counseling a family that their newborn XYY infant might exhibit violent criminal tendencies reinforces the perception that many ethical conundrums in medicine turn out to be problems of correct factual knowledge versus misinformation.

The 20th century was very much a behavioralist century. Nurture was simply assumed to be more important than any genetic contribution. Behavioral psychology was responsible for a number of striking advances in the management of challenging behaviors and several developmental disorders (Lovaas, 2000). But late in the century significant limitations of the contribution of the home environment, early rearing, and nurturance to both typical behavior and behavioral psychopathology and deviance would be documented (Bruer, 1999; Eyer, 1992; McBroom, 1980; Rowe, 1994). The significant contribution of genetic burden to intellectual impairment; schizophrenia; mood disorders; autism; attention-deficit/ hyperactivity disorder; alcoholism; personality disorders; political, religious, and social attitudes; emotional valences (e.g., introversion/extroversion); and even reproductive behavior has been well documented in twin and other studies (Clark & Grunstein, 2000; Plomin, DeFries, McClearn, & McGuffin, 2008).

At first *behavioral phenotypes* referred to characteristic, stereotyped, or reproducible patterns of behavior consistently associated with biological disorders—usually genetic syndromes (Nyhan, 1972). The behaviors could be motor, cognitive, linguistic, or social. Most existing genetic syndromes had a component of intellectual impairment, and many of the atypical behaviors identified were negative. The study of exceptional (in the positive sense) behaviors proceeded more slowly (Obler & Fein, 1988).

Harris (1987) suggested "unlearned behavior disorders" as descriptive for behavioral phenotypes, and the Society for the Study of Behavioural Phenotypes (1990) (http://www.ssbp.co.uk) admitted that "some of the behaviors exhibited by children with biologically based mentally handicapping disorders are organically determined." But behaviors that cluster together in recognizable patterns without any genetic substrate can also be considered "phenotypes." The impact of severe isolation in monkeys (Harlow, 1971) is imitated in children reared in institutions (Provence & Lipton, 1962). The limited emotional repertoire generated by such isolation resembled the features of autism later seen in children internationally adopted from foreign orphanages or other deprivational settings. Several of the most common behavioral phenotypes do not necessarily have any genetic, dysmorphic, or other organ system involvement. On the other hand, attention-deficit/hyperactivity disorder (Accardo, Shapiro, & Capute, 1997), autism spectrum disorders, depression, and schizophrenia all have significant familial and genetic contributions.

INTO THE 21ST CENTURY

In the last decades of the 20th century, as the field of behavioral phenotypes started to be defined, it almost immediately drew some of

the same criticisms that had been leveled at the study of the genetic contribution to intelligence. But current research is using much more complicated end points on both sides of the equation: on the one hand, single genes have been replaced by complex arrays of single-nucleotide polymorphisms and copy number variants, and, on the other hand, "intelligence" has been replaced by endophenotypes composed of increasingly smaller and simpler behavioral units, such that any determinism is neatly concealed in an extremely complex set of interactions.

The first edition of O'Brien and Yule's *Behavioral Phenotypes* in 1995 focused on a major division between genetic syndromes associated with significantly higher verbal intelligence and those associated with higher nonverbal or performance intelligence scores. In the 19th century human handedness had given rise to an interest in cortical asymmetry that led to many studies of right brain/left brain differences in the 20th century (Dimond & Blizard, 1977). But it was not until the late 20th century that IQ tests began to emphasize more than verbal intelligence. Dysfunctional brains could also exhibit right/left differences, and sometimes the right/left differences were the dysfunction. This right/left, verbal/performance distinction was downplayed by O'Brien (2002) but remains especially useful in dealing with autism spectrum disorders: the greater the verbal/performance discrepancy—with lower verbal skills—the more likely the patient is to exhibit some behaviors characteristic of autism.

The clinical depiction of behaviors associated with specific genetic syndromes preceded the revolutionary advances in mapping the human genome, but it provides a rich basis for further research into the complex associations between gene complexes and behavioral profiles. This "new genetics" included suggestions of a behavioral determinism that was greeted at first with suspicion if not discomfort (Dykens, 1995). With progress in the delineation of behavioral phenotypes, cautions are now raised with regard to potential methodological weaknesses (Dykens & Hodapp, 2007):

- Developmental changes across the life span (such changes would reflect the complex interaction of physical, emotional, educational, and social factors over time)

- Gender differences (the impact of which would be as much social as endocrinological/physical; see, e.g., Philippe et al., 2008)

- Positive behaviors (this would probably include the multiple gene sites associated with higher intelligence; see, e.g., Obler & Fein, 1988)

The question of nature (genetic or biological determinism) versus nurture (the impact of family, education, environment, social class, and so forth) is an ancient one that simply cannot be resolved by scientific fact. The interpretation of the "facts"—starting with their selection, going through their measurement, all the way to the most detailed statistical manipulation of their interpretation—will always be subject to the philosophical presuppositions that one brings to the table. Mottram (1952) presented a comprehensive set of philosophical responses to physiological determinism that have not been superseded.

PARENT PERSPECTIVE

When children develop slowly, misbehave, or behave in an atypical manner, families want to know why—and what to do about it. In many cases, parents (and teachers) adopt as their first explanation that the child is lazy, stupid, or deliberately oppositional. Usually this explanation is wrong. Parents are—almost by nature—nurturists; like most adults they believe in the "I think I can, I think I can" school of motivational psychology. This preference for nurture is reflected in the popularity of dairy-free, gluten-free, and other specialized diets, chelation for lead and mercury, avoidance of recom-

mended vaccines, and a wide variety of other alternative therapies whose main attraction is the avoidance of "drugs." This mind-set is a modern continuation of magical belief, in which science is perceived as cold, unfeeling, and, even if effective, only effective at too great a price.

Understanding that there are significant biological, physiological, and genetic contributions to these challenging behaviors does not create a self-fulfilling prophecy or a situation of despair in the face of genetic determinism. Rather, it gives families and the professionals working with them tools to better influence outcomes by selecting the most appropriate behavioral/environmental interventions. Relieving parents of the guilt of causing challenging behaviors releases a great deal of energy that can now be channeled into effective management. If the nurturist option strikingly resembles ancient alchemical medicine, the genetics of behavioral phenotypes seems, in turn, to be a reincarnation of astrology—something many people relish as a hobby or peripheral interest, but become very nervous about when it gets too close to home (Accardo & Accardo, 2008). Functional behavior assessments can generate powerful plans for positive behavioral supports. These will be most effective with behaviors that are triggered and/or rewarded by environmental variables. Some challenging behaviors that have their origin more in neurochemical differences that are genetically controlled may be better addressed by psychopharmacology. Understanding the steps from the base pair to the behavior may still allow the use of positive behavioral supports to effectively interrupt the linkage. All genetic determinism operates only in a specific environment—one that can be changed.

REFERENCES

Accardo, P.J., & Accardo, J.A. (2008). A medical history of developmental disabilities. In P.J. Accardo (Ed.), *Capute and Accardo's neurodevelopmental disabilities in infancy and childhood:*

Vol. II. The spectrum of neurodevelopmental disabilities (3rd ed., pp. 1–14). Baltimore: Paul H. Brookes Publishing Co.

Accardo, P.J., Shapiro, B.K., & Capute, A.J. (Eds.). (1997). *Behavior belongs in the brain.* Baltimore: York Press.

Alexander, F., & French, T.M. (1948). *Studies in psychosomatic medicine.* New York: Ronald Press.

Bateson, W. (1909). *Mendel's principles of heredity.* Cambridge: Cambridge University Press.

Black, E. (2003). *The war against the weak: Eugenics and America's campaign to create a master race.* New York: Four Walls Eight Windows.

Bruer, J.T. (1999). *The myth of the first three years.* New York: Free Press.

Chase, A. (1980). *The legacy of Malthus: The social costs of the new scientific racism.* Urbana: University of Illinois Press.

Clark, W.R., & Grunstein, M. (2000). *Are we hardwired? The role of genes in human behavior.* New York: Oxford University Press.

Dimond, S.J., & Blizard, D.A. (Eds.). (1977). *Evolution and lateralization of the brain* (vol. 299). New York: Annals of the New York Academy of Sciences.

Down, J.L.H. (1866). Observations on the ethnic classification of idiots. *London Hospital Reports, 3,* 25.

Down, J.L. (1990). *Mental affections of childhood and youth.* Oxford: Blackwell Scientific Publications/MacKeith Press. (Original work published 1887)

Dykens, E.M. (1995). Measuring behavioral phenotypes: Provocations from the "new genetics." *American Journal of Mental Retardation, 99,* 522–532.

Dykens, E.M., & Hodapp, R.M. (2007). Three steps toward improving the measurement of behavior in behavioral phenotype research. *Child and Adolescent Psychiatric Clinics of North America, 16,* 617–630.

Eyer, D.E. (1992). *Mother-infant bonding: A scientific fiction.* New Haven, CT: Yale University Press.

Forrest, D.W. (1974). *Francis Galton: The life and work of a Victorian genius.* New York: Taplinger.

Galton, F. (1979). *Hereditary genius.* London: Julian Friedmann. (Original work published 1892)

Gellis, S.S., & Feingold, M. (Eds.). (1968). *Atlas of mental retardation syndromes: Visual diagnosis of facies and physical findings.* Washington, DC: U.S. Department of Health, Education, and Welfare.

Harlow, H. (1971). *Learning to love.* New York: Ballantine.

Harris, J.C. (1987). Behavioral phenotypes in mental retardation: Unlearned behaviors. *Advances in Developmental Disorders, 1,* 77–106.

Herrnstein, R.J. (1973). *I.Q. in the meritocracy.* Boston: Little, Brown & Company.

Herrnstein, R.J., & Murray, C. (1994). *The bell curve: Intelligence and class structure in American life.* New York: Free Press.

Itard, J.M.-G. (1962). *The wild boy of Aveyron* (G. Humphrey & M. Humphrey, Trans.). New York: Appleton-Century-Crofts. (Original work published 1801/1806)

Jencks, C. (1972). *Inequality.* New York: Basic Books.

Jensen, A.J. (1973). *Educability and group differences.* New York: Harper & Row.

Jensen, A.L. (1980). *Bias in mental testing.* New York: Free Press.

Kanner, L. (1964). *A history of the care and study of the mentally retarded.* Springfield, IL: Charles C Thomas.

Kent, L., Evans, J., Paul, M., & Sharp, M. (1999). Comorbidity of autistic spectrum disorders in children with Down syndrome. *Developmental Medicine and Child Neurology, 41,* 153–158.

Kretschmer, E. (1970). *Physique and character: An investigation of the nature of constitution and of the theory of temperament* (W.J.H. Sprott, Trans.). New York: Cooper Square. (Original work published 1936)

Lane, H. (1977). *The wild boy of Aveyron.* London: George Allen & Unwin.

Lejeune, C. (2008). *Life is a blessing: A biography of Jérôme Lejeune* (M.J. Miller, Trans.). San Francisco: Ignatius Press.

Lovaas, O.I. (2000). Experimental design and cumulative research in early behavioral intervention. In P.J. Accardo, C. Magnusen, & A.J. Capute (Eds.), *Autism: Clinical and research issues* (pp. 133–161). Timonium, MD: York Press.

Maccoby, E.M. (Ed.). (1966). *The development of sex differences.* Stanford, CA: Stanford University Press.

Maccoby, E.M., & Jacklin, C.N. (1974). *The psychology of sex differences.* Stanford, CA: Stanford University Press.

McBroom, P. (1980). *Behavioral genetics.* Washington, DC: U.S. Government Printing Office.

Mottram, V.H. (1952). *The physical basis of personality.* Baltimore: Penguin.

Newton, M. (2002). *Savage girls and wild boys: A history of feral children.* New York: St. Martin's Press.

Nyhan, W.L. (1972). Behavioral phenotypes in organic genetic disease. *Pediatric Research, 6,* 1–9.

Obler, L.K., & Fein, D. (Eds.). (1988). *The exceptional brain: Neuropsychology of talent and special abilities.* New York: Guilford Press.

O'Brien, G. (Ed.). (2002). *Behavioural phenotypes in clinical practice.* Cambridge: MacKeith Press.

O'Brien, G., & Yule, W. (Eds.). (1995). *Behavioural phenotypes.* Cambridge: MacKeith Press.

Philippe, A., Boddaret, N., Vaivre-Douret, L., Robel, L., Danon-Boileau, L., Malam, V., et al.

(2008). Neurobehavioral profile and brain imaging study of the 22q13.3 deletion syndrome. *Pediatrics.* Retrieved from www.pediatrics.org/cgi/doi/10.1542/peds.2007-2584

Plomin, R., DeFries, J.C., McClearn, G.E., & McGuffin, P. (2008). *Behavioral genetics.* New York: Worth.

Provence, S., & Lipton, R.C. (1962). *Infants in institutions.* New York: International Universities Press.

Rowe, D.C. (1994). *The limits of family influence: Genes, experience, and behavior.* New York: Guilford Press.

Scarr, S. (Ed.). (1981). *Race, social class, and individual differences in I.Q.* Hillsdale, NJ: Lawrence Erlbaum Associates.

Sheldon, W.H. (1954). *Atlas of men.* New York: Gramercy.

Smith, D.W. (1967). Compendium of shortness of stature. *Journal of Pediatrics, 70* (3, Part 2, Suppl.), 463–519.

Spitz, H.H. (1986). *The raising of intelligence: A selected history of attempts to raise retarded intelligence.* Hillsdale, NJ: Lawrence Erlbaum Associates.

Tobach, E., & Proshansky, H.M. (Eds.). (1976). *Genetic destiny.* New York: AMS Press.

Tytler, G. (1982). *Physiognomy in the European novel.* Princeton, NJ: Princeton University Press.

Wechsler, J. (1982). *A human comedy: Physiognomy and caricature in 19th century Paris.* Chicago: University of Chicago Press.

Witkin, H.A., Mednick, S.A., Schulsinger, F., Bakkestrøm, E, Christiansen, K.O., Goodenough, D.R., et al. (1977). Criminality in XYY and XXY men. *Science, 193,* 547–556.

Wright, L., Schaefer, A.B., & Solomons, G. (1979). *Encyclopedia of pediatric psychology.* Baltimore: University Park Press.

Yanet, H.J. (1957). Classification and etiological factors in mental retardation. *Journal of Pediatrics, 50,* 226–230.

Index

Page numbers followed by *f* indicate figures; those followed by *t* indicate tables.

AAC, *see* Augmentative and alternative communication
ABA, *see* Applied behavior analysis
ABC assessments, *see* Antecedent-behavior-consequence (ABC) assessments
Aberrant Behavior Checklist, 36, 42, 55
Acetylcholinesterase inhibitors, 193*t*, 207
ADHD, *see* Attention-deficit/hyperactivity disorder
ADIS-P, *see* Anxiety Disorders Interview Schedule–Parent
Adjustment disorders, 186–187
ADOS, *see* Autism Diagnostic Observation Schedule
Aggressive behavior
  causes, 187
  consequences of, 218
  examples, 136*t*
  fragile X syndrome (FXS), 38
  Lesch-Nyhan syndrome, 10
  pharmacologic intervention, 197, 204–205
Alcohol-related birth defects, *see* Fetal alcohol syndrome (FAS)
Alcohol-related neurodevelopmental disorders, *see* Fetal alcohol syndrome (FAS)
Allele size mosaicism, 30
Alleles, 243–244
Alpha-adrenergic agents, 193*t*, 194, 206
Alprazolam, 203
Amantadine, 208
Amitiptyline, 200
Amphetamine, 191, 192*t*
Aneuploidy, 246, 248
Angelman syndrome, 72, 73*f*, 248–249, 250*t*, 253
Animal models
  fragile X syndrome (FXS), 39–40, 281
  limitations of, 5
Antecedent-behavior-consequence (ABC) assessments, 141, 144*f*
Antidepressants, 192*t*, 200–202, 225
Antiepileptics, 193*t*, 203, 204–205
Antipsychotic medications, *see* Neuroleptic medications
Anxiety disorders
  comorbidity with speech-language disorders, 165
  correlation with problem behavior, 222
  diagnostic considerations, 190, 222
  FMR1 premutations, 38
  fragile X syndrome (FXS), 34, 36
  pharmacologic intervention, 200–201, 202, 203–204, 208

prevalence, 204
  Williams syndrome, 90–91, 92–93, 92*t*
Anxiety Disorders Interview Schedule–Parent (ADIS-P), 91
Anxiolytics, 193*t*, 203–204
Apoptosis, 258
Applied behavior analysis (ABA)
  behavioral interventions, 223–224
  overview, 136–139, 137*t*, 138*t*, 219–220
  see also Functional behavioral assessment (FBA)
Aripiprazole, 196, 197
ASDs, *see* Autism spectrum disorders
Atomoxetine, 191, 192*t*, 194, 195
Attention-deficit/hyperactivity disorder (ADHD)
  brain development, 6–7
  comorbidity with speech-language disorders, 165
  diagnosis of in neurogenetic disorders, 7, 7*t*
  FMR1 premutation, 32–33
  fragile X syndrome (FXS), 33–34
  intermediate phenotypes, 270
  pharmacologic intervention, 191, 194–196, 198, 206, 207, 225
  prevalence, 191
  Williams syndrome, 90–92, 91*f*
Atypical antipsychotics, 192*t*, 225
Augmentative and alternative communication (AAC), 172–173
Autism Behavior Checklist, 42
Autism Diagnostic Observation Schedule (ADOS), 55, 88–89, 89*t*
Autism spectrum disorders (ASDs)
  behavior problems, 219
  catatonia, 159
  characteristics and endophenotype, 71, 72–74
  cognitive abilities, 57*t*, 268, 296
  compared to Prader-Willi syndrome (PWS), 71–72, 75–78, 283
  component phenotypes, 265–269
  covariates, 272
  diagnosis and classification, 7, 265–269
  Down syndrome, 281–282
  FMR1 premutation, 32–33
  fragile X syndrome (FXS), 34–37, 35*f*, 37*f*, 40–43, 44, 280–281
  genetics, 74, 76–77, 258
  heritability, 76–77, 268
  heterogeneity of, 74
  intermediate phenotypes, 270
  literacy skills, teaching, 171–172

Autism spectrum disorders (ASDs)—*continued*
    pharmacologic intervention, 195, 197, 199,
        201–202, 205, 206, 207, 226
    self-injury, 75–76
    Smith-Magenis syndrome (SMS), 22–23
Avoidance behavior, 230

Bacterial artificial chromosome (BAC) clones,
    254
Baseline exaggeration in psychiatric diagnosis,
    157
Behavior modification, *see* Applied behavior
    analysis (ABA)
Behavior problems
    behavioral interventions, 223–224, 228, 231
    case examples, 186–188
    comorbid conditions, 165–170, 187
    consequences of, 135, 218
    Cornelia de Lange syndrome, 146
    definition, 185–186
    diagnosis and classification, 186–190, 186*t*
    Down syndrome, 55–60, 57*t*, 58*f*, 59*f*, 60*f*, 147
    examples, 136*t*, 186–188
    fetal alcohol syndrome (FAS), 125–126
    FMR1 mutations/premutations, 38, 100
    fragile X syndrome (FXS), 100, 104, 106*f*
    integrative treatment approach, 217–218,
        226–231
    Lesch-Nyhan syndrome, 4, 10, 145–146
    neurofibromatosis type 1 (NF1), 101, 104, 106*f*
    pharmacological intervention, 208–211,
        224–226, 228–231
    Prader-Willi syndrome (PWS) and autism,
        146–147
    prevalence, 218–219
    reinforcement of, 138–139, 138*t*
    research challenges, 135–136
    Rett syndrome, 147
    risk factors and determinants, 135, 218, 219
    Smith-Magenis syndrome (SMS), 22–23
    speech-language disorders and, 165–170
    Williams syndrome, 104, 106*f*
    see also Functional behavioral assessment
        (FBA)
Behavioral checklists, 140–141
Behavioral phenotypes
    definitions, 4–5, 53, 116, 164, 277
    developmental trajectories, 5–7
    distinguished from psychiatric phenotypes,
        278–279
    historical perspective, 3, 293–297
    nurture versus nature controversy,
        295–296, 297
    psychiatric diagnosis and, 156–158
    research considerations, 5, 5*t*
    see also specific syndromes and disorders
Benzodiazepine, 193*t*, 203, 204

Beta-endorphin, 227
Bipolar disorder, 190, 197
Birt-Hogg-Dube syndrome, 18*f*
Brain development
    attention-deficit/hyperactivity disorder
        (ADHD), 6–7
    Down syndrome, 55, 63–64, 65*f*
    importance of studying, 7
    see also Cognitive abilities
Bupropion hydrochloride, 192*t*, 200
Buspirone, 193*t*, 203, 204

Carbamazepine, 204, 205
Caregiver interviews, 139–140
CDKL5 gene, 250*t*, 251
CELF-IV, *see* Clinical Evaluation of Language
    Fundamentals-IV
CGH arrays, *see* Comparative genomic
    hybridization arrays
Charcot-Marie-Tooth syndrome, 18*f*
Child Behavior Checklist, 36
Chlorpromazine, 196
Chromosomes
    7q11.23 deletion, 253
    15q11-13 deletion, 72, 73*f*, 74, 76, 248–249
    17p11.2 deletion, 15, 17–19, 18*f*, 19*f*
    22q11 deletion, 246–247, 253
    overview, 246–249, 247*t*
Citalopram, 200
Clinical Evaluation of Language Fundamentals-
    IV (CELF-IV), 87
Clomipramine, 200, 201, 202
Clonazepam, 202, 203, 204
Clonidine, 194, 206
Clozapine, 158, 196
CNVs, *see* Copy number variants
Cognitive abilities
    adaptive behavior and, 104, 105*f*, 107*f*, 108,
        109–110
    autism spectrum disorders (ASDs), 57*t*,
        268, 296
    behavior problems and, 57–58, 57*t*, 59–60,
        59*f*, 60*f*
    Down syndrome, 57–60, 57*t*, 58*f*, 59*f*, 60*f*,
        99, 109
    fetal alcohol syndrome (FAS), 115, 120–125
    FMR1 premutations, 38
    fragile X syndrome (FXS), 30–31, 33, 100,
        102–104, 103*f*, 107–111, 107*f*
    neurofibromatosis type 1 (NF1), 101, 102,
        103*f*, 104, 107–108, 107*f*
    Smith-Magenis syndrome (SMS), 21–22
    Williams syndrome, 82, 83–86, 84*f*, 93, 100,
        102–104, 103*f*, 107–111, 107*f*
    see also Intellectual disabilities
Cognitive disintegration in psychiatric
    diagnosis, 157

Communication disorders, *see* Speech-language disorders

Comparative genomic hybridization (CGH) arrays, 254–255, 255*t*

Component phenotypes, 265–269, 266*t*

Conduct disorders
literacy and, 171
pharmacologic intervention, 199, 204–205

Congenital hypothyroidism, 154

Copy number variants (CNVs), 253

Cornelia de Lange syndrome, 135, 146, 148, 148*t*

Cortisol levels, 37, 42

Covariates, 266*t*, 271–272

Cytoplasmic FMR1-interacting protein 1 (CYFIP1), 43

DBC, *see* Developmental Behavior Checklist

Deletions, chromosomal, *see* Chromosomes

Depression
case example, 159
correlation with problem behavior, 222
diagnostic considerations, 155, 160, 190, 222
fragile X syndrome (FXS), 34
pharmacologic intervention, 225

Desipramine, 200

Developmental Behavior Checklist (DBC), 6

Developmental trajectories
phenotype expression and, 6–7
research considerations, 5*t*

Dexmethylphenidate, 191

Dextroamphetamine, 191, 192*t*, 194, 195, 208

Diagnostic and Statistical Manual of Mental Disorders, Fourth Edition, Text Revision (DSM-IV-TR)
autism spectrum disorders (ASDs), 7, 35, 266
Down syndrome, 56
fragile X syndrome (FXS), 33, 35
schizophrenia, 155–156
use of for diagnostic classification, 189, 221–222, 263, 278

Diagnostic and Statistical Manual of Mental Disorders, Second Edition, 263

Diagnostic and Statistical Manual of Mental Disorders, Third Edition, 263

Diagnostic Manual–Intellectual Disability (DM-ID), 6, 155, 160, 189–190

Diagnostic overshadowing, 157, 208, 210, 279

Diazepam, 203

Differential Ability Scales (DAS), 83–85, 84*f*

DiGeorge syndrome, 246–247

Direct observation of behavior, 141, 142*f*, 143

Disorder
disconnect from etiology, 263
distinguished from phenotype, 264–265

Disruptive behavior disorder, *see* Behavior problems

Divalproex sodium, 205

DM-ID, see Diagnostic Manual–Intellectual Disability

DNA, 244–245, 244*f*, 249–252

Donepezil, 207

Dopamine dysfunction
anxiety and, 203
attention-deficit/hyperactivity disorder (ADHD), 191
behavior problems and, 196
Down syndrome, 60
Lesch-Nyhan syndrome, 10–11
phenylketonuria (PKU), 8

Down syndrome
behavior problems, 55–60, 57*t*, 58*f*, 59*f*, 60*f*, 147
behavioral phenotype, 3, 54, 55–57, 281
brain development, 55, 63–64, 65*f*
case examples, 188
cognitive abilities, 57–60, 57*t*, 58*f*, 59*f*, 60*f*, 99, 109
comorbidity with autism, 281–282
etiology, 53–54
functional behavior analysis/assessment, 55, 147, 148, 148*t*
genetics, 246, 250*t*, 281
historical perspective, 3, 295
neuroanatomic findings, 60–64
pharmacologic intervention, 199, 202
prevalence, 281
research questions, 54–55, 64–65
social phenotype, 281–282
stereotypy, 56–59, 58*f*, 59*f*, 62–63

Drugs, *see* Pharmacological management; *specific medications*

DSM-IV-TR, see Diagnostic and Statistical Manual of Mental Disorders, Fourth Edition, Text Revision

Dual diagnosis, 189–190, 221–222
see also Psychiatric diagnosis

Dyslexia, 119

EBC, *see* Eye-blink conditioning

ELN gene, 250*t*

End-state phenotypes, 116

Endophenotypes, 71–72, 116, 266*t*, 269–271

Environmental factors
behavior problems and, 209
nurture versus nature controversy, 295–296, 297
phenotype expression and, 6
role in genetic variation, 258–259

Epigenetics, 259

Escitalopram, 200

Executive function
across neurogenetic disorders, 6, 126
fetal alcohol syndrome (FAS), 115, 121–123, 126

Eye-blink conditioning (EBC), 118–119

FAIF, *see* Functional Analysis Interview Form
FAS, *see* Fetal alcohol syndrome
FAST, *see* Functional Analysis Screening Tool
FBA, *see* Functional behavioral assessment
FCT, *see* Functional communication training
Fetal alcohol syndrome (FAS)
   behavior problems, 125–126
   behavioral phenotype/endophenotype, 6,
      116–117, 116*f*, 127
   cognitive abilities, 115, 120–125
   eye-blink conditioning (EBC), 118–119
   neuroimaging studies, 117–118
   pharmacologic intervention, 195
   sensory and perceptual impairments, 119–120,
      123–124
Flip 'n Talk (FNT) communication system,
   172–173, 177–178, 177*f*
Fluorescent in situ hybridization (FISH), 248,
   255*t*
Fluoxetine, 200, 201–202, 203
Fluphenazine, 196
Fluvoxamine, 200, 201
FMR1 gene
   mutation, 29–30, 29*f*, 37–39, 99–100, 250*t*,
      252, 279
   premutation, 32–33, 37–39, 43–44, 279
   see also Fragile X syndrome (FXS)
FMRP, *see* Fragile X mental retardation protein
FNT, *see* Flip 'n Talk communication system
Fragile X mental retardation protein (FMRP),
   29–30, 39, 40, 43
Fragile X mutation, *see FMR1* gene; Fragile X
   syndrome (FXS)
Fragile X syndrome (FXS)
   adaptive behavior, 104, 105*f*, 107*f*, 108,
      109–110
   animal models, 39–40, 281
   anxiety disorder and, 204
   autism spectrum disorders (ASDs) and, 34–37,
      35*f*, 37*f*, 40–43, 44, 280–281
   behavior problems, 100, 104, 106*f*, 110
   behavioral analysis, 148, 148*t*
   behavioral phenotype, 4, 6, 33–37, 109,
      279–280
   cognitive abilities, 30–31, 33, 100, 102–104,
      103*f*, 107–111, 107*f*
   dysmorphic features, 30, 99–100
   genetics, 29–30, 29*f*, 37–39, 99, 250*t*, 252, 279
   neurobiological characteristics, 30–31, 31*t*,
      39–43
   pharmacologic intervention, 195, 205
   prevalence, 30, 279
   primary ovarian insufficiency (POI), 32, 33*t*
   social phenotype, 280
   tremor/ataxia syndrome (FXTAS), 31–32, 32*t*,
      39–40, 43–44
Fragile X–associated tremor/ataxia syndrome
   (FXTAS), 31–32, 32*t*, 39–40, 43–44

Frontal-subcortical circuits, 60–63, 61*f*
Functional Analysis Interview Form (FAIF), 140
Functional Analysis Screening Tool (FAST), 140
Functional analysis, *see* Functional behavioral
   assessment (FBA)
Functional behavioral assessment (FBA)
   behavioral interventions, 223–224, 228
   classification of problem behavior, 220–221
   compared with functional analysis, 139
   of genetic disorders, 144–147
   integration with pharmacologic interventions,
      208–209, 224–231
   methods, 139–141, 142*f*, 143–144, 144*f*
   overview, 136–137, 139
   relationship to psychiatric diagnosis, 220–221,
      222–223, 227–228
Functional communication training (FCT),
   223–224, 226
FXS, *see* Fragile X syndrome
FXTAS, *see* Fragile X–associated tremor/ataxia
   syndrome

G protein-coupled receptor 155 (GPR155), 43
GABA modulation, 197, 203
GABA receptor genes, 72, 76, 77–78
Gabapentin, 204
Genetic principles
   basic concepts, 243–246
   chromosomes, 246–249, 247*t*
   DNA, 244–245, 244*f*, 249–252
   epigenetics, 259
   genetic errors, 245
   heterogeneity, 74, 250–251, 264, 271
   historical perspective, 243, 258
   single versus multiple genetic variants,
      264–265
   see also specific syndromes and disorders
Genetic techniques
   comparative genomic hybridization arrays,
      254–255, 255*t*
   copy number variants (CNVs), 253
   fluorescent in situ hybridization (FISH),
      248, 255*t*
   high-resolution banding, 246, 255*t*
   hybridization, 244, 248, 255*t*
   linkage analysis, 255–258
   mass spectrography, 251
   polymerase chain reaction (PCR) detection,
      251
   single-nucleotide polymorphism mapping,
      252–253
   Southern blotting, 251, 255t
Genetically informative phenotypes
   component phenotypes, 265–269, 266*t*
   covariates, 266*t*, 271–272
   intermediate phenotypes, 266*t*, 269–271
Genotype, definition, 265

Glutamate modulators, 193*t*, 197, 207–208
Guanfacine, 194, 206

Halo effect, 141
Haloperidol, 196
Hearing loss
    chromosome 17 region, 18*f*
    fetal alcohol syndrome (FAS), 119
    Williams syndrome, 90*f*, 92
High-resolution banding techniques, 246
Homozygosity linkage analysis, 258
Hybridization genetic techniques, 244, 248, 255*t*
Hyperuricemia, 10
Hypothalamic-pituitary-adrenocortical (HPA)
    axis, 37, 40, 41, 42
Hypoxanthine-guanine
    phosphoribosyltransferase (HPRT) deficiency,
    see Lesch-Nyhan syndrome

*ICD, see International Classification of Diseases and
    Related Health Problems*
IgA deficiency, 18*f*
Imipramine, 200
Intellectual disabilities
    behavior disturbance and, 185
    behavior problems, 219
    diagnosis considerations, 6, 155, 160, 189–190
    diagnostic shadowing, 157, 208, 210, 279
    historical perspective, 293–296
    see also Cognitive abilities; *specific syndromes and
        disorders*
Intellectual distortion in psychiatric diagnosis, 157
Intermediate phenotypes (endophenotypes),
    71–72, 116, 266*t*, 269–271
Internalizing behaviors, 58–59, 58*f*, 59*f*
*International Classification of Diseases and Related
    Health Problems (ICD)*, 189, 221–222
IQ, *see* Cognitive abilities; Intellectual disabilities

Karyotypes, 246

Lamotrigine, 204, 205
Language disorders, *see* Speech-language
    disorders; Specific language impairment
LCRs, *see* Low copy repeats
Learning disabilities, *see* Cognitive abilities
Lesch-Nyhan syndrome
    assessment and diagnosis, 11
    behavior problems, 135
    behavioral phenotype, 3, 4, 5, 10, 145
    etiology, 10–11
    functional behavior assessment, 145–146
    neurobiological characteristics, 8–9, 9*t*, 145
    neuroimaging, 11

prevalence, 9
treatment, 11–12
Levetiracetam, 204, 205
Linguistic disabilities, *see* Speech-language
    disorders
Linkage/association analysis (genomic), 255–258
Linkage disequilibrium, 257–258, 259
Literacy, language and, 171–172
Lithium, 193*t*, 204, 205, 225
Lod score analysis, 257
Lorazepam, 159, 203
Low copy repeats (LCRs), 253

Mania, 197, 204–205
MAS, *see* Motivation Assessment Scale
Mass spectrography, 251
MECP2 gene, 250–251, 250*t*
Medications, *see* Pharmacological management;
    specific medications
Meiosis, 245
Melatonin, 23–24, 24*f*, 193*t*, 207
Memantine, 208
Mendelian genetics, 243–244, 264
Methylation mosaicism, 30
Methylphenidate, 191, 192*t*, 194, 195, 198
Midazolam, 203
Milieu speech-language treatment approach, 170
Mitosis, 245
Mood disorders
    diagnostic considerations, 190
    fragile X syndrome (FXS) and, 34
    pharmacologic intervention, 200–202, 208
Mood stabilizers, 193*t*, 204–205
Mosaicism, 30, 245, 247*t*, 249, 252
Motivation Assessment Scale (MAS), 140
Moya Moya disease, 25
Mullen Scales of Learning, 85
Mutations, *see* Chromosomes; Genetic principles

Naltrexone, 193*t*, 207, 226–227
National Human Genome Research Institute,
    243, 253–254
National Institutes of Health
    applied behavior analysis (ABA), 223
    intellectual disability and behavior
        disturbance, 185
National Parent Survey (fragile X syndrome), 34,
    36, 38
Nefazodone, 200
Neurobehavioral disorder, definition, 263
Neurofibromatosis type 1 (NF1)
    adaptive behavior, 104, 105*f*, 107*f*, 108
    behavior problems, 101, 104, 106*f*, 110
    cognitive abilities, 101, 102, 103*f*, 104,
        107–108, 107*f*, 110
    genetics, 101

Neuroleptic medications, 192*t*, 196–200,
    225, 230
NF1, *see* Neurofibromatosis type 1
NIPA1 gene, 72, 73*f*
NIPA2 gene, 72, 73*f*
Noncontingent reinforcement intervention, 224
Nonparametric linkage analysis, 257
Nonsocial reinforcement, 138, 138*t*, 220, 230
Noradrenergic receptors, 200
Norepinephrine, 194, 203
Norepinephrine reuptake inhibitors, 191, 192*t*,
    194, 195
Nortriptyline, 200

Observation of behavior, 141, 142*f*, 143
Obsessive-compulsive disorder (OCD), 160, 200,
    203, 225
Olanzapine, 196, 198
Oligonucleotides, 254
Onychotillomania, 21
Operant conditioning
    effects of medications on, 229–230
    examples, 137*t*
    in functional communication training (FCT),
        223–224
    neuroanatomic findings, 229
    overview, 137–139, 138*t*, 219
Opioid atagonists, 193*t*, 207
Oxcarbazepine, 204

Paliperidone, 196
Parametric linkage analysis, 257
Paroxetine, 200
Partial fetal alcohol syndrome, *see* Fetal alcohol
    syndrome (FAS)
PCR, *see* Polymerase chain reaction
Peabody Picture Vocabulary Test (PPVT), 86–87
Pervasive developmental disorder (PDD), 34–35,
    89, 195–196, 197, 207
Pharmacological management
    dosing and monitoring, 209–211
    effects on operant processes, 229–230
    general principles, 208–211
    integration with behavioral interventions,
        208–209, 224–231
    polypharmacy, 156
    psychiatric diagnosis and, 156, 189–190,
        228–229
    selection approaches, 209
    self-injury, 75–76
    see also Antidepressants; Anxiolytics; Mood
        stabilizers; Neuroleptic medications;
        Stimulant medications
Phenotypes
    definitions, 265, 277
    psychiatric phenotypes, 277–279

    see also Behavioral phenotypes; Genetically
        informative phenotypes; Social phenotypes
Phenylketonuria (PKU), 8
Phobias, 200, 203
Pica, 136*t*, 218
PKU, *see* Phenylketonuria
Pleiotropy, 264
POI, *see* Primary ovarian insufficiency
Polymerase chain reaction (PCR), 249, 250*t*,
    251, 257
Posttraumatic stress disorder (PTSD), 158, 160
PPVT, *see* Peabody Picture Vocabulary Test
    (PPVT)
Prader-Willi syndrome (PWS)
    behavioral phenotype, 4, 6, 71
    compared with autism, 71–72, 75–78, 283
    endophenotype, 72, 73*f*
    functional behavior assessment, 146–147
    genetics, 72, 73*f*, 76–77, 248–249, 250*t*, 253
    heritability, 76–77
    pharmacologic intervention, 199, 202
    physical characteristics, 72, 282
    prevalence, 282
    self-injury, 75–76, 146–147
    social phenotype, 282–283
Pragmatic language disorders, 164–165
Prefrontal-subcortical circuits, 60–63, 61*f*, 62*f*
Prelinguistic milieu teaching, 171
Primary ovarian insufficiency (POI), 32, 33*t*
Propranolol, 202
Proteomics, 246
Psychiatric diagnosis
    case examples, 158–159
    challenges and limitations, 155–158, 159–160,
        221–222
    comorbidities, 165–170, 208
    correlation with problem behavior, 222–223
    dual diagnosis, 189–190, 221–222
    hetergeneity of, 164
    historical perspective, 153–154
    importance of accuracy, 154–155, 232
    neurodevelopmental disabilities and, 154–155,
        156–158, 159–161
    prevalence of psychiatric disorders, 221
    recommendations, 159–161
    relationship to behavioral analysis, 220–221,
        227–229
    speech-language disorders and, 165–170
Psychiatric phenotypes, 277–279
Psychopharmacology, *see* Pharmacological
    management; Psychotropic medications
Psychosis, correlation with problem behavior, 222
Psychosocial masking in psychiatric
    diagnosis, 157
Psychotropic medications
    general principles, 208–211
    integration with behavioral interventions,
        224–231

medication classes, 192*t*–193*t*
purposes of, 190–191
see also Antidepressants; Anxiolytics; Mood
   stabilizers; Neuroleptic medications;
   Stimulant medications
PTSD, *see* Posttraumatic stress disorder
PWS, *see* Prader-Willi syndrome

Questions About Behavioral Function
   (QABF), 140
Quetiapine, 196

RAI1 gene, 15, 17–18
Rating scales, 141
Reading difficulties, 171–172
Recombination, 245
Reinforcement
   in applied behavior analysis (ABA), 137–139,
      137*t*, 138*t*, 220
   in speech-language treatment, 170
Research needs
   Down syndrome, 54–55, 64–65
   methodological considerations, 5*t*, 297
   speech-language treatment, 179–180
Responsivity education, 171
Rett syndrome, 147, 250–251, 250*t*, 256
Riluzole, 208
Risperidone, 196, 197, 198, 199, 225–226, 230

SBFE, *see* Stanford-Binet Intelligence Scale,
   Fourth Edition
Scatterplot assessments, 141, 142*f*
Schizophrenia
   diagnostic criteria, 155–156, 160
   pharmacologic intervention, 197, 208
   velocardiofacial syndrome (VCFS), 286–287
SEBDs, *see* Social-emotional-behavioral disorders
Selective serotonin reuptake inhibitors (SSRIs),
   192*t*, 197, 200, 201, 202, 203, 225
Self-injurious behaviors
   Cornelia de Lange syndrome, 146
   examples, 136*t*
   fragile X syndrome (FXS), 38
   Lesch-Nyhan syndrome, 10, 145–146
   pharmacologic intervention, 197, 199, 225,
      226
   Prader-Willi syndrome (PWS) and autism,
      75–76, 146–147
   reinforcement and, 138, 138*t*
   Smith-Magenis syndrome (SMS), 283, 284
   treatment, 225, 226–227
Sensory impairments
   and behavior problems, 219
   hearing loss, 18*f*, 90*f*, 92, 119
   vision impairment, 90*f*, 100, 119, 120, 123–124

visual-spatial processing, 83–86, 84*f*, 100, 109
Serotonin, 197, 200, 201, 203
Sertraline, 200, 202
Short Sensory Profile (SSP), 90, 90*f*
Shprintzen syndrome, *see* Velocardiofacial
   syndrome (VCFS)
Single-nucleotide polymorphisms (SNPs),
   252–253
Sjogren-Larson syndrome, 18*f*
Sleep disorder
   fetal alcohol syndrome (FAS), 119
   pharmacologic intervention, 207
   Smith-Magenis syndrome (SMS), 20*f*, 23–24,
      24*f*, 146
Smith-Magenis syndrome (SMS)
   behavior problems, 22–23, 135, 146
   behavioral phenotype, 19–21, 146, 283
   cognitive abilities, 21–22
   dysmorphic features, 16*f*, 17*f*, 283
   functional behavior assessment, 146
   genetics, 15, 17–19, 18*f*, 19*f*, 283
   neurobiological characteristics, 15–17, 16*t*,
      17*t*, 24–26
   prevalence, 15, 283
   sleep disorder, 20*f*, 23–24, 24*f*, 146
   social phenotype, 284
   speech-language development, 22
   treatment, 23
SNPs, *see* Single-nucleotide polymorphisms
Social Approach Scale, 37
Social phenotypes
   Down syndrome, 281–282
   fragile X syndrome (FXS), 280–281
   Prader-Willi syndrome (PWS), 282–283
   Smith-Magenis syndrome (SMS), 284
   Turner syndrome, 284–285
   velocardiofacial syndrome (VCFS), 286–287
   Williams syndrome, 285–286
   see also Social skills and traits
Social reinforcement, 138, 138*t*, 220, 230
Social skills and traits
   autism spectrum disorders (ASDs), 73, 77
   as diagnostic criteria, 7
   fetal alcohol syndrome (FAS), 124–125
   fragile X syndrome (FXS), 34–37, 35*f*, 37*f*,
      40–43
   psychosocial masking in psychiatric
      diagnosis, 157
   speech-language disorders and, 166–167, 173
   Williams syndrome, 88–89, 89*t*
   see also Social phenotypes
Social-emotional-behavioral disorders (SEBDs)
   comorbidity with speech-language disorders,
      165–170, 178–179, 188–189
   definition, 165
   see also Psychiatric diagnosis
Somatic illnesses, 188
Southern blotting, 251, 255*t*

Specific language impairment
   autism spectrum disorders (ASDs), 77
   social skills and, 166
Speech-language disorders
   augmentative and alternative communication
      (AAC), 172–173
   autism spectrum disorders (ASDs), 73, 77
   and behavior problems, 219
   case example, 173–178
   comorbidity with social-emotional-behavioral
      disorders (SEBDs), 165–170, 178–179,
      188–189, 223–224
   covariates, 271
   diagnostic considerations, 163–165, 167
   fetal alcohol syndrome (FAS), 123
   fragile X syndrome (FXS), 33
   functional communication training (FCT),
      223–224, 226
   literacy and, 171–172
   neuroimaging, 169
   performance variability, sources of, 170
   professional training needs, 179
   research needs, 179–180
   Smith-Magenis syndrome (SMS), 22
   social skills and, 166–167, 173
   specific language impairment, 77, 166
   terminology and definitions, 164–165
   treatment approaches, 170–173
   Williams syndrome, 82–83, 86–89, 89t, 93–94
SSP, see Short Sensory Profile
SSRIs, see Selective serotonin reuptake inhibitors
Stanford-Binet Intelligence Scale, Fourth Edition
   (SBFE), 101–102
Stereotypic behaviors
   Down syndrome, 56–59, 58f, 59f, 62–63
   Lesch-Nyhan syndrome, 4
   pharmacologic intervention, 225, 226
   Rett syndrome, 147
   Smith-Magenis syndrome (SMS), 16t, 17t, 21
Stimulant medications, 191, 192t, 194–196, 225,
   229–230

TBX1 gene, 250t
Test for Reception of Grammar (TROG), 88
Translocations, chromosomal, 247, 247t
Trazodone, 200
Triazolam, 203
Tricyclic antidepressants, 192t, 200
Trinucleotide repeats, 251–252
Trisomy 21, see Down syndrome
TROG, see Test for Reception of Grammar
Tuberous sclerosis, 6

Turner syndrome, 284–285
Twin studies, 263–264, 267
Tyrosine, 8

UBE3A gene, 72, 77–78, 250t
Uniparental disomy, 248, 249

VABS, see Vineland Adaptive Behavior Scales
Valproic acid, 204, 205
Velocardiofacial syndrome (VCFS), 247, 250t,
   286–287
Venlafaxine, 200
Vineland Adaptive Behavior Scales (VABS), 102
Vision impairment
   fetal alcohol syndrome (FAS), 119, 120,
      123–124
   Williams syndrome, 83–86, 84f, 90f,
      100–101, 109

Wechsler Adult Intelligence Scale (WAIS), 83
Wechsler Intelligence Scale for Children
   (WISC), 83, 121
Williams syndrome
   adaptive behavior, 104, 105f, 107f, 108,
      109–110
   behavior problems, 104, 106f, 110
   behavioral phenotype, 4, 109
   cognitive abilities, 82, 83–86, 84f, 93, 100,
      102–104, 103f, 107–111, 107f
   genetics, 81, 94, 100, 248, 250t, 253, 285
   historical perspective, 81–83
   language abilities, 82–83, 86–89, 89t, 93–94
   neurobiological characteristics, 81, 81f,
      100–101, 285
   personality profiles, 89–93, 94
   physical characteristics, 81, 81f, 100
   prevalence, 81, 285
   social phenotype, 285–286
   visual-spatial processing, 83–86, 84f,
      100–101, 109
Williams Syndrome Personality Profile
   (WSPP), 89
WISC, see Wechsler Intelligence Scale for
   Children
WSPP, see Williams Syndrome Personality
   Profile

Ziprasidone, 196